Science and Society in the Sanskrit World

Sir Henry Wellcome Asian Series

VOLUME 21

The titles published in this series are listed at *brill.com/was*

Christopher Zand Minkowski at the Old Indian Institute, University of Oxford, 2020.
SOURCE: CHRISTOPHER Z. MINKOWSKI

Science and Society in the Sanskrit World

Edited by

Christopher T. Fleming
Toke L. Knudsen
Anuj Misra
Vishal Sharma

BRILL

LEIDEN | BOSTON

Cover illustration: Saraswati. Lithograph by Raja Ravi Varma, 1894.

The Library of Congress Cataloging-in-Publication Data is available online at https://catalog.loc.gov
LC record available at https://lccn.loc.gov/2022062026

Typeface for the Latin, Greek, and Cyrillic scripts: "Brill". See and download: brill.com/brill-typeface.

ISSN 1570-1484
ISBN 978-90-04-52231-2 (hardback)
ISBN 978-90-04-53686-9 (e-book)

This book is printed on acid-free paper and produced in a sustainable manner.

To Professor Christopher Minkowski

यन्नेत्रैस्त्रिभिरीक्षते न गिरिशो नाष्टाभिरप्यब्जभूः
स्कन्दो द्वादशभिस्तथा न मघवा चक्षुःसहस्रेण वा ।
सम्भूयापि जगत्त्रयस्य नयनैर्द्रष्टुं न यत्पार्यते
प्रत्याहृत्य दृशौ समाहितधियः पश्यन्ति तत्पण्डिताः ॥

That which eludes
even Śiva's three eyes
Brahmā's eight
Skanda's twelve
Indra's thousand,
What all the eyes in the world
put together still cannot see,
Learned persons behold
shutting their two eyes
mind stilled in concentration.

From the *Saduktikarṇāmṛta* (attributed to Śālikanātha),
translated by SUHAS MAHESH.

Contents

PART 2
Astral and Medical Sciences

PART 3
Statecraft and Jurisprudence

PART 4
Sanskrit Intellectuals and Their Communities

Figures and Tables

Figures

Tables

Publications by Christopher Z. Minkowski

Books and Monographs

Priesthood in Ancient India: A Study of the Maitrāvaruṇa Priest Publications of the De Nobili Research Library 18. Vienna: University of Vienna, 1991.

A Descriptive Catalogue of the Sanskrit Astronomical Manuscripts Preserved at the Maharaja Man Singh II Museum in Jaipur India, compiled by David Pingree from the notes taken by Setsuro Ikeyama, Christopher Minkowski, David Pingree, Kim Plofker, Sreeramula Rajeswara Sarma, and Gary Tubb. Philadelphia: American Philosophical Society, 2003.

Scholar Intellectuals in Early Modern India: Discipline, Sect, Lineage and Community. Routledge: Abingdon, 2015. (Edited, with Rosalind O'Hanlon and Anand Venkatkrishnan).

Articles

"Outlandishness in Sanskrit Commentary," in *Les scolastiques indiennes: genèses, développements, interactions.* Sous la direction d'Émilie Aussant et de Gérard Colas. Paris/Pondichéry, École Française d'Extrême-Orient (Collection Études Thématiques). 2020. 185–202.

"Nīlakaṇṭha Caturdhara and Appayya's Bhāratasārasaṃgrahastotra" *Journal of Oriental Research* 70th Anniversary Special Issue (Kuppuswami Sastri Research Institute) 2018.

"Big Science: Classicism and Conquest," in eds. Benjamin Elman and Sheldon Pollock, *What China and India Once Were: The Pasts that May Shape the Global Future* New York: Columbia University Press, 2018. 199–231. (with Benjamin Elman).

"An Early Modern Account of the Views of the Miśras," *Journal of Indian Philosophy* 46 (2018) 889–933. DOI 10.1007/s10781-018-9368-0

"Nīlakaṇṭha Caturdhara's Advaita Vedānta," in ed. Jonardon Ganeri, *The Oxford Handbook of Indian Philosophy*, Oxford: Oxford University Press, 2017. DOI 10.1093/oxford hb/9780199314621.013.39. ISBN 978-01-99-31462-1

"Maryādām ullaṅghya: the Boundaries of Interpretation in Early Modern India," in eds. A. Grafton and G. Most, *Canonical Texts and Scholarly Practices: A Global Comparative Approach*, Cambridge University Press, 2016. 90–109.

"Nīlakaṇṭha's Teachers and Gurus, Part 1: Lakṣmaṇa Paṇḍita," *Wiener Zeitschrift Kunde Sudasiens*, 55 (2013–2014), 33–76. DOI: 10.1553/wzks2013–2014s33

"Social History in the Study of Indian Intellectual Cultures?" *South Asian History and*

Culture 6.1 (2015) 1–9. (With Rosalind O'Hanlon and Anand Venkatkrishnan) DOI: 10.1080/19472498.2014.969006

"Apūrvaṃ Pāṇḍityam: On Appayya Dīkṣita's Singular Life," *Journal of Indian Philosophy*. 44.1 (2016) 1–10. DOI: 10.1007/s10781-014-9230-y

"Appayya's Vedānta and Nīlakaṇṭha's Vedāntakataka," *Journal of Indian Philosophy*. 44.1 (2016) 95–114. DOI: 10.1007/s10781-014-9252-5

"Learned Brahmins and the Mughal Court: the Jyotiṣas," in eds. Vasudha Dalmia and Munis Faruqui, *Religious Interactions in Mughal India*. New Delhi: Oxford University Press, 2014, 102–134.

"The Meaning and Design of the Mahābhārata in Nīlakaṇṭha's Commentary," in eds. F. Grimal and C. Vielle, *Proceedings of the Pondicherry Conference on Commentary*, Forthcoming.

"The Ṛtuvarṇana Chapter of Jñānarāja's Siddhāntasundara: Text, Translation, and Notes," *Gaṇita Bhāratī*, 33.1–2 (2011) 53–87. (With Toke Knudsen.)

"Seasonal Poetry as Science: Ṛtuvarṇana in Some Astronomy Treatises," *Gaṇita Bhāratī*, 33.1–2 (2011) 89–93.

"The Aśvistuti and the Enigma of the Mahābhārata," in eds. Silvia D'Intino and Caterina Guenzi, *Aux Abords de la Clarière: Ètudes indiennes et comparées en l'honneur de Charles Malamoud* (Paris: Brépols, 2012) 79–92.

"Advaita Vedānta in Early Modern History," in eds. Rosalind O'Hanlon and David Washbrook, *Religious Cultures in Early Modern India: New Perspectives*. Special Volume of *South Asian History and Culture* 2.2 (2011) 205–231. DOI: 10.1080/19472498.2011.553493

"Where the Black Antelope Roam: Dharma and Human Geography in India," in eds. Kurt A. Raaflaub and Richard Talbert, *Geography, Ethnography, and Perceptions of the World in Ancient Civilizations*, (Oxford: Blackwell, 2010) 9–31.

"Sanskrit Scientific Libraries and their Uses: Examples and Problems of the Early Modern Period," in eds. F. Bretelle, C. Proust, *Looking at it from Asia: The Processes that shaped the sources of the History of Science* Boston Studies in Philosophy of Science 206 (2010) 81–114.

"Nīlakaṇṭha's Mahābhārata," *Seminar* 608 (2010) 32–38.

"I'll Wash Out Your Mouth With My Boot: A Guide to Philological Argument in Early Modern Banaras," in ed. Sheldon Pollock, *Epic and Argument in Sanskrit Literary History: Essays in Honor of Robert P. Goldman*. (Delhi: Manhar, 2010) 117–141.

"What Makes People Who They Are? Pandit Networks and the Problem of Livelihoods in Early Modern Western India," *Indian Economic and Social History Review*, 45.3 (2008) 381–416. (With Polly O'Hanlon)

"Nīlakaṇṭha and His Historical Context," *Vostok* (*Oriens*—Moscow Academy of Sciences) No. 4 (2008) 37–49.

"Meanings Numerous and Numerical: Nīlakaṇṭha and Magic Squares in the *Ṛgveda*," in eds L. Kulikov and M. Rusanov, *Indologica T. Ya. Elizarenkova Memorial Volume* (Moscow: Russian State University for the Humanities, 2008) 315–328.

"The Study of Jyotiḥśāstra and the Uses of Philosophy of Science," *Journal of Indian Philosophy*, 36 (2008) 587–597.

"Why Should We Read the Maṅgala Verses?" in ed. W. Slaje, *The Śāstrārambha: Inquiries into the Preamble in Sanskrit* Abhandlungen für die Kunde des Morgenlandes 62 (2008) 1–24.

"On the Success of Nīlakaṇṭha's Mahābhārata Commentary," in ed. F. Squarcini, *Boundaries, Dynamics and Construction of Traditions in South Asia*, (Florence: Firenze University Press, 2005) 225–252. DOI: 10.7135/UPO9781843313977.011

"Nīlakaṇṭha's Vedic Readings in the Harivaṃśa Commentary," in ed. Petteri Koskikallio, *Epic, Khilas, and Purāṇas: Continuities and Ruptures. Proceedings of the Third Dubrovnik Conference on the Sanskrit Epics and Purāṇas, September 2002*. (Zagreb: Croatian Academy of Sciences and Arts, 2005) 411–433.

"Nīlakaṇṭha's Instruments of War: Modern, Vernacular, Barbarous," *Indian Economic and Social History Review*, 41 (2004) 365–385.

"A Nineteenth Century Sanskrit Treatise on the Revolution of the Earth: Govinda Deva's *Bhūmibhramaṇa*," *SCIAMUS* 5 (2004) 199–224.

"On Sūryadāsa and the Invention of Bi-directional Poetry (*vilomakāvya*)," *Journal of the American Oriental Society*, 124.2 (2004) 325–333.

"The Vedastuti and Vedic Studies: Nīlakaṇṭha on Bhāgavata Purāṇa X.87," in eds. A. Griffiths and J.E.M. Houben, *The Vedas: Texts, Languages & Ritual. Proceedings of the Third International Vedic Studies Workshop*, (Groningen: Egbert Forsten, 2004) 125–142.

"Competing Cosmologies in Early Modern Indian Astronomy," in eds. Charles Burnett, Jan Hogendijk, and Kim Plofker, *Ketuprakāśa: studies in the history of the exact sciences in honor of David Pingree* (Leiden: Brill, 2004) 349–385.

"Astronomers and their Reasons: Working Paper on Jyotiḥśāstra," *Journal of Indian Philosophy* 30 (2002) 495–514.

"The 'Bhūgolavicāra': A Cosmological Manuscript from Jaipur," in ed. G.U. Thite, *Subhāṣiṇī: Dr. Saroja Bhate Felicitation Volume* (Pune: Prof. Dr. Saroja Bhate Felicitation Committee, 2002) 250–263.

"Nīlakaṇṭha Caturdhara's *Mantrakāśīkhaṇḍa*," *Journal of the American Oriental Society*, 122.2 (2002) 329–344.

"The Interrupted Sacrifice and the Sanskrit Epics," *Journal of Indian Philosophy*, 29 (2001) 169–186.

"Nīlakaṇṭha Caturdhara and the Genre of Mantrarahasyaprakāśikā," in ed. Y. Ikari, *Proceedings of the Second International Vedic Workshop*, Kyoto, forthcoming.

"The Paṇḍit as Public Intellectual: The Controversy over Virodha or Inconsistency in the Astronomical Sciences," in ed. Axel Michaels, *The Pandit: Traditional Scholar-*

ship in India (Festschrift P. Aithal). Heidelberg South Asian Studies XXXVIII (New Delhi: Manohar Publications, 2001) 79–96.

"I Veda," ("The Vedas,") in ed. Sandro Petruccioli, *Storia della Scienza*, (Rome: Enciclopedia Italiana, 2001) 2: 709–715.

"Testi per i Rituali Solenni (Śrautasūtra)," in ed. Sandro Petruccioli, *Storia della Scienza*, (Rome: Enciclopedia Italiana, 2001) 2: 756–761.

"Raccolte di Aforismi sui Rituali Domestici (Gṛhyasūtra)," in ed. Sandro Petruccioli, *Storia della Scienza*, (Rome: Enciclopedia Italiana, 2001) 2: 766–768.

"Nīlakaṇṭha's Cosmographical Comments in the Bhīṣmaparvan," *Purāṇa* 42 (2000) 24–40.

"More Notes on the Cipher in Sanskrit Manuscripts," *Berliner Indologische Studien* 11/12 (1998) 159–171.

"School Variation in the Text of the Nivids," in *Inside the Texts Beyond the Texts*, ed. M. Witzel. Harvard Oriental Series Opera Minora v. 2 (Cambridge, Mass.: Harvard University Press, 1997) 167–184.

"Nividica," *Berliner Indologische Studien* 9 / 10 (1996) 225–236

"Some Lost Verses for the Aptoryāma," *Berliner Indologische Studien* 8 (1995) 132–157.

"*evám, evá* and Ācārya Gāṇagāri." *Journal of the American Oriental Society* 115.3 (1995) 388–400.

"An Omen Reader and a Cypher in Vedic Manuscripts." in *Vāman Śāstrī Bhāgavat Felicitation Volume*, ed. Saroja Bhate. (Poona: 1994) 122–127.

"Snakes, *sattras* and the Mahābhārata." in ed. Arvind Sharma, *Essays in the Mahābhārata*, Brill's Indological Library I. (Leiden: E.J. Brill, 1991) 384–400.

"Janamejaya's *sattra* and Ritual Structure." *Journal of the American Oriental Society* 109.3 (1989) 401–420.

"The Rathakāra's Eligibility to Sacrifice," *Indo-Iranian Journal* 32 (1989) 177–194.

"The *udumbara* and its Ritual Significance." *Wiener Zeitschrift für die Kunde Südasiens* 33 (1989) 5–23.

Selected Reviews

David Pingree, *Pathways into the Study of Ancient Sciences: Selected Essays*. *Journal of the American Oriental Society* 138.4 (2018) 917–918.

Wendy Doniger, *Redeeming the* Kāmasūtra, *London Review of Books*, vol. 40, no. 12, 21 June, 2018. 35–36.

Charles Allen, *Aśoka: the Search for India's Lost Emperor*. *Times Literary Supplement*, 4 January, 2013. (5727) 11

Klaus Karttunen and Petteri Koskikallio (ed.): *Vidyārṇavavandanam. Essays in honour of Asko Parpola*. *Asiatische Studien / Études Asiatiques* 59.2 (2005) 628–634.

David Pingree, *Catalogue of Jyotiṣa Manuscripts in the Wellcome Library: Sanskrit astral and mathematical literature*. *Medical History* 49 (2005) 391–392.

Laurie L. Patton, *Myth as Argument: The Bṛhaddevatā as Canonical Commentary*. History of Religions 43 (2004) 263–265.

Thomas Trautmann, *Aryans and British India*. *Journal of Interdisciplinary History* 29.1 (1998) 166–168.

T. Elizarenkova, *Language and Style of the Vedic Ṛṣis*. WZKS 41 (1997) 234–236.

J.C. Heesterman, *The Broken World of Sacrifice*. *JAOS* 116.2 (1996) 341–344.

Sebastian Carri, *Contribution of Bhaṭṭa Bhāskara to Vedic Exegesis*. *IIJ* 32 (1989) 144–147.

Personal Reflections on the Intellectual Life of Christopher Z. Minkowski

Recollections from Ithaca

Christopher Z. Minkowski taught at Cornell University in Ithaca, New York (U.S.A.) from 1989 to 2006. Hired first as an Assistant Professor, Minkowski quickly rose through the faculty ranks at Cornell, promoted to Associate Professor in 1995, and to Professor in 2002. Chris Minkowski's 2006 move to the University of Oxford was a blow to many of us at Cornell, who appreciated greatly Chris' intellectual and collegial qualities. However, we could only felicitate him on the appointment as Boden Professor of Sanskrit, a post with a most illustrious lineage, further brightened by Chris Minkowski's many contributions. Having read much Dorothy Sayers in my younger days, I was delighted to picture Chris at Balliol College.

In the Department of Asian Studies at Cornell University, Chris Minkowski's personal office stood directly opposite the Department administrative suite. Chris thus occupied a prime piece of real estate. This was much to the benefit of colleagues in the Department, since Chris generously kept an "open door policy," as it is known in the States. Especially for those of us who were junior-ranked faculty in the Department at the time, Chris Minkowski's open door was a real boon. We were learning the ropes, in the Department and in the University; we needed to talk things through, learn the ways of the institution, develop our strategies for navigating the world into which we had been hired. Chris was unfailingly generous and approachable, serving as an informal mentor, and he encouraged us to anticipate life after tenure, and to craft our intellectual and institutional futures. As Robin McNeal put it, "Chris Minkowski brought a wise and gentle eye to matters under consideration; he combined insight with self-reflexivity." Academic spaces are not terribly well populated by people who combine rich intellectual talent, insight, and self-reflection. At Cornell, and elsewhere, Chris' presence has been remarkable.

In his office chatting over lunch or tea, and in well-judged (neither too brief, nor inconveniently long) hallway conversations, Chris Minkowski manifested wide-ranging intellectual interests, as well as his skill at adapting conversations to the strengths and interests of his interlocutors, in a spirit of inquiry as well as collegiality. As our colleague Brett de Bary remarked, "What always struck me was the ease with which Chris could come up with a truly interesting topic or observation that bridged our interests. They were often meta-level reflections which had to do with the way he was formulating a certain topic for a class, or how to answer students' questions, but he always asked with an eye to prising out underlying assumptions, and an eye to broader methodological debates." With some colleagues he might speak about approaches to translation and literary interpretation, with another ancient intellectual history and cosmology, with others how to conceptualize "early modernity," and so on. There was a hint of mischief in Chris' eyes sometimes; he would choose topics playfully.

Over the years, chatting in offices, hallways, and on the long and dramatic walking paths that connected a northern parking area with the library, Chris contributed to the gradual development of mutual intelligibility and intellectual trust among colleagues working across vastly different geographic, topical, and methodological domains. This was no small feat in a department aiming to address (in teaching and research) much of Asia, from deeply pre-modern to contemporary times. Among the unintended consequences of Chris' intellectual vitality and congeniality were hiring decisions increasingly made on the basis of stronger mutual understanding among departmental colleagues. Chris Minkowski's contributions to intellectual connectivity within the department helped prepare the ground for a new doctoral program called *Asian Literature, Religion, and Culture*, uniting all of the Department's professorial faculty. While this doctoral program was inaugurated after Chris had relocated to Oxford, it bears important traces of his presence. Moreover, the intellectual history of South Asia remains of central interest to the department, and a key sub-field within the *Asian Literature, Religion and Culture* doctoral program under the talented stewardship of Larry McCrea, who joined Cornell University after Chris Minkowski's departure. Larry and Chris have collaborated for years, including in the ongoing research project, "The Age of Vedānta: A Social History of Knowledge in South Asia," undertaken by an international scholarly team.

Chris' own intellectual interests are dynamic and far-ranging. During his years at Cornell Chris moved from work in Vedic materials to the study of astronomy and mathematics. He also became strongly engaged by research on the intellectual history of India during the last centuries of the second millennium A.D. During his Cornell days, Chris Minkowski's work in history of science and cosmology sometimes engaged explicitly with collaborative studies of late precolonial intellectual history. One dimension of this was Chris' participation in the collaborative research project on *Sanskrit Knowledge Systems on the Eve of Colonialism*, led by Sheldon Pollock. Chris' research in these areas was a boon for me as a junior colleague, offering a valuable conversation partner as I worked to frame my second monograph about Buddhist monastic intellectual projects of 19th-century Laṅkā (now Sri Lanka). At that time, I was just finding my way from Buddhist Studies into more comparative studies in intellectual and institutional history. Chris Minkowski was equally at home speaking with

Robin McNeal about the intellectual and ritual histories of China. Chinese history was increasingly on Chris' mind owing to a then-emergent collaboration with Ben Elman (Princeton University). More recently, their collaboration is manifested in a comparative research project on the Comparative History of Philology in Early Modern Asia.

Chris' readiness to think comparatively, combined with his generous collegiality, made him a good fit for leadership of Cornell University's South Asia Program, where he served as Associate Director from 1992 to 1995, and Director from 1998 to 2002. The South Asia Program is a university-wide interdisciplinary program supporting faculty and graduate (= post-graduate) students in research on South Asia. Moreover, the Program typically brings to Cornell multi-million dollar grants from the U.S. federal government to support the teaching of South Asian languages for undergraduate and graduate students. Dan Gold recalls colleagues' appreciation of Chris Minkowski as Director of the South Asia Program. Chris was thought to be an efficient administrator, also well able to integrate Cornell's wider South Asia studies community. Perhaps he approached the South Asia Program directorship from the perspective he later shared with Robin McNeal: one should accept administrative posts such as department Chair (= Head) and program Director "because of what you learn."

Chris Minkowski is remembered not only in the corridors of the Department of Asian Studies, but also in those of the departments of Classics and Linguistics where Chris had strong connections to faculty members Hayden Pelliccia, Alan Nussbaum, and Jay Jasanoff (now at Harvard University). Hayden Pelliccia (Department of Classics) and Sheila Haddad (Executive Staff Assistant, and subsequently Administrative Manager, Department of Linguistics) also shared with Chris an enjoyment of life in the semi-rural environs of the Trumansburg region, located across Cayuga Lake from Cornell University. Chris combines bookishness with an appreciation of the natural world and vigorous physical exercise. Thus, it was no surprise that he resided on a country property offering many occasions to be outdoors, while also providing plenty of solitude. Here Hayden Pelliccia was a friendly neighbor. On more than one occasion, the tortoises of Robin McNeal's household made their summer home on Chris' property; there was perhaps an exchange of propitiousness in this encounter. The foodies among us relished that fact that Chris' residence was located near one of the area's better restaurants, and also near a popular spot for locally prepared ice cream. Occasionally we continued our conversations with Chris in these pleasant locations away from campus.

At present, Chris Minkowski remains a Visiting Scholar in the Department of Asian Studies at Cornell. We hope that Cornell's excellent Kroch Asia Library, and the harmonious landscapes of the Finger Lakes region, will call him back to us more often in years to come. Perhaps Chris' collection of Jimi Hendrix records (still safe at Robin's house) will deepen the call.

Anne M. Blackburn
Department of Asian Studies
Ithaca, NY USA

Śāstradīpaka: Christopher Minkowski and the Boden Professorship

Before arriving in Oxford to interview for the Asoke Kumar Sarkar research fellowship at Balliol College, I had only known Christopher Minkowski as the Advaita guy. His article "Advaita Vedanta in Early Modern History" had changed the way I understood what was possible in the field. The article patiently mapped out the relationships, families, text traditions, and research agendas that constituted the social and intellectual history of Advaita Vedanta in the sixteenth and seventeeth centuries. I had always been interested in the lives of scholars, but it wasn't until reading this essay that I learned that you could actually write about them. Many years and conversations later, I know why he could. For Chris, ideas are inseparable from the people who articulate them. I see it in the way he speaks of his teachers, his fellow students, and even his college friends. Their personalities—their gruff demeanors, rumpled vests, bristling beards and pointed opinions—are present in and through his representation of their thoughts. We often talk about the life of the mind as if it were the mind that mattered, when it's really the life.

For all of Balliol's charms, I was always grateful for the other American in the room, who would be just as amused by the British social codes that went over my head. I was equally glad to have someone just as attentive to the history of the Boden Professorship. Chris' inaugural lecture expanded the ambit of Indology from its classical corner to the global stage, as he reviewed how knowledge of and from the non-West traveled around the world. But he was also ready to entertain conversations about the very local nature of Sanskrit learning at Balliol. While the promotion of Sanskrit studies at Balliol in the nineteenth century was more show than substance, India as part of the imperial imaginary was nevertheless central to the broader process of educational reform at the University of Oxford. Sir Monier Monier-Williams recorded some of this history, in remarkably emotional detail, in his still-unpublished autobiography in the Bodleian's archives. Some of my favorite conversations with Chris were about the benighted Monier-Williams and the ill-fated Indian Institute, once at the very center of the city, a royal elephant atop its weathervane. While it sometimes felt as if Sanskrit were another curio, like the many trinkets of empire that dotted the Pitt Rivers Museum, Chris ensured that its history would not be forgotten, most recently facilitating the acquisition of papers and portraits from the Monier-Williams family.

During my time at Oxford, I was lucky enough to add my notes to the work that Chris had begun with Polly O'Hanlon on social history in the study of Indian intellectual cultures. The core of my doctoral work studied the scholarly career of a family of Maharashtrian Brahmins in Banaras, one of the many who featured in Chris and Polly's work. They sent me to the Chandra Shum Shere collection of Sanskrit manuscripts in the Bodleian Library. Since the manuscripts had been bought wholesale from Banaras over a century ago, they contained material traces of the family networks Chris had charted out. There was a distinct pleasure in being able to share with Chris the experience of making notes on manuscript marginalia, ownership stamps, and scribal signatures that would never feature in our written work. In addition to participating in two iterations of the Oxford Early Modern South Asia workshop, I had the good fortune to work with Chris on another collaborative research project called the "Age of Vedanta." We met in various locations around the world—Delhi, Toronto, Vancouver, Stanford—to read together and share writing. Chris had the unparalleled ability to draw the group out of the minutiae of the text in order to ask the questions with the broadest implications. His essay for the volume will usher in yet another landmark research agenda on traditions of scholarship in late colonial India.

Chris' tenure as the Boden Professor will be remembered for his foundational studies in the history of Sanskrit scholarship, a fitting contribution to this foundational post. From astrology to Advaita, there are few śāstras he has not studied. His studies are distinguished by their attention to the authors of śāstric works, whether it be singular personalities like Nīlakaṇṭha Caturdhara or Anantakṛṣṇa Śāstrī, or entire generations of scholars in early modern India. He has helped us understand what goes into the making of a Sanskrit scholar, and has in turn exemplified those qualities.

Anand Venkatkrishnan
Assistant Professor,
University of Chicago Divinity School

Introduction

Christopher T. Fleming

> It seems fitting for a book in honor of Chris Minkowski that he is probably one of a handful of people in the world who could enjoy and understand all the chapters equally … because they cover a range of topics rarely mastered or known equally to a single person (Minkowski being an exception) … What you get, therefore, is something like four to five mini-books in one.
>
> Anonymous reviewer of Festschrift manuscript

∴

The epigraph above foregrounds the difficulty of editing a Festschrift for Professor Christopher Z. Minkowski, the erstwhile Boden Professor of Sanskrit at Balliol College, Oxford. Over the course of his forty-plus-year career, Professor Minkowski has written about, collaborated with leading scholars of, and supervised students within, *inter alia*, the disciplines of Vedic studies, astral sciences, the epics, Vedānta, and Sanskrit jurisprudence. Take, for example, his three books: *Priesthood in Ancient India: A Study of the Maitrāvaruṇa Priest*; *A Descriptive Catalogue of the Sanskrit Astronomical Manuscripts Preserved at the Maharaja Man Singh II Museum in Jaipur, India*; and *Scholar Intellectuals in Early Modern India: Discipline, Sect, Lineage and Community*.[1] These volumes attest to Minkowski's expertise in intricate Vedic philology, abstruse exact science, and complex Sanskrit knowledge systems.

Similarly, the essays in this Festschrift reflect the depth and breadth of the expertise of Professor Minkowski's students and colleagues, covering Vedic

1 Christopher Z. Minkowski, *Priesthood in Ancient India: A Study of the Maitrāvaruṇa Priest*, Publications of the De Nobili Research Library 18 (Vienna: University of Vienna, 1991); *A Descriptive Catalogue of the Sanskrit Astronomical Manuscripts Preserved at the Maharaja Man Singh II Museum in Jaipur, India*, compiled by David Pingree from the notes taken by Setsuro Ikeyama, Christopher Minkowski, David Pingree, Kim Plofker, Sreeramula Rajeswara Sarma, and Gary Tubb (Philadelphia: American Philosophical Society, 2003); and *Scholar Intellectuals in Early Modern India: Discipline, Sect, Lineage and Community*, edited with Rosalind O'Hanlon and Anand Venkatkrishnan (Routledge: Abingdon, 2015).

 | DOI:10.1163/9789004536869_002

zoology, eighteenth-century global astronomy, scholarly commentary, and juridical science, to name a few. Perhaps the reader, like the anonymous reviewer, might approach the essays in this volume as five distinct 'mini-books' and thus discern in them evidence of Professor Minkowski's status as an anachronism: a vestige, borne from intellectual acumen, of the once common, but now uncommon, capacity of great minds to operate felicitously within many fields of intellectual endeavour.[2] The effect of reading this volume in this manner, might be, then to find a section within one's area of competence, bemoan the hyper-specialization of Sanskrit studies, label Minkowski *sui generis*, and quote the Bard's words that "he was a man, take him for all in all, I shall not look upon his like again."[3]

1 Christopher Z. Minkowski: An Intellectual History

This paean, of course, risks reducing Professor Minkowski to "an absolute *Iohannes fac totum* (a Jack-of-all-trades)" and viewing the suturing thread of this Festschrift as the serendipitous vicissitudes of his scholarly career.[4] The contributions from Anne Blackburn and Anand Venkatkrishnan in Part 1 of this volume offer a mixture of praise and biography, and, before turning to the *thematic* unity of this Festschrift and of Minkowski's oeuvre, I attend to their *biographical* unity. Professor Minkowski began his career as a Vedicist, studying under Daniel H.H. Ingalls at Harvard University (1975–1986) and produced a monograph on the Maitrāvaruṇa Priest in 1991 as well as a brace of exceptional articles (that prefigured his later work on the Sanskrit epics and their commentaries) on the relationship between the *Mahābhārata* and the structure of Vedic rituals.[5] Over the course of time, Minkowski pursued other interests,

2 For a history of polymathism and an analysis of how "the further acceleration of knowledge has led to increased specialisation and to an environment that is less supportive of wide-ranging scholars and scientists," see Peter Burke, *The Polymath: A Cultural History from Leonardo da Vinci to Susan Sontag* (New Haven, CT: Yale University Press, 2020).

3 *Hamlet*, Act 1, Scene 2.

4 The attack on Shakespeare appears in Robert Greene's *Groats-worth of Witte* (1592). See *Groats-worth of Witte*, ed. G.B. Harrison, *The Bodley Head Quartos* (1923; rpt. New York: Barnes & Noble, 1966), pp. 45–46. For a wider history of the term, see Allen Carroll, "*Johannes Factotum* and Jack Cade," *Shakespeare Quarterly* 40.4 (1989), pp. 491–492.

5 For the monograph, see note 1, above. For the brace of articles, see "Snakes, Sattras and the Mahābhārata," in *Essays in the Mahābhārata*, ed. Arvind Sharma (Leiden: Brill, 1991), pp. 384–400; and "Janamejaya's Sattra and Ritual Structure," *Journal of the American Oriental Society* 109.3 (1989), pp. 401–420. Also see, "The Interrupted Sacrifice and the Sanskrit Epics," *Journal of Indian Philosophy* 29 (2001), pp. 169–186.

but he retained his interest and expertise in Vedic literature: he co-taught Vedic Sanskrit at Oxford with Elizabeth Tucker and her successor, John Lowe (their essays on tetanus in the *Artharaveda* and sandhi in the *Ṛgvedapadapāṭha* celebrate him in this volume); produced a series of articles on sundry aspects of Vedic studies; and supervised Barbora Sojková's brilliant dissertation on Vedic bovine zoology.[6]

While employed as a Sanskrit lecturer in the Department of Classics at Brown University (1985–1987), Minkowski came under the influence of David Pingree, the *primus inter pares* amongst scholars of the exact sciences in Sanskrit, and developed remarkable expertise in Sanskrit astronomy (Jyotiḥśāstra). Minkowski played an instrumental role in the compilation of Pingree's monumental *Descriptive Catalogue of the Sanskrit Astronomical Manuscripts Preserved at the Maharaja Man Singh II Museum in Jaipur, India* and he published widely on technical aspects of Sanskrit astronomy as well as the comparative history of science.[7] Building from David Pingree's work, Minkowski took a

6 For Minkowski's contribution to Vedic studies, see "The *Udumbara* and Its Ritual Significance," *Wiener Zeitschrift für die Kunde Südasiens* 33 (1989), pp. 5–23; "The Rathakāra's Eligibility to Sacrifice," *Indo-Iranian Journal* 32 (1989), pp. 177–194; "An Omen Reader and a Cypher in Vedic Manuscripts," in *Vāman Śāstrī Bhāgavat Felicitation Volume*, eds. Saroja Bhate and Madhav Deshpande (Poona: Vaidika Samshodhana Mandala, 1994), pp. 122–127; "School Variation in the Text of the Nivids," in *Inside the Texts Beyond the Texts*, ed. M. Witzel, Harvard Oriental Series Opera Minora Vol. 2 (Cambridge, MA: Harvard University Press, 1997), pp. 167–184; "Raccolte di Aforismi sui Rituali Domestici (Gṛhyasūtra)," in *Storia della Scienza*, ed. Sandro Petruccioli (Rome: Enciclopedia Italiana, 2001), pp. 766–768; "I Veda," ("The Vedas,") in ibid., pp. 709–715; "Testi per i Rituali Solenni (Śrautasūtra)," in ibid., pp. 756–761; "Nīlakaṇṭha Caturdhara's *Mantrakāśīkhaṇḍa*," *Journal of the American Oriental Society* 122.2 (2002), pp. 329–344; "The Vedastuti and Vedic Studies: Nīlakaṇṭha on Bhāgavata Purāṇa x.87," in *The Vedas: Texts, Languages & Ritual. Proceedings of the Third International Vedic Studies Workshop*, ed. A. Griffiths and J.E.M. Houben (Groningen: Egbert Forsten, 2004), pp. 125–142; and "Meanings Numerous and Numerical: Nīlakaṇṭha and Magic Squares in the *Ṛgveda*," in *Indologica T. Ya. Elizarenkova Memorial Volume*, ed. L. Kulikov and M. Rusanov (Moscow: Russian State University for the Humanities, 2008), pp. 315–328. Also see Barbora Sojková's "When Cows Sacrifice" in this volume and "Animals in Vedic Literature" (unpublished D. Phil. thesis, Balliol College, Oxford, 2022).

7 For the catalogue, see note 1, above. See Clemency Montelle and Kim Plofker's essay in this volume for Minkowski's contributions to that project: "Chris produced a significant part of the manuscript cataloguing notes for the astronomical material (in addition to personally couriering a portable photocopier from New York to Jaipur—a fairly challenging mission with the technology of a quarter-century ago!—to facilitate its study). The manuscripts described in the resulting catalogue … furnish the main subject of our discussion in this modest tribute to Chris's many accomplishments in the study of Sanskrit literature and knowledge systems." For Minkowski's essays on technical problems in Indian astronomy, see "A Nineteenth Century Sanskrit Treatise on the Revolution of the Earth: Govinda Deva's Bhūmibhramaṇa," SCIAMUS

particular interest in a subgenre of Jyotiḥśāstra literature devoted to the reconciling mathematical inconsistencies between Purāṇic and Siddhāntic cosmological models.[8] In this context, Minkowski encountered, in what evolved into a paradigmatic change in his career trajectory, the intellectual oeuvre of the seventeenth-century Sanskrit polymath Nīlakaṇṭha Caturdhara, when he examined Caturdhara's *Saurapaurāṇikamatasamarthana*.[9] Minkowski discovered that Nīlakaṇṭha furnishes an elaborate discussion of cosmology in his famous commentary on the *Mahābhārata*, the *Bhāratabhāvadīpā*'s (Lamplight on the True Meaning of the Bhārata) section on the *Bhīṣmaparvan*. While untangling the threads of these arguments, Minkowski developed a longstanding interest in Nīlakaṇṭha's exegetical efforts in relation to the *Mahābhārata* as well as his other sundry works.[10] Minkowski and Vishal Sharma's current Lever-

---

5 (2004), pp. 199–224; "Learned Brahmins and the Mughal Court: the Jyotiṣas," in *Religious Interactions in Mughal India*, ed. Vasudha Dalmia and Munis Faruqui (New Delhi: Oxford University Press, 2014), pp. 102–134; and "The Ṛtuvarṇana Chapter of Jñānarāja's Siddhāntasundara: Text, Translation, and Notes," *Gaṇita Bhāratī*, 33.1–2 (2011), pp. 53–87 (with Toke Knudsen). For comparative work, see "The Study of Jyotiḥśāstra and the Uses of Philosophy of Science," *Journal of Indian Philosophy* 36 (2008), pp. 587–597; "Astronomers and Their Reasons: Working Paper on Jyotiḥśāstra," *Journal of Indian Philosophy* 30 (2002), pp. 495–514; "Sanskrit Scientific Libraries and their Uses: Examples and Problems of the Early Modern Period," in *Looking at it from Asia: The Processes that Shaped the Sources of the History of Science*, ed. F. Bretelle and C. Proust (Dordrecht, etc.: Boston Studies in Philosophy of Science 265, 2010), pp. 81–114; and "Big Science: Classicism and Conquest," in *What China and India Once Were: The Pasts that May Shape the Global Future*, ed. Benjamin Elman and Sheldon Pollock (New York: Columbia University Press, 2018), pp. 199–231 (with Benjamin Elman).

8 See, for example, "The Paṇḍit as Public Intellectual: The Controversy over Virodha or Inconsistency in the Astronomical Sciences," in *The Pandit: Traditional Scholarship in India*, ed. Axel Michaels (New Delhi: Manohar Publications, 2001), pp. 79–96; and Minkowski's contribution to David Pingree's Festschrift, "Competing Cosmologies in Early Modern Indian Astronomy," in *Ketuprakāśa: Studies in the History of the Exact Sciences in Honor of David Pingree*, ed. Charles Burnett, Jan Hogendijk, and Kim Plofker (Leiden: Brill, 2004), pp. 349–385. Minkowski's article builds on Pingree's, "The Purāṇas and Jyotiḥśāstra: Astronomy," *Journal of the American Oriental Society* 110.2 (1990): 274–280.

9 Unsurprisingly, for details on Nīlakaṇṭha's life and works, see Minkowski, "'Nīlakaṇṭha's Teachers and Gurus, Part 1: Lakṣmaṇa Paṇḍita'," *Wiener Zeitschrift Kunde Sudasiens* 55 (2013–2014), pp. 33–76; "On the Success of Nīlakaṇṭha's Mahābhārata Commentary," in *Boundaries, Dynamics and Construction of Traditions in South Asia*, ed. F. Squarcini (Florence: Florence University Press, 2005), pp. 225–252; and "Nīlakaṇṭha and His Historical Context," *Vostok Oriens—Moscow Academy of Sciences* 4 (2008), pp. 37–49.

10 "Nīlakaṇṭha's Cosmographical Comments in the Bhīṣmaparvan," *Purāṇa* 42 (2000), pp. 24–40. For Nīlakaṇṭha on the *Mahābhārata*, see "The Meaning and Design of the Mahābhārata in Nīlakaṇṭha's Commentary," in *Proceedings of the Pondicherry Conference on Commentary*, ed. F. Grimal and C. Vielle (forthcoming); "Nīlakaṇṭha's Vedic Readings in the

hulme Trust-funded project, "The Mahabharata's Supporting Texts in Literary and Cultural History," aims to produce two new monographs and a series of articles that

> investigate ... the cultural moment in late-medieval and early modern India when the genre of epic commentary emerged, and examines how premodern scholastic, religious and literary communities used these texts to debate the epic's meaning.[11]

Nīlakaṇṭha's extensive writings on Vedānta (broadly, 'theology') provided another inflection point in Professor Minkowski's academic career.[12] As Anand Venkatkrishnan notes in his laudatory contribution to this volume,

> Before arriving in Oxford ... I had only known Christopher Minkowski as the Advaita guy. His article "Advaita Vedanta in Early Modern History" ... patiently mapped out the relationships, families, text traditions, and research agendas that constituted the social and intellectual history of Advaita Vedanta in the sixteenth and seventeeth centuries. I had always been interested in the lives of scholars, but it wasn't until reading this essay that I learned that you could actually write about them ... For Chris, ideas are inseparable from the people who articulate them. I see it in the way he speaks of his teachers, his fellow students, and even his college

---

Harivaṃśa Commentary," in *Epic, Khilas, and Purāṇas: Continuities and Ruptures. Proceedings of the Third Dubrovnik Conference on the Sanskrit Epics and Purāṇas, September 2002*, ed. Petteri Koskikallio (Zagreb: Croatian Academy of Sciences and Arts, 2005), pp. 411–413; "Nīlakaṇṭha's Instruments of War: Modern, Vernacular, Barbarous," *Indian Economic and Social History Review* 41 (2004), pp. 365–385. For Nīlakaṇṭha's Vedic works, see, note 6, above.

11 For a description of the project and its outputs, see https://www.orinst.ox.ac.uk/the-mahabharatas-supporting-texts-in-literary-culture-and-history. For an excellent monograph on the ending of the Mahābhārata by Minkowski's erstwhile student, Naama Shalom, see *Re-ending the Mahābhārata; A Study of the Rejection of Justice* (*Dharma*) *in the Sanskrit Epic* (Albany: State University of New York Press, 2017).

12 According to Minkowski's "Nīlakaṇṭha Caturdhara's Advaita Vedānta," in *The Oxford Handbook of Indian Philosophy*, ed. Jonardon Ganeri (Oxford: Oxford University Press, 2017), "Nīlakaṇṭha wrote at least four independent works on Vedānta: the *Determination of the Nonduality of Śiva* (*Śivādvaitanirṇaya*), the *Examination of the Three Jewels* (*Ratnatrayaparīkṣā*), the *Soap Nut for Vedānta* (*Vedāntakataka*), and the *Essence of the Six Systems* (*Ṣaṭtantrīsāra*), probably in that order ... he [also] wrote an Advaitin commentary on the "Praise by the Vedas" (*Vedastuti*) chapter of the *Bhāgavata Purāṇa* ... [and] the *Vedānta through Vedic Verses* (*Mantraśārīraka*) ... comments on verses from the *Rigveda* so as to disclose the topical sections (*adhikaraṇa*) of the *Brahma-sūtras* in their running order."

> friends … We often talk about the life of the mind as if it were the mind that mattered, when it's really the life.

Minkowski's seminal "Advaita Vedanta in Early Modern History" offers a representative example of his significant contributions to the scholarly study of Vedānta—contributions that continue in participation in the 'Age of Vedānta' project and that are reflected in Jonathan Duquette's contribution to this Festschrift.[13]

Given its relatively late emergence as a core area of his research and its comparatively humble percentage in his oeuvre, Vedānta's close association with Professor Minkowski is somewhat ironic. Nevertheless, Minkowski's contribution to Vedānta, as Venkatkrishnan notes, typifies his fifth area of competence: the intellectual history of Sanskrit knowledge systems in early modern India. In 2001, Professor Minkowski joined Sheldon Pollock's National Endowment for the Humanities-funded project, 'Sanskrit Knowledge Systems on the Eve of Colonization' (SKSEC), with the stated aim of investigating, "the nature of the 'knowledge systems' or scholarly disciplines in India on the eve of colonial rule."[14] The 'knowledge systems', sometimes described in the project grant as 'sciences' included: "linguistic analysis [Vyākaraṇa], hermeneutics [Mīmāṃsā], logic-epistemology [Nyāya], literary theory [Alaṅkāraśāstra], moral-political discourse [Dharmaśāstra], and astral science [Jyotiḥśāstra] [along with] medicine [Āyurveda] and Vedic ritualism …"[15] In the following decades, scholars associated with this project, including Professor Minkowski, produced

13 "Advaita Vedānta in Early Modern History," in *Religious Cultures in Early Modern India: New Perspectives*, ed. Rosalind O'Hanlon and David Washbrook. Special Volume of *South Asian History and Culture* 2.2 (2011), pp. 205–231; and "Appayya's Vedānta and Nīlakaṇṭha's Vedāntakataka," *Journal of Indian Philosophy* 44.1 (2016), pp. 95–114. The Age of Vedānta was a multi-year research project funded in part by the Social Sciences and Humanities Research Council of Canada. The project was spearheaded by Ajay Rao and Lawrence McCrea and brought together a number of Sanskritists and historians of religion, including Professor Minkowski. Together, the project argues that there was a period in India (c. 1500–1800) where Vedānta became the premiere discipline through which Hindu theology would be articulated. This work is, in some sense, a continuation of Minkowski's work on Advaita Vedānta in that it shows on a large scale that this discipline was not frozen in time, but transformed by agents, institutions, and patronage.

14 From the proposal, see http://www.columbia.edu/itc/mealac/pollock/sks/proposal.html#workplan. Also see Sheldon Pollock, "Introduction: Working Papers on Sanskrit Knowledge-Systems," *Journal of Indian Philosophy* 30.5 (2002), pp. 431–439.

15 Ibid. "Forms of Sanskrit science without close analogy in the West, such as hermeneutics or language analysis, disappeared as creative modes of thought in the face of what were vaguely comparable European forms (in this case, historical textual criticism, on the one hand, and descriptive grammar on the other)."

an astonishing body of research that not only examined the complex, interdisciplinary nature of scholastic Sanskrit writings from India's early modern period (c. 1500–1800), but also how "discipline, sect, lineage, and community" as well as political factors shaped the lives and writings of various authors.[16]

My own journey to Oxford to study early modern Sanskrit jurisprudence was the result of attending a conference, held at Saint Antony's College in 2013, organized by Professors Minkowski and Rosalind O'Hanlon: "Discipline, Sect, Lineage, and Community: Scholar Intellectuals in India c. 1500–1800." I was fascinated by Minkowski and O'Hanlon's microhistories of groups of Mahārāṣṭrian scholars of Sanskrit jurisprudence (Dharmaśāstra), such as Kamalākarabhaṭṭa (active 1610–1640) and Anantadeva (seventeenth-century), who intervened in the social life of the Marāṭhā Empire.[17] The takeaway from the conference, articulated beautifully in Minkowski, O'Hanlon, and Venkatkrishnan's succinct introduction to the resulting edited volume, was that the history of the intellectual ideas of early modern indigenous Sanskrit scholars were inexorably intertwined with the social lives of those same scholars.[18] Minkowski and his collaborators' (especially Rosalind O'Hanlon, who has provided an essay for this volume) contribution steers scholarly focus towards a strand of intellectual

16 Christopher Minkowski et al., "Social History in the Study of Indian Intellectual Cultures?" *South Asian History and Culture*, 6.1 (2015): 1–9, 2. The primary research output of the project was, Sheldon Pollock, ed., *Forms of Knowledge in Early Modern Asia: Explorations in the Intellectual History of India and Tibet, 1500–1800* (London: Duke University Press, 2011). For a general account of early modernity, see John Richards, "Early Modern India and World History," *Journal of World History* 8.2 (1997), pp. 198–203. For intellectual history in India, see Pollock, "Is There an Indian Intellectual History? Introduction to 'Theory and Method in Indian Intellectual History,'" *Journal of Indian Philosophy* 36.5–6 (2002): 431–439. Also see Sheldon Pollock, "New Intellectuals in Seventeenth Century India," *Indian Economic and Social History Review* 38.1 (2001), pp. 3–31.

17 See, for example, "What Makes People Who They Are? Pandit Networks and the Problem of Livelihoods in Early Modern Western India," *Indian Economic and Social History Review* 45.3 (2008), pp. 381–416 (with Polly O'Hanlon). This article is part of a longer collaboration with O'Hanlon. See her extensive works on the social history of Vārāṇasī and Mahārāṣṭra: "Letters Home: Banaras Pandits and the Maratha Regions in Early Modern India," *Modern Asian Studies* 44.2 (2010), pp. 201–240; "The Social Worth of Scribes: Brahmins, Kayasthas and the Social Order in Early Modern India," *Indian Economic Social History Review* 47.4 (2010), pp. 563–595; "Speaking from Śiva's Temple: Banaras Scholar Households and the Brahman 'Ecumene' of Mughal India," *South Asian History and Culture* 2.2 (2011), pp. 253–277; "Contested Conjunctures: Brahman Communities and 'Early Modernity' in India," *American Historical Review* 116.2 (2013), pp. 765–787; and "Discourses of Caste over the Longue Durée: Gopīnātha and Social Classification in India, ca. 1400–1900," *South Asian History and Culture* 6.1 (2015), pp. 102–129 (with Gergely Hidas and Csaba Kiss).

18 See Minkowski et al., "Social History."

history that "understood texts as active expressions of language, or speech acts … [which] made interventions in a broader discourse or milieu within which they participated."[19] This 'Skinnerian' spirit of contextualism, or, how "a particular document or pronouncement [may be] situated in a biographical, social, political and literary context rich enough to enable an inference to be drawn about the nature of the illocutionary intervention the document embodies" animates Minkowski's sundry essays on early modern Sanskrit scholiasts.[20]

In short, one unifying facet of this Festschrift is biographical: the essays illuminate the multifaceted Sanskritic intellectual endeavors that piqued Professor Minkowski's interest during his long and illustrious career. This volume, therefore, helps to dispel his typecasting as 'the Vedānta guy' or 'the scholar of Nīlakaṇṭha,' as an erstwhile Vedacist, or even as a scholar of Sanskrit astronomy. His expertise in one field informs his expertise in another field, and the reader of this Festschrift will, it is to be hoped, garner a greater appreciation for Minkowski's intellectual development. Furthermore, the deep interconnectedness of his disciplinary expertise suggests that, unlike a Jack-of-all-trades, Minkowski is a master-of-many. While this biographical account of the essays in the present volume serves one purpose of a Festschrift admirably—that is, praising a mentor, teacher, and colleague in a way that highlights their contributions to their fields—we are still left with the task of providing a *thematic* account of the volume. While I recognize that the Festschrift's diversity of disciplines makes this a challenging task, it enables us to deconstruct two meth-

---

19 Anand Venkatkrishnan, "Mīmāṃsā, Vedānta, and the Bhakti Movement" (unpublished PhD dissertation, Columbia University, 2015), pp. 13–14.

20 For the use of contextualism in intellectual history in early modern India, see Jonardon Ganeri, "Contextualism in the Study of Indian Intellectual Cultures," *Journal of Indian Philosophy* 36.5 (2008), pp. 551–562, 553. Also see Quentin Skinner, *Visions of Politics*, Vol. 1: *Regarding Method* (Cambridge: Cambridge University Press, 2002). For Minkowski's writings on early modern Sanskrit knowledge systems (excluding Astronomical, Vedānta, and Dharmaśāstric works mentioned above), see, "Outlandishness in Sanskrit Commentary," in *Les Scolastiques Indiennes: Genèses, Développements, Interactions. Sous la direction d'Émilie Aussant et de Gérard Colas* (Paris/Pondichéry: École Française d'Extrême-Orient, Collection Études Thématiques, 2020), pp. 185–202; "An Early Modern Account of the Views of the Miśras," *Journal of Indian Philosophy* 46 (2018), pp. 889–933; "Maryādām Ullaṅghya: the Boundaries of Interpretation in Early Modern India," in *Canonical Texts and Scholarly Practices: A Global Comparative Approach*, ed. A. Grafton and G. Most (Cambridge: Cambridge University Press, 2016), pp. 90–109; "Apūrvaṃ Pāṇḍityam: On Appayya Dīkṣita's Singular Life," *Journal of Indian Philosophy* 44.1 (2016), pp. 1–10; "I'll Wash Out Your Mouth With My Boot: A Guide to Philological Argument in Early Modern Banaras," in *Epic and Argument in Sanskrit Literary History: Essays in Honor of Robert P. Goldman*, ed. Sheldon Pollock (Delhi: Manhar, 2010), pp. 117–141.

odological terms whose assumed meanings sometimes impede the pursuit of interdisciplinary research in Sanskrit studies: science and society.

2 Science and Society in the Sanskrit World

This Festschrift bears the imprimatur of Brill's Sir Henry Wellcome Asian Series, a series that strives "to make available fine editions of the medical and scientific classics of Asia." At first blush, perhaps, only the third section of the present volume, covering the 'Astral and Medical Sciences,' matches the themes of previous volumes in the Wellcome Series—which examine, *inter alia*, Jyotiśāstra manuscripts in the Wellcome Library, therapeutic uses of opium in ninth-century Bagdad, and Mongol-era Chinese dietary medicine.[21] Put another way, the reader might ask why this Festschrift, in which ostensibly non-'scientific' essays on Vedic philology, statecraft, and theology out-proportion medical and mathematical essays in a three-to-one ratio, is included in the Wellcome Series, *honoris causa*, merely on the basis of one section on the astral and medical sciences? The answer, of course, hinges on the range of meanings of the word 'science.'

In this volume, I employ the word 'science' in a twofold sense. The first sense is the common, knockabout use of science as the rigorous, mathematically informed theorization of the natural world through the use of testable experimentation and empirical evidence.[22] Seminal, twentieth-century historians of science, such as Herbert Butterfield, A. Rupert Hall, and Thomas Kuhn crafted what Andrew Cunningham and Perry Williams have described, critically, as the 'big picture' narrative of the 'scientific revolution' and its aftermath.[23] The standard narrative, as Cunningham and Williams explain,

---

21 David Pingree, *Catalogue of Jyotiṣa Manuscripts in the Wellcome Library: Sanskrit Astral and Mathematical Literature* (Leiden: Brill, 2004); Selma Tibi, *The Medicinal Use of Opium in Ninth-Century Baghdad* (Leiden: Brill, 2006); Paul Buell and Eugene Anderson, eds. and trans., *A Soup for the Qan: Chinese Dietary Medicine of the Mongol Era As Seen in Hu Sihui's Yinshan Zhengyao: Introduction, Translation, Commentary, and Chinese Text*, 2nd revised and expanded edition (Leiden: Brill, 2010). Many thanks to Eric M. Gurevitch (Vanderbilt) for his immensely helpful comments regarding the history of science.

22 For contemporary definitions of science, see John Heilbron, *The Oxford Companion to the History of Modern Science* (New York: Oxford University Press), pp. vii–x: "modern science is a discovery as well as an invention. It was a discovery that nature generally acts regularly enough to be described by laws and even by mathematics; and required invention to devise the techniques, abstractions, apparatus, and organization for exhibiting the regularities and securing their law-like descriptions." Also see E.O. Wilson, "The Natural Sciences," in his *Consilience: The Unity of Knowledge* (New York: Vintage, 1999), pp. 49–71.

23 Andrew Cunningham and Perry Williams, "De-Centring the 'Big Picture': 'The Origins of

enshrines the belief that the scientific revolution of the late sixteenth- and early seventeenth-centuries marked "the true beginning of the modern world and the abandonment of the ancient and medieval world."[24] Most people possess a cursory awareness of the canonical events that marked this "transformation of human thought into a form close to that of 'scientific knowledge' as defined by twentieth-century philosophers of science": Nicolaus Copernicus' heliocentric cosmology in *De Revolutionibus Orbium Coelestium* (1543); William Harvey's mechanistic cardiac biology in *Exercitatio Anatomica de Motu Cordis et Sanguinis in Animalibus* (1628); Galileo Galilei's observational defense of Copernican Heliocentrism in *Dialogo Sopra i due Massimi Sistemi del Mondo* (1632, Latin trans. 1635); and Newton's classical mechanics in *Philosophiæ Naturalis Principia Mathematica* (1687).[25]

One problem with this narrative, as Cunningham, Williams, and others have argued in detail, is that it involves an anachronistic projection of an essentialized, transhistorical model of 'science' as "a universal human enterprise … [comprising] all respectable knowledge about the natural world" onto historical phenomena. Harvey's mechanistic model of the heart, for example,

> once seen as the exemplary case of the physiological application of mechanical ideas … is now generally accepted to have been the result of an essentially Aristotelian investigation into the 'final cause' of the heart's motion and structure, into the ways in which this organ served the purposes of the soul.[26]

Likewise, Newton, who saw himself as a 'natural philosopher,' rather than a 'scientist' employed "theology, mysticism, alchemy and biblical chronology" as part of "a coherent attempt to reach a deeper understanding of the Christian God by studying His Creation."[27]

Modern Science' and the Modern Origins of Science," *The British Journal for the History of Science* 26.4 (1993), pp. 407–432. Also see Herbert Butterfield, *The Origins of Modern Science 1300–1800* (London: G. Bell & Sons, 1949); A. Rupert Hall, *The Scientific Revolution 1500–1800: The Formation of the Modern Scientific Attitude* (London: Longmans, Green and Co., 1954); and Thomas S. Kuhn, *The Copernican Revolution: Planetary Astronomy in the Development of Western Thought* (Cambridge, MA: Harvard University Press, 1957).

24 Ibid., p. 408.

25 Ibid., p. 412.

26 Ibid., 413. Also see Andrew Cunningham, "William Harvey: The Discovery of the Circulation of the Blood," in *Man Masters Nature: 25 Centuries of Science*, ed. Roy Porter (London: BBC Books, 1987), pp. 65–76.

27 Ibid., 414. Also see J.E. McGuire and P.M. Rattansi, "Newton and the 'Pipes of Pan'" *Notes*

Furthermore, as Fritz Staal notes,

> Newton expressed the principles of his Principia by drawing figures, employing tables and abbreviations, but without using the simple equations that we have been taught in school.[28]

Newton's idiosyncratic mathematical language, reminiscent in some respects of the notational strategies explored by Takanori Kusuba in his analysis of Nārāyaṇapaṇḍita's '*meru*' diagrams in this Festschrift, required a series of interlocutors, including Leonhard Euler (1707–1783) and Joseph Lagrange (1736–1813).[29]

Rejecting the ahistorical conception of science as a "timeless logic ... embodying absolute moral values of freedom, rationality and progress," Cunningham and Williams date "the invention" of science, as an "historically contingent ... [knowledge system] embodying the vales, aims and norms of a particular social group" to 1760–1848.[30] Adopting a Skinnerian approach, they argue that, during the "Age of Revolutions" in Western Europe, a radical transformation of the term science occurred: the Latin word *scientia* (knowledge), which had, previously, referred to a wide variety of knowledge systems, came to refer "*only* to the sciences of nature."[31] The process of inventing science—of delimiting modern from premodern meanings of the word science—developed in conjunction with economic, political, and social transformations: the industrial revolution, the rise of the middle class, the creation of nation states, secu-

and Records of the Royal Society 21.2 (1966), pp. 108–143; and Simon Schaffer, "Godly Men and Mechanical Philosophers: Souls and Spirits in Restoration Natural Philosophy," *Science in Context* (1987), pp. 55–85. For the distinction between science and natural philosophy, see Roger French and Andrew Cunningham, *Before Science: The Invention of the Friars' Natural Philosophy* (London: Routledge, 1996); and David Cahan, ed., *From Natural Philosophy to the Sciences: Writing the History of Nineteenth-Century Science* (London: University of Chicago Press, 2003).

28 Fritz Staal, "The Sanskrit of Science," *Journal of Indian Philosophy* 23.1 (1995), pp. 73–127, 75. For example, Newton expresses his second law, $f = ma$ (force is the product of mass and acceleration), in Latin prose: mutationem motis proportionalem esse vi motrici impressae, et fieri secundum lineam rectam qua vis ilia.

29 Ibid., pp. 76–77. Also see C. Truesdell, "Reactions of Late Baroque Mechanics to Success, Conjecture, Error, and Failure in Newton's *Principia*," in his *Essays in the History of Mechanics* (Berlin: Springer-Verlag, 1968), pp. 138–183; and John Fauvel, "Newton's Mathematical Language," in *The Foundations of Newtonian Scholarship*, ed. Richard Dalitz and Michael Nauenberg (London: World Scientific), pp. 145–159.

30 Cunningham and Williams, "De-Centring," p. 418.

31 Ibid., p. 422.

larization, and European imperialism.[32] Key events in this sequence include Laplace's desacralization of Newtonian physics in the *Traité de mécanique céleste* (1798–1825); the founding of the *Gesellschaft Deutscher Naturforscher und Ärzte* (1822) in the wake of the Kantian revolution in knowledge; and, in the English case, either the founding of University College, London as a secular institution (1826) or William Whewell's coining the term, 'scientist' in 1834.[33]

Historicizing the invention of 'science' enables us to tackle its vexed relationship to imperialism and to garner a more nuanced approach to using science as a standard of comparison when examining the history of similar knowledge systems in non-Western cultures. These eighteenth- and nineteenth-century definitional projects were often—perhaps always—used to separate the West from the Rest, by using 'science' (in the singular with a single history and methodology) to serve as a mark of Western modernity and to justify colonial domination of ostensibly backwards communities.[34] Comparative histories of science must dispense with an 'evolutionary view' in which non-Western cultures (and, indeed, pre-modern Western cultures) developed imperfect, defective versions of 'true' scientific knowledge, epitomized by a phrase from Al-Biruni's *Taḥqīq mā li-l-Hind* (redeployed against the *Ain-i-Akbari* by H.T. Colebrooke in the late

32 Ibid., pp. 425–426.

33 For the Académie des sciences during the Age of Revolutions and Laplace, see Charles Gillispie, *Science and Polity in France: The Revolutionary and Napoleonic Years* (Princeton, NJ: Princeton University Press, 2004); and *Pierre-Simon Laplace, 1749–1827: A Life in Exact Science* (Princeton, NJ: Princeton University Press, 1997). For Germany, see Denise Phillips, *Acolytes of Nature: Defining Natural Science in Germany, 1770–1850* (Chicago: University of Chicago Press, 2012). For England, the situation was complicated by a Christian, Tory conservatism at Oxbridge that did not subside until the 1860s. See, *inter alia*, J.B. Morrell, "Brewster and the early British Association for the Advancement of Science," in *Martyr of Science: Sir David Brewster 1781–1858*, ed. A.D. Morrison-Low and J.R.R. Christie (Edinburgh: Royal Scottish Museum, 1984), pp. 25–29; Richard Yeo, *Defining Science: William Whewell, Natural Knowledge and Public Debate in Early Victorian Britain* (Cambridge: Cambridge University Press, 1993); and Sydney Ross, "*Scientist*: The Story of a Word," *Annals of Science* 18.2 (1962), pp. 65–85.

34 See Brett Bennett and Joseph Hodge, eds., *Science and Empire Knowledge and Networks of Science across the British Empire, 1800–1970* (Basingstoke: Palgrave Macmillan, 2011); Marwa Elshakry, "When Science Became Western: Historiographical Reflections," *Isis* 101.1 (2010), pp. 98–109; John Gascoigne, *Science in the Service of Empire: Joseph Banks, the British State and the Uses of Science in the Age of Revolution* (Cambridge: Cambridge University Press, 1998); Michael Dodson, "Translating Science, Translating Empire: The Power of Language in Colonial North India," *Comparative Studies in Society and History* 47.4 (2005) pp. 809–834; and Zaheer Baber, *The Science of Empire: Scientific Knowledge, Civilization, and Colonial Rule* (Albany: State University of New York Press, 1996).

eighteenth-century), "a dunghill, in which a pearl or two lie hid."[35] Nevertheless, as Fritz Staal notes,

> historians of mathematics write about ancient mathematics from the point of view of contemporary mathematics ... [and] I believe ... that we cannot but accept what historians of science like Joseph Needham have laid down: 'To write the history of science we have to take modern science as our yardstick.'[36]

Similarly, Peter Dear, Professor Minkowski's former colleague at Cornell notes that, "acknowledging that science is not ... a natural kind, while ... recognizing that ... 'science' is culturally very real" can perhaps allow us to avoid essentialism while "speaking coherently about ... knowledge enterprises, such as the mathematical sciences, that had a career that spread across the Eurasian continent."[37] Clemency Montelle and Kim Plofker's analysis of trigonometric tables in eighteenth-century Jaipur exemplifies this Eurasian history of scientific development. Furthermore, employing a Western normative definition of science also enables us to see some of the scientific aspects of essays in this Festschrift that are not in the section devoted to the astral and medieval sciences. For example, Elizabeth Tucker's analysis of tetanus' causes and treatments in the *Atharvaveda* or Barbora Sojková's taxonomic analysis of bovine evolution in late Vedic texts presuppose modern, scientific concepts of

---

35 E.W. Sachau, trans., *Alberuni's India: An Account of the Religion, Philosophy, Literature, Geography, Chronology, Astronomy, Customs, Laws and Astrology of India about* A.D. *1030*, Vol. 1 (London: Trübner & Co 1888), pp. 24–25: "(Hindu scientific theorems) are ... devoid of any logical order, and ... I can only compare their mathematical and astronomical knowledge to a mixture of pearls ... and dung, ... Both kinds of things are equal in their eyes since they cannot raise themselves to the methods of a strictly scientific deduction." For Colebrooke's version of the aspersion, see Daniel Ingalls, "On the Study of the Past," *Journal of the American Oriental Society* 80.3 (1960), pp. 191–197, 192.

36 Staal, "Sanskrit of Science," p. 95. See Joseph Needham, *Science and Civilization in China*, Vol. 5: *Chemistry and Technology, Pt. Ill: Spagyrical Discovery and Invention: Historical Survey, from Elixirs to Synthetic Insulin* (Cambridge: Cambridge University Press, 1976), p. xxix. Needham's seminal history of science in China has been the subject of an extraordinary amount of reflecting over the years. For analyses of the 'Needham Question', see Florence Hsia and Dagmar Schäfer, "History of Science, Technology, and Medicine: A Second Look at Joseph Needham," *Isis* 110.1 (2019), pp. 94–99.

37 Peter Dear, "What Is the History of Science the History Of?: Early Modern Roots of the Ideology of Modern Science," *Isis* 96.3 (2005), pp. 390–406, 406. Dear attributes this position to Minkowski.

medicine or zoology that are etic to the ancient Sanskritic sources that they examine. Thus, science, as radically *etic*—from the Sanskritic perspective—serves as a useful, if not *exclusive*, yardstick to open up new vistas in this volume.

I also employ science in a second, more general sense, borrowed from Professor Minkowski as much as from Sheldon Pollock: "disciplines of systematized rational inquiry."[38] Under this rubric,

> a variety of terms drawn from the language of our own day are used to designate these subjects, such as "disciplines," "knowledge systems," "sciences," "arts," "learned traditions," and their variants.[39]

Contemporary historians and philosophers of science cast doubt on the ostensible uniformity of the "scientific method" and argue, either for the "disunity" of science as a discipline or for a more pluralist model of "the sciences"—formal, modern knowledge systems such as mathematics, physics, medicine, and so forth.[40] These disputes highlight the fact that, even within the narrow parameters of contemporary Western science, there has been a paradigm shift toward the study of knowledge systems—of the sciences that comprise modern science. If the "view that all scientific knowledge could be reduced to a single unified science in the form of general causal laws" is untenable, then, as Lorraine Daston argues, scholars ought to approach the history of science as a history of organized methods of *knowledge*.[41] As Daston notes, this inclusivist approach,

> licenses historical exploration of all corners of the contemporary *mappemonde* of the disciplines, embracing the humanities and social sciences as well as the sciences (and turning the translation of *Wissenschaft* or *sciences humaines*—or *physis* or *ziran* or *shizen*—from a headache into a research topic).[42]

---

38 Elman and Minkowski, "Big Science," p. 200.

39 Ibid., p. 200.

40 Cunningham and Williams, "De-Centring," p. 415. See, for example, Paul Feyerbend, *Against Method: Outline of an Anarchistic Theory of Knowledge* (London: Verso, 1978) and Peter Galison and David Stumps, eds., *The Disunity of Science: Boundaries, Contexts, Power* (Palo Alto, CA: Stanford University Press, 1996).

41 Lorraine Daston, "The History of Science and the History of Knowledge," *KNOW* 1.1. (2017), pp. 131–154.

42 Ibid., pp. 142–143. Daston's approach also flips the script on the perennial debate concerning the 'scientific' nature of the social sciences wherein questions of quantitative reason-

This broad formulation of the semantic range of the word 'science' enables us to identify a deeper connection between all the essays in this volume. Each author, from their own disciplinary background—Western and non-Western alike—presents a methodical, rational account of various phenomena from within the parameters of an intellectual discipline—their 'science.' This may seem too capacious: all academic writing would, presumably fall under this broad definition; but adopting it enables us to make a series of observations about the ostensibly unitary nature of science and about the ostensibly non-unitary nature of the sciences. The end result, therefore, is that this volume's secondary use of the term science emphasizes the plurality of the sciences—some emic to India, others etic to India—that, collectively, share the common feature of being organized, rational systems of inquiry. In essence, we may say that science (itself an historically contingent amalgamation of various sciences) is a science in the sense that it is one of many formalized knowledge systems.

In fact, this approach reflects the longstanding premodern usage of the term, 'science.' In medieval England, science, from the Latin *sciencia* (knowledge), was often used

> as a general term for knowledge as such, as in 'for God of sciens is lord' (1340), and this use was still active in Shakespeare: … hath not in natures mysterie more science then I have in this Ring. (All's Well That Ends Well, v, iii).[43]

Also, in English parlance,

> science became more generally used, often interchangeably with art, to describe a particular body of knowledge or skill: "his science Of metre, of rime and of cadence" (Gower, 1390); "thre Sciences … Divinite, Fisyk, and Lawe" (1421); "Liberal Sciencis … fre scyencis, as gramer, arte, fisike, astronomye, and otheris" (1422).

---

ing or predictive accuracy predominate. See, *inter alia*, Gunnar Myrdal, "How Scientific are the Social Sciences?" *Journal of Social Issues* 28.4 (1972), pp. 151–170; Karin Knorr-Cetina, "Social and Scientific Method or What Do We Make of the Distinction Between the Natural and the Social Sciences?" *Philosophy of the Social Sciences* 11.3 (1981), pp. 335–359; and Rein Taagepera, *Making Social Sciences More Scientific: The Need for Predictive Models* (Oxford: Oxford University Press, 2008).

43 Raymond Williams, "Science," *in Keywords: A Vocabulary of Culture and Society* (Oxford: Oxford University Press, 1985).

The medieval English usages reflect the country's Latin heritage and the classical, hierarchical model of the seven-fold lower arts—the *Trivium* (grammar, logic, and rhetoric) and the *Quadrivium* (arithmetic, geometry, music, and astronomy)—and the three higher sciences (medicine, jurisprudence, and, especially, theology).[44] The claim, attributed to Aquinas—and so often appropriated, uncritically by scholars in other disciplines—that theology is the 'queen of the sciences' serves as a timely reminder that the *least* scientific discipline from the perspective of modern science is certainly scientific from the perspective of Daston, Minkowski, and others.[45]

This broad approach to the sciences as organized knowledge systems enables us to see commonalities between, say, Stephanie Jamison's excellent linguistic analysis of *cvi* (a grammatical formation with the meaning 'become/make X') suffixes in the *Rig Veda* and Jonathan Duqette's survey of early modern Vedāntins' theological debates concerning the nature of Hindu deities. Both essays revolve around the application of rational, scientific systems.

The classical Sanskrit tradition articulates its own *emic* theory of organized knowledge and, even as I argue that this Festschrift is scientific in a Western, normative sense, I must acknowledge that it is also, and perhaps more so, scientific in an Indian, Sanskritic sense. How does the Sanskrit tradition theorize systematic, scientific knowledge? As Sheldon Pollock notes, "*Science, systematic knowledge, scholarship, learning* (as well as *rule* and even *scripture*) would

---

44 For classical models of the sciences, see James A. Weisheipl, "Classification of the Sciences in Medieval Thought," *Medieval Studies* 27 (1965), pp. 54–90; Frédéric Goubier and Irène Rosier-Catach, "The Trivium in the 12th Century," in *A Companion to Twelfth-Century Schools*, ed. Cédric Giraud (Leiden: Brill, 2019), pp. 141–179; and Edward Grant, *The Foundations of Modern Science in the Middle Ages: Their Religious, Institutional, and Intellectual Contexts* (Cambridge: Cambridge University Press, 1996).

45 While Aquinas argues for the primacy of theology among all other sciences, this specific expression, attributed to Aquinas does not actually appear in his *Summa Theologiae* (Question 1, Article 5), Aquinas notes that, "Other sciences are called the handmaidens of this one [theology]: 'Wisdom sent her maids to invite to the tower' [Proverbs 9:3]: Sed contra est quod aliae scientiae dicuntur ancillae huius, Prov. 9,3: misit ancillas suas vocare ad arcem." For Aquinas' theory of science, see Geoffrey Turner, "St Thomas Aquinas on the "Scientific" Nature of Theology," *New Blackfriars* 78.921 (1997): 464–476. For usages of the phrase in the exact and social sciences, see Michael Atiyah, "Mathematics: Queen and Servant of the Sciences," *Proceedings of the American Philosophical Society* 137.4 (1993), pp. 527–531; Paul Samuelson, "Credo of a Lucky Textbook Author," *Journal of Economic Perspectives* 11.2 (1997), pp. 153–160, 159; and Dick Houtman, "The Disenchantment of the World and the Authority of Sociology: How the Queen of the Sciences Lost Her Throne," in *Science Under Siege: Contesting the Secular Religion of Scientism*, ed. Dick Houtman, Stef Aupers, and Rudi Laermans (Basingstoke: Palgrave Macmillan, 2021), pp. 37–64.

all be legitimately translated by the Sanskrit word *śāstra*."[46] This begs the question: "what exactly is *śāstra*, and how does it relate to other, kindred concepts, such as *jñāna* (and *vijñāna*) and *vidyā* (all variously translated as knowledge, learning, scholarship .... and science?)"[47]

In a seminal essay on this question, Pollock invokes a Sanskrit proverb that encapsulates the common thread linking the multivalent usage of the word śāstra in Sanskrit literature, "clearing up all uncertainties and revealing the imperceptible, *śāstra* is the all-seeing eye—to lack it is to be truly blind."[48] So, śāstra is a vehicle for accurate knowledge about the world and, in the orthodox Brāhmaṇical worldview, a knowledge system's accuracy (its śāstricity) hinged on it capacity to reveal (invariably in Sanskrit) immutable cosmic archetypes anchored in the Vedas. In his *Nyāyamañjarī*, the ninth-century CE Kashmiri logician Jayantabhaṭṭa argues that,

> All sciences have existed, precisely like the vedas, from the first creation. People, however, ascribe them to one or another human author who has sought to abbreviate or expand them.[49]

Beyond this contingence of its epistemological validity on its connection to Vedic ontology, śāstra—artisanal, philosophical, or religious—"has a relatively formalized and public character ... [with] an unmistakable aspect ... of systematicity, stability, and repetition."[50]

One canonical, and relatively early (fifth-century) formulation of the śāstric sciences is the fourteen 'sources of knowledge' (*Vidyāsthāna*-s) that appear in *Yājñavalkyasmṛti* (1.3): the four Vedas; the six 'limbs of the Vedas' (*Vedāṅga*-s)—language analysis (Vyākaraṇa), etymology (Nirukta), phonetics (Śikṣā), ritual (Kalpa), metrics (Chandas), and astronomy (Jyotiḥśa); and jurisprudence (Dharmaśāstra), scriptural hermeneutics (Mīmāṃsā), logic (Nyāya), and his-

46 Sheldon Pollock, "The Languages of Science in Early Modern India," in *Forms of Knowledge in Early Modern Asia: Explorations in the Intellectual History of India and Tibet, 1500–1800*, ed. Sheldon Pollock (Durham, NC: Duke University Press, 2011), pp. 19–48, 21.

47 Ibid., p. 21.

48 Sheldon Pollock, "The Idea of Śāstra in Traditional India," in *Shastric Traditions in Indian Arts* ed. Anna Dallapiccola (Stuttgart: Steiner Verlag; Wiesbaden: GMBH, 1989), pp. 17–26, 17: anekasaṃśayocchedi parokṣārthasya darśakam / sarvasya locanaṃ śāstram yasya nasty andha eva saḥ //.

49 Sheldon Pollock, "The Theory of Practice and the Practice of Theory in Indian Intellectual History," *Journal of the American Oriental Society* 105.3 (1985), pp. 499–519, 516.

50 Pollock, "The Idea of Shastra," p. 19.

tory/mythology (Itihāsapurāṇa).[51] This scheme contrasts, seemingly, with a four-fold taxonomy of the sciences as the Vedas, politics (Daṇḍanīti), epistemology (Ānvīkṣikī), and economics (Vārttā) that appears in, among other places, the *Mānavadharmaśāstra*.[52] Jayantabhaṭṭa reconciles these two lists by distinguishing between śāstras which serve worldly goals (*dṛṣṭārtha*) and those which serve transcendent goals (*adṛṣṭārtha*).[53] Alternatively, the *Viṣṇupurāṇa* (3.6.28–29) solves this issue by raising the number of sciences to eighteen (including politics, medicine, drama, and weaponry).[54] Other authors categorized śāstric knowledge differently, but their task and rubric remained constant: accounting for secular disciplines such as erotology (Kāmaśāstra), elephantology (Gajaśāstra), and architecture (Śilpaśāstra) by distinguishing between the transcendent or secular purposes of various śāstras—even if, a priori, every śāstra, *qua śāstra*, reveals knowledge of a pre-existing transcendent reality.[55] As in the western tradition, intellectuals writing in Sanskrit also debated the hierarchy of the sciences, and Patañjali, in his *Vyākaraṇamahābhāṣya*, claimed that Vyākaraṇa was foremost of the six Vedāṅgas (*pradhānam ca ṣaṭsu aṅgeṣu vyākaraṇam*).[56] By the seventeenth-century, scholars writing in Sanskrit boasted, ubiquitously, of having mastered the triple-science (*Vidyātraya*) of grammar, hermeneutics, and epistemology (*Pada-Vākya-Pramāṇa*) which should be taken as a reflection on the centrality of these disciplines to the pursuit of organized knowledge.[57]

Professor Minkowski's scholarly work—and many of the essays in this volume—reflect this emic, Sanskritic model of scientific thought. Some essays, such as Patrick Olivelle's analysis of violence (*hiṃsā*) in Dharmaśāstra, provide careful analyses of phenomena according to the parameters of a particular śāstric discipline. Of course, the series of essays on the astral and medical sciences are deeply śāstric in this sense as well. Robert Goldman and Deven Patel's

---

51 Pollock, "Theory of Practice," p. 502. See *Yājñavalkyasmṛti* 1.3: purāṇanyāyamīmāṃsādharmaśāstrāṅgamiśritāḥ / vedāḥ sthānāni vidyānāṃ dharmasya ca caturdaśa //3//.

52 Ibid., p. 502 fn 17. See Arthaśāstra 1.2.11, Raghuvaṃśa 3.30, Yājñavalkya 1.310, and Manu 7.43: traividyebhyas trayīṃ vidyāṃ daṇḍanītiṃ ca śāsvatīm /ānvīkṣikīṃ cātmavidyāṃ vārtārambhāṃś ca lokataḥ //7.43//.

53 Pollock, "The Idea of Shastra," pp. 21–23.

54 Ibid., pp. 23–24.

55 See Pollock, "Languages of Science," p. 22: "As the Yājñavalkyasmṛti expresses it, 'no śāstra exists other than the Veda-śāstra; every śāstra springs from it' na vedaśāstrād anyat tu kiṃcic chāstraṃ hi vidyate / niḥsṛtaṃ sarvaśāstraṃ tu vedaśāstrāt sanātanāt //."

56 *Patañjali's Vyākaraṇamahābhāṣya*, Vol. 1, ed. F. Kielhorn (Bombay: Government Central Book Depot, 1880), p. 1.

57 Pollock, "Languages of Sciences," p. 23.

essays attest to the centrality of śāstric knowledge systems (such as Vyākaraṇa and Dharmaśāstra) as hermeneutical devices in Sanskrit commentarial tradition. In Goldman's estimation, Nīlakaṇṭha Caturdhara's grammatical acumen enables him to exonerate the epic's heroes' names of sundry moral and criminal charges deftly, while in Patel's analysis, nuanced grammatical transformations of the root text are one of the hallmarks of commentary as a genre. John Lowe's essay on unusual sandhi forms in Śākalya's *Ṛgvedapadapāṭha* is deeply scientific in an emic, Sanskritic way: it employs an early Vedāṅga (Vyākaraṇa) to garner a deeper appreciation for the meaning of the Veda itself.

In short, *Science and Society in the Sanskrit World* ought not to be viewed, as the reviewer suggested, as a series of independent mini-books. Rather, a core feature of the volume is its deeply scientific character. On one level, the volume offers a substantial number of scientific essays that examine Indian phenomena through the lenses of mathematics, astronomy, medicine, and zoology. On another level—one that problematizes modern, Western-normative views of science as a monolithic category—the volume's essays employ a variety of etic and emic scientific knowledge systems—sciences and śāstras—to examine a host of textual and historical phenomena. It is fitting then, that Sarasvatīdevī, the goddess of learning, knowledge, music and, *inter alia*, science, would grace the cover of a Festschrift the celebrates the diversity and depth of knowledge in the Sanskrit tradition generally and in Professor Minkowski personally.

So much for science, but what about this volume's other core theme, *society*? O'Hanlon's essay, on the lawsuit of the Brāhmaṇa widow, Gotmāi, in early modern Mahārāṣṭra, invites us to attend to the ways in which the production and application of formal systems of knowledge were imbricated in the lived experiences of individuals within specific historical and cultural milieux. As noted above, a 'Skinnerian' approach to intellectual history animates Professor Minkowski's collaborative work "Sanskrit Knowledge Systems on the Eve of Colonialism" and many of the essays in this Festschrift.[58] At its core, the aim of this project was to shift the focus of Sanskrit studies away from 'the old philology,' which traced a history of linguistic and textual change, *sans* ideas, and away from the 'old Orientalism' which excavated a series of timeless, transcendent ideas *sans* history.[59] Instead, contemporary scholars of Indian intellectual history focus on

58 For the Skinnerian school of intellectual history, see note 20, above.

59 Pollock, "Is there an Indian Intellectual History," p. 534.

> understanding change in the Indian thought world ... in relation to broader processes of disciplinary formations (comprised of language communities, forms of knowledge, text genres, and the like), society, polity, institutions ... [etc.][60]

In many respects this Skinnerian approach to Sanskrit knowledge systems parallels developments in the history of science. After all, Cunningham and Williams' attribution of the emergence of science to the Age of Revolutions (1760–1848) was predicated, methodologically, on Skinner and Dunn's emphasis on the need to "respect and to seek to understand the terms and categories in which the past people we study actually thought of and performed their work."[61] If Sanskritic knowledge systems, like science, cannot be viewed as an "absolute moral criteria or [as a] timeless logic ... [and] if these things do not derive from some transcendent realm, then they ... require historical explanation in the usual way."[62]

Sanskrit philology, in the sense of careful readings of texts, plays a crucial role in crafting an intellectual history of Sanskritic thought that attends to social realities. While every essay in this volume displays the linguistic skills of its author, and while some essays, such as those of Thomas Trautmann, Frederick Smith, and Johannes Bronkhorst *describe* classical Indian society through a careful reading of Sanskrit texts, a few, such as Jonathan Duquette's, exemplify the core feature of intellectual historical work: that ideas (in Vedānta, Mīmāṃsā, Dharmaśāstra, Jyotiḥśāstra, etc.) and social realities (sect, lineage, polity, patronage, etc.) are mutually constitutive. Using Sanskrit texts to describe aspects of Indian society offers a valuable contribution to Indology. Ideas, however, are shaped by society, and in turn, shape society. This principle, of course, is a key facet of Minkowski, O'Hanlon and Venkatkrishan's edited volume and collaborative research. Duquette's essay acknowledges the intimate relationship between theological debates concerning the nature of Hindu deities and pedagogical political networks in the Kāverī delta in South India during the early modern period. Similarly, Clemency Montelle and Kim Plofker's analysis of trigonometric tables in eighteenth-century Jaipur reconstructs not just the mathematical details of complex astronomical calculations, but also the social networks and relationships—between Maharaja Sawai Jai Singh II (r. 1700–1743), the Mughal court, Jyotiḥśāstrins, and the Jesuit order that influenced those calculations. Furthermore, Montelle and Plofker reveal

60 Ibid., p. 534.
61 Cunningham and Williams, "Decentring the Big Picture," p. 419.
62 Ibid.

the intimate interconnections between the history of astral science and political history.

Professor Minkowski produced an unusually prodigious and exceptionally wide-ranging body of research in the field of Sanskrit studies. While the previous section of this Introduction has highlighted the biographical and chronological continuity of his intellectual history—following one interdisciplinary thread after another—this section has excavated the thematic continuities between those disciplines. Professor Minkowski's books and articles, along with the essays in this volume are deeply scientific. They display a careful, philologically precise attentiveness to the rules and categories of formal knowledge systems, including the exact sciences, theology, jurisprudence, linguistics, and commentary. Nevertheless, they are all scientific. Furthermore, the essays in this volume reflect Professor Minkowski's careful focus on the relationship between scientific endeavor and broader social realities. Reading scientific texts philologically enables us to gain greater clarity on the historical conditions of Indian society. At other times, we can see how historical conditions shaped and were shaped by intellectual disciplines. In short, I hope that, in reading this volume in its entirety, our readers will enjoy the contributions individual authors make within a series of specialized Sanskrit knowledge systems and garner a deeper appreciation for the unifying scientific aspects that made these knowledge systems so integral to classical and contemporary Indian society.

3 Summary of the Essays

This Festschrift's introductory section contains a brace of essays from Professor Minkowski's colleagues at Cornell and Oxford, Anne Blackburn and Anand Venkatkrishnan, respectively. The section aims to flesh out the skeletal outline of Professor Minkowski's CV and list of publications and to paint a portrait of his personality as a scholar, colleague, and mentor. Blackburn outlines Professor Minkowski's manifold contributions to the Department of Asian Studies during his seventeen-year tenure there. Blackburn provides an account of his many collaborations and administrative roles, but, also, offers insight into Minkowski as an individual: a tendency towards insight and self-reflexivity; a hint of playful mischief in his eyes; a propensity for casual, meta-level observation; and a combination of "bookishness with an appreciation of the natural world and vigorous physical exercise." Blackburn enables us to appreciate Minkowski's life of the body and mind in Cornell and "in the semi-rural environs of the Trumansburg region, located across Cayuga Lake from

Cornell University." Venkatkrishnan, in turn, proffers perspective on Professor Minkowski's time as the Boden Professor of Sanskrit at Balliol College, Oxford.

This pair of personal pieces in the first section attest to a broader overlap between the biographic and thematic frames for our volume. Professor Minkowski's pursuit of various Sanskrit knowledge systems over his long career was in no small measure the product of his personality and his professional relationships. His research, in turn, reflects this biographical theme in its adherence to the principle that study of specialized Sanskrit knowledge systems (the sciences) is most illuminative when viewed through the lens of social history.

Part One, "The Vedic World," revolves around the polestar of Vedic philology and contains essays from Barbora Sojková, Elizabeth Tucker, Stephanie Jamison, and John Lowe. The first two essays employ philology in conjunction with modern biological science to provide insight into ancient Vedic Society. Sojková examines the *Gavāmayana* ("the path/course of cows"), a myth concerning a year-long Soma ritual performed by bovines that "occurs in six parallel versions in five Vedic texts ... two Black Yajurveda Saṃhitās (TS 7.5.1.1–2, TS 7.5.2.1–2, and KS 33.1) ... [and] the Ṛgveda and Sāmaveda Brāhmaṇas (AB 4.17, JB 2.374, PB 4.1)." In these narratives, the bovines' ritual actions result in the formation of a peculiar type of cow, the *tūpara*, "a peculiar word which is understood by the later Sanskritic tradition as a term for a hornless animal." Sojková argues that, to the contrary, the tūpara should be identified with the Zebu (*Bos indicus*), a bovine with distinctive, upwardly turned horns. Sojková employs a mixture of comparative philology (looking at Indo-European sources), detailed studies of the Vedic Saṃhitās themselves, and contemporary studies of historical zoology. In matching Vedic philology with contemporary zoology, Sojková makes a compelling case for rethinking the received understanding of the tūpara.

Like Sojková, Tucker approaches her analysis of tetanus in the *Atharvaveda* with an eye towards linking classical philology with contemporary biology. Tucker observes that, although, "the Atharvaveda (AV) has often been dismissed as unscientific, a collection of spells and magic formulae ... [nevertheless] ... it contains poems which show accurate physiological observations ... [and] it is the bunching of [these physiological] terms that indicates medical knowledge." Tucker provides an excellent edition of fourteen-stanza composition that is peculiar to the Paippalādins, PS 11.2. A mysterious word, *viśara* (attested in only one other text), appears seven times in the *Caturdaśarcakāṇḍa*. Tucker draws on an impressively comprehensive review of related terms in Vedic materials to argue that *viśara* should be understood as a form of tetanus (lockjaw). Moreover, *viśara* and related terms appear to have been closely linked with arrow wounds (something that would, from a modern medical perspective, too, be seen as a vector for tetanus). Tucker is careful to emphasize that the hymn doesn't display

> any [modern, medical] conception of a bacterium *clostridium tetani* which released toxins that affected the nervous system, but my claim is that this sūkta shows good observation of symptoms which belong together and some understanding of their possible sources.

Tucker concludes that, if her analysis is correct, the *Caturdaśarcakāṇḍa* provides evidence that the symptoms of tetanus and at least one of its causes were recognized in India some centuries before Hippocrates.

The second set of essays in the first part of the Festschrift, from Stephanie Jamison and John Lowe, exemplify the discipline of historical linguistics (as applied to the *Rigveda*). Jamison examines *Rigveda* X.28, a twelve-verse dialogue hymn, dedicated to Indra, intricate in its structure and unique in the RV for its use of abbreviated animal fables for instruction. Jamison argues that, in X.28.12ab, the phrase *suśámī abhūvan* ought to be taken as a potentially second usage of the cvi-construction in the *Rigveda*.[63] Jamison begins with the claim that, "the *śam* in *suśámī* belongs instead with the etymologically and semantically independent indeclinable *śám* 'weal, luck', which is extremely common in the RV." She observes that Sāyaṇa, the famed, fourteenth-century commentator on the *Rigveda*,

> treats *suśámī* as if it were a nominative plural bahuvrīhi (*sukarmāṇaḥ* 'having good [ritual] work') characterizing the subject, which seems to me tantamount to analyzing *suśámī abhūvan* as a *cvi* construction.

Jamison then turns to the wider Indo-European world and concludes that, in short, *pace* the

> *opinio communis* about the origin of the [*cvi*] construction … the hefty representation of the construction in technical vocabulary, including

63 As Jamison notes, "The *cvi* construction is a well-known feature of Classical Sanskrit, treated by Pāṇini (5.4.50, 51; 7.4.26, 32) as well as in the standard Western Sanskrit grammars (e.g., Whitney 1889: 401–402; Renou 1968: 150–152). In this construction a noun stem acquires the invariant suffix *-ī* and is then construed with a form of √*bhū* 'become' or √*kr̥* 'make', with the combinatorial sense 'become/make X', of the type *stambhī bhavati* 'becomes a post' to *stambha-* 'post'. As in the example just given, the derivational base is often an *-a*-stem, with substitution of *-ī* for the stem vowel. All of this is, of course, generally known and uncontroversial."

> agriculture, and in lower-register items suggests that the construction developed not in order to predicate abstractions, but as a technical and slangy shorthand, a way to make verb phrases out of concrete nouns that lacked clearly associated verbal roots—in other words as an alternative way of making denominatives—perhaps originally popular in vernacular speech.

Yes, Jamison's analysis is highly technical, and will undoubtedly thrill those with the prerequisite philological skills to appreciate their subtleties, but I enjoyed Jamison's remarkably accessible insight into the social world of Rigvedic India—capturing a sense of ancient Sanskrit slang.

John Lowe's essay attends to three apparently inexplicable instances of *parigraha* (marking a word by placing *iti* after it, and then repeating the word after the *iti*) in relation to the word han in Śākalya's padapāṭha to the *Rigveda* (at RV 5.29.4d, 5.31.1d, and 10.99.6d). Śākalya's padapāṭha divides "the original, continuous 'saṃhitāpāṭha' into separate words (padas) ... [and also] undertook to analyse compounds words into two constituent parts." Lowe notes that Śākalya's core project, the formal linguistic analysis and preservation of the text of the *Rigveda*, represents "the beginnings of vyākaraṇa, and hence of śāstric and scientific thought in India." In relation to these anomalous parigraha's in the padapāṭha Lowe introduces two possibilities: 1) that "Śākalya simply forgot to apply his rules consistently in a number of cases" and, as a result, there was "an overapplication of otherwise consistently applied rules to cases which ordinarily should not have been marked"; or 2) "that there was indeed a rationale underlying the analysis that comes down to us," such as that "the reading of this pāda may have been a matter of debate, and the itikaraṇa [the use of parigraha] of the verb may have been intended to confirm Śākalya's interpretation over some other contemporary reading." Lowe weighs both options and concludes, following a skillful analysis of related phenomena in the *Rigveda*, that,

> the parigraha of han was Śākalya's means of ensuring the correct transmission of verse-final han, in the face of an apparent tendency to weakened or reduced pronunciation of short, unaccented words in that position ... [and] it is possible, though of course unprovable, that Śākalya was even aware of contemporary recitations of the RV which treated han in these cases as ha.

The beauty of Lowe's essay, besides its philological rigor, is its assumption that the padapāṭha, by virtue of its scientific nature (formal analysis according to internally consistent rules and methods) entails an internal—if not always apparent—logic. Lowe not only uncovers that internal logic—resolving ambiguities—but also provides insight into the social world of the padapāṭha, in which, perhaps there existed "specific controversies of the recitation to which we no longer have access." The science of language thus offers a lens with which to illuminate the social.

Part Two of this Festschrift, "Astral and Medical Sciences," covers, *inter alia*, the disciplines of Jyotiḥśāstra and Āyurveda. Although this section covers the most formally scientific (in the sense of modern science), it also provides much insight into the social worlds which shaped the lives of astral and medical scientists in medieval and early modern India. Takanori Kusuba offers an example of pure mathematics, taken from Nārāyaṇapaṇḍita's 1356 treatise the *Gaṇitakaumudī* (Moonlight of Mathematics). While discussing mathematical rules that seemed peculiar too him, Nārāyaṇa introduces what contemporary high school pupils might describe, bemoaningly, as a 'word-problem' involving the calculation of sequences:

> A cow gives birth to a calf every year, and the calf becomes mature in three years. Tell [me], wise man, how many offspring will one cow have in twenty years?

Kusuba explains that Nārāyaṇa offers two methods for solving the problem: 1) something we might describe as "a rule for the multiple sum of the natural series $S(n) = 1 + 2 + 3 + \cdots + n$."; and 2) a '*meru*,' although he does not elaborate on the latter. Kusuba turns to the *Karmapradīpika*, Nārāyaṇa's commentary on Bhāskara II's *Līlāvati*, where Nārāyaṇa describes *meru*-s as a type of notational chart for extended calculation. Kusuba applies this technique to the *Gaṇitakaumudī*'s bovine problem and arrives at the correct answer, 2745. Kusuba's calculations are clever, of course, but his analysis on India's bevy of notational techniques (much more opaque to modern readers than others) to address narratively stated questions or principles bears characteristic similarity to early modern mathematics. Newton, for example, stated his theorems narratively (and in Latin) and developed idiosyncratic tabular methods that would be largely incomprehensible to contemporary students of calculus. Kusuba's essay remind us that, while mathematic *principles* may be universal, notational *techniques* are highly circumscribed by one's social location.

Setsuro Ikeyama's essay on Nīlakaṇṭha Somayājin's (fifteenth–sixteenth-century) *Jyotirmīmāṃsā* (Investigations on Astronomical Theories) explores one of the core features of what Sheldon Pollock describes as the "language ideology" of Sanskrit scientific thought: that intellectual 'innovation' within the knowledge systems must, invariably, frame itself, on the face of it, as the recovery, or harmonization of theories articulated in an unimpeachable cannon.[64] According to Ikeyama, "Nīlakaṇṭha tried to rectify the astronomical constants not only of other schools [of Indian astronomy] but also of his own tradition, and he insisted that his canon, [the] *Āryabhaṭīya* … [of Āryabhaṭa, c. 476–550] supported his attempts." In the *Jyotirmīmāṃsā*, Nīlakaṇṭha contends that, although the differing canonical tables of India's diverse array of astronomical schools (for example, the Ārya school with the sunrise-epoch system (*audayika*) and the Ārdharātrika school with the midnight epoch (*ārdharātrika*) system, both various proposed and used by Āryabhaṭa) are all valid, nevertheless, they must be revised in order to bring them into alignment with empirical observation (*parīkṣa/parīkṣaṇa*). Toward that end, Nīlakaṇṭha sets out to revise a series of astronomical constants from these schools of Indian astronomy. Ikeyama works deftly though Nīlakaṇṭha's solutions to a series of incongruous astronomical constants between the schools. Ikeyama notes that Nīlakaṇṭha "does not make any judgements about which value is better or which is closer to observations … since his goal is to rectify astronomical constants by observation." Ikeyama helps us to see that scientific innovation occurs within cultural and scriptural constraints. Just as Copernicus and Galileo's postulation of a heliocentric universe (and the mathematics supporting this claim) occurred within a cultural milieu dominated by the principles of biblical scriptural authority, Nīlakaṇṭha's empirical adjustments to the classical Sanskrit astronomical canon assumed the cultural ideology of the Sanskrit classics as an unimpeachable canon.

Clemency Montelle and Kim Plofker's essay on the integration of logarithm concepts from Sanskrit, Muslim, and European astronomy in eighteenth-century Jaipur under the reign of Maharaja Sawai Jai Singh II (r. 1700–1743) excavates a complex, global network of knowledge and power during the early modern period. In the early 1700s, Jai Singh expressed a passionate interest in astronomy in both Sanskrit and Greco-Islamic traditions and assembled "a learned team of Hindu and Muslim scientists with extensive knowledge of Sanskrit, Arabic and Persian text corpora." Jai Singh sought to develop a comprehensive "system of astronomical tables named *Zīj-i Muḥammad Shāh* in honor of the [eponymous] Mughal emperor," Muhammad Shah. With the assistance of "Father Miguel Figuereda of the Jesuit mission at Agra," Jai Singh arranged for "a Portuguese astronomer named Pedro da Silva" to come to Jaipur

64 Pollock, "Languages of Science," pp. 21–22. For the prolemptic dimension of Sanskrit texts, see Ganeri, "Contextualism," pp. 555–561.

"with various instruments and texts," including, "the *Tabulae astronomicae* compiled by the French astronomer-mathematician Philippe de La Hire."[65] Although the 'heliocentrism-agnostic' viewpoint of La Hire's *Tabulae* "proved acceptable … to Jai Singh's Hindu and Muslim geocentrists" (and to Singh's Jesuit knowledge brokers), nevertheless, as Montelle and Plofker relate, the *Tabulae*'s "use of tabulated logarithms of trigonometric function values" caused some challenges for its Jaipur students. Sanskrit and Islamic astronomy utilized trigonometric tables, but they "did not employ the computational aid of composing trigonometric with logarithmic functions that had been devised in early seventeenth-century Europe." La Hire's *Tabulae* relies on "log trig values" even though he "neither described the basic principles of logarithm rules nor provided any general tables of logarithms, either of common numbers or of trigonometric function values." Montelle and Plofker explore a series of unpublished Sanskrit manuscripts in which Jai Singh's chief astronomer, Kevalarāma, attempted to translate the *Tabulae*'s trigometric tables into Sanskrit and transculturate them with the Jyotiḥśāstra tradition. In addition to its lucid account of the mathematical nuances of Kevalarāma's work, Montelle and Plofker reveal how the complex political and social features of eighteenth-century India shaped the contours of astronomical knowledge.

The next essay, by Johannes Bronkhorst, explores the potential influence of changing social factors in the development of the theory of cyclical world ages (*yugas*, or *kalpas*) in classical Brāhmaṇical ideology. Here, the numerical calculation of time is inseparable from the ideology of time. Bronkhorst argues that "the cyclical vision of time came from the eastern region of northern India, from the culture of Greater Magadha," especially in Buddhist and Jaina sources. According to Bronkhorst, Buddhist and Jaina sources, such as the *Sāmaññaphala Sutta* of the *Dīgha Nikāya* and the *Viyāhapannatti*, respectively argue for a cyclical model of time that features extraordinarily long eons (*kalpas*)—even if these early texts attribute such a model to Makkhali Gosāla, the putative founder of Ajivikism. Bronkhorst notes that "a small number of early Brahmanical texts," such as the *Mānavadharmaśāstra* (1.68–70) and the *Mahābhārata* (3.186.18–23), "present a [rather different] picture of cyclical time" that features much smaller eons and that emphasizes the ills of human-

65 For the influential role members of the Society of Jesus in the formation of modern scientific knowledge in Europe, overseas, and in India in particular, see Agustín Udías Vallina, *Jesuit Contribution to Science: A History* (Cham: Springer, 2015); Steven J. Harris, "Jesuit Scientific Activity in the Overseas Missions, 1540–1773," *Isis* 96.1 (2005), pp. 71–79; and Dhruv Raina, "French Jesuit Scientists in India: Historical Astronomy in the Discourse on India, 1670–1770," *Economic and Political Weekly* 34.5 (1993), pp. 30–38.

ity's current epoch. Later Purāṇic literature transformed these smaller epochs into larger eons by revising "human" years into "divine" years (multiplying them by 360) and, as Bronkhorst observes,

> the long Purāṇic world-periods share an important feature with the long world-periods that were current in the culture of Greater Magadha, viz. their great length.

Bronkhorst attributes this calendrical transformation to socio-political forces and notes that, in the *Yuga Purāṇa*, the invasion of the Śakas (second–first-centuries BCE) is taken as evidence of "the approaching end of an era (*yugānta*)" and as "an indication that a new *kṛtayuga* will begin soon." Although, at "this time ... certain Brahmins ... who suffered most from the repeated invasions that struck northwestern India, thought that ... the end of the present world-period—was near," nevertheless, when,

> presumably under the reign of the Guptas [fourth–sixth centuries CE], this fear no longer looked justified, a reinterpretation—or rather: a recalculation—of the four *yugas* put the end of the world at a safe distance ... [and] this brought these cycles closer to what must have been their source of inspiration: the time-cycles of Greater Magadha.

The final essay in the section on astral and medical sciences, by Dominik Wujastyk, provides a translation of the first chapter of the *Śārīrasthāna* (chapter on the body) of the *Carakasaṃhitā*, an Āyurvedic encyclopedia compiled by Caraka in the first or second-centuries of the common era.[66] Collated from ten manuscripts, the translation provides a skeletal overview of the physical, psychological, and spiritual anatomy of a human being: its constituents (*dhātu*); the soul (*bhūtātman*); sensation (*vedanā*); the mind (*manas*); the organs of sense (*indriya*) and of action (*karmendriya*); past and future; disease (*vyādhi*); and yoga and liberation (*mokṣa*). Wujastyk's contribution is doubly scientific in the sense that it employs a phylogenetic text-tree to produce a collated Sanskrit text to translate and it covers medicine, a traditionally 'scientific' field. Nevertheless, Caraka's anatomy is as metaphysical as it is medicinal. The *Śārīrasthāna*

66 For the *Carakasaṃhitā*, see Karin Preisendanz, "Text Segmentation, Chapter Naming and the Transmission of Embedded Texts in South Asia, with Special Reference to the Medical and Philosophical Traditions as Exemplified by the *Carakasaṃhitā* and the *Nyāyasūtra*," in *Pieces and Parts in Scientific Texts*, ed. Florence Bretelle-Establet and Stephane Schmitt (Cham: Springer, 2018), pp. 159–220.

maps out the functioning components of the human body with a clear teleological eye toward liberating the human soul. Physical disease leads to and stems from spiritual malaise. If, consequently, the reader were inclined to view Caraka's anatomy as comprised by its spirituality, they would be reminded, I think, to remember that Harvey's theory of the heart as a pump was the result of an "essentially Aristotelian investigation ... into the ways in which this organ served the purposes of the soul."[67]

Part Three, "Statecraft and Jurisprudence," adopts a social scientific perspective, with a series of essays tackling legal, political, and economic questions according to the text traditions of Sanskrit jurisprudence (Dharmaśāstra) and political science (Arthaśāstra). These essays query the ways in which Sanskrit sciences conceived of social phenomena. Patrick Olivelle's essay explores how

> [classical] Brahmanical moral theologians, heirs to the Vedic sacrificial religion, found themselves in a quandary resulting from a value system inherited from Buddhism and other śramaṇic traditions and anchored in the principle of *ahiṃsā*, not killing or injuring living beings.

The quandary is that, although *ahiṃsā* appears as a general, universal ethical good in many places in the Brāhmaṇical tradition (see, for example, *Mahabhārata* (13.117.37–38); MDh 10.63; 11.222; 12.83; YDh 1.122; 3.213; ViDh 2.16; BDh 1.8.2; 3.1.27; 3.10.13; VaDh 4.4; 30.8), nevertheless, many other places in the Brāhmaṇical tradition endorse violence toward animals and humans in the specific contexts of Vedic sacrifice and combat. As Olivelle notes, the *Mānavadharmaśāstra* (5.39) declares that "in sacrifice killing is not killing" (*yajñe vadho 'vadhaḥ*). Olivelle turns to the text-tradition of ritual hermeneutics (Mīmāṃsā) and explores how Mīmāṃsā notions of Vedic injunctions (*vidhi*) and exceptions (*apavāda*-s) provided a hermeneutical strategy for Sanskrit jurists to resolve the dilemma of violence and nonviolence. Although *ahiṃsā* is "an *utpattividhi*, an originative injunction (*utsarga*) of broad applicability," nevertheless, injunctions to perform animal sacrifices or to engage in armed conflict are exceptions (*apavāda*-s) to this broad injunction. Later Dharmaśāstric commentaries, such as the *Manubhāṣya* of Medhātithi (ninth–eleventh-century) and the *Vacanamālā* (a late sub-commentary on the *Bālakrīḍā* of Viśvarūpa) of Rāmaśaurya, as Olivelle demonstrates, take this logic of general rule and exception even further. To the extent that one's congenital dharma (*svadharma*) of

67 See note 26, above.

caste and class (*varṇāśrama*) enjoins one to live as a hunter or to kill humans as part of a royal duty to punish evildoers, there is no 'violence,' per se, in following those injunctions. As Olivelle notes, "*svadharme vadho 'vadhaḥ*, 'within the dharma specific to one, killing is not killing.'" Here we see a classic example of the science of jurisprudence: the careful application of the formal rules of a knowledge system to gingerly reconcile two competing strands of logical interpretation.

The next essay, by Valters Negribs, examines the science of statecraft (Arthaśāstra) in the *Mahābhārata*. Negribs discusses the 'ascetic' virtues of ministers (*amātya*-s), chaplains (*purohita*-s) and advisor-chaplains (*mantripurohita*-s) in two sections of the *Mahābhārata*: Bhīṣma's advice to Yudhiṣṭhira in the *Śāntiparvan* (adhyāyas 80–86 and 116–119); and Vidura's warning to Dhṛtarāṣṭra in *Udyogaparvan*. As others have noted, the *Mahābhārata*, like the *Mānavadharmaśāstra* (and other Dharmasmṛtis) appropriates the political scientific language of the *Kauṭilīyārthaśāstra* and resituates it within the post-Mauryan, Brāhmaṇical political theology of Varṇāśramadharma.[68] Negribs argues that, in the specific context of royal advisers as described in the *Mahābhārata*, Kauṭilīyārthaśāstric language concerning ministers' qualifications (intellect, pedigree, loyalty, etc.) is suffused with Śrāmaṇic moral virtues (learnedness, moral virtue, etc.). Negribs notes, that, although "it might be a mere coincidence that the *Mahābhārata* talks about royal advisors using the same terms that the Buddhist tradition uses to talk about learned and virtuous monks," nevertheless, it is also "plausible that it points to a mutual interaction and possible competition for patronage." The *Mahābhārata* emphasizes that these 'ascetic' virtues, whatever be their historical and religious origins, comprise, essential qualifications for royal advisers. Sometimes, as in the case of Vidura and in fable of the Jackal who became a ministers (12.112), moral virtues compensate for a lack of noble pedigree. In other cases, such as in Bhīṣma's warning to Yudhiṣṭhira concerning the dangers of avarice (12.152), pedigreed, learned ministers without ascetic virtue can destroy a kingdom. Negribs' essay suggests that the *Mahābhārata*'s science of statecraft bears the indelible imprint of a Brāhmaṇical response to the social prestige of renunciants under the Mauryan regime: ministers should be learned like Kauṭilya and be moral like an ascetic.

Fred Smith's essay examines the geographical and psychomental relationships between the village (*grāma*) and the wilderness (*araṇya*) in four case

68 See, for example, Mark McClish, "The Dependence of Manu's Seventh Chapter on Kauṭilya's Arthaśāstra," *Journal of the American Oriental Society* 134.2 (2014), pp. 241–262.

studies: 1) the description of Yudhiṣṭhira's horse sacrifice (*aśvamedha*) in the *Āśvamedhikaparvan*, the fourteenth book of the *Mahābhārata*; 2) Yudhiṣṭhira's absorption of Vidura's essence (after the latter's death) in the fifteenth book of the *Mahābhārata*; 3) the transference, as inheritance, of 'possessed' identities, such as Arjuna, Draupadī, Kuntī, and so forth, between participants in Pāṇḍav līlā-s (folk performances of local *Mahābhārata* narratives) in the remote Himalayan region of Garhwal; and 4) "complex mechanisms of possession and counter-possession" at the healing centers of Bālājī and Chottanikkara in Rajasthan and Kerala, respectively. Smith draws on Charles Malamoud's seminal work (concerning first millennium BCE Vedic ritual texts), which

> contrasts the differences between *grāma* (village) and *araṇya* (forest): each has its own spaces; its own domesticated and non-domesticated animals, reified in ritual ordering, including choice of sacrificial animals; their orderly predictability and unpredictable wildness; their mechanisms of peacefulness opposing the unbridled threat of violence; and the householder stage of life (*gṛhasthāśrama*) contrasted with retirement and renunciation (*vānaprasthāśrama* and *saṃnyāsāśrama*).[69]

For Smith (reading Malamoud), the distinction between village and forest, which appears ubiquitously in Sanskrit social scientific literature (such as Dharmaśāstra and Arthaśāstra), was, and is, "as psychological and sociological as it was strictly environmental." Assuming this psychological dimension to social-scientific categories, Smith avers that "internal constructions of space condition notions of external space" and, furthermore, he proceeds to examine how,

> Liminal spaces or buffer zones such as well-tended forests or sanctioned dissociative experience, including certain forms of possession, helped mediate the physical and emotional space between *grāma* and *araṇya* in India centuries and millennia ago, as they continue to do today.

Smith's case studies explore a variety of liminal environments (both psychical and geographical) which serve as "buffer zones for resolving the crises that arise from the confrontation of *grāma* and *araṇya*." Yudiṣṭhira's horse sacrifice enacts sovereignty over the grāma by domesticating the resources of the

---

69 Charles Malamoud, *Village and Forest in the Ideology of Brahmanic India, Cooking the World: Ritual and Thought in Ancient India*, trans. David White (Delhi: Oxford University Press, 1996), pp. 74–91.

araṇya. Similarly, Yudhiṣṭhira's absorption of Vidura's spirit in the wilderness augments the formers' control over the Kuru realm. In the Himalayan Pāṇḍav Līlā, "the surprising rite of transference of an alternate identity as Arjuna" by an "actor ... to his son through an act of possession" is part and parcel of a larger process of "merging ... the present with the historical presence of the *Mahābhārata*" and, thereby, "bringing greater depth and purpose to their lives." Finally, at the healing shrines of Bālājī and Chottanikkara, "the unpredictable wildness of the forest," represented by fearsome demons possessing the pilgrims, is "sanctioned within the confines of these otherwise orthodox and domesticated Hindu temple complexes" and, ultimately, domesticated through the deities' acts of counter-possession. In short, Smith provides a poignant, indigenously Indian counterpart to the longstanding theory of a nature/culture divide that lies at the heart of contemporary anthropology as a social-scientific tradition.[70]

The final essay in this section, by Thomas Trautmann, discusses the concept of "forest people" (*aṭavī*-s, *āṭavika*-s, *vanacara*-s) in the *Kauṭalīyārthaśāstra* from a normative, political scientific standpoint. Trautmann works, very carefully, through a series of Arthaśāstric passages in order to recover how the text, on its own terms, understood the nature and utility of individuals who lived in the forest. Trautmann argues that Kauṭilya's "naming [of people] by means of an environmental zone" reflects a broader Arthaśāstric and Āyurvedic contrast between *grāma* and *araṇya*. Although the settled *grāma* "is normative and supplies the point of view from which people of the forest are seen occupying a landscape that is the opposite of the ideal for humans," nevertheless, the forest and its human and nonhuman inhabitants are essential for the functioning of the idealized kingdom that Kauṭilya describes. Trautman identifies three themes wherein the forest people play an important role in Sanskrit political science: 1) "as servants of the king in connection with forests supplying forest produce and elephant forests"; 2) as a source of military strength; and 3) as agents in interstate conflict, "especially as it concerns the succession to kingship."

In book 2 (2.2.5), Kauṭilya notes that a king should establish produce forests wherein forest-dwellers (*vanacara*-s) are "brought into the royal service by the

70 For the nature/culture divide, see, David Greenwood and William Stini, *Nature, Culture, and Human History* (New York: Harper and Row, 1977). For feminist, intersectional approaches to the same issue, see Ynestra King, "Healing the Wounds: Feminism, Ecology, and Nature/Culture Dualism," in *Gender/Body/Knowledge: Feminist Reconstructions of Being and Knowing*, ed. Alison Jaggar and Susan Bordo (New Brunswick: Rutgers University Press, 1989), pp. 115–141.

proffer of a wage." The same model follows (2.5.6.) in the case of elephant forests (*gajavana*-s). Trautman notes that, because of elephants' "sluggish rate of reproduction and the late age at which they became sufficiently mature for war (twenty years, according to the *Arthaśāstra*)" a king *required* extensive tracts of uncultivated land (and the assistance of forest people) in order to maintain units of war-elephants. This reveals that, in ancient Sanskrit political science, the *grāma* and *araṇya* were, to some extent, mutually interdependent: the world of the *grāma*, with its kings, temples, and armies required the forest, with its elephants, resources, and natives, in order to thrive, while the forest required royal protection in order to stave off the onslaught of development. Furthermore, the *Arthaśāstra* (9.2.13) describes six types of troops, with forest troops (*aṭavī*-s) described as being the most unreliable. Trautman notes that, according to Kauṭilya, their "propensity ... to plunder the remote regions of the kingdom," forest troops are highly useful in specialized circumstances such as guiding the king's army through the wilderness or countering an opposing king's forest troops. Finally, Trautman examines the "nature of the polities of forest people" from Kauṭilya's perspective. Kauṭilya offers a "grudging semi-recognition that forest peoples are organized into polities" and views the chief of the forest people as having "sky-high" political importance, especially as an ally to a prince in disfavor. Consequently, Trautman notes, that, although "we cannot hope to find texts of ancient India in which forest people speak for themselves, and from their own point of view," we can glean a fairly comprehensive sense of how forest people were viewed from *within* the classical Sanskrit political science.

Part Four of this Festschrift, "Sanskrit Intellectuals and Their Communities," attends to relationship between specific intellectuals who wrote in Sanskrit and the social worlds in which they lived. This section exemplifies Venkatkrishnan's observation that, for Professor Minkowski, "ideas are inseparable from the people who articulate them" and the communities in which those people lived. The first brace of essays, by Robert Goldman and Deven Patel, examines the social mores behind and the phenomenology of commentary, respectively. Robert Goldman's essay explores several medieval and early modern commentators' analyses of passages from the *Rāmāyaṇa* and the *Mahābhārata* wherein the elaborate use of hermeneutical tools (drawn from Dharmaśāstra, Mīmāṃsā, Vyākaraṇa, and other śāstric discplines) enables the commentators to bring the epic heroes' ostensibly errant behavior into alignment with the social mores of classical Indian society. A key observation from Goldman is that these commentators "deploy to the fullest their lexical knowledge and grammatical skills" in order to demonstrate that "despite the apparent surface meaning of their texts, its heroes and heroines do not deviate from two central social imperatives, deference to their elders and unwavering adherence to the truth." In essence, Goldman demonstrates that commentary (*bhāṣya*, *ṭīkā*, etc.) is a decidedly scientific endeavor to the extent that its practitioners employ the methods of several formal knowledge systems. This scientific endeavor, however, was deployed in support of the preservation of the normative social values of the commentators' communities.

Take, for example, the reunion of Rāma, Lakṣmaṇa, and Bharata in the *Rāmāyaṇa*'s *Yuddhakāṇḍa* (6.115.33). Goldman notes that Vālmīki "seems to describe an older brother paying homage to his junior," something of an anathema in the social world of the *Rāmāyaṇa*'s commentators. Consequently, commentators, including Nāgeśabhaṭṭa, Maheśvaratīrtha, Govindarāja, Satyatīrtha, and Śivasahāya deploy a series of grammatical and hermeneutical principles to interpret away this violation of social mores. Goldman reveals that, "what might strike the western reader as an essentially trivial narrative verse" was, for "all of these deeply learned scholars," a necessary occasion to "expend … much intellectual energy and ingenuity" in defense of the proposition that these epic characters did not "in any way act counter to the social imperative." Likewise, Goldman turns to "the opening four *adhyāya*s of the *Mahābhārata*'s fourth book, the *Virāṭaparvan*," wherein the Pāṇḍavas (and their wife, Draupadī) must live incognito for a year. The issue, for the *Mahābhārata*'s commentators is that the Pāṇḍavas *appear* to *lie* to King Virāṭa—claiming that they are former servants of Yudhiṣṭira—in order to work, disguised, as servants in Virāṭa's court. As Goldman notes,

> the tales the Paṇḍāvas and Draupadī tell the king would appear clearly to violate the categorical imperative of truthfulness … and that, in the commentators' view, is something that can never be ascribed to these godlike men and their divine wife. Therefore, they make it their mission … to demonstrate, through the deployment of their considerable, even encyclopedic, knowledge of the *kośas, vyākaraṇa*, and the broader intellectual culture of which they are defenders, that each of the Pāṇḍavas' statements about themselves and their proposed occupations are, in fact true, if phrased in such a way as to escape the understanding of all but the most learned …

Goldman deftly outlines Nīlakaṇṭha Caturdhara's intricate reasoning in depth and he highlights similar arguments from Arjunamiśra, the author of the *Virodhārthabhañjanī*, Caturbhujamiśra, and Sarvajñanārāyaṇa, that, collectively reinterpret each of the Pāṇḍavas' and Draupadī's deceptive "résumés" to Virāṭa as equivocal, almost Jesuitical, statements that, while fulfilling their primary function of obtaining employment as servants, are, nevertheless, factually true.[71] Contrary to what one might expect, the commentators display "their greatest grammatical knowledge and lexical ingenuity" not to manifest the true meaning of archaic words or difficult syntactical passages, but, rather, to argue, against the clear, denotative meaning of the text. Grammatical aptitude enables commentators to show "that these noble figures, so central to the culture's image of elite manhood and womanhood, cannot possibly be guilty of violating the strict code of truthfulness."

While Goldman provides several examples of commentary *in action*, the next essay, by Deven Patel, offers a phenomenology of commentary as a scientific discipline. Patel argues that Sanskrit commentary (*ṭīkā*), "is arguably *the* signature of learned and literate culture in India for the past one thousand years, much as *glossa* was in the European Middle Ages."[72] Patel notes that early Indologists, like some indigenous commentators such as Bhoja, bemoan the tendency of *ṭīkās* to remain silent on grammatically difficult passage and to wax poetic in places where the authors take a polemical stance. The issue, of course, is that *ṭīkā*, as a genre, does not map onto Western expectations of either commentary or translation. Patel, however, argues that

> the very same markers of distinction for variable commentarial and editorial acts hold for forms of *translation* as well: choices are made on the basis of the reader's intellectual project (historicist, pragmatic, semiotic, etc.), their attitudes toward the text (fixed and semantically invariant, somewhat unstable and polysemous, totally open and reader-dependent), and the hermeneutic strategies found to be acceptable and useful to enact a specific interpretive vision.

Consequently, by shifting our understanding of the *ṭīkā* to "a special kind of translational act" we can historicize "ancient and medieval reading practices" and appreciate *ṭīkā* as a genre—on emic terms—and the specific hermeneutical projects of individual commentators. Thus, according to Patel,

71 The pejorative term 'Jesuitical' for excessive equivocation has intriguing roots in the oppression of Catholics in post-Reformation England. See, for example, Janet Halley, "Equivocation and the Legal Conflict over Religious Identity in Early Modern England," *Yale Journal of Law & Humanities* 3.33 (1991), pp. 33–52.

72 For a similar typology for Latin *glossa*, see Rita Copeland, "Gloss and Commentary," in *The Oxford Handbook of Medieval Latin Literature*, ed. Ralph Hexter and David Townsend (Oxford: Oxford University Press, 2012), pp. 171–191.

> in resisting a hasty conflation of *ṭīkā* with translation and bracketing the convention to assume it to be an unproductive commentary or merely a helpful gloss, a provisional position might understand the *ṭīkā* as delivering a close-to-the-text reading that makes more elaborate readings possible.

A *ṭīkā*, unlike the more commentarial *bhāṣya* or a briefer, direct translation, serves as a touchstone for future readers. A ṭīkā's ability to disturb "the source out of its slumber and yet itself remaining sketchy and elusive," along with its comparative brevity makes its composition ideal "for pedagogical reasons and for justifying credentials in institutional contexts." Patel surveys Jain, Buddhist, and Mīmāṃsaka hermeneutic techniques before turning to Nīlakaṇṭha Caturdhara's ourvre: the *Bhāratabhāvadīpā* and his four *Mantrarahasyaprakāśa* works. Patel notes that

> Nīlakaṇṭha's readings seem idiosyncratic because he himself completes one of the many generative interpretive routes that a *ṭīkā* on a source text can inspire, while most authors of *ṭīkā* rarely exceed their pedagogical goals to merely make available those readings.

Patel's essay, in providing a typology of *ṭīkās* and in querying Nīlakaṇṭha's place within that typology, helps us to appreciate the contours of commentary as a scientific endeavor as well as the idiosyncrasies of individual commentators who practiced that discipline.

The final brace of essays in this Festschrift, by Jonathan Duquette and Rosalind O'Hanlon, foregrounds the social politics behind intellectual life in early modern India. Duquette examines the work of "four scholars of Advaita Vedānta active between the seventeenth and nineteenth centuries, who belonged to a dynamic intellectual network centred around the Kāverī delta in South India": Bodhendra Sarasvatī (1610–1692); Ayyaṇṇa Dīkṣita (c. 1750); Tyāgarāja Śāstrī (1815–1903); and Rāmasubrahmaṇya Śāstrī (1840–1922). Duquette argues that, *pace* Andrew Nicholson's position in *Unifying Hinduism*, "we find the first genuine attempts to elaborate an 'all-inclusivism' akin to that of neo-Advaita, particularly in relation to the notion that there is a supreme reality underlying gods and goddesses" in the works of these four scholars, rather than "in those of Vijñānabhikṣu, Madhusūdana Sarasvatī or other late medieval scholars."[73] Duquette acknowledges that modern, "'neo-Advaita' thinkers such as Sarve-

73 Andrew Nicholson, *Unifying Hinduism* (New York: Columbia University Press, 2010).

palli Radhakrishnan came to regard Advaita Vedānta as the culminating point of Hinduism" on the basis of *some* Advaita-Vedāntins' "inclusivistic tendencies," but he argues that

> Advaita Vedāntins were not unanimous on the question of God as we entered the modern period. Although they all belonged to the same 'school', and thus shared the same basic metaphysical premises about the non-dual nature of the world, self, and Brahman, the four pandits examined in this study did not agree on the ultimate nature of the godhead nor on which deity, if applicable, ought to be considered supreme.

Duquette opens his intellectual history of inclusivism in Advaita Vedānta with "a paradigm shifting moment" that "occurred in the sixteenth century when Appayya Dīkṣita [1520–1593] challenged the dominant [Vaiṣṇava] theological model of the period, Viśiṣṭādvaita Vedānta." In his *Śivatattvaviveka* (Inquiry into the Śiva-principle), Appayya "targets Śrīvaiṣṇava theologians of Vedānta" and argues that "Śiva is the controller (*niyantṛ*) of all beings, *including the deities of the trimūrti*—Brahmā, Hari (= Viṣṇu) and Hara (= Śiva)." Nevertheless, in his later, mature works, such as the *Ānandalaharī* ('Waves of Bliss'), Appaya shifted towards "a more comprehensive view of God" and conceded that, "whether one worships Viṣṇu or Śiva ... should be of no importance given their core belief that Brahman is the non-dual reality of everything." In the following centuries, Duquette's quartet of paṇḍitas responded to Appayya's controversial work. In his *Hariharādvaitabhūṣaṇa* and *Hariharabhedadhikkāra*, Bodhendra, "the 59th pontiff of the Śaṅkarite *maṭha* in Kāñcīpuram," draws on Appayya's inclusivist arguments to defend the position that, "there is essentially no difference between Viṣṇu and Śiva and that both deities are aspects of the same reality—a reality he refers to as Harihara." Likewise, Ayyaṇṇa Dīkṣita "argues for the identity of Śiva and Viṣṇu, but he does so primarily with a focus on the *Brahmasūtras*." Ayyaṇṇa endorses Appayya's inclusivism, but he adds the caveat that, "*all* religious texts ... including the so-called 'heterodox' (*nāstika*) traditions of Buddhism, Jainism ... ultimately teach non-duality."

Not all Advaitins, however, endorsed Appayya's inclusivist tendencies. Tyāgarāja Śāstrī took up Appayya's pro-Śaiva arguments from the *Śivatattvaviveka*, while Rāmasubrahmaṇya Śāstrī developed a pro-Viṣṇu rejoinder to the *Śivatattvaviveka* that incorporated some of Appayya's argumentative strategies. Duquette's broader argument is that "socio-religious concerns can play a key role even in what appears to be purely philosophical matters." All four paṇḍitas that Duquette examines "belonged to a network of scholars centered around the Kāverī delta in South India" and their polemical intellectual debates were intimately shaped by local political, social, and pedagogical ecologies. In short, the inclusivist Advaita tendencies that intrigued Radhakrishnan and other new-Advaitins and that has become representative of Hinduism, writ large, was, to some extent, born as a result of competition—for patronage, to defend one's maṭha, to endorse one's chosen deity, and so on—amongst a discrete set of scholars working in a highly specific location in southern India.

Rosalind O'Hanlon closes this Festschrift with a detailed micro-history of Gotmāi Kāle, the "widow of a Brahman family based in the *taraf* or district subdivision of Chiplun town in the *subāh* or province of Dabhol in Maharashtra's Konkan littoral," who lived during the mid-seventeenth and early eighteenth-centuries. Gotmāi, as the scion of a once illustrious Brāhmaṇa clan, found herself "drawn into a prolonged struggle to retain control of the family's prestigious hereditary office as district accountants." O'Hanlon explores Gotmai's efforts to navigate the complex networks of family, tradition and the Marāṭha bureaucracy in her battle "against distant relatives and outsiders bent on usurping the office and the valuable rights attached to it." Gotmāi's family held the right of *deśakulkarṇi*, the hereditary office "of district-level accountant" who "oversaw revenue collection and the framing of accounts for the *pargaṇā* … a geographical unit of anything between 20 and 100 villages" within the Marāṭha Empire. According to a 1715 *karīnā*, a "lengthy factual statement usually made to a *majālis* or judicial assembly meeting to hear a local dispute," the office of *deśakulkarṇi* "had been bestowed on one Harī Krishna," an ancestor of "Kālopant, the husband of Gotmāi." Kālopant was the victim of a violent murder and, when Kālopant and Gotmāi's son, Antājī Kālo "himself died, leaving Gotmāi alone with her daughter-in-law, now also herself a widow," Gotmāi arranged for "Harbājī Dātīr, the father of Gotmāi's now widowed daughter in law" to exercise the office of *deśakulkarṇi* on the Kāle family's behalf.

After Gotmāi and Dātīr fell out, Gotmāi hired two newcomers, "Timājī Bhāskar and his brother Gangājī" to exercise the office of *deśakulkarṇi*. The Bhāskar brothers used wider political events—Afzal Khān's 1659 raid of the Marāṭha territory—to secure a personal *ḍaula* (a record of offices and estates) from the Bijapur Sultan and, thereby, to usurp the office of *deśakulkarṇi* for themselves. In order to save herself and her family, Gotmāi "retired to the prominent temple of Paraśurām in Chiplun, took up a seat in front of the god, and embarked on a public fast" which, in the course of time, "brought another local official, Antājī Bāvājī, the *mazūmdār* or keeper of revenue accounts at Chiplun, to enquire into her situation." The *karīnā* notes that Gotmāi likened her situation to that of "Draupadī, heroine of the epic *Mahābhārata*, rescued by the god Krishna himself from the malign Kaurava prince Dushashana as he attempted to disrobe her in public." Eventually, Mahādjī, the *subedār* of Dabhol appointed "Vinājī Balāl, son of Gotmāi's unlettered kinsman Bālo Vināyak" to the office of *deśakulkarṇi*. O'Hanlon deftly describes how Gotmāi operated strategically within the bureaucratic and Dharmic mores that circumscribed her as a woman in early modern Mahārāṣṭra. Gotmāi could not, according to Dharmaśāstric norms, adopt a son, so she had to find kinsmen to exercise the office of *deśakulkarṇi*. She could, nevertheless, leverage her personal virtue as a Brāhmaṇa widow to stage "a major drama in the local temple" and to persuade the *subhedār* to come to her aid. If Duquette, Patel, and Goldman reveal how scholarly pursuits in Dharmaśāstra, Vyākaraṇa, Vedānta and other Sanskrit sciences shaped early modern social and political milieux, O'Hanlon illuminates the agency of women whose lives were circumscribed the normative standards of those social contexts.

Bibliography

Primary Sources:

Alberuni's India: An Account of the Religion, Philosophy, Literature, Geography, Chronology, Astronomy, Customs, Laws and Astrology of India about A.D. *1030*, Vol. 1., ed. & trans. E.W. Sachau, (London: Trübner & Co 1888).

Groats-worth of Witte, ed. G.B. Harrison, *The Bodley Head Quartos* (1923; rpt. New York: Barnes & Noble, 1966).

Patañjali's Vyākaraṇamahābhāṣya, Vol. 1, ed. F. Kielhorn (Bombay: Government Central Book Depot, 1880)

Summa Theologiæ: Latin Text and English Translation, Introductions, Notes, Appendices, and Glossaries, ed. & trans. Thomas Gilby, (Cambridge: Cambridge University Press, 2006).

Secondary Sources

Atiyah, Michael. 'Mathematics: Queen and Servant of the Sciences'. *Proceedings of the American Philosophical Society* 137.4 (1993): 527–531.

Baber, Zaheer. *The Science of Empire: Scientific Knowledge, Civilization, and Colonial Rule*. Albany: State University of New York Press, 1996.

Bennett, Brett and Joseph Hodge, eds. *Science and Empire Knowledge and Networks of Science across the British Empire, 1800–1970*. Basingstoke: Palgrave Macmillan, 2011.

Burke, Peter. *The Polymath: A Cultural History from Leonardo da Vinci to Susan Sontag*. New Haven, CT: Yale University Press, 2020.

Butterfield, Herbert. *The Origins of Modern Science 1300–1800*. London: G. Bell & Sons, 1949.

Cahan, David, ed. *From Natural Philosophy to the Sciences: Writing the History of Nineteenth-Century Science*. London: University of Chicago Press, 2003.

Carroll, Allen. '*Johannes Factotum* and Jack Cade'. *Shakespeare Quarterly* 40.4 (1989): 491–492.

Copeland, Rita. 'Gloss and Commentary'. In *The Oxford Handbook of Medieval Latin Literature*, eds. Ralph Hexter and David Townsend, 171–191. Oxford: Oxford University Press, 2012.

Cunningham, Andrew. 'William Harvey: The Discovery of the Circulation of the Blood'. In *Man Masters Nature: 25 Centuries of Science*, ed. Roy Porter, 65–76. London: BBC Books, 1987.

Cunningham, Andrew & Perry Williams. 'De-Centring the 'Big Picture': 'The Origins of Modern Science' and the Modern Origins of Science'. *The British Journal for the History of Science* 26.4 (1993): 407–432.

Daston, Lorraine. 'The History of Science and the History of Knowledge'. *KNOW* 1.1. (2017): 131–154.

Dear, Peter. 'What Is the History of Science the History Of?: Early Modern Roots of the Ideology of Modern Science'. *Isis* 96.3 (2005): 390–406.

Dodson, Michael. 'Translating Science, Translating Empire: The Power of Language in Colonial North India'. *Comparative Studies in Society and History* 47.4 (2005): 809–834.

Elshakry, Marwa. 'When Science Became Western: Historiographical Reflections'. *Isis* 101.1 (2010): 98–109.

Fauvel, John. 'Newton's Mathematical Language'. In *The Foundations of Newtonian Scholarship*, eds. Richard Dalitz & Michael Nauenberg, 145–159. London: World Scientific, 2000.

Feyerbend, Paul. *Against Method: Outline of an Anarchistic Theory of Knowledge*. London: Verso, 1978.

French, Roger and Andrew Cunningham. *Before Science: The Invention of the Friars' Natural Philosophy*. London: Routledge, 1996.

Galison, Peter and David Stumps, eds. *The Disunity of Science: Boundaries, Contexts, Power*. Palo Alto, CA: Stanford University Press, 1996.

Ganeri, Jonardon. 'Contextualism in the Study of Indian Intellectual Cultures'. *Journal of Indian Philosophy* 36.5 (2008): 551–562.

Gascoigne, John. *Science in the Service of Empire: Joseph Banks, the British State and the Uses of Science in the Age of Revolution*. Cambridge: Cambridge University Press, 1998.

Gillispie, Charles. *Science and Polity in France: The Revolutionary and Napoleonic Years*. Princeton, NJ: Princeton University Press, 2004.

Gillispie, Charles. *Pierre-Simon Laplace, 1749–1827: A Life in Exact Science*. Princeton, NJ: Princeton University Press, 1997.

Goubier, Frédéric and Irène Rosier-Catach. 'The Trivium in the 12th Century'. In *A Companion to Twelfth-Century Schools*, ed. Cédric Giraud, 141–179. Leiden: Brill, 2019.

Grant, Edward. *The Foundations of Modern Science in the Middle Ages: Their Religious, Institutional, and Intellectual Contexts.* Cambridge: Cambridge University Press, 1996.

Greenwood, David and William Stini. *Nature, Culture, and Human History*. New York: Harper and Row, 1977.

Hall, A. Rupert. *The Scientific Revolution 1500–1800: The Formation of the Modern Scientific Attitude*. London: Longmans, Green and Co., 1954.

Halley, Janet. 'Equivocation and the Legal Conflict over Religious Identity in Early Modern England'. *Yale Journal of Law & Humanities* 3.33 (1991): 33–52.

Harris, Steven J. "Jesuit Scientific Activity in the Overseas Missions, 1540–1773," *Isis* 96.1 (2005): 71–79.

Houtman, Dick. 'The Disenchantment of the World and the Authority of Sociology: How the Queen of the Sciences Lost Her Throne'. In *Science Under Siege: Contesting the Secular Religion of Scientism*, eds. Dick Houtman, Stef Aupers, and Rudi Laermans, 37–64. Basingstoke: Palgrave Macmillan, 2021.

Heilbron, John. *The Oxford Companion to the History of Modern Science*. New York: Oxford University Press.

Hsia, Florence & Dagmar Schäfer. 'History of Science, Technology, and Medicine: A Second Look at Joseph Needham'. *Isis* 110.1 (2019): 94–99.

Ingalls, Daniel. 'On the Study of the Past'. *Journal of the American Oriental Society* 80.3 (1960): 191–197.

King, Ynestra. 'Healing the Wounds: Feminism, Ecology, and Nature/Culture Dualism'. In *Gender/Body/Knowledge: Feminist Reconstructions of Being and Knowing*, ed. Alison Jaggar and Susan Bordo, 115–141. New Brunswick: Rutgers University Press, 1989.

Knorr-Cetina, Karin. 'Social and Scientific Method or What Do We Make of the Distinction Between the Natural and the Social Sciences?'. *Philosophy of the Social Sciences* 11.3 (1981): 335–359.

Kuhn, Thomas S. *The Copernican Revolution: Planetary Astronomy in the Development of Western Thought*. Cambridge, MA: Harvard University Press, 1957.

Malamoud, Charles. *Village and Forest in the Ideology of Brahmanic India, Cooking the World: Ritual and Thought in Ancient India*, trans. David White. Delhi: Oxford University Press, 1996: 74–91.

McClish, Mark. 'The Dependence of Manu's Seventh Chapter on Kauṭilya's Arthaśāstra'. *Journal of the American Oriental Society* 134.2 (2014): 241–262.

McGuire, J.E. & P.M. Rattansi. 'Newton and the 'Pipes of Pan'.' *Notes and Records of the Royal Society* 21.2 (1966): 108–143.

Minkowski, Christopher Z. 'Janamejaya's Sattra and Ritual Structure'. *Journal of the American Oriental Society* 109.3 (1989): 401–420.

Minkowski, Christopher Z. 'The Rathakāra's Eligibility to Sacrifice'. *Indo-Iranian Journal* 32 (1989): 177–194.

Minkowski, Christopher Z. 'The *Udumbara* and Its Ritual Significance'. *Wiener Zeitschrift für die Kunde Südasiens* 33 (1989): 5–23.

Minkowski, Christopher Z. *Priesthood in Ancient India: A Study of the Maitrāvaruṇa Priest*. Publications of the De Nobili Research Library 18. Vienna: University of Vienna, 1991.

Minkowski, Christopher Z. 'Snakes, Sattras and the Mahābhārata'. In *Essays in the Mahābhārata*, ed. Arvind Sharma, 384–400. Leiden: Brill, 1991.

Minkowski, Christopher Z. 'An Omen Reader and a Cypher in Vedic Manuscripts'. In *Vāman Śāstrī Bhāgavat Felicitation Volume*, eds. Saroja Bhate and Madhav Deshpande, 122–127. Poona: Vaidika Samshodhana Mandala, 1994.

Minkowski, Christopher Z. 'School Variation in the Text of the Nivids'. In *Inside the Texts Beyond the Texts*, ed. M. Witzel, Harvard Oriental Series Opera Minora Vol. 2, 167–184. Cambridge, MA: Harvard University Press, 1997.

Minkowski, Christopher Z. 'The Interrupted Sacrifice and the Sanskrit Epics'. *Journal of Indian Philosophy* 29 (2001): 169–186.

Minkowski, Christopher Z. 'The Paṇḍit as Public Intellectual: The Controversy over Virodha or Inconsistency in the Astronomical Sciences'. In *The Pandit: Traditional Scholarship in India*, ed. Axel Michaels, 79–96. New Delhi: Manohar Publications, 2001.

Minkowski, Christopher Z. 'Astronomers and Their Reasons: Working Paper on Jyotiḥśāstra'. *Journal of Indian Philosophy* 30 (2002): 495–514.

Minkowski, Christopher Z. 'Nīlakaṇṭha Caturdhara's *Mantrakāśīkhaṇḍa*'. *Journal of the American Oriental Society* 122.2 (2002): 329–344.

Minkowski, Christopher Z. 'Nīlakaṇṭha's Cosmographical Comments in the Bhīṣmaparvan'. *Purāṇa* 42 (2000): 24–40.

Minkowski, Christopher Z. *A Descriptive Catalogue of the Sanskrit Astronomical Manuscripts Preserved at the Maharaja Man Singh II Museum in Jaipur India*. Compiled by David Pingree from the notes taken by Setsuro Ikeyama, Christopher Minkowski, David Pingree, Kim Plofker, Sreeramula Rajeswara Sarma, and Gary Tubb. Philadelphia: American Philosophical Society, 2003.

Minkowski, Christopher Z. 'A Nineteenth Century Sanskrit Treatise on the Revolution of the Earth: Govinda Deva's Bhūmibhramaṇa'. *SCIAMUS* 5 (2004): 199–224.

Minkowski, Christopher Z. 'Competing Cosmologies in Early Modern Indian Astronomy'. In *Ketuprakāśa: Studies in the History of the Exact Sciences in Honor of David Pingree*, eds. Charles Burnett, Jan Hogendijk, and Kim Plofker, 349–385. Leiden: Brill, 2004.

Minkowski, Christopher Z. 'Nīlakaṇṭha's Instruments of War: Modern, Vernacular, Barbarous'. *Indian Economic and Social History Review* 41 (2004): 365–385.

Minkowski, Christopher Z. 'The Vedastuti and Vedic Studies: Nīlakaṇṭha on Bhāgavata Purāṇa X.87'. In *The Vedas: Texts, Languages & Ritual. Proceedings of the Third Inter-*

national Vedic Studies Workshop, ed. A. Griffiths and J.E.M. Houben, 125–142. Groningen: Egbert Forsten, 2004.

Minkowski, Christopher Z. 'Nīlakaṇṭha's Vedic Readings in the Harivaṃśa Commentary'. In *Epic, Khilas, and Purāṇas: Continuities and Ruptures. Proceedings of the Third Dubrovnik Conference on the Sanskrit Epics and Purāṇas, September 2002*, ed. Petteri Koskikallio, 411–413. Zagreb: Croatian Academy of Sciences and Arts, 2005.

Minkowski, Christopher Z. 'On the Success of Nīlakaṇṭha's Mahābhārata Commentary'. In *Boundaries, Dynamics and Construction of Traditions in South Asia*, ed. F. Squarcini, 225–252. Florence: Florence University Press, 2005.

Minkowski, Christopher Z. 'Meanings Numerous and Numerical: Nīlakaṇṭha and Magic Squares in the *Ṛgveda*'. In *Indologica T. Ya. Elizarenkova Memorial Volume*, eds. L. Kulikov and M. Rusanov, 315–328. Moscow: Russian State University for the Humanities, 2008.

Minkowski, Christopher Z. 'Nīlakaṇṭha and His Historical Context'. *Vostok Oriens—Moscow Academy of Sciences* 4 (2008): 37–49.

Minkowski, Christopher Z. 'The Study of Jyotiḥśāstra and the Uses of Philosophy of Science'. *Journal of Indian Philosophy* 36 (2008): 587–597.

Minkowski, Christopher Z. 'I'll Wash Out Your Mouth With My Boot: A Guide to Philological Argument in Early Modern Banaras'. In *Epic and Argument in Sanskrit Literary History: Essays in Honor of Robert P. Goldman*, ed. Sheldon Pollock, 117–141. Delhi: Manohar, 2010.

Minkowski, Christopher Z. 'Sanskrit Scientific Libraries and their Uses: Examples and Problems of the Early Modern Period'. In *Looking at it from Asia: The Processes that Shaped the Sources of the History of Science*, ed. F. Bretelle and C. Proust, 81–114. Dordrecht, etc.: Boston Studies in Philosophy of Science 265, 2010.

Minkowski, Christopher Z. 'Advaita Vedānta in Early Modern History'. In *Religious Cultures in Early Modern India: New Perspectives*, eds. Rosalind O'Hanlon and David Washbrook, 205–231. Special Volume of *South Asian History and Culture* 2.2 (2011).

Minkowski, Christopher Z. "Nīlakaṇṭha's Teachers and Gurus, Part 1: Lakṣmaṇa Paṇḍita". *Wiener Zeitschrift Kunde Sudasiens* 55 (2013–2014): 33–76.

Minkowski, Christopher Z. 'Learned Brahmins and the Mughal Court: the Jyotiṣas'. In *Religious Interactions in Mughal India*, eds. Vasudha Dalmia and Munis Faruqui, 102–134. New Delhi: Oxford University Press, 2014.

Minkowski, Christopher Z. 'Appayya's Vedānta and Nīlakaṇṭha's Vedāntakataka'. *Journal of Indian Philosophy* 44.1 (2016): 95–114.

Minkowski, Christopher Z. 'Apūrvaṃ Pāṇḍityam: On Appayya Dīkṣita's Singular Life'. *Journal of Indian Philosophy* 44.1 (2016): 1–10.

Minkowski, Christopher Z. 'Maryādām Ullaṅghya: the Boundaries of Interpretation in Early Modern India'. In *Canonical Texts and Scholarly Practices: A Global Comparative Approach*, eds. A. Grafton and G. Most, 90–109. Cambridge: Cambridge University Press, 2016.

Minkowski, Christopher Z. 'Nīlakaṇṭha Caturdhara's Advaita Vedānta'. In *The Oxford Handbook of Indian Philosophy*, ed. Jonardon Ganeri. Oxford: Oxford University Press, 2017.

Minkowski, Christopher Z. 'An Early Modern Account of the Views of the Miśras'. *Journal of Indian Philosophy* 46 (2018): 889–933.

Minkowski, Christopher Z. 'Outlandishness in Sanskrit Commentary'. In *Les Scolastiques Indiennes: Genèses, Développements, Interactions. Sous la direction d'Émilie Aussant et de Gérard Colas*, 185–202. Paris/Pondichéry: École Française d'Extrême-Orient, Collection Études Thématiques, 2020.

Minkowski, Christopher Z. 'The Meaning and Design of the Mahābhārata in Nīlakaṇṭha's Commentary'. In *Proceedings of the Pondicherry Conference on Commentary*, eds. F. Grimal and C. Vielle (forthcoming)

Minkowski, Christopher Z. and Benjamin Elman. 'Big Science: Classicism and Conquest'. In *What China and India Once Were: The Pasts that May Shape the Global Future*, eds. Benjamin Elman and Sheldon Pollock, 199–231. New York: Columbia University Press, 2018.

Minkowski, Christopher Z. and Toke Knudsen. 'The Ṛtuvarṇana Chapter of Jñānarāja's Siddhāntasundara: Text, Translation, and Notes'. *Gaṇita Bhāratī*, 33.1–2 (2011): 53–87.

Minkowski, Christopher Z. and Rosalind O'Hanlon. 'What Makes People Who They Are? Pandit Networks and the Problem of Livelihoods in Early Modern Western India'. *Indian Economic and Social History Review* 45.3 (2008): 381–416.

Minkowski, Christopher Z., Rosalind O'Hanlon, & Anand Venkatkrishnan. 'Social History in the Study of Indian Intellectual Cultures?' *South Asian History and Culture*, 6.1 (2015): 1–9.

Minkowski, Christopher Z., Rosalind O'Hanlon, & Anand Venkatkrishnan, eds. *Scholar Intellectuals in Early Modern India: Discipline, Sect, Lineage and Community*. Abingdon: Routledge, 2015.

Morrell, J.B. 'Brewster and the early British Association for the Advancement of Science'. In *Martyr of Science: Sir David Brewster 1781–1858*, eds. A.D. Morrison-Low and J.R.R. Christie, 25–29. Edinburgh: Royal Scottish Museum, 1984.

Myrdal, Gunnar. 'How Scientific are the Social Sciences?' *Journal of Social Issues* 28.4 (1972): 151–170.

Needham, Joseph. *Science and Civilization in China*, Vol. 5: *Chemistry and Technology, Pt. Ill: Spagyrical Discovery and Invention: Historical Survey, from Elixirs to Synthetic Insulin*. Cambridge: Cambridge University Press, 1976.

Nicholson, Andrew. *Unifying Hinduism*. New York: Columbia University Press, 2010.

O'Hanlon, Rosalind. 'Letters Home: Banaras Pandits and the Maratha Regions in Early Modern India'. *Modern Asian Studies* 44.2 (2010): 201–240.

O'Hanlon, Rosalind. 'The Social Worth of Scribes: Brahmins, Kayasthas and the Social

Order in Early Modern India'. *Indian Economic Social History Review* 47.4 (2010): 563–595.

O'Hanlon, Rosalind. 'Speaking from Śiva's Temple: Banaras Scholar Households and the Brahman 'Ecumene' of Mughal India'. *South Asian History and Culture* 2.2 (2011): 253–277.

O'Hanlon, Rosalind. 'Contested Conjunctures: Brahman Communities and 'Early Modernity' in India'. *American Historical Review* 116.2 (2013): 765–787.

O'Hanlon, Rosalind, Gergely Hidas and Csaba Kiss. 'Discourses of Caste over the Longue Durée: Gopīnātha and Social Classification in India, ca. 1400–1900'. *South Asian History and Culture* 6.1 (2015): 102–129.

Phillips, Denise. *Acolytes of Nature: Defining Natural Science in Germany, 1770–1850*. Chicago: University of Chicago Press, 2012.

Pingree, David. 'The Purāṇas and Jyotiḥśāstra: Astronomy'. *Journal of the American Oriental Society* 110.2 (1990): 274–280.

Pollock, Sheldon. 'The Theory of Practice and the Practice of Theory in Indian Intellectual History'. *Journal of the American Oriental Society* 105.3 (1985): 499–519.

Pollock, Sheldon. 'The Idea of Śāstra in Traditional India'. In *Shastric Traditions in Indian Arts* ed. Anna Dallapiccola, 17–26. Stuttgart: Steiner Verlag; Wiesbaden: GMBH, 1989.

Pollock, Sheldon. 'New Intellectuals in Seventeenth Century India'. *Indian Economic and Social History Review* 38.1 (2001): 3–31.

Pollock, Sheldon. 'Introduction: Working Papers on Sanskrit Knowledge-Systems'. *Journal of Indian Philosophy* 30.5 (2002): 431–439.

Pollock, Sheldon. 'Is There an Indian Intellectual History? Introduction to 'Theory and Method in Indian Intellectual History''. *Journal of Indian Philosophy* 36.5–6 (2002): 431–439.

Pollock, Sheldon. 'The Languages of Science in Early Modern India'. In *Forms of Knowledge in Early Modern Asia: Explorations in the Intellectual History of India and Tibet, 1500–1800*, ed. Sheldon Pollock, 19–48. Durham, NC: Duke University Press, 2011.

Preisendanz, Karin. 'Text Segmentation, Chapter Naming and the Transmission of Embedded Texts in South Asia, with Special Reference to the Medical and Philosophical Traditions as Exemplified by the *Carakasaṃhitā* and the *Nyāyasūtra*'. In *Pieces and Parts in Scientific Texts*, eds. Florence Bretelle-Establet and Stephane Schmitt, 159–220. Cham: Springer, 2018.

Raina, Dhruv. 'French Jesuit Scientists in India: Historical Astronomy in the Discourse on India, 1670–1770'. *Economic and Political Weekly* 34.5 (1993): 30–38.

Richards, John. 'Early Modern India and World History'. *Journal of World History* 8.2 (1997): 198–203.

Ross, Sydney. '*Scientist*: The Story of a Word'. *Annals of Science* 18.2 (1962): 65–85.

Samuelson, Paul. 'Credo of a Lucky Textbook Author'. *Journal of Economic Perspectives* 11.2 (1997): 153–160.

Schaffer, Simon. 'Godly Men and Mechanical Philosophers: Souls and Spirits in Restoration Natural Philosophy'. *Science in Context* (1987): 55–85.

Shalom, Naama. *Re-ending the Mahābhārata; A Study of the Rejection of Justice (Dharma) in the Sanskrit Epic*. Albany: State University of New York Press, 2017.

Skinner, Quentin. *Visions of Politics*, Vol. 1: *Regarding Method*. Cambridge: Cambridge University Press, 2002.

Staal, Fritz. 'The Sanskrit of Science'. *Journal of Indian Philosophy* 23.1 (1995): 73–127.

Taagepera, Rein. *Making Social Sciences More Scientific: The Need for Predictive Models*. Oxford: Oxford University Press, 2008.

Truesdell, C. 'Reactions of Late Baroque Mechanics to Success, Conjecture, Error, and Failure in Newton's *Principia*'. In *Essays in the History of Mechanics*, ed. C. Truesdell, 138–183. Berlin: Springer-Verlag, 1968.

Turner, Geoffrey. 'St Thomas Aquinas on the 'Scientific' Nature of Theology'. *New Blackfriars* 78.921 (1997): 464–476.

Udías Vallina, Agustín. *Jesuit Contribution to Science: A History*. Cham: Springer, 2015.

Venkatkrishnan, Anand. 'Mīmāṃsā, Vedānta, and the Bhakti Movement'. Unpublished PhD dissertation, Columbia University, 2015.

Weisheipl, James A. 'Classification of the Sciences in Medieval Thought'. *Medieval Studies* 27 (1965): 54–90.

Williams, Raymond. 'Science'. In *Keywords: A Vocabulary of Culture and Society*. Oxford: Oxford University Press, 1985.

Wilson, E.O. *Consilience: The Unity of Knowledge*. New York: Vintage, 1999.

Yeo, Richard. *Defining Science: William Whewell, Natural Knowledge and Public Debate in Early Victorian Britain*. Cambridge: Cambridge University Press, 1993.

PART 1

The Vedic World

CHAPTER 1

When Cows Sacrifice: The Meaning of Vedic *tūpara*

Barbora Sojková

1 Introduction

Animals, and especially domestic livestock, played a major role in ancient societies from the earliest times, and the Vedic culture is by no means an exception. It is therefore not surprising that domestic animals dominated the poetic language of Vedic literature, being the most common source of metaphors and similes [Jamison 2009, p. 197], and that they figured in many mythological accounts in both the Vedic Saṃhitās and Brāhmaṇas.

In this paper, we will re-examine one of these animal myths, found in six versions throughout the Vedic *śākhās*. The mythical narrative in question tells a story of times when cows did not have horns (and hooves). To obtain them, they decided to organize a year-long *soma* ritual, so called *sattra*, known since then as *gavāmayana* (literally "the path, course of cows"). After ten months, the cows acquired their desired horns but a group of them decided to continue with the ritual activity for the rest the year, as planned. These cows became known as *tūpara*, a peculiar word which is understood by the later Sanskritic tradition as a term for a hornless animal. In the context of this myth, however, this meaning does not make much sense. In what follows, I will reconsider the evidence of the mythical narrative and other Vedic sources to understand what could be signified by the word. Although this paper is called "When Cows Sacrifice", I will not focus on the ritual actions of the *gavāmayana*; what interests me here are the cows themselves.

2 The Narrative

The mythical narrative under consideration occurs in six parallel versions in five Vedic texts. Three versions are found in two Black Yajurveda Saṃhitās (TS 7.5.1.1–2, TS 7.5.2.1–2, and KS 33.1), whereas the other three are from the Ṛgveda and Sāmaveda Brāhmaṇas (AB 4.17, JB 2.374, PB 4.1).[1] Despite their

1 Throughout the paper, I am using the standard abbreviations of Vedic texts (for the list, see

 | DOI:10.1163/9789004536869_003

diverse *śākhā* affiliations, the passages are very close both in terms of grammar and narrative. The close affinity is further emphasized by their shared introductory formula or incipit, "the cows held a *sattra*":

KS 33.1 *gāvo vai sattram āsata*
TS 7.5.1.1 *gā́vo vā́ etát sattrám āsata*
TS 7.5.2.1 *gā́vo vā́ etát sattrám āsata*
AB 4.17 *gāvo vai satram āsata*
JB 2.374 *gāvo vā etad agre sattram āsata*
PB 4.1 *gāvo vā etat sattram āsata*

Apart from the JB which adds the word *agre*, used frequently by the JB to introduce mythical episodes that happened *in illo tempore*, the incipits of the passages are almost completely similar. This suggests not only the affinity of the mythical versions but also a considerable age of the myth. As Stephanie Jamison [1997, p. 127] has argued, many Vedic myths are introduced with an almost identical incipit, even though the main narrative can be rendered differently in various texts and *śākhās*. The shared incipit suggests that the myth in question is rather old: it, or at least its part, must have existed in a precise verbal form before the Vedic tradition divided into the respective *śākhās*.

To make my argument easier to follow, I will quote the JB version in full as an example of the mythical narrative. The JB is certainly not the oldest version of the story; in fact, it might even be the youngest. It is, however, the longest account and despite a few changes in the narrative, it provides us with some additional details about the ritual. Following the JB example, I will point out the differences in the respective versions and treat the textual crux of the passage, the word *tūpara* (usually translated as "hornless"). Let us look at an example of the myth in its Jaiminīya version:

JB 2.374[2] *gāvo vā etad agre sattram āsatānnādyam avarurutsamānās | tā daśame māsy udatiṣṭhan sarvam annādyam avārutsmahīti manyamānās | tā etāś śṛṅgiṇīs | tāsāṃ tvā ivābruvan sarvān evaitān dvādaśa māsas*

section "Abbreviations"). All translations are mine unless otherwise stated. In translations, parentheses are used to indicate which specific Sanskrit word is translated. Brackets are used to insert words which are not present in the Sanskrit original. In Sanskrit passages, curly brackets are used when a word is emended; an explanation of the emendation is provided in a footnote.

2 I am citing the critical edition of Murakawa [2010, pp. 5–7] and not the standard edition of Vira & Chandra [1954]. For the sake of consistency between the passages, I use *daṇḍas* where Murakawa uses full stops.

> *samāpayāmeti | tā atiprāyuñjata | tāsāṃ dvādaśe māsi śṛṅgāṇi prāvartanta | tā etās tūparās | tasmāt sattriṇo dvādaśe māsy api śikhāḥ pravapante | gavāṃ hi tarhy anurūpā bhavanti | tā udatiṣṭhann arātsmety āptvā sattram āptvā sarvam annādyaṃ | tasmād etās sarvān dvāsaśa māsa uttiṣṭhanti {prartvāpi}*[3] *vārṣikāv api śaiśirau* | [...]
>
> In the beginning, the cows held a *sattra* in this way, wishing to obtain the food to eat (*annādyam*). They, finishing in the tenth month, thought: "We have obtained all the food to eat." These are the horned ones. But some of them said: "We will complete all twelve months." They went on performing the *prayoga* past [the ten months]. Their horns got curved (*prāvartanta*)[4] in the twelfth month. They are the *tūpara* [cattle]. Therefore, those who hold a *sattra* (*sattrin*) shave off the hair towards the region of their hair-knots (*śikhā*) [i.e. in order to have the *śikhā* in the front of their head] in the twelfth month. Because then, they become like the cows in form. They finished (the *sattra*), (saying) "we have obtained [the the food to eat]". Having obtained the *sattra*, having obtained all [of] the eating of food. Therefore, they [i.e. the *tūpara* cattle] stand up [even] for the duration of all the twelve months, having roamed widely even during the two rainy [months], even in the two winter [months]. [...]

As we have seen in the passage above, the cows are said to hold the *sattra* in order to obtain *annādya*, "food to eat".[5] Other passages, however, reveal their

---

3 This is a problematic form which also occurs in the passage TS 7.5.1.1–2. Murakawa [2010, p. 7] reads *partvā*, whereas Caland [1919; see also 1927, p. 253] emends *partvā* to *prartvā*, followed also by Vira & Chandra [1954]. Hoffmann [1992, p. 785] opposes Caland and argues that we would ideally expect **prártya*. He also states that *prartvā* is only Caland's emendation without any sufficient textual support. Gerhard Ehlers notes in his unpublished JB edition (courtesy of Professor Michael Witzel) that Hoffmann's assertions are not correct as *prartvā* occurs in two mss. He keeps the form despite the problems with the reading as it ultimately makes the most sense in the context. I have decided to follow Caland and Ehlers in my translation as I find their reasoning convincing.

4 In his translation of the PB, Caland [1931, p. 43] argues that it is unlikely that the verb *prāvartanta* means "they fell off", as it is sometimes suggested, e.g. by Keith [1920, footnote 2, p. 208]. He argues that it should be translated as "they became curved", "round", or "circular", comparing it with words such as *pravarta* "a circular ornament", or *pravṛtta* "round". I agree with Caland's opinion and use this translation.

5 *annādya* is a peculiar compound used throughout the Vedic corpus, and is especially common in the JB cosmology. In this mythical narrative, JB is not the only text using this concept; we also find it in the PB 4.1. The compound is a merism formed from two participles of the verb √*ad-, atti,* "to eat": the past passive participle *anna-* and the gerundive (i.e. the future pass-

true motivation: they desired to obtain horns, and in the case of the AB, hooves as well.

> TS 7.5.2.1 *gā́vo vā́ etát sattrám āsatāśṛṅgā́ḥ satī́ḥ śṛ́ṅgāṇi síṣāsantīs* […] |
>
> The cows, being hornless and desiring horns, held a *sattra*.
>
> AB 4.17 […] *śaphāñ chṛṅgāṇi siṣāsantyas, tāsāṃ daśame māsi śaphāḥ śṛṅgāṇy ajāyanta* |
>
> [The cows held a *sattra*], desiring hooves and horns. In the tenth month, their hooves and horns came into being.

After ten months, the cattle obtained the desired horns; in a couple of passages, these cows are called "the horned ones" (JB 2.374 *śṛṅgiṇīs*, AB 4.17 *śṛṅgiṇyo*). As we have seen in the JB passage, a smaller group of the cows (or half of them, as said by TS 7.5.2.1 and KS 33.1), decided to go on with the ritual. At the end of the year-long period, their horns became "curved" (*prāvartanta*). This happened due to their lack of "ritual confidence"[6] (*aśraddhayā*) as is suggested by the AB. Both of the YV Saṃhitā passages also mention the concept of *śraddhā*:

> KS 33.1 *tāsāṃ śṛṅgāṇi prāvartanta śraddhayā vāśraddhayā vā, tā imā yās tūparās* |
>
> Their horns got curved either due to their confidence or a lack thereof. These are the *tūpara* [cattle].

ive participle) *adya-*. There have been many attempts to translate it as a unit: for example, Jamison & Witzel [2003, p. 63] suggest "stored food", Olivelle [1998, pp. 490–491, n. 3.17] "food-stuff", and Weber-Brosamer [1988] "Nahrungsmittel", i.e. any substance that may be used as food. According to Jamison & Witzel [2003, p. 63], it is not a mere food but an abstract power, or substance which is ritually exchanged among god and men. In this paper, I am following Jones' [2017, pp. 78–80] translation "food to eat". The compound also frequently occurs in an alliterative figura etymologica *annam annādyam* which Jones [ibid.] translates with a somewhat redundant phrase as "food, food to eat". He also suggests that we could understand this phrase as a merism inherited from the IE poetic canon (cf. Watkins 2005, p. 45) and render it perhaps as "grain [or other specific staple food] and food generally", although this would "eliminate the participial, morphological, and phonological redundancy of the etymological figure" [Jones 2017, pp. 79].

6 Jamison and Witzel [2003, p. 65] note, following Köhler [1973], that the usual translation of *śraddhā* as "faith" is erroneous and argue that it "originally referred to 'confidence' in the efficacy of the ritual, i.e. [ritual] ability to motivate counter-gifts."

All the passages unanimously call this group of animals *tūpara*. As suggested above, this is most likely the key to the whole passage, and will be dealt with later. In the BYV passages, the *tūpara* cattle are also said to be the ones that obtained the "liquid nourishment", *ūrj*.[7] This is perhaps a YV variant of *annādya*, "food to eat", which we saw earlier in the JB. The other Sāmavedic Brāhmaṇa, the PB, retains the notion of "food to eat", although it does not mention it nearly as often as is the case in the JB:

> **PB 4.1** *tāsāṃ dvādaśasu māḥsu śṛṅgāṇi prāvartanta, tāḥ sarvam annādyam āpnuvaṁs, tā etās tūparās* |
>
> In the twelfth months, their horns got curved. They obtained all the food to eat. These are the *tūpara* [cattle].
>
> **TS 7.5.2.2** *ubháyyo vā́vá tā́ ārdhnuvan yā́ś ca śṛ́ṅgāṇy ásanvan yā́ś córjam avā́rundhata* |
>
> Both are indeed satisfied: those who gained horns, and those who obtained the liquid nourishment (*ūrj*).

By far the most interesting point of these passages is their descriptions of the behavior of the *tūpara* cattle. In the JB passage, we saw that the *tūpara* cattle was said to keep outside even during the rainy and cold seasons. This is confirmed by other passages:

> **TS 7.5.1.2** *tásmāt tūparā́ vā́rṣikau mā́sau {prártvā}*[8] *cara{n}ti*[9] |
>
> Therefore, the *tūpara* [cattle] are active, having roamed widely during the two rainy months.
>
> **KS 33.1** *tasmād dvādaśa mās{a}s*[10] *tūparāḥ prertvarīś*[11] *caranti, daśa śṛṅgiṇīr*

7 Cf. Minkowski [1989, p. 10].

8 See the explanation of the emendation above, n. 4.

9 The reading *carati* (3rd singular, present of √*car-*) does not make sense in conjunction with the nominative being in plural. Based on the parallel passage in KS 33.1, I am reading *caranti*.

10 I am reading *māsas* instead of *māsās*, following Caland [1927, p. 253].

11 As Caland [ibid.] notes, this form is most likely connected with the mysterious form *prártvā* found in the TS and JB versions, and treated above in n. 4.

Therefore, the *tūpara* [cattle] are active, moving about for twelve months; the horned [cattle are active] for ten [months].

AB 4.17 *tasmād u tāḥ sarvān ṛtūn prāptvottaram*[12] *uttiṣṭhanty, ūrjaṃ hy asunvan* |

Therefore, they [i.e. the *tūpara* cattle] stand upright during all the seasons, for they pressed out the liquid nourishment (*ūrj*).

3 The Meanings of *tūpara*

The descriptions of the behavior of the *tūpara* cattle, found in the majority of the passages, bring us to the core of this paper: the word *tūpara* itself. The word is ordinarily translated as a "hornless [animal]", or specifically a "hornless goat". As suggested from the outset, I doubt that it is the original meaning of it, or at least not in this mythical passage. In fact, I am not the first one to express such a doubt: Willem Caland [1931, p. 43] noted in his translation of the PB that "a *tūpara* animal is [...] not only a hornless animal, but also one with small, inward bent horns" and Karl Hoffmann [1992, pp. 782–783] understood it to mean "cow with deformed horns" ("Kuh mit mißgebildeten Hörnern"). Unfortunately, neither Caland nor Hoffmann provided any explanation for their reasoning for such claims, and we will therefore have to test their assertions. To do so, let us consider the evidence of the narrative itself once again, followed by an exposition of the etymology and textual history of the word *tūpara*.

If we accept the translation "hornless", the narrative seems rather inelegant. All the passages tell us that the cows decided to follow a year-long ritual in order to obtain horns. All of the cows acquired horns after ten months but those who persevered became *tūpara*, on top of obtaining the more practical goals of the ritual, such as the "food to eat" or the "liquid nourishment". We would expect that the cattle which actually finished up the challenge of the year-long *sattra* in full obtained something for their effort. Hornlessness would be a return to the pre-ritual state, a negation of the goal of the ritual. Moreover, every iteration of the mythical story describes the different behavior of the *tūpara* cattle, as we have seen above. They are said to move about the whole rainy season and winter, unlike the horned cows which are moving outside only for ten months

12 Instead of *prāptvā*, we would expect the absolutive form *prāpya*. Perhaps, this could be a reflection of the form *prártvā* found in the TS and JB?

of the year. As far as I am aware, the lack of horns does not cause any change of behavior in cattle. On this ground, "hornless" is an unlikely meaning.

The longer, ritualistically-oriented JB passage provides an indispensable detail regarding the sacrificers' hairstyle which could help us understand the word *tūpara*. The sacrificers, observing a strict *dīkṣita* lifestyle, do not cut their hair for the whole duration of the ritual, i.e. the whole year. At the end of the ritual period, they shave in a specific way:

> **JB 2.374** *tasmāt sattriṇo dvādaśe māsy apiśikhāḥ*[13] *pravapante | gavāṃ hi tarhy anurūpā bhavanti |*
>
> Therefore, those who hold the *sattra* shave off [their hair] towards the region of their hair-knots (*apiśikhā*), [i.e. in order to have the *śikhā* in the front of their head]. Then, their form corresponds to that of cows.

This passage suggests that the sacrificers keep a tuft of their hair in the very front of the head, mirroring the *tūpara* cows whose horns "became curved" (*prāvartanta*) at the end of the year-long *sattra*.[14] If the cows were to be hornless, we would expect the sacrificers to follow them and shave their hair completely. To me, this is a clear internal evidence of the narrative that we are not dealing with hornless cattle but rather with cattle which has different shape of horns, a possibility which we will treat in detail below.

Stepping aside from the narrative, let us look at the word *tūpara* itself, starting with its meaning and possible etymology. Monier-Williams understands the word as "hornless", specifically a "hornless goat", perhaps following the usual meaning found in later Sanskrit literature starting with the late Brāhmaṇas and

13 *Apiśikhāḥ* is a curious word. Its morphology, i.e. *api* + *śikhā-*, reminds of a few Vedic words for body parts, such as *apikarṇa*, "close to ear" (ṚV 6.48.16), or *apikakṣa*, "region of the shoulder blade, armpit" (ṚV 4.40.4, 10.134.7). On the later example, see Jamison [1987, p. 84]. Professor Minkowski (personal communication, July 2020) suggested to me that the prefix *api-* might be comparable to the Classical prefix *abhi-*. I am following this suggestion in my translation, using the prepositional phrase "towards the region of".

14 As Ehlers [2003], suggests, the *pūrvaśikhā* hair-fashion might be alluded in this myth. The *pūrvaśikhā* style is explained by Mahadevan and Staal [2003, pp. 5–6] who suggest that "there were in the main two different waves of Vedism arriving in South India at two different periods of history: the first is represented by the *pūrvaśikhā* Brahmans with their fronted top-knots and the second by the *aparaśikhā* Brahmans, their topknots toward the back of their heads, making a pony tail." Although Mahadevan's and Staal's theory is not widely accepted, I find Ehlers' following argument and connection with this myth convincing. For illustration of the *pūrvaśikhā* hair-style see the cover photo of the paper by Mahadevan and Staal [2003].

Śrauta Sūtras. This sense is supported by Mayrhofer [*KEWA* Vol. 1, pp. 518–519] who also adds the meaning "blunt". According to him [ibid.], the word has a few variants: *tūvara* or *tūbara*, referring to a "hornless bull", "beardless man", or an "eunuch", and *tūvaraka* or *tūbaraka*, a MBh derogatory term for a "castrated person", "eunuch". Following Turner [*CDIAL*, p. 348] who catalogues the word under **thubba*, "defective", Mayrhofer [*EWA* Vol. 1, p. 661] further connects the word with modern Indian variants such as the Bengali words *thubaṛā*, or *thobṛā*, "blunt" which, however, do not seem to be plausibly associated with the Sanskrit word *tūpara*.[15]

All of these variants show a great degree of flexibility in the second labial consonant; they fluctuate between *-v-*, *-b-*, and *-p-*. As F.B. Kuiper [1948, p. 74] argues, the variants with *-v-* and *-b-* are usually understood to be later, Middle Indian equivalents of *tūpara* but they might equally be the original forms. Kuiper [ibid., pp. 74–75] also suggests that the word is unlikely to be of an Indo-European etymology but Proto-Munda, with the initial *t* standing for *ṭ*. He compares it to a number of lexemes, such as Santali *tope*, "short, to cut off", *thopro*, "low, short brushwood", or *ṭhe̱ble̱*, "low, small, dwarwish", or Mundari *ṭupā, ṭumpu*, "blunt", or *ṭhepā*, "short in stature, dwarf". This would point at a possible original meaning of the word "small", "little", or "cut off". He [ibid., p. 75] argues: "[…] the root *ḍa-wa* may develop the meanings ("cut off, stripped" >) "solitary, bereft" and, on the other hand, ("cut off" >) "stunted, small, dwarwish"". He also notes that other Sanskrit words describing hornless animals might have had a similar development:[16]

---

15 According to Lubomír Ondračka, the Bengali words are most likely not derived from the Sanskrit *tūpara* as the initial *th-* in Bengali come from Old Indo-Aryan *sth-*. Moreover, the meaning of the Bengali words such as *thubaṛā* and *thobṛā* does not seem to mean a "blunt" object but rather a "dull", "old", or "doddery" person, or somebody who is feeling "flattened". Thanks to Lubomír Ondračka (personal communication, 25 April 2021) for very useful suggestions.

16 With the breadth of Sanskrit vocabulary in mind, it is hard to argue that *tūpara* cannot mean "hornless" because other words with the same meaning exist in Sanskrit. It is, however, useful to consider them and the context in which they appear. One of these is the word *śama*, inherited from the PIE * *k̑em-*, "hornless" (cf. Engl. *hind*, or Grk. κεμάς "young deer"; see Mallory and Adams [2006, p. 137]) and attested only in two ṚV hymns, 1.32.15 and 1.33.15. In the former instance, Indra is said to be "the king […] of the horned and the hornless" (*índro* […] *rā́jā śámasya ca śṛṅgíno*), a merism which possibly describes all the *paśus* (the hornless being human and horse, the horned the rest of them). Another term for a hornless animal, mentioned also by Kuiper [1948, pp. 74–75], is the word *kūṭa*. This word, used as infrequently as *śama*, occurs e.g. in AV 12.4.3, in a list of blemished cows: "hornless, lame, crippled, one-eyed" (*kūṭáyāsya* […] *śloṇáyā* […] *baṇḍáyā* […] *kāṇáyā* […]). Similar list is found in ŚBM 3.3.1.16, where the *soma* cow is supposed to be of an unblemished

> Parallel developments are, e.g., Skr. *kūṭá-* "not horned": *kuṇṭha-* "blunt" from *ga-ḍa*, S. *ḍonḍo* "with very small horns, bent backwards" from *ḍa-ḍa* "to shave, cut off", Skr. *muṇḍin-* "hornless": S. *muṇḍṛa daṅgra* "a hornless bullock", *muṇḍṛa* "having his hair cropped or shaved", *moṇḍgoc'* "bare, hornless, polled, to shave or cut off" from *wa-ḍa*. Cf. also the various meaning of Engl. *to poll* ("to crop the hair, cut off the top, cut off horns").[17]
>
> KUIPER ibid., p. 75

The possible non-Indo-European etymology, together with the instability of the second consonant shows that we might be dealing with a rather unclear term with a highly colloquial meaning. The colloquialism is especially present in the variants *tūvaraka* and *tūbaraka* which are formed with the widespread Indo-Iranian colloquial suffix *-ka*. As shown by Stephanie Jamison [2009], this suffix is predominantly associated with negative or lower-class features and frequently found in animal names, especially those which are small and obnoxious, such as *vṛścika*, "scorpion", *maṇḍūka*, "frog", or *maśaka*, "stinging insect". The *-ka* suffix in the word *tūbaraka/tūparaka* shows its pertinence to the sphere of everyday life; it is not, I believe, a particularly high-brow word.

The colloquiality of the word might be also a reason why the meaning of the word is not fully set, at least not in its early history. All of the listed meanings point at a state of defectiveness, a state in which an object lacks its defining feature. It describes a cow or a goat which is hornless, despite it being a characteristic mark of the species; a man who is beardless or castrated, both signs of the lack of masculinity (cf. TS 2.1.1.5); or a *yūpa*, "sacrificial post" which should be pointy but is blunt (TB 1.3.7.2, ĀpŚS 18.1.). Perhaps, instead of translating the word as "hornless" in all cases, the translation "cut off" or "trimmed" would be more fitting, or more specifically, "the one whose characteristic feature has been cut off" or "polled". Although being rather wordy, this translation might be preferable as it also encompasses the sense "hornless" which could be the most possible meaning in most of the ritual passages dealing with sacrificial animals, aside of the myth in question.

---

body: it should not be "tailless, hornless, one-eyed, earless, with any other specific mark, or seven-hoofed" (*syād avaṇḍākūṭākāṇākarṇālakṣitāsaptaśaphā*). Lastly, we find the word *aviṣāṇa* which occurs, as far as I know, only in a single instance, ŚB 5.1.3.8: "Man is indeed the closest to Prajāpati, as he is *tūpara*, hornless; Prajāpati is indeed *tūpara*, hornless." (*puruṣo vai prajāpaternediṣṭhaṃ so 'yaṃ tūparo 'viṣāṇastūparo vā aviṣāṇaḥ prajāpatiḥ*). Its positive counterpart, *viṣāṇa*, is analogously rare, and occurs a few times in the Vedas, e.g. in the ṚV 7.1.8.7 as a name of a non-Vedic tribe.

17 In this quote, "S." stands for Santali and "Skr". for Sanskrit.

4 The Textual History of *tūpara*

To understand the word *tūpara* in context, let us consider a few of its instances in the Vedic canon.[18] The word occurs more than forty times; of these, more than half are in the Vedic Saṃhitās.[19] Many instances are, moreover, parallel passages or verbatim quotations, such as our mythical narrative. Missing in the ṚV, *tūpara* is first found in the Śaunaka recension of the AV, in a poorly understood hymn which Whitney [1905, p. 651] labels "to conquer enemies":[20]

> **AVŚ 11.9.22** *yé ca dhī́rā yé cā́dhīrāḥ párāñco badhirā́ś ca yé | tamasā́ yé ca tūparā́ átho bastābhivāsínaḥ | sárvāṃs tā́ṃ arbude tvám amítrebhyo dṛśé kurūdārā́ṃś ca prá darśaya ||22||*
>
> Both those who are wise and those who are not wise, [those who] have left and those who are deaf, the dark ones, and moreover the "polled ones" (*tūpara*) who bleat like a billy-goat (*bastābhivāsin*); O Arbudi, make our enemies see all of those and show [them] the lofty [ones].

Throughout the hymn, Arbudi is pleaded to make the enemies see various, more or less frightening phenomena, such as apsarases accompanied by dogs, four-tusked and blood-faced creatures, serpents, or forest trees. The beings described in the aforementioned verse as *tūpara* are clearly on considered to be negative as well: they are preceded by creatures connected with darkness (*tamasa*) and possibly described as "bleating like a billy goat" (*bastābhivāsin*). Especially the later adjective is suggestive of the possible negative association of the word *tūpara*, as the AV commonly attributes unpleasant animal noises to dangerous animals.[21] Interestingly, when mentioning animals, the hymn 11.9

18 For the sake of brevity, I will be translating the word *tūpara* as "polled" rather than "one whose characteristic feature has been cut off" in the following passages.

19 AV 11.9.22; KauS 49.1–10; KS 13.1–4, 19.8, 24.7, 33.1, 45.2, 48.3; MS 2.5.1, 2.5.9, 3.13.2, 3.13.13; TS 2.1.1.4–5, 2.1.5.7, 2.1.6.5, 5.5.1.1–2, 5.5.23.1, 5.5.24.1, 5.6.11–20.1, 7.5.1.2, 7.5.2.2; VSM 24.1, 24.15, 24.15; PB 4.1.2; AB 4.1.2; JB 2.262., 2.371, 2.374; TB 1.2.5.3, 1.3.7.2, 3.8.20.4, 3.8.23.1, 3.9.22.4; ŚBM 5.1.3.7–8, 6.2.2.2, 6.2.2.6, 6.2.2.7, 6.2.2.15, 9.5.1.57, 13.2.2.2, 13.5.1.13, 13.5.2.10.

20 The word is altogether missing from the Paippalāda recension. Thanks for this information are due to Elizabeth Tucker (email communication, 20th March 2021).

21 Cf. AV 8.6.10, *yé śā́lāḥ parinṛ́tyanti sāyáṃ gardabhanādínaḥ* [...] *tā́n oṣadhe tváṃ gandhéna viṣūcī́nān ví nāśaya*, "Those who dance around houses in the evening, braying like a donkey [...], make them disappear in different directions with your smell, o herb." It is, however, absolutely unclear whether the word *bastābhivāsin* is a qualifier of *tūpara* or a separate entity. Thanks to Elizabeth Tucker for this suggestion (email communication, 20th March 2021).

limits itself to wild creatures which gives an indication of the early association of the word *tūpara* with the sphere of wilderness.

In the Yajurvedic Saṃhitās, the word occurs relatively frequently, predominantly to describe the appearance of *paśu*s, domestic animals. Overwhelmingly, it is found in lists of animals to be sacrificed, such as this passage from the TS:

> **TS 5.6.20.1** *somaindrā babhrulalāmās tūparāḥ* |
>
> The white-spotted "polled" ones are for Soma and Indra.

This example comes from a long list of *aśvamedha* animal victims, found in TS 5.6.11–20.1. This highly formulaic passage enumerates ten sets of eighteen animals fastened to the sacrificial stakes, categorizes them according to color (red, dappled, white, with white features, spotted, tawny) or bodily features (humped, long-eared, short-eared, etc.), and ascribes them to a particular deity. The *tūpara* animals are mentioned as the last ones in each section, as if their contribution to the ritual was the least important, perhaps due to their bodily "blemish".

In the following instance, also a from a formulaic list of animal victims, the VS associates the *tūpara* animal with a horse and a gayal. This is a rather common animal trio, frequently found together elsewhere in the Vedic canon:[22]

> **VS 24.1** *aśvas tūparo gomṛgas te prājāpatyāḥ* |
>
> A horse, a "polled" one, and a gayal; these belong to Prajāpati.

The association of both horse and *tūpara* with the god Prajāpati is also comparatively frequent (in more detail, I will treat it below). Prajāpati himself is said to be *tūpara* in the ŚB (ŚBM 6.2.2.2. *tūparo hí prajā́patiḥ*), and so is a human being:

> **ŚBM 5.1.3.8** *púruṣo vaí prajā́pater nédiṣṭhaṃ só 'yáṃ tūparó 'viṣāṇás* |
>
> A man is indeed closest to Prajāpati, in that he is "polled" (*tūpara*), hornless (*aviṣāṇa*).

Here, the ŚB provides an explanation of the word *tūpara* with the term *aviṣāṇa*, "without horns". This embedded Brāhmaṇa gloss would suggest that by

22 Cf. MS 3.13.2, TB 3.8.20.4, 3.8.23.1; ŚB 13.2.2.2, 13.5.1.13, 13.5.2.10.

the time of the ŚB, the word *tūpara* is understood predominantly as "hornless".

Prajāpati and human beings are not the only ones that are said to be *tūpara* in the Brāhmaṇa cosmology. A goat, as said above, is another one of them, and really becomes the prototypical *tūpara* by the time of the Śrauta Sūtras. Here, we see an example of a *tūpara* goat from the TS:

> **TS 2.1.1.4** *táto 'jás tūparáḥ sám abhavat* |
>
> Then, the "polled" (*tūpara*) goat came to life.

Lastly, horse is also described as *tūpara*, again at the time of the later Vedic prose. This brings to mind an interesting passage, found a few times in the Vedic prose, which enumerates characteristic features of *paśus*. Here, horse is described as the prototypical hornless animal, as opposed to the rest of the livestock (but man) which are horned.[23]

> **JB 2.371**[24] *śveto lomaśas tūparo lapsugy,*[25] *anyatodaṃś catuṣpād, yac chveto lomaśas tad avīnāṃ rūpaṃ, yat tūparas tad aśvānāṃ, yal lapsugī tat puruṣāṇāṃ, yad anyatodaṃs tad gavāṃ, yad ajas tad ajānām* |
>
> The [domestic animals are] white, woolly, "hornless", bearded, with incisors in the lower jaw, [and] four-footed. That which is white [and] woolly has a form of sheep; that which is "hornless" of a horse, that which is

---

23 Mallory and Adams [2006, p. 137] argue that in certain IE languages, especially in Balto-Slavic, horse is the prototypical hornless livestock, in contrast to cattle, the horned livestock: Slavic words for "cow" are derived from PIE **ḱóru-*, "horn", see OCS *krava*, Cz. *kráva*, Rus. *koróva*, cf. OPrus. *camnet*, or Grk. *κεραός* 'horned'. Mallory and Adams base this interpretation on the term **ḱem-*, "hornless" which they understand as the source of OCS *koņь*, (cf. Rus. *kon'*, Cz. *kůň*). This interpretation is, however, highly improbable; first of all, in Slavic, we would expect the palatovelar turn into sibilant, i.e. **ḱem* > **sem-*. Moreover, I am not aware of any extant poetic figures in Slavic that would suggest the opposition of "horned" and "hornless" cattle (such as the merism "horned and hornless" in ṚV 1.32.15, see n. 17 above). The interpretation of Mallory and Adams seems, therefore, rather improbable, albeit very attractive for my argument here. Thanks to Jan Bičovský and Jiří Dynda for the useful suggestions on Slavic linguistics (email communication, 26 April 2021).

24 Cf. TS 2.1.1.5, 5.5.1.1–2; JB 2.262.

25 This is an unexplained reading of the relatively rare word *lapsudin*, "bearded", which also occurs in JB 2.262. See also the commentary of Caland [1919, p. 208, n. 3], Schrapel [1970, p. 6], and Ehlers [1988, p. 67].

bearded of men, that which has incisors in the lower jaw is of cattle, that which is goaty[26] of goats.

The late-Brāhmaṇa association of the word *tūpara* with humans, horses, and goats and the ŚB gloss *aviṣāṇa* both indicate that towards the end of the Brāhmaṇa period, the word must have already meant "hornless". I suggest, however, that in the earlier layer of the late YV Saṃhitās and early Brāhmaṇas, especially in the passages pertaining to our myth, we could perhaps understand it as a term for an animal with a different horn morphology to common cows, perhaps wild or semi-domesticated cattle of some sort. Our mythical narrative is an obvious port of call for this suggestion, with all the passages unanimously using the verb *prāvartanta*, "they became curved", to describe the horns and the descriptions of behavior of the cows and the *yajamāna*'s *śikhā* style. The association with the wilderness is present in its earliest occurrence in the AV where the poet mentions many species of wild animals but not a single domestic one.

An additional evidence for treating *tūpara* as the name of a different species or a breed of cattle rather than an adjective describing another animal could be the relatively common association of *tūpara* with horse and gayal (*gomṛga*). Especially the presence of the gayal, is peculiar here, as it is a rather marginal animal species in the Vedic literature.[27] Gayal is a large, wild or semi-domesticated species of cattle from South and Southeast Asia characterized by large horns with a considerable curvature inwards [Ahrestani 2016, p. 4]. I argue that the association of *tūpara* with horse and gayal is not incidental—on one hand, we have the hornless horse, and on the other, wild cattle with the largest and the curviest horns in the region. Perhaps, the morphology of the *tūpara* horns which are described as "bent" (*prāvartanta*) is somewhere in between these two species.

5 The Cattle and Their Horns

This brings us to the last section of this paper. Here, we will consider the evidence of zoology which, I believe, can help us identify what kind of animal is our mythical narrative talking about when it mentions *tūpara*. My hypothesis

26 The word *aja* could be also possibly understood as "unborn". My translation here follows both Caland and Keith, translators of the other parallel passages.

27 Gayal (*Bos frontalis*), also known as *mithan, methne*, or *mithun*, is largely similar to a gaur (*Bos gaurus*, the largest living bovid), although smaller in every aspect, and found only in hillier parts of the Indian subcontinent (Bhutan, Assam, Nepal, and the Chittagong Hills).

is that *tūpara* signifies a wild or semi-domesticated animal, perhaps cattle species, which is horned but has a different shape of horns than the "common" domestic cow. The horns, I believe, are the point of comparison which distinguishes between the two groups, the *śṛṅgiṇī*s and the *tūparas*. Unlike Hoffmann [1992, pp. 782–783], however, I do not think that the horns are defective; rather, they are just differently shaped. Moreover, if we accepted Kuiper's suggestion [1948, pp. 74–75] about the Munda origin of the word, *tūpara* would be an animal which must have been present in India before the arrival of the Indo-Aryans and known and named by the local communities. Last clue is the behavior of the *tūpara* animals presented by the myth; the *tūpara*s are said to be active outdoors during monsoon and winter months, whereas the *śṛṅgiṇī* cows cannot withstand the elements.

First of all, I suggest that when meeting a different species of cattle, it makes sense that the Indo-Aryans would describe it in terms of their horns. Looking at the earliest poetry of the ṚV, horns seem to be an important characteristic feature of both wild and domestic cattle. Animals and deities alike are described with epithets such as "sharp-horned" (*tigmáśṛṅga*; e.g. ṚV 6.16.39) or "ample-horned" (*bhū́riśṛṅga*; e.g. ṚV 1.154.6). Bulls and buffaloes are said to sharpen their horns (e.g. ṚV 8.60.13 *śíśāno vṛṣabhó* [...] *śṛ́ṅge*), and Indra is said to be the king of horned and hornless beings (ṚV 1.32.15 *índro* [...] *rā́jā śámasya*[28] *ca śṛṅgíṇo*). We can trace this development all the way to the Brāhmaṇas where wild beasts are described with the same words, as "sharp-horned" (e.g. AB 6.18 *yas tigmaśṛṅgo vṛṣabho na bhīma iti*). Clearly, horns are an important characteristic feature of cattle for the ancient Indian tradition.

Horns are a type of cranial appendages which consist of a bony horn core covered by a keratin sheath, and in most species of the family Bovidae, they exist in both sexes. In the Linnaean system, all cattle are essentially defined by them. The Indian subcontinent is a habitat to a number of cattle species, domesticated, semi-domesticated, feral, and wild. Amongst the wild species, these are the water buffalo (*Bubalus bubalis*, in Sanskrit called *mahiṣa*), living in the wet forests, marshes, and grasslands from Nepal to Central India;[29] the gaur (*Bos gaurus*, in Sanskrit *gaura*) whose habitat are the forested areas all

Some believe that it is a semi-domesticated form of the *gaur*, whereas others consider it to be a distinct species. See Van der Geer [2008, p. 84], cf. Ahrestani [2016].

28 Note the inherited term *śama* for "hornless" cattle, commentary on it above, n. 17 and n. 24.

29 The buffalos' horns are very different from the rest of the Asiatic wild cattle: they are very thick and heavy at the base, with a triangular cross section, and marked with vertical ridges. All the cattle of the *Bos* family have oval-shaped horns. [van der Geer 2008, p. 120].

over the Indian subcontinent; the gayal (*Bos frontalis*, alias *gomṛga*), possibly a semi-domesticated form of the gaur;[30] and the yak (*Bos grunniens*, in Sanskrit *camara*) which lives in the high mountains delineating the subcontinent. All of these, however, have regular Sanskrit names that occur relatively frequently in the Vedic literature (excluding the yak which, as far as I am aware, occurs only from the MBh onwards) which makes them unlikely candidates for being our *tūpara*.

Among the domesticated and semi-domesticated cattle species, there are two species of the "common" cow in India, the taurine cattle (*Bos taurus*) and the zebu (*Bos indicus*), domesticated locally in South Asia. These two species largely differ in terms of horn-shape and ability to withstand the somewhat harsh elements of the Indian subcontinent. Since our mythical narrative does not mention any other difference between the *śṛṅgiṇī*s and *tūpara*s than their behavior and difference in horns, we could suspect that it might be talking about the two different species, breeds, or variants of cattle. In that case, these two would be an obvious choice.

The taurine cattle began being domesticated from aurochs (*Bos primigenius*), wild cattle with large, long horns, in the area of Fertile Crescent in western Asia around 10,500 years ago. In the now extinct aurochs,[31] horns were carried by both sexes which would suggest "that the original form of domestic cattle was horned as well" [Schafberg & Swalve 2015, p. 62]. As Schafberg & Swalve [ibid., p. 65; cf. Zeuner 1963, p. 211] suggest, there were differences in horn size and shape from the ancient times. The ancient Egyptians, for example, must have kept long-horn, short-horn, and polled animals, as it is suggested by the archaeological and visual evidence.

The South Asian species of cattle, zebu (*Bos indicus*), was most likely domesticated independently on the taurine cattle from a separate, now extinct ancestor, *Bos namadicus*, in the Neolithic period. The earliest evidence of the domestication of the zebu is from Merhgarh in the Bolan Pass, eastern Pakistani Baluchistan, around 6000 BCE [Meadow 2003; Patel & Meadow 2017; cf. Bökönyi 1997]. Zebu cows and bulls are characteristic for their large hump, dewlap, elongated eyes, and drooping ears—traits which help them tolerate the stress

30 As Marleen Felius [2016, p. 29] states: "Gayal and gaur have the same mtDNA, Y-chromosomal DNA [...] and karyotype. Introgression from wild gaur bulls is supposed to take place frequently, although gayal has different horns and a more friendly character."

31 The last auroch was hunted down in a marshy, deciduous Polish forest of Jaktorów in 1627. On history and archaeology of aurochs see Wright [2013] and Zeuner [1963, pp. 201–210], on aurochs in Indian sculpture see Van der Geer [2008, pp. 111–114].

of extreme weather conditions better.[32] In comparison with the taurine cattle, the zebu has a much more slender built and even a different bone structure, including an elongated skull [Grigson 1973; cf. van der Geer 2008, pp. 88–89]. As is the case in the taurine cattle, there are breeds of zebu that have longer and shorter horns, or a complete lack thereof [Bökönyi 1997, p. 648].[33]

Despite the fact that they were domesticated in specific regions (the Fertile Crescent and the area of the Indus Valley respectively), both of the species dispersed widely throughout the Eurasia. The zebu, highly adapted to a warmer climate, migrated both westwards and arrived to the warmer regions of Mesopotamia in ca. 4000 BP.[34] Eastwards, it migrated to China from about 3000 BP onward [Felius 2016, p. 45]. Taurine cattle reached the Central Europe by 7500 BP and quickly spread out throughout Europe, whilst eastwards, it reached China between 5000 BP and 4000 BP [Felius ibid., pp. 40–44]. The archaeological evidence for taurine cattle in the Indian subcontinent is not fully sufficient but since the taurine species migrated with people all the way to the east of Asia, we can presume their presence in India as well. Moreover, the soapstone seals from the Indus Valley show the possible coexistence of humped and humpless cattle [Felius ibid., p. 45]. Taking the movement of the Indo-Europeans into consideration, it is more than probable that the Indo-Aryans would have encountered both species of cattle; however, they would be perhaps more familiar with the humpless, taurine breeds, populating the Central Asian planes. Today, taurine cows coexist with zebu (and the numerous zebu-taurine hybrids) in the Indian subcontinent, especially due to colonialism, as multiple modern taurine dairy breeds were brought by the British colonists to encourage dairy production on the large scale [Govindrajan 2018, p. 72].

Although we have ruled out the possibility that *tūpara* means "hornless" in the earlier layer of the language, or our mythical narrative, let us briefly comment on hornlessness in cattle. An evidence of polled (i.e. dehorned) taurine cattle is attested in archaeological data from the central and eastern Europe, western Asia, and north Africa since the early Neolithic. Cattle was both selectively bred to be polled and dehorned artificially. This is perhaps not surprising, as working with hornless cattle has always had clear advantages for farmers. The existence of hornless taurine cattle is further confirmed by the textual and

32 On the versatile adaptations of various cattle breeds and species to climatic conditions, see Felius [2016, p. 11].

33 Regarding polled zebus, Auld [1887b, pp. 886–890] mentions an anecdotal evidence of white hornless zebu cows used solely for traction in 19th century India.

34 Nowadays, the westernmost habitat of a pure-bred zebu is in Azerbaijan [van der Geer 2008, p. 89].

visual material from the ancient Mediterranean civilizations which domesticated cattle, including the Egyptians and the Roman Empire [Schafberg & Swalve ibid., p. 65]. These were presumably used for different purposes, analogously to the situation in premodern Europe: long-horned animals for traction, short-horned and polled animals for dairy farming. In zebus, there does not seem to be sufficient archaeological data to suggest that zebu cattle were bred to be hornless or were artificially polled, at least not at the time the late Vedic Saṃhitās. Even though there is not enough archaeological evidence for the existence of hornless zebu cattle in Vedic India, they could have, nevertheless, existed naturally on the basis of spontaneous mutation. The "polled allele", responsible for the lack of horns is, after all, a dominant trait when it occurs in ruminants.[35] Unfortunately, the absolute proportion of polled cattle in the taurine cattle is difficult to ascertain, and the situation is even more unclear in the zebu.

What is more interesting for our argument than the sole existence of polled cattle in India is, however, the morphology of cow horns. As said above, horns are the defining feature of bovids, both taxonomically and visually, and must have been an important characteristic mark even for the ancient Indian tradition, as suggested by the breadth of metaphors in the ṚV and further Vedic literature. The Linnaean taxonomical tradition of the 19th and early 20th century classified different species of cattle predominantly on the basis of the shape of their horns [Felius et al. 2011, p. 665]. Based on the morphology of horns—the long-horned, middle-horned, short-horned, and polled cattle [e.g. Youatt 1834]—it was proposed that the various breeds of taurine cattle actually had different ancestors [Felius et al. 2011, p. 667, Table 1]; an idea which no longer holds any traction in the scientific debate. Regardless their length, however, the horns of taurine cattle are usually down-turned, with the tips of the horns lying at the level of the crest, or higher. Upturned horns are not completely absent but they are certainly rarer in the *Bos taurus* [Zeuner: 1963, p. 206]. In contrast, the horns of the zebu cattle are visibly upright. Sándor Bökönyi summarizes the shape of zebu horns thus:

> The horns vary in size and form. Both long-horned and short-horned zebus exist and hornless animals also occur. Their horn form basically dif-

35 See Schafberg & Swalve [2015, pp. 63–65] on the genetics of hornlessness: "[…] polledness is autosomal dominantly inherited with polled being dominant over horned. But even after the bovine genome was completely decoded in 2009 […] and additional insights were provided by molecular biology, it proved to be difficult to identify the causal mutation […]. Polled animals appear among smaller ruminants, too […]."

> fers from that of taurine cattle: their typical direction is lateral or upright with a backward tendency (this is the so-called "auchenokeratos" type) and in profile they are well behind the plane of the face.
>
> BÖKÖNYI 1997, p. 648

The difference of the horn morphology between the taurine and zebu cattle is remarkable. As noted above, however, this is a rather general observation; there are many individual breeds of both of the species with different horn lengths.

As for the distinct seasonal behavior of the *śṛṅgiṇī* and *tūpara* cattle reported by the myth, I suggest that this could also be accounted for by the difference between the two species. Zebu cows are usually reported to be much sturdier than taurine cattle [Felius 2016, p. 28]. For inspiration, we could take into consideration the multispecies anthropological research of Radhika Govindrajan [2018] who reports the differences in behavior and social interpretation of taurine and zebu cattle in contemporary Uttarakhand. Govindrajan reports that the local villagers strongly distinguish between the taurine species ("Jersey cows") and the *pahari*, mountainous breed of the zebu. The "Jerseys", albeit gentle animals, more productive in terms of milk, do not seem to withstand the elements and the mountainous feed. They are reported to behave differently when milked and even being "*bekar* (useless) when it came to *puja*" [Govindrajan ibid., p. 84]. The humped *pahari* cows, on the other hand, slender-bodied and temperament, can resist the high temperature differences, including the harsh mountainous winters. Throughout Govindrajan's accounts, the *pahari* cows are repeatedly said to have ritual *shakti*, "divine power" [Govindrajan ibid., p. 66]. This is, of course, an ethnographic evidence from contemporary India and we cannot base our understanding of an ancient myth on it. It shows, however, that people are partial to different species and breeds of cattle.

Although anecdotal, Govindrajan's ethnographic research indicates another important feature of the zebu cattle that we have not considered yet; its slender built. In comparison with almost all breeds of taurine cattle, zebus are very small and light; some of them even have miniature variants [Felius 2016, p. 61]. Returning to the etymology of *tūpara* yet again, we are reminded of Kuiper's comparanda such as Santali *tope*, "short, to cut off", *ṭhe̱ble̱*, "low, small, dwarwish", or Mundari *ṭhepā*, "short in stature, dwarf". The word *tūpara*, however, most likely does not apply to the built of the cows in our myth but to their horns; otherwise, the interpretation of *tūpara* cattle as "small" would be very tempting.

6 Conclusion

Despite the sparse descriptions of the *tūpara* cattle, I propose that the myth is not talking about "hornless" cattle but about a different cattle species, breed, or variant. The locally domesticated zebu, with their distinctive upright horns, slender body, and resistance from the extreme weather conditions, are an obvious fit for our mythical narrative.

As noted above, the myth is shared in parallel passages throughout the Vedic canon. Its incipit, shared across the *śākhā* boundaries, suggest its relatively old age. Since the word *tūpara*, a crucial term used by every single of the parallel passages, first occurs only in the AV, we can suspect that the myth must have been composed only when the Indo-Aryans arrived in the Indian subcontinent (if we accepted Kuiper's suggestion that the word to was of Munda origin, the Indo-Aryans would have to moved considerably eastwards by the time they composed this myth). I suggest, with a pinch of salt, that we could understand the mythical story to portray a time when the semi-pastoral Indo-Aryans met a new species of cattle, perhaps semi-domesticated or feral (judging from the AV association with wilderness), with a different bodily characteristics and behavior. Meeting such a strange, yet familiar animal, they had to account for it in their symbolic system: such a hardy animal with a different shape of horn must have surely deserved the body by a prolonged ritual action. Without a doubt, however, the word *tūpara* was assimilated to the Indo-Aryan system by the end of the Brāhmaṇa period, and started being used as a term for any animal without horns.

Acknowledgements

This paper is a modified version of my talk *When Cows Sacrifice*, first presented at the 231th Annual Meeting of the American Oriental Society (online, 13th March 2021). I have greatly benefited from the helpful comments and suggestions of many people in the audience. Apart from them, I would like to thank numerous people for their generous help with this article: Professor Christopher Minkowski for first reading the mythical narrative with me, Lubomír Ondračka for suggestions regarding the Bengali comparanda of *tūpara* and his kind proofreading of this paper, Elizabeth Tucker for helpful commentaries on the AV passages, and Jan Bičovský and Jiří Dynda for recommendations on the horse and cow in Slavic languages and PIE. Thanks are also due to Christopher Fleming and the other editors of this volume for inviting me to contribute. All errors and inaccuracies are my own.

Abbreviations

The Vedic Corpus

AB	Aitareya Brāhmaṇa
ĀpŚS	Āpastamba Śrautasūtra
AVŚ	Atharvaveda Saṃhitā, Śaunakīya
BYV	Black Yajurveda (= KS, TS, and MS)
JB	Jaiminīya Brāhmaṇa
KS	Kāthaka Saṃhitā
KauS	Kauthuma Saṃhitā
MS	Maitrāyaṇī Saṃhitā
PB	Pañcaviṃśa Brāhmaṇa
ṚV	Ṛgveda Saṃhitā
ŚBM	Śatapatha Brāhmaṇa, Mādhyaṃdina
ŚBK	Śatapatha Brāhmaṇa, Kaṇva
TB	Taittirīya Brāhmaṇa
TS	Taittirīya Saṃhitā
VS	Vājasaneyi Saṃhitā
VSM	Vājasaneyi Saṃhitā, Mādhyaṃdina

Other Abbreviations

BP	before present
CDIAL	*Comparative Dictionary of Indo-Aryan Languages*
Cz.	Czech
EWA	*Etymologisches Wörterbuch des Altindoarischen*
Grk.	Greek
KEWA	*Kurzgefaßtes etymologisches Wörterbuch des Altinidischen*
OCS	Old Church Slavonic
OPrus.	Old Prussian
PIE	Proto-Indo-European
Rus.	Russian

Primary Sources

Aitareya Brāhmaṇa:

Aufrecht, T., ed. *Das Aitareya Brāhmaṇa mit Auszügen aus dem Commentare von Sāyaṇācārya und anderen Beilagen*. Bonn: Marcus, 1879.

Atharvaveda Saṃhitā:

Roth, R., and W.D. Whitney, eds. *Atharva Veda Sanhita.* 2 vols. Berlin: Ferd. Dümmler, 1855.

Bloomfield, M., trans. *Hymns of the Atharva-veda: Together with extracts from the ritual books and commentaries.* Oxford: Clarendon Press, 1897.

Whitney, W.D., trans. *Atharva-veda-saṃhitā.* two vols. Delhi: Motilal Banarsidass, 1855. Reprint.

Jaiminīya Brāhmaṇa:

Vira, Raghu, and Lokesh Chandra, eds. *Jaminiya Brahmana of the Samaveda.* Nagpur: Lokesh Chandra, 1954.

Bodewitz, H.W., trans. *Jaiminīya Brāhmaṇa I, 1–65.* Leiden: Brill, 1973.

Bodewitz, H.W., trans. *The Jyotiṣṭoma Ritual: Jaiminīya Brāhmaṇa I, 66–364.* Leiden: Brill, 1990.

Caland, Willem, trans. *Das Jaiminīya-Brāhmana in Auswahl.* Verhandelingen Der Koninklijke Akademie van Wetenschappen Te Amsterdam, Afdeeling Letterkunde; Nieuwe Reeks, d. 19, No. 4. Amsterdam: Johannes Müller, 1919.

Kāṭhaka Saṃhitā:

von Schroeder, L., ed. *Kāṭhakam: Der Saṃhitā der Kaṭha-Śākhā.* 3 vols. Leipzig: Brockhaus, 1900–1910.

Maitrāyaṇī Saṃhitā:

von Schroeder, L., ed. *Maitrāyaṇī Saṃhitā.* 4 vols. Leipzig: Brockhaus, 1881–1886.

Pañcaviṃśa Brāhmaṇa:

Chandra, Lokesh, ed. *Pañcaviṃśa-Brāhmaṇa with the Commentary of Sāyaṇa.* New Delhi: Sharada Rani, 1981.

Caland, Willem, trans. *Pañcaviṃśa Brāhmaṇa: The Brahmana of twenty five chapters.* Calcutta: Baptist Mission Press, 1931.

Ṛgveda Saṃhitā:

van Nooten, B.A., and G.B. Holland, eds. *Rig Veda: A Metrically Restored Text With an Introduction and Notes.* Cambridge, MA: Harvard University Press, 1994.

Jamison, S.W., and J.P. Brereton, trans. *The Rigveda: The Earliest Religious Poetry of India.* 3 vols. Oxford; New York: The University of Texas, South Asian Institute—Oxford University Press, 2014.

Śatapatha Brāhmaṇa, Kāṇva:

Caland, Willem, and Raghu Vira, eds. *The Śatapatha Brāhmaṇa in the Kāṇvīya recension.* Delhi: Motilal Banarsidass, 1998. Reprint.

Swaminathan, C.R., ed. *Kāṇvaśatapathabrāhmaṇam.* 4 vols. New Delhi: Indira Gandhi National Centre for the Arts, 1994–2001.

Śatapatha Brāhmaṇa, Mādhyaṃdina:

Weber, Albrecht, ed. *The Çatapatha-Brâhmana in the Mâdhyandina Çâkhâ with Extracts from the commentaries of Sâyaṇa, Harisvâmin and Dvivedaganga.* Varanasi: Chowkhamba Sanskrit Series Office, 1964. Reprint.

Eggeling, J., trans. *The Śatapatha Brāhmaṇa: According to the text of the Mādhyandina School.* 5 vols. New Delhi: Low Price Publications, 1996. Reprint.

Taittirīya Brāhmaṇa:

Śāstri, Mahadeva, ed. *The Taittirīya Brāhmaṇa with the Commentary of Bhaṭṭabhāskaramiśra.* 4 vols. Mysore: Government Branch Press, 1908–1921.

Taittirīya Saṃhitā:

Keith, Arthur B., trans. *The Veda of the Black Yajus School entitled Taittiriya Sanhita.* 2 vols. Cambridge, MA: Harvard University Press, 1914.

Weber, Albrecht, ed. *Die Taittirīya-Saṃhitā.* 2 vols. Leipzig: Brockhaus, 1871–1872.

Vājasaneyi Saṃhitā:

Weber, Albrecht, ed. *The Vâjasaneyi-Sanhitâ in the Mâdhyandina-and the Kâṇva-Çâkhâ with the Commentary of Mahîdhara.* Berlin; London: F. Dümmler—Williams; Norgate, 1852.

Secondary Sources

Ahrestani, Fashid. "Bos Frontalis and Bos Gaurus (Artiodactyla: Bovidae)." *Mammalian Species* 50, no. 959 (2018): 34–50.

Auld, R.C. "Hornless Ruminants." *The American Naturalist* 21, no. 8 (1887): 730–746.

Bökönyi, Sándor. "Zebus and Indian Wild Cattle." *Anthropozoologica* 25–26 (1997): 647–654.

Caland, Willem. "Vedica." *Acta Orientalia* 5 (1927): 252–253.

Feelius, Marleen, Peter A. Koolmees, Bert Theunissen, European Cattle Genetic Diversity Consortium, and Johannes Lenstra. "On the Breeds of Cattle—Historic and Current Classifications." *Diversity* 3 (2011): 660–692.

Felius, Marleen. "On the Breeds of Cattle: Their History, Classification and Conservation." PhD thesis, Utrecht University, 2016. https://dspace.library.uu.nl/handle/1874/328463.

Govindrajan, Radhika. *Animal Intimacies: Interspecies Relatedness in India's Central Himalayas.* Chicago; London: The University of Chicago Press, 2018.

Grigson, Caroline. "Bos Indicus and Bos Namadicus and the Problem of the Autochthonous Domestication in India." In *Recent Advances in Indo-Pacific Prehistory: Proceedings of the International Symposium Held at Poona, December 19–21, 1978*, edited by V.N. Misra and P. Bellwood, 425–428. New Delhi: Oxford & IBH, 1985.

Grigson, Caroline. "The comparative craniology of Bos taurus L., B. indicus L., B. primigenius Boj. and B. namadicus Falc." PhD thesis, University of London, 1973.

Hoffmann, Karl. "Vedica." In *Aufsätze zur Indoiranistik*, edited by Johanna Narten and Sonja Glauch, 3:767–800. Wiesbaden: Reichert, 1992.

Jamison, Stephanie W. "Formulaic Elements in Vedic Myth." In *Inside the Texts, beyond the Texts: New Approaches to the Study of the Vedas: Proceedings of the International Vedic Workshop, Harvard University, June 1989*, edited by M. Witzel, 127–138. Harvard Oriental Series: Opera Minora, Vol. 2. Cambridge, MA: Harvard University Press, 1997.

Jamison, Stephanie W. "Linguistic and philological remarks on some Vedic body parts." In *Studies in Memory of Warren Cowgill (1929–1985): Papers from the Fourth East Coast Indo-European Conference Cornell University, June 6–9, 1985*, edited by Calvert Watkins, 67–91. Berlin; Boston: De Gruyter, 2003. First edition: 1987.

Jamison, Stephanie W. "The Function of Animals in the Rig Veda, RV X.28, and the Origins of Story Literature in India." In *Penser, Dire et Représenter l'animal Dans Le Monde Indien*, edited by Nalini Balbir and Georges-Jean Pinault, 97–218. Bibliothèque de L'École Des Hautes études. Sciences Historiques et Philologiques (1995); T.345. Paris: Champion, 2009.

Jamison, Stephanie W., and Michael Witzel. "Vedic Hinduism." In *The Study of Hinduism*, edited by Arvind Sharma, 65–113. Columbia, SC: University of South Carolina Press, 2003. Written in 1992/95, only partially updated.

Jones, Michael Brattus. "Agriculture and Religion in Ancient India." PhD thesis, The University of Texas, 2017.

Köhler, Hans-Werbin. *Śrad-dhā in der vedischen und altbuddhistischen Literatur*. Wiesbaden: FSteiner, 1973.

Kuiper, Franciscus Bernardus Jacobus. *Proto-Munda Words in Sanskrit*. Amsterdam: Noord-Hollandsche UitgMij, 1948.

Mahadevan, Thennilapuram P., and Frits Staal. "The Turning-Point in a Living Tradition Somayāgam 2003." *Electronic Journal of Vedic Studies* 10 (2016): 1–29.

Mallory, J.P., and Douglas Q. Adams. *The Oxford Introduction to Proto-Indo-European and the Proto-Indo-European World*. Oxford: Oxford University Press, 2006.

Murakawa, Akiko. "Das Gavamayana-Kapitel im Jaiminiya-Brahmana: The Gavamayana Chapter of Jaiminiya-Brahmana." PhD thesis, Freie Universität Berlin, 2010.

Olivelle, Patrick. *The Early Upaniṣads: Annotated Text and Translation*. New York; Oxford: Oxford University Press, 1998.

Schafberg, R., and H.H. Swalve. "The History of Breeding for Polled Cattle." *Livestock Science* 179 (2015): 54–70.

Schrapel, Dieter. "Untersuchung der Partikel iva und anderer lexikalisch-syntaktischer Probleme der vedischen Prosa, nebst zahlreichen Textemendationen und der kritischen Übersetzung von Jaiminīya-Brāhmaṇa 2, 371–373 (Gavāmayana 1)." PhD thesis, University of Margburg, 1970.

Turner, Ralph Lilley. *A comparative dictionary of Indo-Aryan languages*. London: Oxford University Press, n.d. https://dsal.uchicago.edu/dictionaries/soas/. [Last accessed: 29 August 2022.]

van der Geer, Alexandra. *Animals in Stone: Indian Mammals Sculptured Through Time*. Leiden; Boston: Brill, 2008.

Weber-Brosamer, Bernhard. *Annam: Untersuchungen zur Bedeutung des Essens und der Speise im vedischen Ritual*. Rheinfelden: Schäuble, 1988.

Wright, Elizabeth. "The history of the European aurochs (Bos primigenius) from the Middle Pleistocene to its extinction: an archaeological investigation of its evolution, morphological variability and response to human exploitation." PhD thesis, University of Sheffield, 2013.

Youatt, W. *Cattle, Their Breeds, Management and Diseases*. London: Baldwin; Cradock, 1834.

Zeuner, Frederick Everard. *A History of Domesticated Animals*. London: Hutchinson, 1963.

Zeuner, Frederick Everard. *Prehistory in India: Four Broadcast Talks on Early Man*. Poona: Deccan College Postgraduate; Research Institute, 1951.

CHAPTER 2

A sūkta of the Paippalāda Atharvaveda about the Symptoms and Causes of Tetanus

Elizabeth Tucker

The Atharvaveda (AV) has often been dismissed as unscientific, a collection of spells and magic formulae, the word-book of sorcerers. However, it contains poems which show accurate physiological observations, e.g. to do with pregnancy and birth (Śaunakasaṃhitā (ŚS) 1.11, Paippalādasaṃhitā (PS) 11.1); or closely observed details of illness, e.g. patterns of recurrent fever (ŚS 7.116, ŚS 1.25.5), fever that occurs at different times of the year (ŚS 5.22.13). This paper, which I have great pleasure in contributing to the volume in honor of Christopher Minkowski, will argue that some very ancient pre-Āyurvedic Sanskrit science has been overlooked because obscure terms (usually nouns) can refer to symptoms rather than identifiable illnesses and it is the bunching of terms that indicates medical knowledge. In illustration I will attempt to analyse a hitherto unstudied 14-stanza composition that is peculiar to the Paippalādins, PS 11.2.

The text of the complete sūkta is given in the Appendix to this article. The edition that I have prepared is based on three palm leaf manuscripts from Oḍisha (Orissa) written in Oriya script (Ku_2 and JM from Central Oḍisha and Pa from Northern Oḍisha).[1] In addition I have taken into account the testimony of Ma and Mā (from Northern Oḍisha) employed by Dipak Bhattcharya (1997), some of whose readings have to be inferred from his edition. The Kashmiri Birchbark manuscript written in Śāradā, which represents a very different Paippalāda tradition, has been re-read from the facsimile published by Bloomfield and Garbe (1901),[2] and its readings compared with those of the Oriya palm leaf manuscripts.[3]

1 High resolution photos of these palm leaves have been kindly made available to me by Arlo Grifftihs.

2 A valuable transcription of PS 11 from this facsimile was published in *JAOS* by Barrett (1926), but it requires some corrections.

3 Following the editing principles advocated by Karl Hoffmann (1968).

 | DOI:10.1163/9789004536869_004

PS kāṇḍa 11 is the caturdaśarcakāṇḍa. In 7 out of the 14 stanzas of PS 11.2 (mostly in the d pādas) there is the formula: *viśaram nāśayāmi te* "I destroy *viśara-* for you, I make *viśara-* disappear for you" (also with metrically equivalent forms of the same verb *nāśayāti, nāśayantu*). What is *viśara-*? It is a word which is only attested in one other place (ŚS 2.4.2 = PS 2.11.2), where it appears in a list. Conjectures about the meaning of *viśara-* in its single ŚS occurrence are mostly based on 'Sāyaṇa's' gloss *śarīraviśaraṇa-*: 'tearing pain' (Bloomfield 1897:284) 'something crushing or tearing to pieces' (Whitney 1905: 42), 'tearing or contorting pain' (Zysk 1985, 1996:55). This is not the only puzzle about PS 11.2, as there are other obscure terms, and the final verse (11.2.14) tells not *viśara-* but *viṣkandha-* to go away. Also the same composition contains stanzas which list the different parts of a bow and arrow (11.2.4, 11.2.13) or pay homage to the divine arrow (11.2.5).

Identical lists of the parts of the bow and arrow are found in an AV sūkta "Against the poison of a poisoned arrow" (PS 5.8.3–4;[4] cf. ŚS 4.6.5–6). Since the Kashmiri birchbark manuscript consistently reads *viṣam* 'poison' in PS 11.2 as the object of *nāśayāmi* in the repeated formula, Barret (1926: 261–262) suggested that this noun should be read throughout this sūkta even in places where the Kashmiri manuscript has *viśaram*. But all the Oriya manuscript evidence goes against this, and *viṣam* 'poison' does not fit the metre.

The verb employed in the formula does not further understanding as the transitive present stem *nāśaya-* has objects of many types in the AV: nouns of known meaning which designate identifiable illnesses, e.g. *takmán-* 'fever' (ŚS 5.22.13cd); other nouns of unknown meaning which the context suggests are illnesses / disabilities, e.g. *kṣetriyá* (ŚS 3.7.3d); and this verb can be used of getting rid of all manner of things /beings, e.g. *yé śā́lāḥ parinṛ́tyanti sāyáṃ gardabhanādínaḥ / ... tā́n ... viṣūcī́nān ví nāśaya* "Those who dance around the dwellings in the evening making donkey-noises ... make them disappear in different directions" (ŚS 8.6.10).

The first stanza of PS 11.2 presents some difficulties of reading,[5] but it suggests by its reference to *yakṣma-* that the whole poem is to be recited to remove illness:

4 Edited and translated by Lubotsky (2002:50–51).

5 The first stanzas of AV hymns often contain an esoteric statement (cf. e.g. PS 6.11.1, ŚS 4.1.1) which adds to the difficulty of understanding the text transmitted by the manuscripts here (see Appendix for details of the readings). Negative *(a)bhūtir* was suggested by Bhattacharya (1997). I am grateful to Diwakar Acharya for suggesting *aśás-* (RV +), but the positive root noun

**aśaso yamanaṃ *śaso 'bhūtir yakṣmam ajījanat |*
imam sahasrabhāg indro viśaraṃ nāśayāti te ||

"Non-Thriving (*ábhūti-*) has given birth to Disease the constrainer *of the reviler, *of the praiser (or "*of the one without chants, *of the one who chants"). May Indra who allots a thousand (forms of aid) destroy this viśara for you."

The second stanza distinguishes in its pāda b types of *viśara-* which arise spontaneously or have an external cause:

PS 11.2.2
yaḥ kār$_{i}$yo yaś ca kṛtaḥ svayaṃjā uta hār$_{i}$yaḥ |
devā ind$_{(a)}$rajya$_{i}$ṣṭhā viśaraṃ nāśayantu te ||

"The *viśara-*, which is to be made (in the future) and which has been made, (which is) born of its own accord and has to be acquired (from external sources), let the gods with Indra as their chief destroy for you."

In PS 11.2.10 *viśara-* appears amid other obscure terms, but some of these can provide clues:

viṣkandhaṃ tvad⁺ apiśaraṃ viśaraṃ ⁺pṛṣṭ$_{i}$yāmayam |
*jambhaṃ hanugrābhaṃ śaṅkhaṃ tān *ajenyāṁ ajāmasi ||*

"From you (we drive) *viṣkandha-*, ⁺*apiśara-* (*apaśara-*?), *viśara-*, ⁺rib-pain, *jambha-*(,) the jaw-seizer, the 'conch', we drive those *which are to be driven."

A parallel first hemistich occurs as PS 2.11.2ab, in a sūkta about the medicinal properties of the Jaṅgiḍa-plant:

---

without a prefix is not attested elsewhere (Scarlata 1999:528–532), although here a nonce form based on the negative term may have been created. Previously I had considered reading:
**asato yamanaṃ *sato bhūtir yakṣmaṃ ajījanat |*
imaṃ sahasrabhāg indro viśaraṃ nāśayāti te ||
"Growth gave birth to Disease the constrainer *of the non-existent, *of the existent." The opposition *ásat* ... *sát* is found in a number of AV compositions (e.g. ŚS 10.7.10, and in converse order PS 5.2.2, ŚS 5.6.1), as well as RV 10.

jaṅgilo jambhād viśarād *viṣkandhād abhiśocanāt* |
maṇis sahasravīr̥yaḥ *pari ṇaḥ pātu viśvataḥ* ||

"Jaṅgiḷa soll uns vor dem zerreissenden Krampf, vor dem quälenden Viṣkandha, das Amulett mit tausend Kräften, rudherum beschützen."

ed. and trans. ZEHNDER 1999:46

This PS 2.11.2 stanza exactly corresponds to ŚS 2.4.2, which contains the only Śaunaka occurrence of *viśara-*. But Whitney (1905:42) took all the nominal forms in pādas ab as nouns in an asyndetical list:

"From *jambhá*, from *viśará*, from *víṣkandha*, from scorching (*abhiśócana*) let the *jaṅgiḍá*, the amulet of thousand-fold valiance (*-vīryà*), protect us about on every side."

Narten (1965:262–264) derived the noun *jambhá-* attested here from a root *jambh* 'to show teeth, snap teeth' (as opposed to root *jambh* 'to smash, to demolish'). Cf. *jámbha-* m. 'teeth', Greek γόμφος 'peg'. She suggested that the agent noun *jambhá-*, literally 'teeth-shower, teeth-snapper', can refer to the baring of the teeth while the jaws are locked as the result of tetanus ('lockjaw'), cf. ŚS 8.1.16 *mā́ tvā jambháḥ sáṃhanur … vidat* "Let not the teeth-barer with jaws together find you"; and, according to this interpretation, PS 11.2.10c *jambha- hanugrābha-* would mean 'the teeth-barer that seizes the jaws' or (as two nouns) 'the teeth-barer, the jaw-seizer' (*jambhá- hanugrābhá-*).

víṣkandha- occurs 8 times in ŚS and the whole of ŚS 2.4 = PS 2.11 is aimed at its destruction by means of the Jaṅgiḍa-plant. Suggestions about its meaning have ranged from rheumatism (Weber 1873:13, 141) to a demon (Bloomfield 1897:282–283) while others have refused to guess (Whitney 1905:17, 42–43, etc.; Filliozat 1975:106–107). Zysk (1985, 1996:54–55) accepts Weber's literal meaning 'tearing the shoulders apart' ('die Schultern ausenander ziehend'); and, because *sáṃskandha-* occurs in the same stanza as *víṣkandha-* at ŚS 19.34.5 = PS 11.3.5, he conjectures "the Vedic poet-healer could have been describing the spasmodic contortions of the body of one suffering from the internal disease commonly known as tetanus or lockjaw." In addition to the word's juxtaposition with *jambhá-* in two AV sūktas, this hypothesis is also supported by a prayer to Soma (PS 1.46.3) that a donor who has been *viddha-* (wounded in battle according to Renou 1964–1965: 439) should be *aviṣkandha-*. However, the premise that the word contains *skandhá-* m. 'shoulder' needs to be modified as the thematic stem does not have this meaning in the AV. It is only found in the plural (never dual!) and appears to refer to bones, probably the vertebrae of the neck and spine (cf. *skandhā́ḥ* beside *pṛṣṭī́r* 'ribs' in ŚS 10.9.20, 10.2.4). This fits the spasms of severe tetanus where the back bends to an extreme extent (opisthotonos): hence *víṣkandha-* perhaps literally 'parting the vertebrae', or (if an original Bahuvrīhi, turned into a name) 'whose vertebrae are asunder'.

The formulae where *viśara-* is the object of *nāśaya-* in PS 11.2 indicate that it is a noun, and so Zehnder's translation as an adjective in PS 2.11.2 is unlikely to be correct. But here *jaṅgiḷo jambhād viśarād viṣkandhād abhiśocanāt … pari ṇaḥ pātu* could show the literary figure of an augmented triad (see West 2004 on this Indo-European stylistic feature) and *abhiśocana-* could be an adjective 'scorching, torturing' qualifying the third noun in the asyndetic group *jambha- viśara- viṣkandha-*. This grouping suggests that *viśara-* belongs to the same semantic sphere as *jambha-* and *viṣkandha-* (cf. Narten 1965:262–264; Bloomfield 1897:283); in addition in PS 11.2.3a it receives the epithet ⁺*vijambha-*[6] 'revealing the teeth'. Even if we are unsure of its precise meaning we can conjecture that *viśara-* refers to another symptom of tetanus.

This makes it possible to understand the role of the PS 11.2 stanzas about bows and arrows. PS 11.2.3 says the mother of *viśara-* is the arrow, the father the bow:

> *viśarasya ⁺vijambhasya-*　　*iṣur mātā dhanuṣ pitā |*
> *adityāḥ putrā ādityā*　　*viśaraṃ nāśayantu te ||*

> "Of *viśara-* ⁺revealing the teeth, the mother is the arrow, the father the bow. The sons of Aditi, the Ādityas, let them destroy *viśara-* for you."

PS 11.2.4 talks about making it disappear from various parts of bows and arrows:

> *dhanvano jyāyā iṣ$_{u}$vā*　　*apaskambhasya bāh$_{u}$voḥ |*
> *apāṣṭhāc chṛṅgāt kurmalād*　　*viśaraṃ nāśayāmi te ||*

> "From the bow, from the bow-string, from the arrow, from the two arms of the handle, from the barb, from the horn, from the shaft, I make *viśara-* disappear for you."

viśara- must have been viewed as something that could be brought into a person by an arrow (the type that was described as *hārya-* in PS 11.2.2b), and in

6 The Oriya manuscripts read *vijṛmbhasya*, the Kashmiri *vijyambhasya*, but emended to *vijambhasya* by Zehnder (1999: 46). A derivative from root *jṛmbh* 'gape, yawn' is unlikely in the context of the whole composition.

this respect it was on a par with the poison of a poisoned arrow. A detailed enumeration of the different parts of the bow and arrow (as in PS 11.2.4 and 11.2.13) was necessary to ensure that nowhere is left where *viśara-* can hide and then enter the wound of the person who is shot. Since an arrow wound was considered so dangerous the arrow is reverenced and placated (with flattery?) in PS 11.2.5, which has an RV parallel (RV 6.75.15) from a riddling hymn about different weapons.

alavatī ruruśīrṣṇ̥y *atho yasyā ayo mukham* |
devyai parjanyaretasa *iṣave kr̥ṇutā namaḥ* ||

"The deer-headed one with a scorpion's tail and whose mouth is metal, make reverence to the divine arrow (sprung) from Parjanya's seed."

I am not suggesting that there was any conception of a bacterium clostridium tetani which released toxins that affected the nervous system, but my claim is that this sūkta shows good observation of symptoms which belong together and some understanding of their possible sources. For identification of the symptom indicated by *viśara-* we are reduced to etymology just as the 'Sāyaṇa' commentary was when it explains *viśara-* as *śarīraviśaraṇa-*. An etymological connection with the root *śr̥̄* 'to crush' receives some support from the juxtaposition of *viśara-* with +*apiśara-* (or *apaśara-*) in PS 11.2.10ab where the poet seems to be playing with different prefixed forms of the same verbal noun.[7] In view of the accentuation transmitted for ŚS 2.4.2 *viśará-* may be an agent noun, literally 'the one who crushes to pieces, the one who smashes apart', cf. the uses of the compound verb *vi-śr̥̄*:

ŚS 12.3.18d *mā́ taṇḍulám̐ ví śarair*[8] *devayántam* "Do not crush to pieces the god-seeking rice-grain."
RV 3.53.17bc *méṣā́ ví varhi mā́ yugám̐ ví śāri* "Let the chariot-pole not be torn apart, let the yoke not be smashed apart."

I suggest it refers to the severe pain[9] associated with the other symptoms mentioned in the stanzas PS 11.2.10 and PS 2.11.2 = ŚS 2.4.2, which we have compared

7 +*apiśara-* is the more likely reading because the verb compound *api-śr̥̄* is attested in Vedic but *apa-śr̥̄* is not. See further below.
8 cf. *aśarait* ŚS 6.32.2, 6.66.2 and the discussion by Narten 1964:257–258.
9 Lubotsky (2002:50) translates *viśara-* as 'piercing power' when discussing PS 11.2.4, which is about its destruction from parts of an arrow, but this meaning does not fit other stanzas in PS 11.2.

above, i.e. *jambhá- hanugrābhá-* and *víṣkandha-*. What would be particularly feared would be the devastating pain associated with the muscular spasms which started from the jaws and then spread.

Can all the obscure items that appear in the longer list in PS 11.2.10 be accounted for as symptoms that can appear together through what we would call tetanus? To repeat this stanza:

viṣkandhaṃ tvad +apiśaraṃ *viśaraṃ +pṛṣṭ$_i$yāmayam* |
jambhaṃ hanugrābhaṃ śaṅkhaṃ *tān *ajenyāṁ ajāmasi* ||

The Oriya manuscripts' puṣṭyāmayam (Ku$_2$ Pa Ma, Mā) in pāda b is almost certainly to be emended to +pṛṣṭ$_i$yāmayam (cf. Bhattacharya 1997, Zehnder 1999: 46). *pṛṣṭyāmayín-* 'having rib-pain /a stitch in the side' occurs once in the RV (RV 1.105.18 in a simile about a craftsman who suddenly has to straighten himself because he experiences pain); and the noun occurs again at PS 11.3.10 in a list of ailments that the Jaṅgiḍa-plant is asked to nullify (close to *āśarīka-* and *viśarīka-*, two hapax legomena which appear to be related to *viśara-*). In the PS 11.2.10b context, it could be the symptom of lung problems, as pneumonia or pulmonary embolisms are now recognized as potential complications of tetanus.

śaṅkha- literally 'conch' following *jambhá- hanugrābhá-* in pāda c is puzzling and I can only offer a very tentative suggestion. The only other evidence for an ailment called *śaṅkha-* 'the shell, the conch' comes from Āpastamba GS.7.18.3–4 where a ritual to cure a boy afflicted by this disease (*śaṅkhinaṃ kumāraṃ*) is described. The commentary explains the curious name: *śaṅkho nāma grahaḥ kumārāṇām bhayaṃkaraḥ yena gṛhītaḥ śaṅkhavan nadati* "A terrible illness of children is called *śaṅkha*, whereby the one affected resounds like a conch shell". In the context of PS 2.11.10c could it be connected with the tightening of the vocal cords (laryngospasm), which is another complication of tetanus? This might cause the sufferer to cough in a strange way.

+*apiśara-* (suggested by the Kashmiri birchbark's āpiśaraṃ, but the Oriya manuscripts have *apaśaraṃ*[10]) 'the crusher-in, the smasher-in' would paral-

10 However, the Oriya manuscripts' apaśaraṃ 'the crusher away' could represent the same word that is found in PS 11.2.7a. As sibilants are frequently confused the Oriya manuscripts' apasaro in the latter stanza is likely to represent *apaśaro*, with the same second element as *viśara-*. Otherwise it would be hard to understand why this stanza is included in this composition. The stanza appears to be about an +*apaśara*-illness that is sexually transmitted and causes impotence:

yas tvā straiṇād +apaśaro *yaḥ puṃso adh$_i$yāruhat* |
āklāntaṃ saṃklāntaṃ snāva *tad u te kalpayāmasi* ||

lel *viśara-* 'the crusher to pieces, the smasher apart' both morphologically and semantically.

Cf. ŚS 4.3.6 *mūrṇā́ mr̥gásya dántā ápiśīrṇā u pr̥ṣṭáyaḥ* "crushed are the beast's teeth, also smashed in are its ribs."

A nominal derivative from the compound *api-śr̥̄* 'to smash in' could refer to the bone fractures which can result from the extreme muscular spasms of tetanus.

The following sūkta in PS, 11.3–4 = ŚS 19.34–35, which praises the Jaṅgiḍa-plant, may cautiously be claimed to provide some support for my interpretation of 11.2.10. One stanza refers to the 'fashioned reed' = arrow and seven *visrasaḥ*, a plural root noun that is probably connected with *vi-sraṃs* 'to fall apart, to collapse', hence 'failings, debilities'.

> PS 11.3.3 cf. ŚS 19.34.3
> *arasaṃ kr̥trimaṃ nāḍam* *arasāḥ sapta visrasaḥ* |
> *apeto jaṅgiḍāmatim* *iṣum asteva sādhaya* ||
>
> "(May the Jaṅgiḍa make) the fashioned reed sapless, the seven debilities sapless. Away from here, O Jaṅgiḍa, send off[11] helplessness, as an archer an arrow."

Possibly these 'debilities' are the seven symptoms named in PS 11.2.10 and they form a sort of canonical list; but to get a total of seven here it would be necessary to take *hanugrābha-* as a noun rather than as a qualifier of *jambha-*. Hence PS11.2.10 could be translated literally "We drive from you the convulsion, the smasher-in, the crusher-to-pieces, the rib-pain, the teeth-barer, the jaw-seizer, the conch-cough."

The Jaṅgiḍa-plant as a means of cure connects PS 11.3–4 with 11.2. Compare the role of the Rishis in 11.2.6 with that in 11.4.1.

"(If) the +*apaśara*-illness has mounted upon you which (comes) from business with women, which (comes) from a male, we arrange that flaccid, impotent tendon for you." In the AV the neuter noun *straíṇa-* is a euphemism for 'sex with women', cf. PS 6.22.3 = ŚS 4.34.2; *snā́van-* can mean 'penis': this sūkta uses *jyā́-* for 'bowstring'.

11 *apa-sādhaya-* corresponds exactly in meaning to the transitive present *apa-sédha-* from *sidh* 'to repel', whose object in the RV is normally *rákṣāṃsi, srídhaḥ, ámīvām*, etc., and once *ámatim* (RV 10.76.4). It is a previously unattested form which must represent a new clearly characterised transitive stem, probably created on the analogy of *sādhaya-* 'makes succeed' because the weak grade of *sidh-* 'to repel' (e.g. absolutive *apa sidhya* ŚS 8.2.7) coincided with the weak grade of *sādh* 'to succeed' (e.g. RV 1.18.7 *sídhyati*).

PS 11.2.6
yām atrayo aṅgiraso *gotamā vīrudhaṃ viduḥ* |
tayā bharadvājaḥ kaṇvo *viśaraṃ nāśayāti te* ||

"The plant which the Atris, the Aṅgirasas, the Gotamas know, with that may Bharadvāja, Kaṇva destroy *viśara* for you."

PS 11.4.1 = ŚS 19.35.1
indrasya nāma gṛhṇanta *ṛṣayo jaṅgiḍaṃ daduḥ* |
devā yaṃ cakrur bheṣajaṃ *agre viṣkandhadūṣaṇam* ||

"Pronouncing the name of Indra[12] the seers have given (us) the Jaṅgiḍa, which the gods in the beginning have made a *viṣkandha*-spoiling remedy."

References to medicinal plants or amulets made from plants as cures for all sorts of ills are common-place in the AV, but more unusual and interesting are two stanzas in PS 11.2 about mantra-recitation as a cure, in other words verbal remedies.

PS 11.2.8
vāta ivābhraṃ cyāvayāmi *yakṣmaṃ te tanuvas pari* |
vācā yac cakrus te guru *tad ṛcā laghu kṛṇmasi* ||

"I make disease move from your body as the wind a cloud. What they have made difficult/ painful for you by means of an imprecation that we make easy by means of a *ṛk*-stanza."

vāc- here contrasts with *ṛc-*. It is fairly frequent for *vácas* to mean 'spell' or 'imprecation' in the AV, and the root noun *vā́c-* also occurs in contexts where its meaning must be 'recited spell' or have an even more pejorative meaning,[13] e.g.

PS 7.8.6c–e
nāmagrāhād vāco heḍād **ikṣitād ghoracakṣasaḥ*
śarma te varma kṛṇmasi ||

12 PS 2.11.12 refers to an Indra-given plant (*vīrudh-*), which is almost certainly the Jaṅgiḍa, as PS 11.3.8–9 states that Indra gave it *vīrya-* and *ojman-*.

13 For pejorative AV use of vocabulary which is employed in a more elevated sense in the RV, cf. AV *kṛ́tvan-* 'performer of spells' versus RV 'performer of rituals', *gṛtsí-* 'wise woman, witch' (PS 11.3.2) versus RV adj. *gṛ́tsa-* 'wise' (of gods).

> "We make for you a cover, an armor, against name-grasping, against [bad] speech, against anger, against the glance, against the evil eye."
>
> ed. and trans. GRIFFITHS 2009:329

This is followed by the striking stanza PS 11.2.9 which has the normal closing formula *viśaraṃ nāśayāmi te* located in Pāda b:

tenāham ṛtod$_i$ yena *viśaraṃ nāśayāmi te* |
yathā na satyaṃ puruṣaḥ *sadā vaditum arhati* ||

> "With that truth-speech I destroy *viśara-* for you, as a man (suffering from *viśara-*) is not able always to speak a truth (i.e. to perform a *satyakriyā*)."

In my opinion, this stanza describes a situation where the patient ('*puruṣaḥ*' in pāda c) is unable to perform a *satyakriyā* 'Act of Truth' himself, but the reciter does so on his behalf and thereby removes the pain. *satyakriyā* is the general Sanskrit term which has been defined as "the magical use of a formal declaration of truth to accomplish a miracle" (Thompson 1998: 145). Starting from the later RV, the formula introducing an 'Act of Truth' is *téna satyéna* (e.g. RV 1.21.6ab, Damayantī at MBh 3.57.17, etc.). But the term in the earlier parts of the RV is *ṛténa* (*ṛtá-* n. 'Truth, true speech': see Lüders 1959: 505–506; Watkins 1979: 629–631) and it is this old term that must be reflected by *ṛtodyena* in the present PS stanza. The later phrase with the same technical meaning, *satyam ... vad-* (cf. Pali *saccavādin-* 'Truth speaking') is used here in connection with the patient. The stylistically more elevated term *ṛtodya-* in pāda a may refer back to the *ṛc-* mentioned in the last pāda of 11.2.8, as a mantra taken from the RV may be supposed to contain the necessary declaration of truth (about the gods, cosmos, etc.) that the reciter utters in order to destroy the patient's *viśara-* . The patient's inability[14] to speak a truth may be explained by the fact that the patient cannot speak at all, if *viśara-* represents the pain of tetanus whose other symptoms include paralysis of the mouth and jaw (*jambha-*, *hanugrābha-*).

If AV sūktas which contain lists or which move with bewildering rapidity from one ailment name to another are talking about co-occurring symptoms, as I have argued for PS 11.2, we may be able to interpret other poorly under-

14 The sense 'to be able' (usually with negative 'is not able') is found for *árhati* e.g. at ŚS 10.1.26 *ná tvā níkartum arhati* "it (the witchkraft) is unable to bring you down"; TS 5.2.5.1 *pakṣī́ bhavati ná hy àpakṣáḥ pátitum árhati* "he becomes winged, for without wings he cannot fly".

stood AV 'medical' compositions in similar fashion. Much more work needs to be done, but there are one or two sūktas which point in such a direction themselves, e.g.

> ŚS 5.22.11–12
> *mā́ smaitā́nt sákhīn kuruthā* *balā́saṃ kāsám udyugám* |
> *mā́ smā́to'rvā́ṅ aíḥ púnas* *tát tvā takmann úpa bruve* ||
>
> "Do not make these your companions- *balā́sa-*, cough,[15] *udyugá-*. Do not come back again from there in this direction; for that I call upon you, O fever."
>
> *tákman bhrā́trā balā́sena* *svásrā kā́sikayā sahá* |
> *pāpmā́ bhrā́tr̥vyeṇa sahá* *gáchāmúm áraṇaṃ jánam* ||
>
> "O fever, together with your brother *balā́sa-*, with your sister slight cough (?), with your cousin *pāpmá-*, go to yonder foreign people."
>
> trans. after WHITNEY 1905:261

There does not appear to be any theoretical basis to Atharvavedic medicine such as that developed in later Āyurveda. Yet if my interpretation of PS 11.2 is roughly correct, it provides evidence that the symptoms of tetanus and at least one of its causes were recognized in India some centuries before Hippocrates.[16] PS 11.2.10 with its six or seven items may indicate that a greater number of symptoms were linked together in Vedic India.

Appendix: Paippalāda Saṃhitā of the Atharvaveda, 11.2

Sigla Used in This Edition of PS 11.2

- **K** single Kashmiri Birchbark ms., in Śāradā script
- **Or** all palm leaf mss. known for PS 11, in Oriya script
- Ku_2 palm leaf from Central Orissa (Oḍisha) in Oriya script
- JM palm leaf from Central Orissa in Oriya script

---

15 I am grateful to Shree Nahata for pointing out that *kāsa-* means 'cough' in *Suśruta*, and it is a case where the later meaning may legitimately be projected back.

16 Hippocrates *Aphorisms 2* makes the connection between wounds and convulsions and *Aphorisms 6* introduces the term *tetanos*.

Pa palm leaf from Northern Orissa in Oriya script
Ma Mā Oriya mss. from N. Orissa read by Dipak Bhattcharya (1997). Readings inferred from this edition are enclosed in brackets [...]

* indicates conjectural emendation.
\+ indicates readings not attested as such in any manuscript but which can be inferred and which only demand small orthographical changes.

Note: the full apparatus criticus is not shown.
Variants are shown only where there is doubt about the reading.

11.2.1 Only PS
**aśaso yamanaṃ *śaso 'bhūtir yakṣmam ajījanat |*
imam sahasrabhāg indro viśaraṃ nāśayāti te ||

"Non-Thriving (*ábhūti-*) has given birth to Disease the constrainer *of the reviler, *of the praiser (or *of the one without chants, *of the one who chants). May Indra who allots a thousand (forms of aid) destroy this viśara for you."

*aśaso yamanaṃ *śaso] atsasoyamanaṃsaso Ku$_2$ (Ma), asasoyamanaṃ saso JM, aṃtsasoyamanaṃ saso Pa, atasaso yamanaṃso Mā, ahaṃ saso yamanaṃ saso **K**

11.2.2 Only PS
yaḥ kār$_i$yo yaś ca kṛtaḥ svayaṃjā uta hār$_i$yaḥ |
devā ind$_{(a)}$rajya$_i$ṣṭhā viśaraṃ nāśayantu te ||

"The viśara, which is to be made (in the future) and which has been made, (which is) born of its own accord and has to be acquired (brought from external sources), let the gods with Indra as their chief destroy for you."

11.2.3 Only PS
viśarasya +vijambhasya- iṣur mātā dhanuṣ pitā |
adityāḥ putrā ādityā viśaraṃ nāśayantu te ||

"Of viśara revealing the teeth, the mother is the arrow, the father the bow. The sons of Aditi, the Ādityas, let them destroy viśara for you."

+vijambhasya-iṣur] vijṛmbhasyeṣur Ku$_2$ JM Pa (Ma), vijṛmbhasyeṣu Mā, vijyambhasyeṣudhar **K**

11.2.4 - b = PS 5.8.3c, c = PS 5.8.4c; cf. ŚS.4.6.5–6
dhanvano jyāyā iṣ$_u$vā *apaskambhasya bāh$_u$voḥ* |
apāṣṭhāc chṛṅgāt kurmalād *viśaraṃ nāśayāmi te* ||

"From the bow, from the bow-string, from the arrow, from the two arms of the handle, from the barb, from the horn, from the shaft, I make viśara disappear for you."

11.2.5 cf. RV 6.75.15
alavatī ruruśīrṣṇ$_i$y *atho yasyā ayo mukham* |
devyai parjanyaretasa *iṣave kṛṇutā namaḥ* ||

"The deer-headed one with a scorpion's tail and whose mouth is metal, make reverence to the divine arrow (sprung) from Parjanya's seed."

The syntactically smoother RV verse (6.75.15) runs:

ā́lāktā yā́ rúruśīrṣṇy *átho yásyā áyo múkham*
idám parjányaretasa *íṣvai devyaí bṛhán námaḥ*

"She who is smeared with poison, with the head of a deer but whose mouth is metal: here is lofty reverence to her who has (received) the semen of Thunder, to the goddess *Arrow*" (trans. Jamison & Brereton 2014: 876–877.)

11.2.6 Only PS
yām atrayo aṅgiraso *gotamā vīrudhaṃ viduḥ* |
tayā bharadvājaḥ kaṇvo *viśaraṃ nāśayāti te* ||

"The plant which the Atris, the Aṅgirasas, the Gotamas know, with that may Bharadvāja, Kaṇva destroy viśara for you."

11.2.7 Only PS
yas tvā straiṇād +apaśaro *yaḥ puṃso adh$_i$yāruhat* |
āklāntaṃ saṃklāntaṃ snāva *tad u te kalpayāmasi* ||

"(If) the +apaśara-illness has mounted upon you which (comes) from business with women, which (comes) from a male, we arrange that flaccid, impotent tendon for you."

straiṇād +apaśaro] straiṇād apasaro **Or**, strīṇā upasado **K**

11.2.8 Only PS
vāta ʼvābhraṃ cyāvayāmi *yakṣmaṃ te tan$_{u}$vas pari* |
vācā yac cakrus te guru *tad ṛcā laghu kṛṇmasi* ||

"I make disease move from your body as the wind a cloud. What they have made difficult for you by means of an imprecation that we make easy by means of a ṛk-stanza."

11.2.9 Only PS
tenāham ṛtod$_{i}$ yena *viśaraṃ nāśayāmi te* |
yathā na satyaṃ puruṣaḥ *sadā vaditum arhati* ||

"With that truth-speech I destroy viśara for you, as the man (suffering from viśara) is not able always to speak a truth (to perform a satyakriyā)."

11.2.10 Only PS
viṣkandhaṃ tvad +apiśaraṃ *viśaraṃ +pṛṣṭ$_{i}$yāmayam* |
jambhaṃ hanugrābhaṃ śaṅkhaṃ *tān *ajenyāṁ ajāmasi* ||

"From you (we drive) viṣkandha, +apiśara, viśara, +rib-pain, jambha (,) the jaw-seizer, the 'conch', we drive those *which are to be driven."

+apiśaraṃ] apaśaraṃ Ku$_2$ Pa (Ma, Mā), āpiśaraṃ K (stanza missing in JM)
+pṛṣṭyāmayam |] puṣṭyāmayam Ku$_2$ Pa Ma, Mā, vṛṣṇyā vayaṃ K
tān *ajenyāṁ] tānajenā Ku$_2$ Pa (Ma, Mā), tvānujayenāṃ K

11.2.11 Only PS
śunam id va oṣadhayo *ni dadhe bheṣajāya kam* |
dhanāyuvaḥ sanāyuvaḥ *puruṣaṃ pārayiṣṇuvaḥ* ||

"I bestow prosperity on you, O herbs, for the purpose of healing, (you who are) seeking wealth, surviving from ancient times (seeking gain?), causing a man to cross beyond (illness)."

11.2.12 Only PS
+ātmatrāṇī devajūtā *vīrudāyuṣ$_{i}$yā kṛtā* |
tayāham indradattayā *viśaraṃ nāśayāmi te* ||

"A plant protecting the self, sped by the gods, made life-prolonging, with that (plant) given by Indra I destroy viśara for you."

+ātmatrāṇī] ātmatrāṇi JM Pa (Ma Mā), āTMatrāṇi Ku$_2$, ātapattrāṇī K

11.2.13 Only PS
alasya v$_i$yañjanasya *veṣṭanasyota parṇadheḥ* |
granther j$_i$yāyā iṣ$_u$vā *viśaraṃ nāśayāmi te* ||

"Of the tail, of the anointing/embellishment, of the band and of the feather-holder, of the knot, of the bow-string, of the arrow, the viśara I destroy for you."

PS 11.2.14 = RV 10.97.13
sākaṃ viṣkandha pra pata *cāṣeṇa kikidīv$_i$yā* |
sākaṃ vātasya dhrāj$_i$ yā *sākaṃ naśya nihākayā* ||

"Fly forth, O viṣkandha, together with the blue jay. Disappear together with the whirl of the wind, together with the storm."

References

Barret, L.C. 1926. The Kashmirian Atharva Veda, Book 11, edited with critical notes. *JAOS* 44, 258–269.

Bhattacharya, Dipak. 1997. *The Paippalāda-Saṃhitā of the Atharvaveda. Volume One, Consisting of the first fifteen Kāṇḍas*. Calcutta: The Asiatic Society.

Bloomfield, M. 1897. *Hymns of the Atharvaveda* (*SBE* vol. 42), Oxford: OUP

Bloomfield, M. and Garbe R. 1901 *The Kashmirian Atharva-Veda (School of the Paippalādas). Reproduced by Chromatography from the Manuscript in the University Library at Tübingen*. 3 Parts. Baltimore: John Hopkins University Press.

Filliozat, J. 1975. *La doctrine classique de la médecine indienne: Ses origines et ses parallèles grecs*. 2eme édition. Paris: École Française d'Extrême Orient.

Lüders, H. 1959. *Varuṇa*. vol. 2 *Varuṇa und das Ṛta*. Göttingen: Vandenhoeck & Ruprecht.

Griffiths, A. 2009.*The Paippalādasaṃhitā of the Atharvaveda, Kāṇḍas 6 and 7: A New Edition with Translation and Commentary*. Groningen: Egbert Forsten.

Hoffmann, K. 1968. Remarks on the New Edition of the Paippalāda-Saṃhitā. *IIJ* 11, 1–10.

Jamison, S. and Brereton, J. 2014. *The Rigveda: the Earliest Religious Poetry of India*. Oxford-New York: OUP.

Lubotsky, A. 2002. *Atharvaveda-Paippalāda, Kāṇḍa* 5: *Text, Translation, Commentary*. (Harvard Oriental Series, Opera Minora, vol. 4), Cambridge, Mass.: Department of Sanskrit and Indian Studies, Harvard University.

Narten, J. 1964. *Die Sigmatischen Aoriste im Veda*. Wiesbaden: Otto Harrassowitz.

Narten, J. 1965. Ai. *jámbha-*, Grk. γόμφος und verwandtes. KZ (= *Zeitschrift für Vergleichende Sprachforschung / Historische Sprachforschung*) 79, 255–264.

Narten, J. 1980. Ved. *āmáyati* und *āmayāvín-*. *StII* 5/6, 153–166.

Renou, L. 1964–1965. Notes sur la version 'paippalāda' de l'Atharvaveda. *Journal Asiatique* 252, 421–450 & 253, 15–42.

Scarlata, S. 1999. *Die Wurzelkomposita im Ṛg-Veda*. Wiesbaden: Ludwig Reichert Verlag.

Thompson, G. 1998. On Truth-Acts in Vedic *IIJ* 41, 125–153.

Watkins, C. 1979. *Is tre fír flathemon*: marginalia to *Audacht Morainn*. In L. Oliver (ed.) *Calvert Watkins: Selected Writings* II, 626–643.

Weber, A. 1873. Zweites Buch der Atharvaveda-Saṃhitâ. *Indische Studien* 13, 129–216. Leipzig: F.A. Brockhaus.

West, M.L. 2004. An Indo-European Stylistic Feature in Homer. In A. Bierl, A. Schmitt & A. Willi (eds.) *Antike Literatur in neuer Deutung* (Festschrift J. Latacz). Leipzig: K.G. Saur.

Whitney, W.D. 1905. *Atharva-Veda Saṃhitā: translated with a critical and exegetical Commentary*. Revised and edited by C.R. Lanman. 2 vols (HOS 7, 8), Cambridge, Mass.: Harvard University Press.

Zehnder, T. 1999. *Atharvaveda-Paippalāda, Buch 2, Text, Übersetzung, Kommentar*. Idstein: Schulz-Kirchner Verlag.

Zysk, K. 1985. *Religious Healing in the Veda*. (Transactions of the American Philosophical Society vol. 75, Part 7). Philadelphia: the American Philosophical Society.

Zysk, K. 1996. *Medicine in the Veda: Religious Healing in the Veda* (Revised edition of Zysk 1985) Delhi: Motilal Banarsidass Publishers.

CHAPTER 3

Another, Unrecognized, *cvi* Construction in the Rigveda, with Cautionary Remarks on the Supposed Origin of the Construction

Stephanie W. Jamison

It gives me great pleasure to offer this small study in honor of Chris Minkowski, in token of the nearly four decades of our association.

The *cvi* construction is a well-known feature of Classical Sanskrit, treated by Pāṇini (5.4.50, 51; 7.4.26, 32) as well as in the standard Western Sanskrit grammars (e.g., Whitney 1889: 401–402; Renou 1968: 150–152). In this construction a noun stem acquires the invariant suffix *-ī* and is then construed with a form of √*bhū* 'become' or √*kr̥* 'make', with the combinatorial sense 'become/make X', of the type *stambhī bhavati* 'becomes a post' to *stambha-* 'post'. As in the example just given, the derivational base is often an *-a*-stem, with substitution of *-ī* for the stem vowel. All of this is, of course, generally known and uncontroversial.

As is also well known, the use of the construction is quite restrained in Vedic, and in particular there are no certain and universally accepted examples in the earliest text, the Rigveda. See Whitney (p. 401): "There is no instance of this in the RV., unless the *ī* of *akkhalīkŕ̥tya-* [*sic*, for *akhkhalī*, swj] ... is to be so explained." Whitney seems to be the first to have suggested, however tentatively, an analysis of this RVic hapax as a *cvi* construction, but it wasn't until Thieme's brief (half-page) treatment of the form (1954: 109 = 1984: 138) that the word received a satisfactory explanation. The form is found in the Frog Hymn, RV VII.103.3, and serves, in the first instance, as a distinctly froggy vocalization (Whitney glosses it as 'croaking'), emphasized by the otherwise anomalous double aspirate *-khkh-*, not ordinarily found in either Sanskrit or Middle Indo-Aryan. One of the controlling images in the Frog Hymn is a pedagogical one, with the frog chorus likened to the antiphonal recitation of text between teacher and students in Vedic education. (See introduction to and translation of this hymn in Jamison-Brereton 2014 ad loc.) Thieme explains *akhkhalī-* as a *cvi* formation, in Middle Indo-Aryan phonological guise, built to *akṣára-* 'syllable'. By the standard MIA phonological rules, *akṣara-* should have yielded **akkhara-*; the double *-khkh-* aspiration is the onomatopoetic froggy touch. Since some form of early MIA was probably the everyday language even in Rigvedic times, it is not surprising that pedagogical vocabulary would appear in this form, not

 | DOI:10.1163/9789004536869_005

in high Vedic Sanskrit. The students are "making syllables," that is, repeating by rote the text recited by their teacher, while being compared to frogs incessantly repeating the ancient Indian equivalent of "ribbit." I have always found Thieme's explanation extremely clever and entirely convincing.

I will now suggest another example of this construction in the RV—though I'm afraid it will take me more than half a page. Like *akhkhalī-kṛ́tya* it involves a play on words. RV X.28 is a twelve-verse dialogue hymn, dedicated to Indra, intricate in its structure and unique in the RV for its use of abbreviated animal fables for instruction. I have treated this hymn in a number of publications, and it has had its share of other commentary.[1] For eleven of its twelve verses it is fiendishly complex; the verbal twists and turns are discussed at length in my online commentary. The two final verses (11–12) are thematically paired: they contrast the fate of bad / non-sacrificers (vs. 11) with that of proper sacrificers (12ab), ending with an appeal to Indra (12cd). In striking contrast to the complexities that mark the rest of the hymn, the first half of vs. 12, describing the good sacrificers, seems flat-footed and clumsily simplistic, esp. pāda a:

X.28.12ab[2] *eté śámībhiḥ suśámī abhūvan, yé hinviré tanvàḥ sóma ukthaíḥ*

The first pāda contains what appears to be an awkward near-pleonastic etymological doubling of instrumentals belonging to the long *ī*-stem *śámī-* 'ritual labor', built to the root √*śam* 'ritually labor' (*śámībhiḥ suśámī*), construed with a finite form of √*bhū* (*abhūvan*) that seems to contribute little semantically. Consider Geldner's translation: "Die sind *durch ihre Opferdienste und guten Dienst* jetzt hochgekommen."[3] Geldner (n. 12a to the translation, ad loc.) gives *abhūvan* something to do by ascribing to it the sense common to √*bhū* in the Brāhmaṇas, namely 'thrive' (his "… sind hochgekommen," Schnaus's "sind …

1 In addition to the introduction to and translation of the hymn in Jamison-Brereton 2014 (ad loc.), see Jamison 2004: 237–249; 2007: 80–83; 2009: 197–218; Brereton-Jamison 2020: 152–153, as well as extensive discussion ad loc. in my online commentary to the RV translation, http://rigvedacommentary.alc.ucla.edu/. Both Oldenberg 1912 and Geldner 1951 (both ad loc.) provide lengthy introductions to the hymn with elaborate interpretations of its content. It is also anthologized and translated in Doniger (1981: 146–148) and translated and commented upon by Schnaus 2008: 203–232. None of these treatments convinces me, but I will not engage with them in detail here.

2 All passages cited are from the Rigveda unless otherwise identified.

3 My italics, as also in the other translations given in this note. Similarly Schnaus (2008: 232): "Diese sind *mit Opferwerken, mit gutem Opferwerk* gediehen." Doniger (1981: 147–148): "Those … have become supreme in their *sacrificial acts* and *devout rituals*" (my italics), but with an unjustified shift in the sense of √*śam* in the second term. Elizarenkova's "Those (men) distinguished themselves by (their) *sacrificial devotion* (and) *generous donation*" (1995: 50) also

gediehen"), but I am not aware of clear examples of this usage in the Rigveda (the other example Geldner suggests in that note should be otherwise interpreted). Therefore, a translation reflecting the RVic usage of √*bhū* and based on an etymological identity between the two *śam*-forms should be the even less inspired "They came to be (joined?) with labors and good labor."

I find it quite improbable that the poet Vasukra, whose three hymns X.27–29 contain some of the most dazzling displays of RVic poetic pyrotechnics, would end this great hymn in such a pedestrian fashion. I therefore suggest that the etymological duplication in *śámībhiḥ suśámī* is only apparent, with the homophony of *śámī(bhiḥ)* and *(su)śámī* actually a pun. The instr. pl. *śámībhiḥ* does indeed belong to the root √*śam* 'ritually labor', but, I suggest, the *śam* in *suśámī* belongs instead with the etymologically and semantically independent indeclinable *śám* 'weal, luck',[4] which is extremely common in the RV.[5]

That the two words are doing different things seems also to be the view of Sāyaṇa, though his treatment is limited to paraphrase: ***śamībhiḥ*** *somayāgakarmabhiḥ* ***suśamī*** *sukarmāṇo* ***'bhūvan***.[6] He glosses *śámībhiḥ* with another instrumental plural specifying that the work is for the Soma sacrifice, but he treats *suśámī* as if it were a nominative plural bahuvrīhi (*sukarmāṇaḥ* 'having good [ritual] work') characterizing the subject, which seems to me tantamount to analyzing *suśámī abhūvan* as a *cvi* construction (though with the same lexical *śam* as the instrumental plural *śámībhiḥ*). (No other translator or commentator has taken note of Sāyaṇa's interpretation, as far as I have found.)

Now in fact *śám* is already regularly construed with and compounded with forms of √*bhū*: see the RVic compounds *śambhū́-* (also *viśvá-śambhū-*), *śambhú-*,[7] and its superlative *śámbhaviṣṭha-*, as well as numerous verb phrases with finite forms of √*bhū*. In all instances the syntagm means "be(come) weal/luck (for X)," with the recipient of the luck in the dative and the subject generally a god or gods (though sometimes just an auspicious object), as in the following examples:

---

arbitrarily fails to reflect the apparent etymological doublets. These two translators obviously felt the need to vary their renderings of the two *śámī* forms, an indirect indication that the text appears on the surface to be inelegantly pleonastic.

4 On the independent etymologies of both see EWA s.v. *ŚAM*[1] and *śám* respectively.

5 According to Lubotsky's *Concordance* (1997, s.v.), there are 161 occurrences, not counting compounds.

6 The glossed forms from the RVic passage are bolded for clarity; their glosses, which follow immediately, are not.

7 On these compounds see Scarlatta 1999: 367–368.

compound: V.42.7 *yáḥ śáṃsate stuvaté śámbhaviṣṭhaḥ*
(Bṛhaspati,) who is most wealful for the one who chants and praises …

verb phrase: VI.74.1 *śáṃ no bhūtaṃ dvipáde śáṃ cátuṣpade*
(Soma and Rudra,) be(come) luck for our two-footed and four-footed.

śám is also construed with finite forms of √*as* in the same general sense as *śám* √*bhū*. In fact an entire hymn (VII.35), quite possibly the most boring hymn in the Rigveda, consists of an unbroken series of predications of *śám*, with the predicating verb alternating between √*bhū*, √*as*, and zero, as in the following verse:

VII.35.5 ***śáṃ*** *no dyā́vāpr̥thivī́ pūrváhūtau,* ***śám*** *antárikṣaṃ dr̥śáye no* ***astu***
śáṃ *na óṣadhīr vaníno* ***bhavantu,*** ***śáṃ*** *no rájasas pátir* ***astu*** *jiṣṇúḥ*

Luck for us Heaven and Earth at the Early Invocation; luck be the Mid-space for us to see.
Luck for us be the plants, the trees; luck be the victorious lord of the dusky realm.

It is also found with √*kr̥* in the sense "make luck for," again with a dative, e.g.:

IV.1.3 *asmábhyaṃ dasma śáṃ kr̥dhi*

For us, o wondrous one (= Agni), make luck.

In the compound *śaṅkara-* this lexeme will have, by accident of nomenclature, a glorious future; it is first attested in a Yajurveda mantra *námaḥ śaṃkarā́ya ca mayaskarā́ya ca* (VS XVI.41 = TS IV.5.8.1 = MS II.9.7 = KS XVII.15), referring to Rudra-Śiva.

But in all these instances the recipient of *śám* is external to the lexeme and expressed in the dative. The subject *is* the luck or *makes* the luck, but is not him/itself lucky: the subject transfers the luck to the dative recipient. In order to indicate that it's the subject that possesses or acquires the luck, I think that Vasukra pressed the emergent *cvi* construction into service, since the predication *śám* √*bhū* was unavailable because it was already well established in a different sense. That the resulting *śám-ī* coincides with the stem of the noun *śámī-* 'labor' and its instrumental singular is, for Vasukra, a happy rhetorical result. By my interpretation then, *śámībhiḥ suśámī* is a punning expression, since the two *śam*-s are unrelated. Rather than the pedestrian doubling assumed by other interpretations, we see here yet another example of Vasukra's poetic artfulness. Hence, my translation of X.28.2a (Jamison-Brereton 2014 ad loc.) "But these (others) *have become very fortunate* [lit. become good fortune] by their ritual labors."

I could at this point declare my work here to be done: I have provided an alternative analysis of the phrase *suśámī abhūvan*, identified a second example of a *cvi* construction in the RV, and salvaged Vasukra's poetic reputation in the process. But to support my claim I need to address several issues: 1) the stem *suśámī-* is found twice elsewhere in the RV as well as in a few mantras in other early Vedic texts; 2) it is generally claimed that the *cvi* construction is built originally and overwhelmingly to *a*-stems, which *śám* obviously isn't; 3) it is generally claimed that the *cvi* construction is inherited (not "emergent" as I termed it above) and that the inherited construction had a particular semantic/functional profile. I will deal with these in turn, the last at greatest length.

The first requires little comment. In a sense the *point* of the *suśámī cvi* construction in X.28.12 is that it mimics the already existing form *suśámī* 'good labor'; this association is signaled by the immediately preceding *śámībhiḥ*. But it is worth noting (though not pursuing here) that the other forms of *suśámī* are not unproblematic. There are two other occurrences in the RV: *suśámī* in VII.16.2 (= VS XV.34, TS IV.4.4.4, etc.) and *suśámi*[8] with short final vowel (in both Saṃhitā and Padapāṭha texts) in V.87.9. Although Grassman identifies them both as instrumentals, in neither case is an instrumental reading certain: both could be modifiers of *yajñáḥ* (VII.16.2) / *yajñám* (V.87.9) or adverbial. Geldner comments on the form ad VII.16.2 "Das Kasusverhältnis ist unklar." A Yajurveda mantra (VS I.15, TS I.1.5.2, KS I.5, etc, with variants in, e.g., MS I.1.6, IV.13.4) also contains a short-vowel *suśámi*, which is most likely adverbial.

The secondary association between my putative *cvi* form *suśámī* and forms of *-śámī-* derived from √*śam* 'labor' will also account for the root accent in the form. Other *cvi* forms in accented texts accent the *-ī* in finite verb phrases (e.g., TS V.3.6.2 *mithunī́ abhavan*, MS IV.1.7 *phalī́ karoti*), though the accent usually falls on the auxiliary in nominalized forms (e.g., RVic *akhkhalī-kŕ̥tya*, AVŚ XI.3.6 *phalī-káraṇa-*). However, immediately preceding *śámībhiḥ* as well as the other occurrences of *suśámī-* would have encouraged root accent in our form, either in the original composition or redactionally introduced.

Another formal consideration gives at least weak support to my *cvi* interpretation: the hiatus in the sequence *suśámī abhūvan*, with the length of the final *-ī* metrically guaranteed. As is shown by the TS phrase just quoted, TS V.3.6.2

8 The uncompounded stem *śámī-* also has several short-*i* *śámi* forms.

mithunī́ abhavan, the *-ī* of the *cvi* construction is pragṛhya, at least in the Taittirīya tradition.[9] However, Oldenberg (1912 ad loc.), who interprets the form as an instrumental, finds the existence of *-ī* before a vowel "irrelevant," and Wackernagel (1896: 321) notes that a number of *-ī* forms (e.g., nom. sg. *pṛthivī́*) are unaltered before vowels, so that the *cvi* formation is not the only possible explanation for the long-vowel hiatus in the Saṃhitā text.

As for the limitation of the construction to underlying *-a*-stems or its supposed origins there, all descriptive accounts must reckon with the fact that, although *a*-stems bases may be overwhelmingly predominant, a variety of other stem-types are found and are among the earliest examples—e.g., KS XIII.6 (2×) *starī-karoti* 'makes barren' to *starī́-* 'barren cow';[10] TS V.2.1.7, VI.1.4.3, ŚB III.1.3.25 *muṣṭī́-karoti* 'makes a fist' to *muṣṭí-* 'fist'. There are certainly well-established *cvi* formations to *-a*-stems among the early examples, e.g., *phalī́ √kṛ* 'hull' vel sim. (AVŚ, AVP, MS) to *phála-* 'fruit', *mithunī́ √bhū* / *√kṛ* 'copulate' / 'make a couple' to *mithuná-* '(sexual) pair(ing)'. Whitney (1889 § 1094) in his discussion of the Classical construction states that "the combinations with a-stems are the immense majority" but points to other stem-types as bases, also in § 1093 concerning the early language. According to Renou (1968 § 116), "En théorie toute forme nominale peut figuerer ici," though "Le seul type productif à date ancienne est celui qui s'attache aux thèmes en *a-*." But given the predominance of *-a*-stems in the lexicon anyway, neither the productivity of the *-a*-stem type nor its numerical advantage should be surprising.

However, for most of those who have pursued the construction into prehistory the connection to the thematic stem has been crucial for their reconstructions, and so the original derivational relationship between the *-a*-stem and the *-ī* of the *cvi* formation has become an article of faith. For a long time the most influential account of the history of the formation was that of Wackernagel (1908). Wackernagel connected the *-ī* of the *cvi* formation with the Latin thematic genitive in *-ī*, and therefore it was important to assert that the Indic construction was also based on the thematic noun: "Unweigerlich muss anerkannt werden, dass die Grundsprache aus den *ŏ-* Stämmen einen Adverbialkasus aus *-ī́* bildete" (1908: 128 = 1955: 1349). Since Wackernagel's explanation has generally been abandoned,[11] I will not treat it further here, beyond noting that it depended on assuming the affiliation of the *cvi* construction with the Sanskrit *-a*-stem.

---

9 Cf. also *śyetī́ akuruta* TS V.5.8.1 = TB I.1.8.3; TS III.4.9.1 *mithunī́ evá karoti*, and Schindler 1980: 387. Though it is not always pragṛhya, at least in ŚB II.1.1.5, etc., *mithuny ènayā (àbhiḥ) syām*.

10 There is no adjective **stara-* 'barren'.

11 Although strongly championed, against Schindler's interpretation, by Schlerath (1988; updated in 1998, published in Schlerath 2000). In the latter Schlerath also criticizes the

The current standard view of the history of the formation was first articulated by Schindler (1980). After dismissing previous explanations of the history of the construction, Schindler presents his own, which also depends on an organic connection between the thematic nominal stem and the *-ī*: "Für die Morphologie ist von der Tatsache auszugehen, dass die Domäne der *cvi*-Bildung die *a*-Stämme sind. Wir konstatierern somit einen Ersatz von **-e/o-* durch **-ī* (älter z.B. **-ih*)" (1980: 389). Schindler starts with a thematic adjective, to which an abstract noun in *-i-* is built. This noun in turn is used as a predicative instrumental in *-ī* with an auxiliary to produce the sense 'be/make with X (noun)' → 'become/make X (adj.)'. The prehistoric example he constructs is **rudhri-h_1 $dheh_1$-, $bhuh_2$-, ($h_1$es)* 'mit Röte machen, werden sein' = 'rot machen, werden, sein' beside **rudhro-* 'red' (p. 393; no accents given). With this analysis Schindler effectively brings the construction into the larger Caland system—though as far as I can find, he never uses the word "Caland" in the article. This basic interpretation of the construction has been developed in a series of publications especially by Irene Balles.[12]

I will raise two objections to this account of the history of the construction: 1) the evidence for the existence of the construction outside Indic is very weak; 2) the theoretical model for its origins and development is at odds with the actually occurring Vedic examples.

To begin with the first, as long as Wackernagel's connection of the *cvi* formation with the Latin thematic genitive prevailed, the construction seemed to be deeply embedded in Indo-European nominal morphology. However, with that view abandoned, the only comparative evidence adduced as far as I know,

---

alternative interpretation of Tichy (1990). Tichy devotes only a few pages (11–13 = 2018: 272–274) of a more general discussion of the methodology and aims of reconstruction (her inaugural lecture at Phllips-Universität Marburg) to the *cvi* construction. She suggests that the *-ī́* is the acc. of the collective **- íh$_2$*, entering into a double accusative construction with 'make'. Since this interpretation requires a transitive construction, she asserts (pp. 12–13 = 274) that the Vedic construction is not found with the verb 'to be', ignoring the examples with √*bhū* (despite citing the very well attested *mithunī́ bhavanti* earlier [p. 11 = 272]). Her proposal is insufficiently worked out to be properly assessed, and her aim is primarily to dispute the old Wackernagel proposal as well as Schindler's (see p. 13 = 274). I will not treat it further here, as it does not seem to have gained any traction in the scholarly literature.

Neither Schlerath nor Tichy questions whether the construction is an inherited one. This is the shared assumption of all scholars who have treated the origin of the construction, whatever the origin they suggest for it, at least as far as I have been able to determine.

12 Balles 2000, 2006, 2008, 2009. Much of her work concerns the theoretical and prehistoric underpinnings of such a construction, with Schindler as starting point, and so much of it is irrelevant to my more limited investigation, as is the evolution of her views in the course of these separate publications.

certainly the only evidence adduced by that very careful and widely knowledgeable scholar Schindler, is an equivocal passage from Younger Avestan: Y 62.2–3, a litany to the sacred fire, a passage also found in the Khorde Avestan in the composite prayer to the fire, Ātaš Niyāyišn (there = sections 8–9). Y 62.2 consists of a series of addresses to the fire with the 2nd sg. opt. *buiiå* 'may you become' preceded by a compound that ends, in Geldner's edition, in *-i*.[13] The first of these (of six) is *dāitiiō.aēsmi buiiå* "may you be with / have the prescribed firewood." The next section, Y 62.3, also consists of a series of parallel statements referring to the fire, again utilizing a form of √*bū*, here an infinitive, preceded by/compounded with a form in *-i* (again, in Geldner's edition); the first of these (of four) is *saoci.buiie ahmiia nmāne* "to be(come) flaming in this house." Since Bartholomae (1895: 148), the *-i* + √*bū* constructions in these two adjacent passages have been compared to the Indic *cvi* construction, a comparison repeated and accepted by, e.g., Benveniste 1935: 64–65 and Schindler 1980: 387. However, there are both functional and formal problems. To begin with, the wishes expressed in section 2 are semantically entirely different from the *cvi* type: they are functionally possessives. The fire is not exhorted to *become* prescribed firewood, but to acquire or possess it. To be sure, section 3 could be considered more conventionally *cvi*-like: "to become flaming" fits the functional template.

On the formal side, the *-i* of Geldner's edition is only one of a variety of final vowels found in the manuscripts. The variants for both sections are meticulously charted by de Vaan in his detailed treatment of this passage (2003: 450, 453–458), in which he rejects the *cvi* interpretation (see esp. 453, 456–457) as well as Geldner's edited final *-i*. See further Redard and Kellens (2013: 11–12), who echo de Vaan's conclusions. Although it might be possible to argue that *-i* is the *lectio difficilior*, with the variants the result of an attempt at normalization, the variability of the final vowel and the marked functional difference between the Avestan examples and the Indic *cvi* forms make the equation highly questionable. Not to mention that even if the fit were better, this short passage in Younger Avestan is a slender basis for reconstructing the construction for Indo-European or even Indo-Iranian—esp. given that the Indic construction does not show any signs of antiquity.[14]

---

13 Given the conventions governing YA orthography, the original / underlying length of the final *-i* is of course unrecoverable.

14 Schindler's attempt to address this issue (p. 393) seems uncharacteristically apodictic: "Das relative späte Auftreten der *cvi*-Bildungen in der ved. Literatur kann nichts als Einwand gegen die hier vorgetragenene Erklärung geltend gemacht werden, auch wenn man die avestischen Parallelen bestreitet: Der Typus gehörte offenbar der Umgangssprache

Nonetheless, the view that the *cvi* construction is not only of Indo-European date but a part of the Caland system, the latter view adumbrated by Schindler and prominent in the work of Balles (see esp. the title of Balles 2009: "The Old Indic *cvi* Construction, the Caland System, and the PIE Adjective"), seems now to be firmly established in the Indo-Europeanist literature. And accompanying this establishment is a semantic/functional model associated with the Caland system: that roots that participate in the Caland system belong to the "property concept" class (color, dimension, etc.), as originally defined by Dixon (1982), and that the source of the *cvi* construction is a predication of Caland forms expressing such property concepts. Although Schindler 1980 does not use the terms "Caland" or "property concept" in his treatment, his constructed example (given above), "be/make with redness" → "be/make red," fits squarely within this model, and it has been developed in somewhat different ways ever since, particularly in the work of Balles. And here we encounter another major stumbling block: the deep disconnect between the model that will supposedly account for the development of the *cvi* construction and the actual *cvi* constructions attested in Vedic.

It is not my intention to investigate the early examples of the *cvi* construction in any detail here, but I have collected a tolerably sizeable catalogue of them. What is striking is that there are *no* clear "property concept" examples among them. Although the extension of the construction outside of its original boundaries would not be surprising—and Schindler (p. 393) constructs a model for such an extension[15]—one might expect to have a few early examples preserved from the original semantic/functional domain, but this is not the case. Instead the examples are stubbornly concrete and tend to cluster in a relatively low and often technical register. Leaving aside the two putative RVic examples discussed above, the construction first really emerges in the Atharvaveda, and by far the best attested example in the AV (primarily AVP) is *phalī* √*kṛ*,[16] to *phála-*

an, und sein nur langsames Eindringen in der Hochsprache ist ein soziolinguistisches Phänomen." Note his "gehörte offenbar der Umgangssprache an," an absolutist declaration without associated support. Although I quite agree that the construction belongs to a lower linguistic register than the usual Vedic discourse, I do not see that this gives us license to reconstruct it for the proto-language in the absence of evidence. Note that Schindler seems to admit that the Avestan evidence is weak.

15 It is worthy of note that his exemplar of the adjectival property-concept model *śuklī́ kṛ* 'make white' (Schindler's accent) isn't textually attested till Classical poetry, but the exemplar of the nominal one supposedly based upon it, *brāhmaṇī́ kṛ* 'make (into) a Brahmin', is found in Vedic, in the (late) ŚB. Or rather, the √*bhū* equivalent: ŚB XIII.4.1.3 *brāhmaṇībhū́ya* 'having become a Brahmin'.

16 Including both nominal and verbal forms: AVŚ XI.3.6, AVP XVI.53.3, XVI.124.2–3, XVII.15.2;

'fruit'. This lexeme doesn't have a straightforward additive meaning 'make (into) (a) fruit', but a technical agriculture sense 'separate grain from chaff', often rendered 'thresh, winnow' vel sim. Elizabeth Tucker (p. c. March 2021) suggests rather 'hull' (remove the tough outer husk of grain), since Vedic already has a number of other words for thresh/winnow. This interpretation is quite appealing, esp. since "hulled barley" is a staple of whole-grain cookery even now—and assuming that barley was the standard early Vedic grain. In any case, whatever the precise meaning, to me the term suggests a complex semantic development in specialized circles of speakers outside the mainstream of high Vedic discourse.

Other agricultural terms are found: e.g., *starī* √*kṛ* to *starī́-* 'barren cow': 'make (into a) barren cow, make barren', found only in one passage in the early BYV Saṃhitā, KS XIII.6, where it is applied to speech (*vāc-*). (As was noted above, the base is not an *-a*-stem.) Although found in somewhat later Vedic, (*su/duṣ*)*matī* √*kṛ* (AB III.38) also belongs to this sphere. Despite appearances *matī* doesn't belong to the root √*man* 'think', but rather with the noun *matyà-* 'harrow', already found in early Vedic (AV, KS, MS, TS). The *cvi* formation derived from it means 'break up clods, smoothe' (originally of fields), and in its sole attestation it is parallel to (*su*/*duṣ*)*kṛṣta-* 'well-/ill-ploughed'. The sense of this combination and its relation with *matya-* is guaranteed both by the presence of *matya-* in the AB passage and by a different *cvi* construction, *samī* √*kṛ*, construed with *matyena* in PB II.9.2: 'make smooth/even with a harrow'.

> AB III.38 *yathā vai* ***matyaṃ*** *evaṃ yajñasya viṣṇus tad yathā duṣkṛṣṭaṃ* ***durmatīkṛtaṃ*** *sukṛṣṭaṃ* ***sumatīkṛtaṃ*** *kurvann iyāt* …
>
> Like a **harrow** (*matya-*), so is Viṣṇu to the sacrifice. Just as one may keep making well ploughed and **well harrowed** what has been ill ploughed and **ill harrowed** …
>
> PB II.9.2. *yathā saptāsthitena* ***matyena samīkaroty*** *evaṃ pāpmānaṃ bhrātṛvyaṃ prarujati.*
>
> Just as one **makes even with a harrow** reposing on seven sharp-pointed (pins), so he breaks to pieces his wicked rival.

---

also in the early BYV Saṃhitā, MS IV.1.17. It is also extremely widely attested in the Brāhmaṇas, likewise in both nominal and verbal forms: e.g., ŚB I.1.4.23, 24; II.4.29, 6.1.8; AB II.9; TB III.3.9.10; KB XVIII.12, etc.

To return to earlier Vedic, another stubbornly concrete example is *muṣṭí* √*kṛ* 'make a fist', to *muṣṭí-* 'fist' (as already noted, not an *-a*-stem) found twice in identical passages in TS (V.2.1.7 = VI.1.4.3), as well as once in the ŚB (III.1.3.25). The same construction is found in Pāli *muṭṭhī katvā* (J VI.331).

One of the best attested Vedic *cvi* constructions is *mithunī́* √*bhū* 'become a couple, have sex' and its less common transitive counterpart *mithunī́* √*kṛ* 'make a couple, cause to mate' found several times in TS, as well as once in KS, and widely in the Brāhmaṇas (AB, JB, ŚB).[17] Cf., e.g., TS V.3.6.2 *tā́ḥ prajā́ḥ … ná **mithunī́ abhavan** tā́su … mithunám adadhāt* "These offspring **did not couple** (/have sex); he established coupling / sex in them." Despite its slightly euphemistic air, this expression seems an example of lower-register plain speaking and certainly does not, at least as I understand the term, express a property concept. Again, we can note a Pāli correspondent: although *mithunī* √*bhū* is not found in Pāli, the semantic equivalent *missī* √*bhū* and √*kṛ* (Skt. *miśra-*, Pāli *missa-* 'mixed') seems to be well established.

A Vedic example that might appear from its gloss to belong to the property concept realm is *tīvrī́* √*kṛ* (ŚB, KB), generally glossed as 'make sharp, strengthen' (*tīvrá-* 'sharp'). But this term too belongs to the domain of *tekhne*. It has a very restricted application, describing the "sharpening" of the ritual drink soma immediately after it has been coagulated (expressed by the likewise technical term *ā́* √*tac*). The adj. *tīvrá-* is already well attested in the RV in technical usage, specialized for a particular type of prepared soma.

Two other examples in later Vedic might also seem to fall into the property concept sphere, but since both are based on already complex nominal deriv-

17

mithunī-bhavantī	KS XXIV.1, KapS 46.5
mithunī́ abhavan	TS V.3.6.2
mithunī-bhavantīḥ	TS V.3.6.2–3
mithunī́ bhavataḥ	TS VI.5.8.6
mithunī́ … karoti	TS III.4.9.1
mithunī … syām	ŚB II.1.1.5, 2.4.15; III.2.1.25 [ŚBK I.1.1.3, 2.4.11; II.7.2.1; IV.2.1.9]
mithunī-bhavati	JB I.351
mithunī-bhavataḥ	ŚB X.5.2.11
mithunī-bhavanti	AB VII.13
mithunī … bhavanti	JB II.166, 439
mithunī-bhavant-	ŚBK V.8.4.10, ŚB X.5.2.2
mithunī-bhaviṣyant-	JB II.166, 439
mithunī-karoti	AiĀr II.3.6; ŚB II.4.4.24; ŚBK I.3.1.26, 27; TB III.3.3.5
mithunī-kārayanti	JB II.405
mithunī-kartar	JB I.300
mithunī-kṛtya	ŚB III.2.1.2, 4; ŚBK IV.2.1.1, 2

ations, there does not seem to be a straight path from the supposed original model to these specialized forms. One is *kālvālī́-kṛta-* 'made bald', once in the ŚB (ŚBM II.2.4.6 = ŚBK I.2.4.3) of the earth. Its supposed base **kālvāla-* is not attested, but it seems related in some not-quite-straightforward way to *-kulva-* 'bald', found once in VS (and later in a śrauta sūtra). Another is *daridrī √kṛ* 'make poor', once in the late ŚB (XI.3.3.5) to the adj *dáridra-*, attested in the sense 'poor' only beginning in the Brāhmaṇas (and there very rarely), ultimately derived from the intensive to *√drā* 'run', originally 'running around (aimlessly?), wandering', vel sim. These are not sturdy foundations on which to base the Caland, property-concept model.

I have not treated all the Vedic examples, even those in early Vedic, though I have discussed all the examples that might be thought to express property concepts. I have left aside even several well-attested examples because their meaning is unclear: *vātī √kṛ* in the AV (a type of health condition, whose precise identity is much discussed—and not important here); *śyetī √kṛ* (TS, also TB, JB) possibly 'congeal' or 'make submissive', not, in any case, connected with *śyetá-* 'white', as is clear from its contexts.

In short, a quick survey of the Vedic evidence provides no support for the current *opinio communis* about the origin of the construction and in fact points in another direction. The hefty representation of the construction in technical vocabulary, including agriculture, and in lower-register items suggests that the construction developed not in order to predicate abstractions, but as a technical and slangy shorthand, a way to make verb phrases out of concrete nouns that lacked clearly associated verbal roots—in other words as an alternative way of making denominatives[18]—perhaps originally popular in vernacular speech. It is worth noting that, in addition to the two Pāli examples I have already cited, the construction is fairly productive in Pāli,[19] which supports the notion of a vernacular origin. The single generally acknowledged RVic example, *akhkhalī-kṛ́tya*, fits this pattern well: it is Middle Indic by phonology[20] and is part of pedagogical vocabulary, suggesting an origin in the domestic life of young children. As for my *suśámī √bhū*, the clever poet Vasukra seized on a developing syntagm type to make his pun, but it is probably not an accident

18 Note that *tīvrī́ √kṛ* is paralleled in the PB (XVIII.5.12, 14, 15) by a nonce denominative *tīvraya-*, made in a more conventional way, in the same type of technical context.

19 See the Pali Text Society Dictionary s.v. *karoti* (p. 198): "It is often compounded with nouns or adjectives with a change of final vowel to ī. … Cp. the same process in connection with bhavati." It should be noted that although the Dictionary always cites it with *-ī*, it is also found with short final *-i* in some manuscripts; see von Hinüber 1991: 11 n. 20. I owe this reference to Diego Loukota, along with valuable discussion of the Middle Indic evidence.

20 Note the "popular" *l* as well as the assimilated sibilant cluster.

that it is found at the end of a hymn specializing in (fragmentary) animal stories, which would have been used to instruct the young in Vedic times, as in later ones. See my speculations on the pedagogical use of the stories in this hymn in Jamison 2009: 216–218.

I have no idea how the construction developed, in particular where the *-ī* came from. But I would maintain that a hard look at the Vedic evidence—and at the lack of Indo-European evidence or of robust Indo-Iranian evidence—can only lead us to the conclusion that the construction developed on Indian soil, as it were, and that however appealing a Caland-system model is in theory, it has no support in the realities of attestation.

Bibliography

Balles, Irene. "Die altindische Cvi-Konstruktion: Alte Deutungen und neue Wege." In *Indoarisch, Iranisch und die Indogermanistik*, edited by Bernhard Forssman and Robert Plath, 25–36. Wiesbaden: L. Reichert, 2000.

Balles, Irene. *Die altindische Cvi-Konstruktion: Form, Funktion, Ursprung*. Münchner Forschungen zur historischen Sprachwissenschaft, Band. 4. Bremen: Ute Hempen Verlag, 2006.

Balles, Irene. "Principles of Syntactic Reconstruction and 'Morphology as Paleosyntax': The Case of Some Indo-European Secondary Verbal Formations." In *Principles of Syntactic Reconstruction*, edited by Gisella Ferraresi and Maria Goldbach, 161–186. Amsterdam: Jon Benjamins, 2008.

Balles, Irene. "The Old Indic cvi Construction, the Caland System, and the PIE Adjective." In *Internal Reconstruction in Indo-European: Methods, Results, and Problems*, edited by Jens E. Rasmussen and Thomas Olander, 1–15. Copenhagen: Museum Tusculanum Press, 2009.

Bartholomae, Christian. "Vorgeschichte der Iranischen Sprachen." In *Grundriss der Iranischen Philologie*, edited by Wilhelm Geiger and Ernst Kuhn, 1–151. Strassburg: Trübner, 1985.

Benveniste, Émile. *Les infinitifs avestiques*. Paris: Adrien-Maisonneuve, 1935.

Brereton, Joel P., and Stephanie W. Jamison. *The Rigveda: A Guide*. New York: Oxford University Press, 2020.

Dixon, R.M.W. *Where Have All the Adjectives Gone? And Other Essays in Semantics and Syntax*. Berlin: Mouton, 1982.

Doniger, Wendy. *The Rig Veda*. London: Penguin, 1981.

Elizarenkova, Tatyana J. *Language and Style of the Vedic Ṛṣis*. Albany, NY: SUNY Press, 1995.

Geiger, Wilhelm, and Ernst Kuhn. *Grundriss der Iranischen Philologie*. Strassburg: Trübner, 1895–1901.

Geldner, Karl Friedrich. *Der Rig-Veda aus dem Sanskrit ins Deutsche übersetzt und mit einem laufenden Kommentar versehen*. 3 vols. Cambridge, MA: Harvard University Press, 1951.

Hinüber, Oskar von. *The Oldest Pali Manuscript: Four Folios of the Vinaya-Pitaka from the National Archives, Kathmandu*. Stuttgart: Steiner, 1991.

Jamison, Stephanie W. "Poetry and Purpose in the Ṛgveda: Structuring Enigmas." In *The Vedas: Texts, Language, and Ritual*, edited by A. Griffith and J. Houben, 237–249. Gronigen: Egbert Forsten, 2004.

Jamison, Stephanie W. "The Function of Animals in the Rig Veda, RV X.28, and the Origins of Story Literature in India." In *Penser, dire et représenter l'animal dans le monde indien*, edited by Nalini Balbir and Georges-Jean Pinault, 197–218. Paris: Champion, 2009.

Jamison, Stephanie W. *The Rig Veda between Two Worlds: Four Lectures at the Collège de France, May 2004*. Collège de France, Publications de l'Institut de Civilisation Indienne, fasc. 74. Paris: de Boccard, 2007.

Jamison, Stephanie W., and Joel P. Brereton. *The Rigveda: The Earliest Religious Poetry of India. An English Translation*. 3 vols. New York: Oxford University Press, 2014.

Lubotsky, Alexander. *A Ṛgvedic Word Concordance*. 2 vols. American Oriental Society, Vols. 82–83. New Haven: American Oriental Society, 1997.

Mayrhofer, Manfred. *Etymologisches Wörterbuch des Altindoarischen*. 3 vols. Heidelberg: Carl Winter, 1986–1996.

Oldenberg, Hermann. *Ṛgveda: Textkritische und exegetische Noten. Siebentes bis zehntes Buch*. Abhandlungen der kgl. Gesellschaft der Wissenschaften zu Göttingen, Philologisch-historische Klassse, vol. XIII.3. Berlin: Weidmann, 1912. Reprint: Göttingen: Kraus, 1970.

Redard, Céline, and Jean Kellens. *La liquidation du sacrifice* (*Y 62 à 72*). Études avestiques et mazdéennes, vol. 5. Paris: de Boccard, 2013.

Renou, Louis. *Grammaire sanscrite*. Paris: Adrien-Maisonneuve, 1968.

Scarlatta, Salvatore. *Die Wurzelkomposita im Ṛg-Veda*. Wiesbaden: Reichert, 1999.

Schindler, Jochem. "Zur Herkunft der altindischen cvi-Bildungen." In *Lautgeschichte und Etymologie: Akten der VI. Fachtagung der Indogermanische Gesellschaft, Wien, 24.–29. September 1978*, edited by Manfred Mayrhofer, Martin Peters, and Oskar E. Pfeiffer, 386–393. Wiesbaden: Reichert, 1980.

Schlerath, Bernfried. "Nochmals zu den cvi-Bildungen." In *Studia Indogermanica et Slavica: Festgabe für Werner Thomas zum 65 Geburtstag*, edited by Peter Kosta, 37–47. Munich: Sagner, 1988.

Schlerath, Bernfried. "Nochmals zu den cvi-Bildungen: Neufassung 1998." In *Kleine Schriften*, 1:398–414. Dettelbach: J.H. Röll, 2000.

Schnaus, Susanne. *Die Dialoglieder im altindischen Rigveda: Kommentar unter besonderer Berücksichtigung textlinguistischer Kriterion*. Philologia, vol. 116. Hamburg: Verlag Dr. Kovač, 2008.

Thieme, Paul. "*Akhkhalīkṛtyā.*" KZ 71 (1954): 109.

Thieme, Paul. *Kleine Schriften*. Wiesbaden: Franz Steiner, 1984.

Tichy, Eva. "Warum und mit welcher Sicherheit rekonstruieren wir heute das Urindogermanische?" *Würzburger Jahrbücher für die Altertumswissenschaft* NF 60 (1990): 5–18. Reprinted in *Kleine Schriften*, ed. Axel Metzger. Pp. 265–282. Bremen: Hempen Verlag, 2018.

Vaan, Michiel de. *The Avestan Vowels*. Amsterdam: Rodopi, 2003.

Wackernagel, Jacob. *Altindische Grammatik, vol. I: Lautlehre*, 1896. Reprint: Göttingen: Vandenhoeck & Ruprecht, 1957.

Wackernagel, Jacob. "Genetiv und Adjektiv." In *Mélanges de linguistique offerts à F. de Saussure*, 125–152. Paris: Champion, 1908. Reprinted in Wackernagel (1955): 1346–1373.

Wackernagel, Jacob. *Kleine Schriften*. 2 vols. Göttingen: Vandenhoeck & Ruprecht, 1955.

Whitney, William Dwight. *Sanskrit Grammar*. 2nd ed. Cambridge, MA: Harvard University Press, 1889.

CHAPTER 4

Hann iti han

John J. Lowe

1 Introduction

The first and foremost śāstra in ancient India was vyākaraṇa, grammar. The beginnings of vyākaraṇa, and hence of śāstric and scientific thought in India, can be seen in the Vedas, the preservation, understanding and analysis of which was the basis of the development of scientific thought in the first millennium BC. The earliest surviving text concerned with grammatical analysis, and hence the earliest monument of the birth of scientific thought in India, is Śākalya's padapāṭha to the *Ṛgveda* (RV). In this short paper I address a problematic feature of the Ṛgvedic padapāṭha, which I first encountered while teaching *Ṛgveda* with Chris Minkowski.

In Śākalya's padapāṭha to the *Ṛgveda*, the word *han*, 3sg. present injunctive 'he struck', appears as *hann iti han* ('parigraha' marking) when it occurs at the end of a verse; this occurs three times, at 5.29.4d, 5.31.1d, and 10.99.6d, for example:

(1) Saṃhitā: *jígartim índro apajárgurāṇaḥ* | *práti śvasántam áva dānavám han* ||
Pada: *jígartim* | *índraḥ* | *apa-járgurāṇaḥ* | *práti* | *śvasántam* | *áva* | *dānavám* | *hann iti han* || (5.29.4d)
Transl.: 'Repeatedly taunting the gulper, Indra smashed the snorting Dānava back and down.' (Jamison and Brereton 2014: 691)

Such 'itikaraṇa' in the RV padapāṭha is ordinarily used to mark final vowels which do not participate in the standard rules of external sandhi ('pragṛhya' vowels, e.g. dual endings in *-e*), or word final *-r* in certain contexts, as discussed in more detail below. The reason for marking *han* in these three instances in this way is not immediately clear, and no explanation is offered by Rastogi (1970) or Jha (1975, 1992), who discuss itikaraṇa in the RV padapāṭha in detail.

In this paper I discuss the parigraha marking of *han* and other words unexpectedly so marked. As a part of this, I present an extended digression on the sandhi of final *-n* in the *Ṛgveda* which grew out of investigating this question; although it is ultimately not clear that this helps us to resolve the parigraha of

 | DOI:10.1163/9789004536869_006

han, the sandhi of final *-n* is an interesting question in its own right, though its patterns have never been presented in detail.

In §2 I present the basic principles of itikaraṇa in the RV padapāṭha, and introduce the cases of exceptional marking. In §3 I discuss in detail the sandhi of final *-n* in the RV. In §4 I return to itikaraṇa, and offer some possible explanations for previously unexplained parigraha markings. In §5 I draw conclusions.

2 Itikaraṇa in the RV padapāṭha

The main analytical features of the RV padapāṭha are well known. The most important feature, from which the pāṭha gets its name, is its division of the original, continuous 'saṃhitāpāṭha', into separate words (padas). In the padapāṭha each word is recited separately, with a pause between each word, and with sandhi as *in pausa*. But this is only the start of the grammatical analysis of the padapāṭha. Beyond word division, Śākalya[1] also undertook to analyse compound words into two constituent parts, separating those parts with an 'avagraha', a term originally referring to a pause of particular length (different from the pause between padas), but now more commonly used for the symbol used in writing to mark this (nagari ऽ, in transliteration in this paper marked with a dash, as in *apa-járgurāṇaḥ* in (1)).[2]

Avagraha is also used to distinguish certain morphological units, notably the case suffixes in *-bh-* and locative plural *-su*, when these occur after athematic stems.[3] For example, at RV 1.1.2 the words *pū́rvebhir ŕ̥ṣibhiḥ* 'by ancient ṛṣis' are rendered in the padapāṭha as *pū́rvebhiḥ ŕ̥ṣi-bhiḥ*.

Besides avagraha, the other analytical marking in the padapāṭha involves the particle *íti*. This itikaraṇa is of two kinds: pragr̥hya marking and parigraha.[4]

---

1 In this paper I refer to the author(s) of the padapāṭha by the name of the scholar to whom the pāṭha is traditionally ascribed, Śākalya, while acknowledging that such an ascription is impossible to prove, and we may in fact be dealing with multiple authors.

2 There are additional details which go beyond the scope of the present discussion. For example, the padapāṭha treats the particle *iva* as compounded with the word it follows.

3 That is, when they occur after a stem which is equivalent to the basic stem with 'external' sandhi applied as appropriate (i.e. sandhi of the type found ordinarily between words, but also found between athematic stems and certain case suffixes). Thus avagraha marks stems which are below the 'word' level but which show external 'word' sandhi (stems which, like full 'words', get the label *pada* in Pāṇini's *Aṣṭādhyāyī*).

4 The terms *avagraha* and *parigraha* refer to the respective types of marking, but *pragraha* is not used to refer to the marking of pragr̥hya words (it is used only in the Taittirīya Prātiśākhya as a synonym of pragr̥hya, i.e. to refer to words so marked).

Pragṛhya marking is the simpler, involving simply the addition of *íti* following a word, with sandhi applied as appropriate. Parigraha involves both the addition of *íti* and the repetition of the word in question after the *íti*. The uses of itikaraṇa in the RV padapāṭha have been described in detail by Rastogi (1970) and Jha (1975, 1992); in the following I provide a brief summary.

Itikaraṇa has three primary functions in the RV padapāṭha. It is used to mark: 1. pragṛhya vowels, that is vowels which do not undergo sandhi even when another vowel follows; 2. visarjanīya(/visarga) (ḥ) which occurs at the end of a word and which represents an original final *-r* as opposed to an original final *-s*; 3. the particle *ī*, which the padapāṭha treats as *īm*.

The first—marking pragṛhya vowels—is relatively unproblematic and is by far the most common use of itikaraṇa in the padapāṭha, occurring over 3,000 times; vowels so marked include:[5] duals in *-e*, *-ī*, *-ū*, such as *índravāyū* at 1.2.4 (pp. *índravāyū íti*); Vedic locatives in *-ī* and *-ū*, such as *sarasī́* at 7.103.2 (pp. *sarasī́ íti*); the Vedic pronominal forms *asmé*, *yuṣmé*, *tvé*, *amī́*, such as *asme* at 8.5.15 (pp. *asmé íti*); and, unexpectedly but consistently, vocatives to *-u* stems, such as *viṣṇo* at 1.156.1 (pp. *viṣṇo íti*).[6]

If the pragṛhya vowel occurs at the end of a compound (or word-*iva* sequence) which is separable in the pp., then parigraha is used; for example, the vocative *kavikrato* at 3.14.7 is given in the pp. as *kavikrato íti kavi-krato*. That is, the two features being marked are marked separately: first the pragṛhya vowel is marked with *íti* following the unanalysed word, and then the compound is presented in analysed form with avagraha as required.

The second major use of itikaraṇa, to mark a final visarjanīya following original *-r*, is considerably less common, being found just over 200 times. The majority of these instances mark r-stem nouns in the vocative (e.g. *pitar*, pp. *pitar íti*, 6.51.57), or the indeclinables *antár*, *púnar*, and *prātár*. Neuter *-r* or *-r/n* stems (e.g. *áhar*, pp. *áhar íti*, 6.9.1) are also marked, and a number of verb forms, such as aorists to roots which in the guṇa grade end in *-r* or in a consonant cluster beginning with *-r*[7] (e.g. *akar*, 2/3sg. of *kṛ* 'do, make', pp. *akar íty akaḥ*, 1.33.15). When original final *-r* remains as *-r* in the saṃhitā, there is no itikaraṇa. With a few problematic exceptions mentioned below, it is only visarjanīya from *-r* which follows a vowel *a*/*ā* which is marked, i.e. only in the context where there is a difference in the sandhi of final *-r* vs. *-s*. This is differ-

5 According to Jha (1975: 50), itikaraṇa marking of pragṛhya vowels is consistent except for one case which Śākalya appears to have missed, the vocative *índrāgnī* at 7.94.8.

6 Also under this heading can be included the particle *u*, which is not pragṛhya in the saṃhitā but is treated as pragṛhya *ūm̐* in the pp.

7 And which will therefore end in *-r* following the simplification of final consonant clusters, if the root is suffixed only with vowel-less suffixes.

ent from the marking of pragṛhya vowels, which are marked regardless of their context.

There are various exceptions and apparent inconsistencies in the use of parigraha instead of pragṛhya marking for visarjanīya, which make it difficult to establish the intended use of parigraha here. As with marking of vowels, parigraha is used when the visarjanīya to be marked appears at the end of a compound, the avagraha to mark the compound appearing in the second iteration of the form following the *íti* (e.g. *pranetaḥ* at 2.28.3, pp. *pranetar íti pra-netaḥ*). According to Jha (1975), parigraha is also found instead of pragṛhya marking when the visarjanīya is part of a verbal form. Yet this does not explain all remaining instances of parigraha with visarjanīya, as Jha (1975: 52–53) notes. Rastogi (1970: 8) gives a slightly different explanation: that it is monosyllabic or disyllabic verbs at the end of a verse or hemistich which get parigraha. But Rastogi too has to admit a few exceptions to this generalization, cases of parigraha instead of pragṛhya marking of both verbs and non-verbs inside verses.

When we find apparent irregularities in a scientific work as old as the padapāṭha, the natural temptation is to assume error or inconsistency on the part of the composer(s) of the work. The composer(s) of the padapāṭha was/were certainly human, and thus liable to err, but it may not be unreasonable for our first assumption to rather be that there was indeed a rationale underlying the analysis that comes down to us. The RV padapāṭha was an extraordinary achievement, not least because it was composed without the aid of writing, based on an original text—the RV saṃhitā—which was also transmitted and studied without the aid of writing. The feats of memory that enabled the saṃhitāpāṭha not only to be preserved and transmitted, but also analysed and reconstructed into the padapāṭha, are such that we should hesitate before assuming that Śākalya simply forgot to apply his rules consistently in a number of cases. Although a complete justification of every analysis offered by the padapāṭha lies beyond the scope of this paper, I want to offer two suggestions which may help with understanding some of the unexpected analyses in the text.

First, it is likely that the padapāṭha is the product of a long process of composition.[8] Even if it was conceived and composed by the genius of one man, it may well have taken a lifetime's work to devise a means of analysing, and then to analyse, every pāda of the RV saṃhitā. Perhaps more likely, the undertaking

8 Jha (1992: 70–74) proposes a number of stages in the development of the padapāṭha, though I see no reason to assume that each feature of the pp. must necessarily be ordered with respect to every other feature in terms of its introduction into the pada analysis. In relation to the present discussion, the most significant claims that Jha makes are firstly that parigraha

was conceived and carried out over a longer time frame, by more than one generation of scholars. It is also likely that the form of the padapāṭha as we have it may have been affected by later redaction, following its initial completion. Granted a perhaps long period of development, it is possible that procedures and principles were altered during the period of composition, and this may explain certain inconsistencies. To take itikaraṇa of visarjanīya as an example, parigraha is unexpectedly found with a few verb forms which are traditionally treated as ending in *-s*, not *-r*. Some of these are 3pls. such as *ūvuḥ* (1.61.8), *guḥ* (3.7.7.), *syuḥ* (×5, only when unaccented, e.g. 1.24.7), which historically ended in **-r̥s* (e.g. Jasanoff 1994: 150); the consonant of the ending *-uḥ* does continue an *-r* historically, via a development **-r̥s* > **-urs* > *-ur*, but in general *-uḥ* from this development is not marked.[9] Besides these 3pls., a few 2sg. optatives are marked with parigraha, e.g. *vaneḥ* at 2.6.1 (pp. *vaner íti vaneḥ*), and two reduplicated 2sgs., *tūtoḥ* (6.26.4d) and *pīpeḥ* (4.16.21b, a pāda repeated with the same marking seven times in the following hymns). This highly sporadic marking of visarjanīya from original *-s* is clearly related to the standard marking of visarjanīya from *-r*, but with no further assumptions can only be explained as an occasional mistake on Śākalya's part, an overapplication of otherwise consistently applied rules to cases which ordinarily should not have been marked.

But let us consider an alternative possibility. If one thinks etymologically and focuses on the two origins of visarjanīya, then there is no valid reason why any visarjanīya from *-s* should be marked in the same way as visarjanīya from *-r*. But Śākalya's concern was not etymological but entirely practical, aimed merely at distinguishing ambiguous sandhi variants. From a purely practical sandhi perspective, words ending in visarjanīya can be divided into two groups: words ending in visarjanīya from *-s* after *a*/*ā*, and everything else: the sandhi of visarjanīya from *-r* is consistent, regardless of the preceding vowel, and is identical to the sandhi of visarjanīya from *-s* following any vowel but *a*/*ā*. If we group the data this way, we can most simply disambiguate the sandhi of visarjanīya by marking one or another group. Conceivably, the original approach taken at an earlier point in the composition of the padapāṭha was to mark all final visarjanīya which show the sandhi of final *-r*, including those from final *-s* after vowels other than *a*/*ā*. Under such an approach, one set of sandhi outcomes for visarjanīya would have been consistently marked with *íti*, in some ways a more

was added later, words now marked with parigraha originally being marked as pragr̥hya, and secondly that the use of itikaraṇa to mark visarjanīya is later than its use to mark pragr̥hya vowels. There is no firm evidence for either claim, though neither is in itself implausible.

9 The final sequence *-uḥ* < **-r̥s* is found in both the 3pl. perfect/aorist ending and the r-stem genitive singular; the latter is never marked.

logical, though less parsimonious, approach than that found in the padapāṭha as we have it.[10] Let us suppose that this was the original approach, but that this approach was changed during the composition of the padapāṭha, such that visarjanīya after a vowel other than *a*/*ā* need not be marked, because there is only one possible sandhi outcome in such cases. Then only visarjanīya after *a*/*ā* which shows the distinctive sandhi of an original *-r* needed to be marked. It seems more likely that, given an existing padapāṭha with all ambiguous visarjanīyas marked, the removal of itikaraṇa was imperfectly carried out, due to the difficulty of altering an established, memorized, text, than that a very clear rule for applying itikaraṇa to words in final *-r* after *a*/*ā* should have been mistakenly applied to a few forms which, like *vaneḥ*, can hardly have been misunderstood as ending in *-r* and which even if they were, did not fit the context for application (i.e. the visarjanīya did not follow *a*/*ā*).[11]

A second consideration is that the recitation of the saṃhitā itself would not have been so invariant or uncontroversial in Śākalya's time as it is in ours. Besides the possible existence of multiple different recensions of the *Ṛgveda*, even the precise form of the recension that survives today would have been subject to some variation. Indeed, Śākalya's padapāṭha was a key part of fixing the form of the *Ṛgveda* (the Śākalya recension, at least) and eliminating textual corruption and variation henceforth.[12] Unfortunately, we have no direct access to the variant readings that Śākalya may have been aware of when the padapāṭha was composed. No other Ṛgvedic recension survives; the closest we have are the verse mantras of the Yajurveda, which are often identical or highly similar to Ṛgvedic pādas or verses, but being part of a different Vedic tradition these cannot necessarily be taken as indicative of the sorts of variation found within the Ṛgvedic tradition itself. We can suppose, however, that in establishing his padapāṭha Śākalya was aware of different readings for some Ṛgvedic verses. Perhaps Śākalya set out to establish, as the fixed form of the *Ṛgveda*, precisely the form that he inherited from his teacher, i.e. the form

10 This would also somewhat parallel the situation with pragṛhya vowels, which are marked even when a consonant follows.

11 All (five) instances of unaccented *syuḥ*, 3pl. opt. of *as* 'be', except one which shows final *-r* in the saṃhitā, are marked with parigraha, suggesting that the marking of this particular form is intentional (whether intentionally retained, or intentionally applied). The lack of accentuation is likely to be a factor, as is the fact that all five instances end a pāda, and four of the five end a verse; these features are also found with several of the unexpectedly marked forms discussed below, including *han*. A full explanation of the marking of *syuḥ* lies beyond the scope of this paper, however.

12 That the surviving recension is named after Śākalya perhaps reflects the importance of the padapāṭha in establishing the fixed form of the recension as we now have it.

that he had himself memorized as a student; or perhaps, Śākalya may have set out to compare competing forms of the text and establish the 'best reading'. But either way, there are likely to be cases where Śākalya's reading set him at odds with other contemporary versions of the saṃhitā. It is therefore also possible that a previously unidentified use of itikaraṇa in the padapāṭha is to mark a particular, potentially controversial, reading, and thereby to resist its replacement by another current reading. So, to us the itikaraṇa of the 2sg. opt. *vaneḥ* at 2.6.1b is not immediately explicable, except with the sorts of hypotheses advanced in the previous paragraph. But in Śākalya's own time, the reading of this pāda may have been a matter of debate, and the itikaraṇa of the verb may have been intended to confirm Śākalya's interpretation over some other contemporary reading. Of course, such an explanation cannot be proven, but as a hypothesis it is preferable to assuming a simple error. This possibility will be considered again in relation to some of the forms discussed below.

Returning now to the uses of itikaraṇa in the *Ṛgveda*, the third regular use is to mark the word *īm* when this appears as *ī* in the saṃhitā. This has no exceptions and does not require further discussion.

Besides these 'regular' uses of itikaraṇa, there are also a number of exceptional cases of parigraha. These include:[13]

(2)
- *gdha*, 3sg. mid. injunctive of *ghas* 'eat' (1.158.5)
- *te*, enclitic gen./dat. pronoun (3.33.10)
- *namasyan*, nom.sg. present participle (1.72.5)
- *pracetā/pracetaḥ*, voc. of *pracetas-* (1.24.14, 7.17.5, 10.83.5)
- *práṇapād*, voc. (8.17.13)
- *ráthyebhiḥ*, instr.pl. of *ráthya-* (1.157.6)
- *vark*, 3sg. inj. of *vṛj* 'twist' (10.8.9)
- *ṣṭaḥ*, 3du. of *as* 'be' (1.61.8)
- *syasva*, 2sg. imper. of *sā* 'bind' (3.4.9, 7.2.9)
- *han*, 2/3sg. present of *han* 'strike, slay' (5.29.4, 5.32.1, 10.99.6)

A number of these have had explanations offered in previous literature. In several cases, there are contextual problems which endanger the correct reading of the form. So, Jha (1975: 53) explains the parigraha of *namasyan* at 1.72.5b as due to the fact that the form is preceded immediately by the word *namasyàm*:

13 To these irregular examples it would be possible to add the itikaraṇa of visarjanīya from original -*s*, discussed above, and indeed Jha (1975) does this for the 2sgs. like *vaneḥ*.

(3) Saṃhitā: *pátnīvanto namasyàṃ namasyan* ||
Pada: *pátnī-vantaḥ* | *namasyàm* | *namasyann íti namasyan* (1.72.5b)
Transl.: 'Along with their wives they did reverence to the one worthy of reverence.' (Jamison and Brereton 2014: 197)

Evidently, the close similarity of *namasyàm* with immediately following *namasyan*, and the inevitable danger of the former influencing the pronunciation of the latter by perseveration, led Śākalya to mark the latter with parigraha. The sandhi of the final *-n* before *íti*, and its repetition following, ensured the survival of that final dental.

A similar explanation must underlie the parigraha of *rā́thyebhiḥ* at 1.157.6d: the form appears at the end of a verse, immediately preceded by the word *rathyā̀*.[14] This is the only form of *rā́thya-*, with long first vowel, in the RV; it may be this difference in vowel length between *rā́thyebhiḥ* and *rathyā̀* which was the feature in danger of being lost in the saṃhitā, and thus the cause of Śākalya's marking it.

The obvious objection to such an explanation is that the padapāṭha itself does not preserve the long vowel of *rā́thya-*, but shortens it to the more familiar *ráthya-*. But this is not the only such case. Similarly, *práṇapāt* at 8.17.13b immediately follows the form *napāt*; again although the retroflexion of *ṇ* is undone in the pp., the danger of losing the *ṇ* of the saṃhitā by perseveration of the dental from *napāt* is the most likely explanation for the parigraha:

(4) Saṃhitā: ... *śṛṅgavṛṣo napāt práṇapāt* ... ||
Pada: *śṛṅga-vṛṣaḥ* | *napāt* | *pránapāt íti prá-napāt* (8.17.13a–b)
Transl.: 'O grandson, great-grandson of Śṛṅgavṛṣ ...' (Jamison and Brereton 2014: 1061)

In both these examples the wording of the padapāṭha would be more comprehensible if we assume that originally one of the two instances of *ráthyebhiḥ* and *pránapāt* in the padapāṭha preserved the long vowel / retroflexion of the saṃhitā. The alterations found in the padapāṭha in both cases are entirely in accordance with its practice of modernizing the saṃhitā reading (presumably in accordance with the contemporary language of the pp. author), but—bearing in mind again the composition of the padapāṭha over a period of time

14 A third parallel case is the parigraha of the 2sg. opt. *madeḥ* at 8.29.16c. While this can be counted along with other cases of parigraha of verbs ending in visarjanīya, like *vaneḥ* discussed above, it is best explained contextually: the end of the pāda runs *máde madeḥ*, the padapāṭha marking protecting this from misrecitation as *máde made*.

and the possibility of redactional changes—it may well be that this practice was a later augmentation to the padapāṭha analysis.[15] It is therefore possible to posit for example an earlier padapāṭha reading *práṇapāt íti prá-napāt* at 8.17.13b; such a reading would be more logical in terms of what the padapāṭha is trying to achieve, but would have been very easy to alter at a later point if a decision was taken that modernized forms should be cited consistently in the padapāṭha.

A slightly more problematic case is the parigraha of the vocative *pracetaḥ/pracetā* at 1.24.14, 7.17.5, and 10.83.5. The first of these, at 1.24.14c, is understandable, since the form shows irregular sandhi in the saṃhitā:

(5) Saṃhitā: ... *asura pracetā rā́jann* ... ||
Pada: *asura* | *praceta íti pra-cetaḥ* | *rā́jan* (1.24.14c)
Transl.: 'O attentive lord and king ...' (Jamison and Brereton 2014: 121)

Instead of the expected sandhi form *praceto* from the vocative *pracetas*, we find *pracetā*, which would reflect the regular sandhi of an r-stem vocative **pracetar* before the following *r*, if such an r-stem existed. Whether the saṃhitā form is to be understood as based on some kind of alteration of *pracetas-* to an r-stem, or whether it is a dialectal or otherwise nonstandard sandhi of final *-as*, Śākalya recognized it as worthy of note. If it were an r-stem, then parigraha would be entirely regular; as it is, the parigraha makes clear that this is not, in fact, an r-stem, but an s-stem vocative.

What is more difficult is the fact that parigraha is also found at 7.17.5a and 10.83.5b, where the saṃhitā form is the unproblematic *pracetaḥ*. One can assume that, given the irregular form at 1.24.14, and given that *pracetaḥ* is in all three saṃhitā passages ambiguous between an s-stem and r-stem, Śākalya marked *pracetaḥ* in these two other instances also to clarify that we are not dealing with an r-stem. However, there are another two instances of this vocative which are unmarked in the pp., at 6.13.3c (saṃhitā form *praceta*) and 10.87.9b. The first is explicable, since the sandhi of the saṃhitā is only possible for an s-stem, so there is no need to clarify. But the second, at 10.87.9b, is no different in the saṃhitā from the form at 10.83.5b, and we would expect it to be marked identically. In this case we must perhaps admit a slip on Śākalya's part.

Another case of parigraha which has not previously been adequately accounted for, but may be best understood in terms of sandhi or phonological

---

15 Jha (1992: 74) takes such modernizations as the latest stage in the development of the padapāṭha, though they certainly precede the *Rkprātiśākhya*, since it is aware of them.

difficulty, is that of *syasva*. This form is 2sg. present middle imperative to *sā/si* 'bind'; it occurs three times in the RV, and is marked with parigraha in two of those, when it ends identical pādas at 3.4.9b and 7.2.9b. The verb in these pādas occurs with the preverb *ví*, which is separated from the verb by one intervening word. Most instances of the root *sā/si* 'bind' occur with one of the preverbs *áva* or *ví*. With *ví* immediately preceding the verb form, its most common position, there is standardly retroflexion, e.g. 2sg. present active imperative *ví ṣya* (×2), 3sg. *ví ṣyatu* (×2), and also once *ví ṣyasva* (1.101.10b).[16] The forms marked with parigraha are the only cases in the RV where forms of the present *ví syati* do not show retroflexion of *syati*.[17] Although the lack of retroflexion is regular, since the *ví* does not directly precede, in the present of this root the sibilant appears in a phonological context—immediately followed by the palatal glide—which naturally favours retraction of the tongue. Thus there is additional pressure with this stem to pronounce the sibilant with a retracted point of articulation, similar to the 'ruki' process which turns alveolar *s* into retroflex *ṣ*. Given this double pressure towards a retracted pronunciation, the correct form of the sibilant may have been liable to be mistaken; the parigraha can therefore be taken as marking, and protecting, the alveolar articulation of the sibilant.

As we have seen, the most common uses of itikaraṇa involve ambiguous sandhi: marking a final segment, either a pragṛhya vowel, or final *-r*, which is in principle susceptible to an alternative interpretation and thus to different sandhi rules. The other 'irregular' uses of itikaraṇa (always parigraha) mark a variety of things, including problematic sandhi, potential mispronunciation due to the surrounding context, and potential misreading. Some forms have not yet been accounted for, besides *han*: *ṣṭaḥ*, *te*, *gdha*, and *vark*; we will address these below. But returning now to the parigraha of *han*, we must first ask which are we most likely to be dealing with: an issue of sandhi, or something more specific to the form and contexts in question. Although there are good reasons to think that sandhi is not the issue in this case, it is true enough that, just like final long vowels and final *-r*, the final segment of *han* shows two different sandhi treatments in the *Ṛgveda* saṃhitā. There is no existing comprehensive treatment of the sandhi of final *-n* in the RV; I therefore turn to this in the next section.

---

16 With *ví ṣyasva* we do not have parigraha in the padapāṭha, although the padapāṭha undoes the retroflexion.

17 There are a couple of other instances of *ví* with forms of *sā/si* which do not show retroflexion, also with other words intervening between preverb and verb, e.g. *ví…sāt* at 5.45.2a and *ví…sitam* at 8.5.9c. There are no such examples with the present stem.

3 Sandhi of Final *-n*

It is well known that metrical irregularities in the *Ṛgveda* can often be resolved by 'restoring' the saṃhitā text to what must represent an earlier and more original form. The most common restoration involves the glides *y* and *v*, which often conceal an earlier vowel *i* or *u* respectively; restoring these vowels resolves numerous metrically defective pādas. For example, the word *mártya-* 'mortal' is always to be read as **mártiya-*; the word *svàr* 'sun' to be read as **súvar*.

While this restoration increases the syllable count for a pāda, there are other possible restorations which do not change the syllable count, but do enable the 'correction' of an irregular or highly rare metrical pattern into a more common metrical pattern.[18] This is particularly clear in the cadence of metrical lines, since all the most common Ṛgvedic metres have one highly regular metrical scheme in the cadence, the violation of which is rare and marked. According to the figures of van Nooten and Holland (1994), 96 % of Jagatī cadences (following their metrical restorations) show the same metrical scheme: HLHLX (heavy-light-heavy-light-anceps), while the next most common Jagatī cadence accounts for less than 2 % of the data. For Triṣṭubh, the majority pattern (HLHX) constitutes 94 % of the data, for Gāyatrī (LHLX) 89 %, and for Anuṣṭubh (LHLX) 80 %, with the second most common pattern never accounting for more than 4 % of relevant cadences. So, if we find a Jagatī cadence which scans HHHLX (a pattern found with only 0.2 % of Jagatī cadences), and if the second syllable of that cadence can be read as light by historical reconstruction, so as to produce an unproblematic HLHLX cadence, then we may be justified in making such a restoration, particularly if the same restoration can also improve the metre when it is possible in parallel contexts as well.

It has been known for a long time that, while the *Ṛgveda* saṃhitā as we have it shows the regular Classical sandhi of final *-n* after a short vowel and before a vowel, i.e. as *-nn*, without exception in all instances, metrical evidence suggests that this exceptionless doubling of final *-n* was not a part of the 'original' form of the hymns. Historically, final *-n* in Sanskrit has multiple origins. In some cases final *-n* derives without change from a final *-n*, such as in the vocative of *n*-stems, the endingless locative of *n*-stems, and the pronominal locative in *-smin*.[19] In other cases, final *-n* derives from a final consonant cluster of which the first consonant was an *n*; by the regular simplification of

18 For example, see Beguš (2015) on the restoration of original word initial **vu-* and **yi-*. It is more controversial whether Proto-Indo-European laryngeals survived long enough as consonants to 'make position' in certain contexts; see Gippert (1997), Kümmel (2014).

19 Although beyond the scope of this paper, it is worth noting that in some cases final *-in*

final clusters in Sanskrit, any consonant following the *-n* was lost. Final *-n* can thus represent an earlier **-nt*, as in the 3pl. secondary verb ending, or the vocative of *nt*-stems, or it can represent an earlier **-ns* or **-nts*, as in the nominative singular of *nt-stems*. It is from the earlier consonant clusters that the sandhi in *-nn* derives, preserving the weight of the original cluster, though not the distinct consonants. At an early period, this sandhi was extended to all cases of final *-n*, regardless of origin. But it has long been known that Ṛgvedic metre can be regularly improved by 'undoing' the *-nn* sandhi for most cases of final *-n* which represent original final *-n*.[20] That is, Ṛgvedic metre preserves evidence for the stage at which the *-nn* sandhi applied only where historically justified.[21]

The clearest account of the patterns found is by Arnold (1905: 141–142), who says:

> Syllables are long by position if the vowel is followed … by *-nn* in the third person plural of secondary tenses and in the nominative singular of participles, but rarely elsewhere.
>
> p. 141

> Amongst forms in which *-nn* represents a double consonant are to be reckoned aorists like *átann* vi 61 9c (2D): the nominative singular in *-ṅṅ* as *sadṛ́ṅṅ* has the same value, for instance in i 94 7a (10). Locatives such as *ájman, tásmin* must everywhere be restored for *-ann, -inn* of the text, and vocatives in *-an, -in* usually for those in *-ann, -inn*: but the final syllable of *maghavann* is always long by position, and *pū́ṣann, rā́jann, vṛ́ṣann* and *sahasāvann* usually, for instance in ii 28 9b (4A), iii 32 1c (8), vii 4 6c (8).
>
> p. 142

As these quotes evidence, the situation is not quite so clear cut as we might hope, particularly with vocatives of *n*-stems. But I am aware of no more detailed assessment of the sandhi of final *-n* in the *Ṛgveda* than those already cited; even van Nooten and Holland (1994: vi) say only that *-nn* sometimes has to be read as *-n*, without even specifying the main patterns.[22]

(both of pronominal locatives, and of vocatives of *in*-stems) is best read as *-i*, with elision of the *-n*; see Arnold (1905: 143–144).

20 For early recognitions and discussions of this see Benfey (1839: 675), Kuhn (1863), Bartholomae (1888: 514 ff.), Wackernagel (1896: I:329–330).

21 As shown by Jamison (2019), the *Ṛgveda* saṃitā occasionally even preserves an original **-nt* as *-nt* or *-nd*, which could survive uncorrected due to misparsing.

22 Likewise Macdonell (1916); the facts are briefly presented also by Gunkel and Ryan (2018: 39–40).

3.1 *Data*

In order to gain a more detailed understanding of the sandhi of final *-n* in the *Ṛgveda*, I analysed all instances of word final *-nn* and *-ṅṅ* in the saṃhitā. For every form, I compared the frequency of the crucial syllable reading light with the frequency of it reading heavy, based on the figures given by van Nooten and Holland (1994: xvii). Metrical studies of this type often focus on cadences alone; for example, Beguš (2015) focused only on trimeter cadences, on the grounds that it is only in trimeter cadences that the metre is sufficiently regular to be probative. But that applies to whole cadences (or openings, or breaks), and not to specific syllables in a particular context. If we look at specific syllables in specific contexts, we find a few other positions which may be metrically significant. In the opening of a Jagatī pāda, for example, the pattern XLLH is found only 78 times, whereas the pattern XHLH is found 2317 times. Given a light third and a heavy fourth syllable, then, the second syllable of a Jagatī pāda is heavy in 96.4% of instances. We find a similar pattern if we compare the patterns XLLL and XLHL in the opening of Gāyatrī pādas (17 vs. 435 instances): given light second and fourth syllables, the third syllable is heavy in 96.24% of instances. I therefore considered final *-nn* in all possible positions, calculating the likelihood of a heavy or light syllable in the manner illustrated.

In total there were 770 instances of final *-nn* or *-ṅṅ* in the *Ṛgveda*. Some data had to be excluded from consideration, however. As van Nooten and Holland (1994) give figures only for the four main metres (Anuṣṭubh, Gāyatrī, Jagatī, Triṣṭubh), any occurrences in metres which do not conform to these standard dimeter or trimeter patterns, for example pādas of 16 or 4 syllables, had to be excluded.[23] In addition, the first and last syllable of every pāda is anceps, i.e. its metrical weight is not significant, thus we can say nothing about final *-n* in these positions.

Removing this data leaves 660 instances.[24] These forms can be subcategorized morphologically:

(6) – Finite verb forms: 136
– Nom./voc. of participles / other *nt*-stems: 266

23 As noted by van Nooten and Holland (1994: vii), 8 and 12 syllable pādas in mixed metres, such as Bṛhatī or Uṣṇih, conform to the same patterns as Gāyatrī and Jagatī pādas. Therefore I excluded only data from pādas which could not be treated as showing the same patterns as one of the four main metres.

24 These are distributed representatively across the ten books: 154 from book 1, 18 from book 2, 43 from book 3, 42 from book 4, 38 from book 5, 44 from book 6, 56 from book 7, 70 from book 8, 74 from book 9, and 121 from book 10.

- Nom. of *añc*-stems (-*ṅṅ*): 10
- Nom. of -*dṛś* compounds (-*ṅṅ*): 4
- Voc. of stems in -*van* or -*vant* (excl. *maghavan*): 12
- Voc. of *maghavan*: 40
- Voc of stems in -*an*: 52
- Voc of stems in -*in*: 25
- Stems ending in -*an* in compound: 2
- Loc of stems in -*an*: 63
- Pronominal locatives in -*smin*: 49
- The word *hiṅ*: 1

For the first four or five categories in (6), the 'original' form would have involved a final consonant cluster, and hence we expect a heavy syllable before a following vowel.[25] The remaining categories (with the exception of *hiṅ*, which has no etymology) all represent a final plain -*n* so should in principle give a light syllable before a following vowel.

The first four categories, where a heavy syllable is uncontroversially expected, make up almost two-thirds of the data, 416 of the 660 tokens.[26] Moreover, there is also a bias towards heavy syllables in terms of the positions in which final -*nn* is found. Five hundred and fifty six of the 660 tokens occur in positions where a heavy syllable is favoured, however slightly; of these, 237 are in positions which show a heavy syllable in at least 90 % of instances, and of these 160 are in positions which show a heavy syllable in at least 95 % of instances. In contrast only 59 tokens occur in positions which show a light syllable in more than 90 % of instances. Overall, this reflects the bias of Ṛgvedic verse towards heavy syllables.

Excluding the problematic vocatives of -*van* and -*vant* stems (including *maghavan*), and excluding forms ending in -*ṅṅ*, since these show no variation, figures 4.1 and 4.2 show the relative distribution of finite and participial forms ending in -*nn* as contrasted with vocatives/locatives in -*an*/-*in*, depending on the likelihood of a light syllable in the given position. Thus 1 on the x-axes represents the positions which are most dependably heavy; those for which the likelihood of a light syllable is 1% or less. Correspondingly, 100 on the

25 The vocative of stems in -*van* or -*vant* were treated together, and separately from other categories, because they show variation and appear to influence each other; they will be discussed below.

26 A fact which may help explain the successful generalization of the heavy sandhi to light contexts.

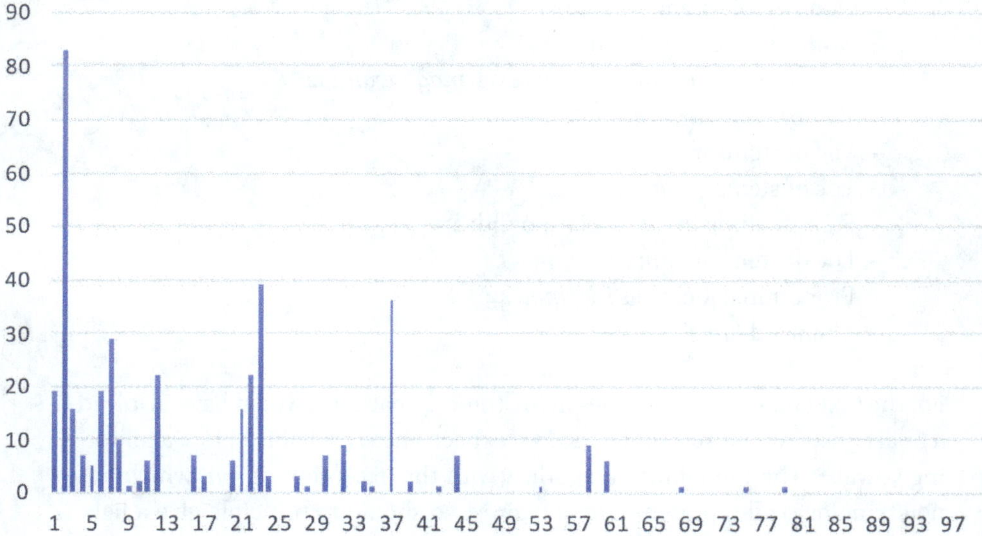

FIGURE 4.1 Distribution of finite/participial forms

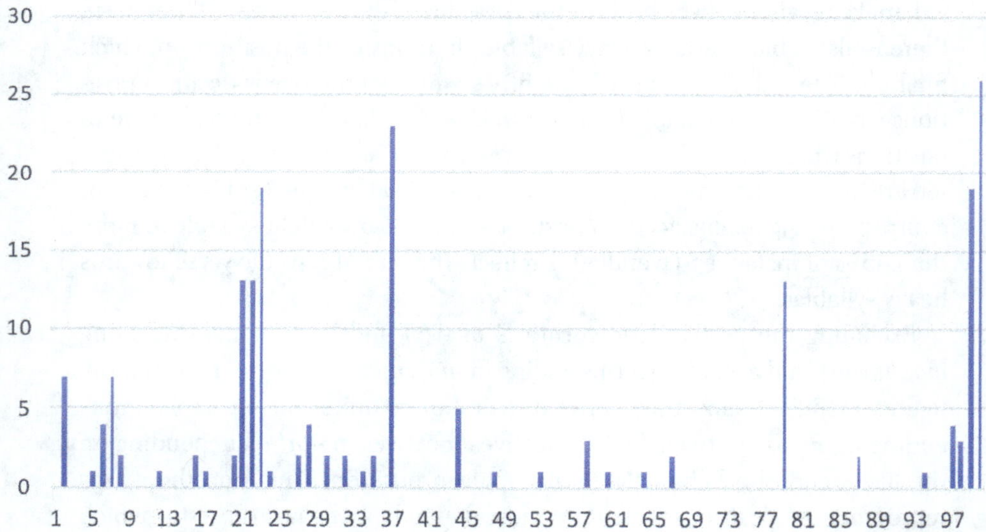

FIGURE 4.2 Distribution of vocative/locative forms

x-axes represents the positions which are most dependably light, i.e. where the likelihood of a light syllable is 99% or more.

It can be seen that, while there is a general tendency for all forms to occur in positions which are around 20–40% likely to be metrically light, reflecting the fact that this is the most common pattern for syllables overall across all Ṛgvedic metres, there is still a clear difference between these two sets of forms in contexts which are below about 15 % likely to be light or above about 75 % likely to be light. As expected, finite and participial forms are much more common in positions which are 85 % or more likely to be heavy, but are never found in positions which are more than 80 % likely to be light, while vocatives/locatives in *-an*/*-in* are common in positions which are above 75 % light, and particularly in positions which are 97 % or more likely to be light.

While finite and participial forms, which historically ended in a consonant cluster, are never found in positions where a light syllable is strongly favoured, the vocatives/locatives, which historically ended in a single consonant, are found in positions where a heavy syllable is strongly favoured. This conforms to the observations of Arnold (1905), given above, who claims that it is vocatives in *-an*/*-in* which must sometimes be read as heavy.

Contrary to Arnold, however, there are a few unambiguous instances of locatives in positions which strongly favour a heavy reading: at 1.114.1c the locative *asmín* occurs in the first syllable of a jagatī cadence, a position which is light in less than 2 % of instances; at 9.86.18c the locative *áhan* likewise occurs in the first position of a jagatī cadence; at 10.6.1a the locative *śárman* occurs in the first syllable of a triṣṭubh cadence, a position which is likewise light in less than 2 % of instances; at 10.115.2a the locative *dán* (from *dám-*) occurs again in the first syllable of a jagatī cadence. Three of these forms are or could be arguable in different ways: the verse 10.6.1a has only ten syllables and is labelled a 'pentad verse' by van Nooten and Holland (1994: 651), so the cadence may not be expected to show the standard triṣṭubh patterning; the form *dán* does represent the locative of *dám-* 'house', but it is phonologically problematic and may in fact represent the inherited genitive **dáns* > *dán* preserved in the phrase *pátir dán*; finally, the locative *áhan* 'on the day' is homophonous with the imperfect of *han* 'slay', so its sandhi could have been influenced by the identical verb form. But even if these three forms can be explained away, the instance of *asmín* at 1.114.1c cannot.

It is also notable that these four forms are all outside the family books. In fact there are no instances of locatives in *-an* or *-smin* in the family books in any position which favours heavy syllables in more than 80 % of instances. As a one-in-five likelihood of being light is by no means statistically unlikely, we can say that the light scansion of locatives in *-an*/*-smin* is without exception in the family books.

Only three vocatives in *-an* or *-in* are found in positions which significantly favour heavy syllables: *sahasin* at 4.11.1a, *śavasin* at 7.28.2b, and *viśvakarman* at 10.81.5b are all in positions which are light in less than 2 % of instances. Vocatives also occur in positions which are only 5–8 % light at: 8.2.17 (5 % light); 1.100.6, 1.181.7, and 10.4.1 (6 % light); 1.42.5,7,8,9, 6.54.6, 8.33.12 and 9.7.1 (7 %

light); 8.2.28 and 8.79.9 (8 % light).[27] These vocatives in 5–8 % positions are all disyllabic pāda-initial vocatives in dimeter verse. It may be that this initial position licensed an emphasis on both syllables of the vocative which encouraged a heavy sandhi of the final syllable.[28] But it is worth bearing in mind that these positions are not as strongly probative as the 2 % positions: between one in every ten or twenty syllables in these positions is uncontroversially light anyway, so it may be that it is not necessary to read all or even any of these vocatives as having a second heavy syllable.[29]

As in the quote from Arnold (1905) above, it is standard doctrine that the vocative *maghavan* always scans with heavy final syllable, despite being unambiguously an *n*-stem.[30] This vocative occurs really only in two positions; ×23 in a three-syllable trimeter break, which shows the third syllable as light in 23 % of instances, and ×13 straddling a two-syllable trimeter break and the first syllable of the cadence, a position where the crucial syllable would be expected to be light in only 2 % of instances. There is one other instance, in the opening of a dimeter pāda (10 % light). So while *maghavan* should certainly be scanned with heavy final syllable in 13 of its occurrences, the other 24 instances are by no means definite, and it is certainly possible that all, or at least some, of the instances in the three-syllable breaks were originally scanned light, as historically expected. So *maghavan* does fit, with its historically expected scansion, into a trimeter line with a three-syllable break, but it must be scanned with final heavy syllable only when the caesura falls after the fifth syllable.

The other stems in *-van* or *-vant* vary. There are three vocatives to uncontroversial stems in *-van-*: *satrādāvan-* at 1.7.6b, *somapāvan-* at 1.55.7a, and *aśva-dāvan-* at 5.18.3d. The final syllables of the latter two occur in strongly light

27 Besides these vocatives, the stem form of *āsán-* 'mouth' occurs a position which is 6 % light at 1.84.16c.

28 Recently, Keydana (2018: 235–236) has discussed this *-nn* sandhi (which he treats as 'lengthening') of final *-n* with disyllabic pada-initial vocatives and tentatively suggested that it may be due to prosodic factors. However he mistakenly took the *-nn* of the saṃhitā as necessarily reflecting a heavy scansion, without considering whether the metre requires a heavy scansion. Thus for example he mentions 14 instances of lengthened *rā́jann*, even though only one of these occurs in a position which strongly suggests a heavy syllable (8.79.9c, second syllable in an Anuṣṭubh line, 8 % light).

29 The forms in question include the vocatives *pū́ṣan*, *rā́jan*, and *vŕ̥ṣan* noted by Arnold. In statistical work it is standard to consider 5 % probability the cut-off for excluding the likelihood of chance: only if a particular distribution has less than 5 % chance of being random ($p<0.05$), can that distribution be considered statistically significant. If we follow this standard, then the 5–8 % light positions can only be considered suggestive, but not probative, of heavy scansion.

30 On the etymology of this word, see Meißner (1993).

positions (>99 % and 97 % respectively), while the latter is in an ambiguous 22 % light position. The vocative *śatāvan* to the *vant*-stem *śatāvant-* occurs in a trimeter break at 6.47.9b, a position which is 68 % light, so almost 1/3 heavy, the expected heavy scansion being therefore unproblematic but not provable. The vocative *sahasāvan* is the most problematic, occurring ×5 in 2 % light positions, ×1 27 %, ×1 34 %, but also ×1 in a position which is >99 % light. The stem for this vocative is traditionally listed as *sahasāvant-* (e.g. in Monier-Williams 1899), but is better analysed as a *van*-stem (so listed by Lubotsky 1997: 1509), originally a root noun compound *sahasā-van-* 'winning with might'.[31] As a *van*-stem the majority occurrence in 2 % light contexts is notable, and suggests that, like *maghavan* (historically also a root noun compound to the same root), the stem has been partly reanalysed as a *vant*-stem. But since one instance must be read as light, the treatment of the form is not consistent.

3.2 *The Sandhi of Final* -n *and* han

The data presented in the preceding section confirms in more detail what was already known: the *Ṛgveda* saṃhitā preserves with few exceptions the historical sandhi distinction between words ending in an original consonant cluster which reduced to *-n*, and words ending in an original plain *-n*. What relevance could this have, if any, for the parigraha of *han*?

As we have seen above, the primary uses of itikaraṇa in the RV padapāṭha are to disambiguate forms with ambiguous sandhi: to mark pragṛhya vowels in distinction from non-pragṛhya vowels, and to mark visarjanīya from original *-r* from visarjanīya from original *-s*. It could therefore have been possible for Śākalya to use itikaraṇa also to disambiguate the two variants of final *-n*. Yet this does not appear to have been the case, presumably because the *-nn* sandhi had already been generalized to final *-n* of all origins by the period of the composition of the padapāṭha. But for the sake of argument, let us push the question to see if there is any possibilty at all of accounting for the parigraha of *han* this way. One could hypothesize that, if the variable sandhi of *-n* was still current in Śākalya's time, then the reason why he did not use itikaraṇa to disambiguate this was because disambiguation was almost never necessary: there are almost no forms which are ambiguous between a form for which final *-n* is expected prevocalically after a short vowel and a form for which final *-nn* is expected in the same context. This contrasts with the case of pragṛhya vowels, where many forms are technically ambiguous (e.g. thematic adjectives, with

31 The form is attested only in the *Ṛgveda* and only in the vocative, so there are no unambiguous forms.

non-pragṛhya masculine locative singular in *-e* and pragṛhya feminine/neuter nominative/accusative dual in *-e*). Forms ending in visarjanīya are rarely ambiguous, but there is one iconic case, namely *káḥ*, which is usually nom.sg.m. of the interrogative pronoun *ká-*, but may also be 2/3sg. root aorist injunctive of *kṛ*.[32]

If this hypothesis held, then we might suppose that Śākalya felt no need to disambiguate the variable sandhi of final *-n*, except in specific cases where such an ambiguity existed. Now our form, unaccented *han*, is one of a very few forms in final *-n* in the RV which could be in principle ambiguous. There is no uncompounded root noun *hán-* attested in the RV, but uncompounded root nouns do exist, and a root noun *hán-* cannot be ruled out in principle.[33] If such a noun did exist, its vocative would be *han*, identical with the unaccented 2/3sg. present injunctive *han* which is marked with itikaraṇa. Moreover, while independent *hán-* is unattested, as the second member of a compound the root noun *hán-* is common, as in the highly frequent *vṛtrahán-* (×39 in the vocative alone); and, naturally, the padapāṭha divides these compounds, marking *han-* as a separate element with avagraha. Now, two of the three instances of *han* marked with parigraha have the object *dānavaṃ* immediately preceding, resulting in a collocation which could conceivably be in danger of misinterpretation as a compound, and thus encouraging the disambiguation of *han* in these cases by means of parigraha.

However, this already tenuous line of argument faces a fatal problem: there is at least one case of genuine ambiguity in the *Ṛgveda*, where both forms are attested, yet neither is marked with itikaraṇa. The ambiguity in question is the form *áhan*, which may be locative of *áhan-* 'day' (attested ×16), or 2/3sg. imperfect of the verb *han* (×28). If Śākalya had been aware of the different sandhi possibilities for final *-n* in the RV, and had chosen to disambiguate unclear cases with itikaraṇa, *áhan* would have been the most important form to disambiguate.[34]

---

32 Although the latter appears only ×3 in the RV with accent (×25 unaccented), the root is one of the most basic roots in the Sanskrit lexicon, and its potential for confusion with the highly frequent interrogative pronoun may have exerted more influence than its actual occurrence warranted. The *Rkprātiśākhya* devotes four rules to correctly distinguishing the two words.

33 Scarlata (1999: 680–697) discusses in detail the attested compounds with *hán-* and emphasizes the lack of *hán-* as an independent element. Following Insler (apud Tucker 2009: 514), the inherited uncompounded agent root noun was replaced in Vedic by the agent noun in *-́tṛ-*, such that the earlier existence and frequency of an independent *hán-* can be inferred from the common agent noun *hántṛ-*.

34 Note that, as mentioned above, one instance of locative *áhan* occurs in a strongly heavy

We must therefore conclude that, although there is a theoretical ambiguity in the sandhi of *han* which could conceivably have been indicated by itikaraṇa, this does not appear a plausible explanation for the itikaraṇa of *han*. We must therefore look for a different explanation.

4 Return to iti-karaṇa

As we have seen, *han* is one of a number of forms which are marked with itikaraṇa in the RV pp. but which do not conform to any of the main categories for which itikaraṇa is employed. The forms discussed above all involved contextual ambiguity or problematic sandhi, where the function of itikaraṇa was to safeguard against misreading or misinterpretation.

Besides *han*, a few forms remain unaccounted for, with no explanation offered in prior literature. Before turning to *han*, let us reconsider these.

4.1 *ṣṭaḥ*

The form *ṣṭaḥ* is marked with parigraha in verse-final position at 1.61.8d. The form is the ruki form (following preceding *pári*) of *staḥ*, 3du. present of *as* 'be'; this is the only 3du. form of this root found in the *Ṛgveda*, beside the phonologically similar and very common 2du. *stháḥ* (×23) and 2pl. *sthá* (×30). It seems most reasonable to attribute the parigraha here to the isolated status of this 3du., and a concern that it could be mistaken for, and replaced by, the commonly encountered 2du. *stháḥ*.[35] There is no reason to attribute its marking to the ruki of the initial cluster (which is undone in the pp.), as suggested by Jha (1975: 53), since this is regularly found, but not marked, with forms of *stháḥ* and *sthá* in the relevant contexts, including in the same verse-final position at 7.43.4d. Thus we have a unique and potentially mistakable form in verse final

---

position, in the cadence of a jagatī pāda (<2% light) at 9.86.18c, being therefore in a sense the instance most in need of disambiguation. There are two other potentially ambiguous forms in the RV. The form *rā́jan* is attested as the vocative of *rā́jan-* 'king'; the form could also be 3pl. injunctive of the verb *rāj*, but this is not found in the RV. The form *dán* functions as the locative of *dám-* 'house', but the same form could theoretically also be nom.sg. of *dánt-* 'tooth', which is not attested in the RV. The latter equation is problematic because, as discussed above, *dán* is not a regular locative for *dám-*, and may represent the original genitive which would go back to a form with a final consonant cluster just like the nom.sg. of *dánt-*.

35 Perhaps, indeed, Śākalya was aware of existing traditions of recitation in which this alteration had already been made.

position, protected by parigraha in the padapāṭha from replacement by a lectio facilior.

4.2 *te*

The parigraha of *te* 'to/for you' at 3.33.10d is perhaps intended to avoid the misinterpretation of the form as part of the preceding word:

(7) Saṃhitā: *máryāyeva kanyā̀ śaśvacaí te* ||
Pada: *máryāya-iva* | *kanyā̀* | *śaśvacaí* | *ta íti te* (3.33.10d)
Transl.: 'I will bend to you like a maiden to her cavalier' (Jamison and Brereton 2014: 514)

Kümmel (2000: 535–536) analyses the form *śaśvacaí* as a thematized perfect subjunctive. The form is 1sg., but the sequence **śaśvacaíte* would be the expected corresponding 3du.[36] This hymn contains several dual verb forms, being addressed to the two rivers Vipāś and Śutudrī; therefore, a 3du. verb form would not have seemed inappropriate, and it is easy to imagine such an error creeping into the recitation. Thus again, we have a case of potential misreading at a verse end, the correct interpretation of which Śākalya reinforces with the use of parigraha.

4.3 *gdha*

The form *gdha* at 1.158.5d is 3sg. middle aorist injunctive of *ghas* 'eat'. Like the preceding two forms, and like *han* when marked, this form is unaccented, and appears at the end of a verse. The verb form *gdha* is phonologically unique: it is the only word in the *Ṛgveda* which begins with two stops.[37] In general, word forms beginning with two stops are extremely rare in Sanskrit, and may have been difficult to pronounce correctly in isolation (just as it is for native speakers of English). In the saṃhitā there is no problem with the form: it follows its preverb, *ápi*, and so the consonant sequence is intervocalic. But it was likely rather unnatural to pronounce a word like *gdha* in isolation, as it had to be in the padapāṭha, risking an incorrect pronunciation such as *gədha*/*gadha*. The parigraha in the padapāṭha can therefore be interpreted as protecting the form of the word by ensuring that it is recited both after a pause *and* in a perfectly pronounceable position, i.e. following the *íti*.

---

36 The perfect middle subjunctive is rare (Kümmel 2000: 40–41), and no such 3du. is attested to any root, but its form can be easily predicted.

37 Excepting the few forms which begin with *cch-*, a sequence which in fact represents an affricate, i.e. a single phoneme with complex articulation (Lipp 2009: 173ff.).

4.4 *vark*

A similar explanation likely holds for *vark*, 2/3sg. root aorist injunctive of *vṛj* 'twist'. This form is also phonologically unique: it is one of the very few Sanskrit words ending in a double consonant, and the only one in the RV ending in -*rk*.[38]

The form *vark* occurs four times in the RV, at 1.63.7c (the only accented instance), 6.26.3b, 8.75.12b, and 10.8.9d. It is marked with parigraha only in the latter case, which is the only occurrence at the end of a verse. As with *ṣṭaḥ*, *te*, and *gdha*, then, we are dealing with a monosyllabic, unaccented word in verse-final position.[39] The instance of *vark* marked with parigraha is also the only 3sg. form; but it seems unlikely that third vs. second person has anything to do with the parigraha: there is no obvious reason why 3sg. *vark* < **vark-t* should be more unstable or liable to misreading than 2sg. *vark* < **vark-s*. The position at the end of the verse is more likely significant. In this final position, the pause following the verse end in the saṃhitā gives no protection to the final -*k*; conceivably, Śākalya employed itikaraṇa here to provide unambiguous protection for the consonant in the padapāṭha (in the first iteration preceding the *íti*).[40]

---

38 The few other attested forms are listed in Wackernagel (1896: I:304–305). Only one other ends in -*rk*: *ū́rk*, nom.sg. of *ū́rj*-, attested in the Vājasaneyī saṃhitā. The rare retention of final consonant clusters is itself interesting and somewhat irregular: while most retentions, including *vark* and one of the other RV forms, (*pary*)*ā́vart* (3sg.aor. from *vṛt* 'turn'), involve a final consonant which is part of the root (and thus conceivably more likely than a suffix to be retained), root final consonants in word final consonant clusters are more usually lost, as in the form *avarīvaḥ*, intensive of *vṛt*, RV 10.129.1c. And the third RV form preserving a final consonant cluster is *dart*, 2/3sg.aor. from *dṝ* 'split', where the final consonant is a suffix. (The only other form not mentioned here which Wackernagel notes as having a final cluster is the form *suhā́rt*, attested in the Atharvaveda.) Retention of final clusters is therefore irregular and sporadic, but our text may conceal a more regular retention of final -*rC* at the period of composition: Wackernagel (1896: I:304–305) notes that final -*ar* < *-*art* never scans light in the RV, and Oldenberg proposed to read -*arr* throughout.

39 As also *syuḥ*, discussed above.

40 The forms ending in -*rt* are never marked with itikaraṇa in the RV padapāṭha. The two occurrences of *vart* (*ā́vart* at 7.59.4c and *paryā́vart* at 10.124.4d) are both in verse-internal position with the -*t* protected in the saṃhitā by the following consonant. Both are 3sg. The form *dárt* is more problematic. At 6.20.10c and 6.27.5d it is clearly 3sg., accented, and in the latter case ends the verse. At 6.20.10c it is verse internal with the consonant protected in the saṃhitā. At 1.174.2b it is best interpreted as 2sg., the -*t* resulting from a redactional alteration of expected *dár*(*r*) on the basis of the identical pāda at 6.20.10c (cf. Oldenberg 1901: 329—who, however, took the form at 6.20.10c as 2sg. also, and the alteration based on the form at 6.27.5d). But 2/3sg. forms of this root without final -*t* are attested: *ā́daḥ* (2sg.) occurs once, and the intensive occurs ×8, *dárdar*/*dardar* (2sg. ×4, 3sg. ×1), *ádardar*/*adardar* (2sg. ×1, 3sg. ×2).

4.5 *han*

Finally, we return to *han*. Parigraha of *han* is found only when the form is unaccented and occurs at the end of a verse; this is the same as with *vark*, and as also found with *gdha* and *ṣṭaḥ* (which never occur in other contexts). We have seen that these forms can all be explained due to their uniqueness, one way or another, and their potential for misreading/misrecitation. While *han* is not a phonologically isolated form, it is yet another unaccented monosyllable marked in verse-final position.

Crosslinguistically, phrase or utterance-final position is a common context for processes of phonetic reduction or loss. The association of unexpected parigraha in the padapāṭha with verse-final position suggests that in the contemporary RV recitation tradition too this position was liable to weaker or less distinct articulation, since it is only in this position that a number of potentially problematic words—mostly unaccented words, moreover—are marked. Thus the form *han* must have been liable to a reduced or weaker pronunciation in this position. The parigraha of *han* is therefore most likely to be a means of protecting the form in this context. In this case, there is also a highly frequent word with which a reduced form of *han* could easily be confused: the particle *ha*, which occurs ×182 in the RV. This particle can occur at the end of a pāda (1.116.14c, 4.23.3c), and though it does not occur in verse-final position in the RV, there is no reason in principle why it might not have done. Thus the parigraha of *han* was Śākalya's means of ensuring the correct transmission of verse-final *han*, in the face of an apparent tendency to weakened or reduced pronunciation of short, unaccented words in that position. Once again, it is possible, though of course unprovable, that Śākalya was even aware of contemporary recitations of the RV which treated *han* in these cases as *ha*.

5 Conclusion

In this paper I have offered a number of explanations for the unexpected parigraha of *han* and other words in the RV padapāṭha. As the context and history of composition of the padapāṭha is now lost to us, all the explanations put forward in this paper must necessarily be considered tentative. But I have worked on the principle that we should seek first a principled rationale behind the markings we find, even where these appear 'irregular', building on the recognition that the composition of the padapāṭha was the product of careful, and highly accurate (though not perfect), linguistic analysis. I have also brought into consideration the fact that the padapāṭha was the product of a perhaps lengthy process of development, meaning that we may find evidence for earlier strata in its composition, and the fact that the padapāṭha must have been composed in a context of varying and competing recitations of the *Ṛgveda*, meaning that the marking of the padapāṭha might conceivably be addressing specific controversies of the recitation to which we no longer have access.

Alongside this, I have provided more complete details on the sandhi of final *-n* in the *Ṛgveda* than has ever previously been done. While this largely confirms previous scholarship, and does not, ultimately, help us to account for the parigraha of *han*, I have shown that some claims of previous scholarship are not necessarily valid, such as the claim that *maghavan* always scans with heavy final syllable.

The RV padapāṭha is more than just a useful aid to reading the *Ṛgveda*, more than just an ancillary edition beside the more fundamental saṃhitāpāṭha. It is an accomplished work of scientific analysis, the earliest significant evidence for, and product of, scientific thought in ancient India. It would undoubtedly repay more extensive, and more detailed, investigation from this perspective; I offer this paper to our honorand as a first tentative step in that direction.

Acknowledgements

I have had the rare privilege of participating in the same class with Chris Minkowski both as a student and, subsequently, as co-teacher. This paper originates from our reading of RV 5.29 in June 2020, and our wondering about the parigraha of *han* at the end of 5.29.4d; though I cannot provide any definite answers, I hope my investigations prove of interest to our honourand. I have learnt so much from Chris about so many things, from understanding the Vedas and Vedic ritual to navigating the modern world of academia, and I offer this paper merely as a small acknowledgement of my inability to repay the debt. I am very grateful to Christopher Haddad, who provided considerable assistance with analysing the data on *-nn*; I also thank Adriana Molina-Muñoz, who extracted the data from the digitized *Ṛgveda* into a useable format. I am also grateful to Stephanie Jamison and to Elizabeth Tucker, who discussed a number of the topics in this paper with me, and to the editors, in particular Christopher Fleming, for inviting me to contribute to this volume. All errors are, or course, my own.

References

Arnold, Edward Vernon (1905). Vedic Metre in its Historical Development. Cambridge: Cambridge University Press.

Bartholomae, Christian (1888). 'Die arische flexion der adjektiva und partizipia auf nt-'. Zeitschrift für vergleichende Sprachforschung 29, pp. 487–586.

Beguš, Gašper (2015). 'A new rule in Vedic metrics'. Journal of the American Oriental Society 135(3), pp. 541–550.

Benfey, Theodor (1839). 'Review of Anthologia sanscritica glossario instructa by Christian Lassen'. Göttingische gelehrte Anzeigen 1839(1), pp. 665–680.

Gippert, Jost (1997). 'Laryngeals and Vedic metre'. In Alexander Lubotsky (ed.), Sound Law and Analogy: Papers in Honor of Robert S.P. Beekes on the Occasion of his 60th Birthday, Amsterdam: Rodopi, pp. 63–79.

Gunkel, Dieter and Kevin M. Ryan (2018). 'Phonological evidence for pāda cohesion in Rigvedic versification'. In Dieter Gunkel and Olav Hackstein (eds.), Language and meter, Leiden: Brill, pp. 34–53.

Jamison, Stephanie W. (2019). 'Hidden in plain sight: Some older verb endings in the Rig Veda'. In Adam A. Catt, Ronald I. Kim, and Brent Vine (eds.), QAZZU warrai: Anatolian and Indo-European studies in honour of Kazuhiko Yoshida, Ann Arbor: Beech Stave Press, pp. 123–129.

Jamison, Stephanie W. and Joel P. Brereton (2014). The Rigveda: The Earliest Religious Poetry of India. An English Translation. Oxford/New York: Oxford University Press.

Jasanoff, Jay H. (1994). 'Aspects of the internal history of the PIE verbal system'. In George E. Dunkel (ed.), Früh-, Mittel-, Spätindogermanisch: Akten der IX. Fachtagung der indogermanischen Gesellschaft vom 5. bis 9. Oktober 1992 in Zürich, Wiesbaden: Reichert, pp. 149–168.

Jha, V.N. (1975). 'iti-karaṇa in the Ṛgveda-Padapāṭha'. Bulletin of the Deccan College Post-Graduate and Research Institute 35(1/2), pp. 49–54.

Jha, V.N. (1992). A linguistic analysis of the Ṛgveda-padapāṭha. Delhi: Sri Satguru Publications.

Keydana, Götz (2018). 'Word-final lengthening in early Vedic'. Journal of South Asian Languages and Linguistics 5(2), pp. 211–240.

Kuhn, Adalbert (1863). 'Sprachliche resultate aus der vedischen metrik'. Beiträge zur vergleichenden Sprachforschung auf dem Gebiete der arischen, celtischen und slawischen Sprachen 3(1 & 4), pp. 113–125, 450–477.

Kümmel, Martin Joachim (2000). Das Perfekt im Indoiranischen: eine Untersuchung der Form und Funktion einer ererbten Kategorie des Verbums und ihrer Weiterentwicklung in den altindoiranischen Sprachen. Wiesbaden: Reichert.

Kümmel, Martin Joachim (2014). 'Laryngeal traces without laryngeals in Vedic metre'. In Jared S. Klein and Elizabeth Tucker (eds.), Vedic and Sanskrit Historical Linguistics: Papers from the 13th World Sanskrit Conference, New Delhi: Motilal Banarsidass, pp. 23–44.

Lipp, Reiner (2009). Die indogermanischen und einzelsprachlichen Palatale im Indoiranischen. Band I: Neurekonstruktion, Nuristan-sprachen, Genese der indoarischen Retroflexe, Indoarisch von Mitanni. Heidelberg: Winter.

Lubotsky, Alexander (1997). A R̥gvedic Word Concordance (2 vol.), volume 82–83 of American Oriental Series. New Haven, Conn.: American Oriental Society.

Macdonell, Arthur Anthony (1916). A Vedic Grammar for Students. Oxford: Oxford University Press.

Meißner, Torsten (1993). 'Zur Etymologie von altindisch maghávan-'. Historische Sprachforschung 106, pp. 42–48.

Monier-Williams, Sir Monier (1899). A Sanskrit-English Dictionary: Etymologically and Philologically Arranged with Special Reference to Cognate Indo-European Languages. New edition, greatly enlarged and improved with the collaboration of E. Leumann, C. Cappeller and other scholars. Oxford: Clarendon Press.

Oldenberg, Hermann (1901). 'R̥V VI, 1–20'. Zeitschrift der deutschen morgenländischen Gesellschaft 55(2), pp. 267–330.

Rastogi, Moti Lal (1970). 'Itikaraṇa in the Rk-Padapāṭha'. Sri Venkateswara University Oriental Journal 13, pp. 1–10.

Scarlata, Salvatore (1999). Die Wurzelkomposita im R̥gveda. Wiesbaden: Reichert.

Tucker, Elizabeth (2009). 'Old Iranian Superlatives in -išta-'. In Werner Sundermann, Almut Hintze, and François de Blois (eds.), Exegisti monumenta: Festschrift in Honour of Nicholas Sims-Williams, Wiesbaden: Harrassowitz, pp. 509–526.

van Nooten, Barend A. and Gary B. Holland (1994). Rig Veda: A Metrically Restored Text. Cambridge, Mass.: Harvard University Press.

Wackernagel, Jacob (1896). Altindische Grammatik. Band I: Lautlehre. Göttingen: Vandenhoeck & Ruprecht.

PART 2

Astral and Medical Sciences

CHAPTER 5

A *Meru* as a Mathematical Tool

Takanori Kusuba

In his *Gaṇitakaumudī* (*Moonlight of Mathematics*) Nārāyaṇa poses a mathematical question concerning the number of cows, which is a reminder of the Fibonacci sequence. The topic is included in the procedure of mathematical series (*śreḍhīvyavahāra*). In his own commentary he said that the question could also be solved by using a figure named *meru*, but he does not explain how. In this article I investigate how he applied the *meru*.

1 Number of Cows: Rule and Example

The *Gaṇitakaumudī* is a complete Sanskrit mathematical work written by Nārāyaṇa Paṇḍita in 1356 CE.[1] The book illustrates the development of Indian mathematics after Bhāskara II and Nārāyaṇa discusses mathematical rules that seemed peculiar to him. A rule with an example on the number of cows in a mathematical series is one of those rules. Example:

> prativarṣaṃ gauḥ sūte varṣatritayena tarṇakī tasyāḥ/
> vidvan viṃśativarṣair gor ekasyāś ca santatiṃ kathaya//[2]
>
> A cow gives birth to a calf every year, and the calf becomes mature in three years. Tell [me], wise man, how many offspring will one cow have in twenty years?

I illustrate this example across eight years. In Figure 5.1, A is a cow that delivers a calf, B, every year. By the eighth year there are eight calves. The first calf (B_1) delivers a third-generation calf (C_{11}) in the fourth year. So B_1 delivers five third-

1 The printed edition of the *Gaṇitakaumudī* (GK) is based on a manuscript preserved by Sudhakara Dvivedi. His son Padmakara discovered the manuscript in his father's collection after the latter's death and published it in 1936 and in 1942—referred to hereafter as GKP1 and GKP2 respectively. Kusuba published a critical edition with an English translation and notes on the GK's last two chapters.

2 GKP1. p. 126.

 | DOI:10.1163/9789004536869_007

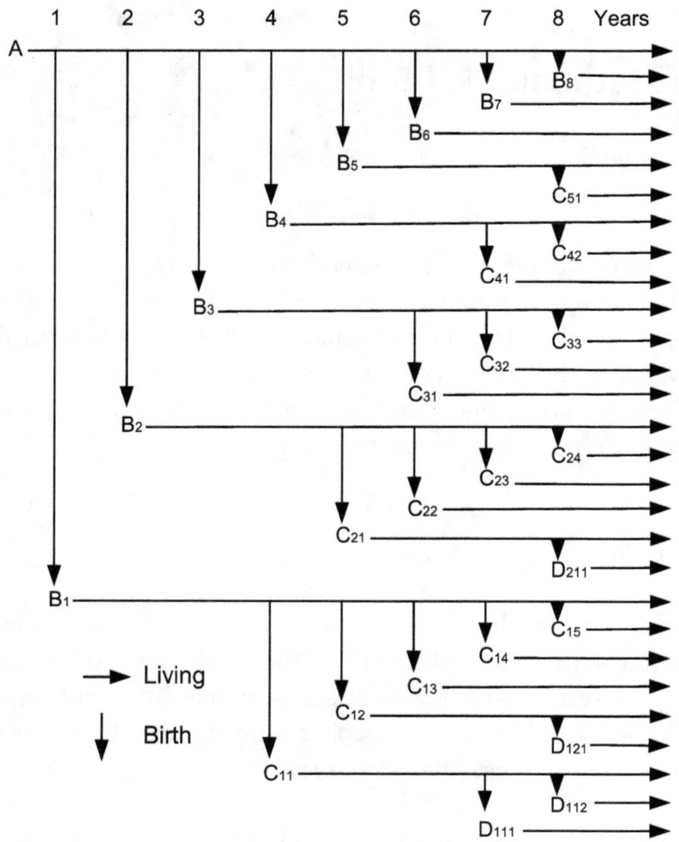

FIGURE 5.1 Family tree of cows

generation cows (C_{11} - C_{15}). Likewise, four cows (C_{21} - C_{24}) are born from B_2. Therefore, the total number of third-generation cows is 5 + 4 + 3 + 2 + 1 = 15. C_{11} delivers two fourth-generation cows (D_{111} - D_{112}) in the seventh and eighth years; C_{12} and C_{21} each bear a calf in the eighth year; four fourth-generation calves are born in the eighth year. Nārāyaṇa describes a rule for this type of example:

> abdās tarṇyabdonāḥ pṛthak pṛthag yāvad alpatāṃ yānti/
> tāni kramaśaś caikādikavārāṇāṃ padāni syuḥ//21//
> ⟨nija⟩nijavāraphalaikyaṃ sābdaṃ rūpānvitaṃ ca gosaṃkhyā/[3]

3 21e saṅkalitānāmaikyaṃ GKP1 p. 126./ 21f ca] tu GKP1. I have used the RORI (Jodhpur) 37782 manuscript (hereafter R) to emend the printed edition. Angled brackets are used to enclose supplement where I suppose that some words have fallen out.

> The years, decreased successively by the years of maturity of a calf until they become small, are terms of the multiple one etc. respectively. The sum of the results from their own multiples, plus the years, increased by unity is the number of cows.

I translate the Sanskrit term *vāra* (multiple) to mean "many times".[4] The rule when applying *vāra* is given as:

> ekādhikavāramitāḥ padādirūpottarāḥ pṛthak te 'ṃśāḥ/
> ekādyekacayaharās tadghāto vārasaṅkalitam//[5]

> [The numbers] beginning with the number of terms and increased by unity, measured by [the number] one more than the multiple, are the numerators respectively, [which have] the divisors beginning with one and increasing by one. The product of these [fractions] is the multiple sum (*vārasaṃkalita*).

This is a rule for the multiple sum of the natural series, $S(n) = 1 + 2 + 3 + \cdots + n$. Let *vāra* be the number of the order m of summation. When $m = 1$ and $n = 6$, $S(6) = 1 + 2 + 3 + 4 + 5 + 6 = 21 = \frac{6\cdot7}{1\cdot2}$. When $m = 2$, $S_2(n) = S(1) + S(2) + S(3) + \cdots + S(n)$. When $m = 2$ and $n = 6$, $S_2(6) = S(1) + S(2) + S(3) + S(4) + S(5) + S(6) = 1 + 3 + 6 + 10 + 15 + 21 = 56 = \frac{6\cdot7\cdot8}{1\cdot2\cdot3}$. When $m = 3$ and $n = 6$, $S_3(6) = S_2(1) + S_2(2) + S_2(3) + S_2(4) + S_2(5) + S_2(6) = 1 + 4 + 10 + 20 + 35 + 56 = 126 = \frac{6\cdot7\cdot8\cdot9}{1\cdot2\cdot3\cdot4}$. In general
$S_m(n) = S_{m-1}(1) + S_{m-1}(2) + S_{m-1}(3) + \cdots + S_{m-1}(n) =$

$$\frac{n}{1} \cdot \frac{n+1}{2} \cdot \frac{n+2}{3} \cdots \frac{n+m}{m+1} = \frac{n(n+1)(n+2)\cdots(n+m)}{1 \cdot 2 \cdot 3 \cdots (m+1)}.$$

I have not found this rule in any mathematical texts before the GK.[6] The commentary on the example is:

> "The setting down: Cow 1. Years 20. Years [for a calf] to be a mature cow, 3. This is the calculation. The 20 years are decreased again and again

4 For example *dvivāra* means twice and *trivāra* three times.

5 GKP1. p. 123.

6 The rule is included in works by a scholar belonging to the Kerala school, Śaṅkara Vāriyar. In his commentary *Yuktidīpikā* (YD) in verse on *Tantrasaṅgraha* based on *Yuktibhāṣā* and in the *Kriyākramakarī* (KK) commentary on the *Līlāvatī*, see YD 2.177–178 (p. 93). 188–189 (p. 94); KK pp. 242–243.

by [the number of] the years the cow takes to mature, 3. The terms of once, twice, thrice, four times, five times and six times are produced

17	14	11	8	5	2
1	2	3	4	5	6

. The once sum of seventeen [years] is 153. The twice [sum] of fourteen [years] is 560. The thrice [sum] of eleven [years] is 1001. The four times [sum] of eight [years] is 792. The five times [sum] of five [years] is 210. The six times [sum] of two [years] is 8. Thus the sums, 153, 560, 1001, 792, 210, 8, are produced. Their total is 2,724. [By Rule 21f] "plus the years, increased by unity" [2,724] is added to the years plus one, 21; the number of cows obtained is 2,745."[7]

Notes:

$$
\begin{array}{lll}
n = 20 - 3 \cdot 1 = 17, & m = 1; & \frac{17 \cdot 18}{1 \cdot 2} = 153 = S(17). \\
n = 20 - 3 \cdot 2 = 14, & m = 2; & \frac{14 \cdot 15 \cdot 16}{1 \cdot 2 \cdot 3} = 560 = S_2(14) \\
n = 20 - 3 \cdot 3 = 11, & m = 3; & \frac{11 \cdot 12 \cdot 13 \cdot 14}{1 \cdot 2 \cdot 3 \cdot 4} = 1001 = S_3(11) \\
n = 20 - 3 \cdot 4 = 8, & m = 4; & \frac{8 \cdot 9 \cdot 10 \cdot 11 \cdot 12}{1 \cdot 2 \cdot 3 \cdot 4 \cdot 5} = 792 = S_4(8) \\
n = 20 - 3 \cdot 5 = 5, & m = 5; & \frac{5 \cdot 6 \cdot 7 \cdot 8 \cdot 9 \cdot 10}{1 \cdot 2 \cdot 3 \cdot 4 \cdot 5 \cdot 6} = 210 = S_5(5) \\
n = 20 - 3 \cdot 6 = 2, & m = 6; & \frac{2 \cdot 3 \cdot 4 \cdot 5 \cdot 6 \cdot 7 \cdot 8}{1 \cdot 2 \cdot 3 \cdot 4 \cdot 5 \cdot 6 \cdot 7} = 8 = S_6(2)
\end{array}
$$

The number of cows of the first generation in 20 years is 20; the number of cows of the second generation, born from the cows of the first generation during the 17 years is $153(= 1 + 2 + 3 + \cdots + 17) = S(17)$; the next result, $560 = S_2(14)$, is the number of cows born from the cows of the second generation over 14 years; and so on. Thus the total number of cows in the twenty years is the sum of these results, of the number of cows born from the original cow, 20, and of the number of the original cow, 1. Thus $153 + 560 + 1001 + 792 + 210 + 8 + 20 + 1 = 2745$.

The commentary ends with the statement: "Alternatively, it is calculated by means of the *meru* defined in the net of numbers [and] metrics. I will explain it later."[8]

7 nyāsaḥ gauḥ 1 varṣāṇi 20 tarṇakīprasavavarṣāṇi 3/ atra karaṇam/ abdā ete 20 (20 ete GKP1) tarṇakīprasavavarṣais tribhiḥ 3 punaḥ punar ūnitā ekadvitricatuṣpañcaṣaḍvārāṇāṃ jātāni padāni

17	14	11	8	5	2
1	2	3	4	5	6

atha saptadaśānām ekavāraṃ saṅkalitam 153/ caturdaśānāṃ dvivāram 560/ ekādaśānāṃ trivāram 1001/ aṣṭānāṃ caturvāram 792/ pañcānāṃ pañcavāram 210/ dvayoḥ ṣaḍvāram 8/ evaṃ jātāni saṅkalitāni 153/560/1001/792/210/8/ (evaṃ om. GKP/ saṅkalitāni om. R.) eṣāṃ yogaḥ 2724/ (yogaḥ] aikyaṃ R.) sābdaṃ rūpānvitam ity abdaviṃśatyā sarūpayā 21 yuto (yutaṃ R) jātā gosaṅkhyā 2745/ GKP1. pp. 126–127.

8 athavāṅkapāśacchandolakṣaṇameruṇā sidhyati/ (-lakṣaṇe R/ -merūṇā GKP) tat purato vakṣye/ GKP1 p. 127.

2 *Meru* in Net of Numbers

At the end of the *Līlāvatī* Bhāskara added a new subject, *aṅkapāśa* and Nārāyaṇa created a *vyavahāra*, modeled on prosody and music. The subjects dealt with can be divided into five patterns according to the method of extension of digits.

Pattern 1. digits unequal to each other, such as 3,7,9 are arranged in a fixed number of places, in this case three.[9]

Pattern 2. digits, some of which are the same, such as 1,1,2,4 are arranged in a fixed number of places, in this case four.

Pattern 3. the number of places and the final digit are fixed, but the sum (*s*) of the digits in each arrangement is variable, shown below.

Pattern 4. the final digit and the sum of the digits are fixed, but the number of places is variable.

Pattern 5. the number of places and the final digit are fixed, but the sum of the digits is variable.

A rule to construct a *meru* or a *sumeru* figure given in *vyavahāra* 13 *aṅkapāśa* is:

> ekādyekottarāḥ kāryā adho 'dhaḥ koṣṭhapaṅktayaḥ/
> sarūpasthānasaṅkhyāś ca tāsv ādyāyāṃ ca rūpakam//36//
> paṅktau likhed dvitīyāyāṃ meror asya tadādime/
> koṣṭhe 'ntimaṃ virūpaṃ ca likhec chṛṅgābhidhaṃ bhavet//37//
> parasmin koṣṭhake rūpaṃ svakoṣṭhordhvasthitaś ca yaḥ/
> śṛṅgaghnas tam adho nyasya vāmakarṇakoṣṭhāṅkayuk//38//
> kramād evaṃ tiryag āśu koṣṭhapaṅktiṣv ayaṃ vidhiḥ/
> sumerukaraṇaṃ prājñaiḥ proktaṃ gaṇitavedibhiḥ//39//[10]

Rows of cells (*koṣṭha*) beginning with one and having a difference of one are to be drawn one below the other; they are counted by the [number of] places plus one. One should write down 1 in the first row among them; then in the first cell of the second [row] of this *meru*, one should write down the final [digit] decreased by one; this is designated as the "horn" (*śṛṅga*). 1 [is written down] in the other cell [of the second row]. [The number] situated above its cell is multiplied by the horn; place that

9 *Līlāvatī* 262. 2,8; 3,9,8; 2,3,4,5,6,7,8,9.

10 Kusuba p. 220; GKP2 pp. 302–301.

1						
1	1					
1	2	1				
1	3	3	1			
1	4	6	4	1		
1	5	10	10	5	1	
1	6	15	20	15	6	1

FIGURE 5.2
meru in $p = 6 \cdot q = 2$

1			
2	1		
4	4	1	
8	12	6	1

FIGURE 5.3
meru in $p = 3 \cdot q = 3$

1			
$q-1$	1		
$(q-1)^2$	$2(q-1)$	1	
$(q-1)^3$	$3(q-1)^2$	$3(q-1)$	1

FIGURE 5.4
General form

[result] below, combined with the number in the cell in the diagonal to the left; in this way place numbers successively in the horizontal rows of cells quickly. This is the procedure. The drawing of the *sumeru* is described [thus] by learned mathematicians.

A *meru* or *sumeru* is formed with respect to the number of places (p), and the final digit (q), and is applied in pattern 3. Let a number at column 1 and row 2, that is $h = (q-1)$, be the horn. Also, let any cell contain x and the cell to the left of it contain y. The procedure for the formation of a *meru* is that the cell vertic-

TABLE 5.1 Extension 1

1	222222	17	222212	33	222221	49	222211
2	122222	18	122212	34	122221	50	122211
3	212222	19	212212	35	212221	51	212211
4	112222	20	112212	36	112221	52	112211
5	221222	21	221212	37	221221	53	221211
6	121222	22	121212	38	121221	54	121211
7	211222	23	211212	39	211221	55	211211
8	111222	24	111212	40	111221	56	111211
9	222122	25	222112	41	222121	57	222111
10	122122	26	122112	42	122121	58	122111
11	212122	27	212112	43	212121	59	212111
12	112122	28	112112	44	112121	60	112111
13	221122	29	221112	45	221121	61	221111
14	121122	30	121112	46	121121	62	121111
15	211122	31	211112	47	211121	63	211111
16	111122	32	111112	48	111121	64	111111

ally below the x is filled with $x \cdot h + y$.[11] The formation is shown in Figures 5.2, 5.3, and 5.4. Nārāyaṇa gives two examples where $p = 6$ and $q = 2$ (Figure 5.2) and $p = 3$ and $q = 3$ (Figure 5.3).[12] A general form is given in Figure 5.4.

11 For any number in the first column $y = 0$.

12 Kusuba p. 238.

TABLE 5.2 Extension 2

1	333	10	332	19	331
2	233	11	232	20	231
3	133	12	132	21	131
4	323	13	322	22	321
5	223	14	222	23	221
6	123	15	122	24	121
7	313	16	312	25	311
8	213	17	212	26	211
9	113	18	112	27	111

The numbers in the last row in Figure 5.2 indicate: among 64 variations as in Table 5.1, the number of variations having no 1s is 1; the number of variations having one 1 is 6; the number of variations having two 1s is 15; the number of variations having three 1s is 20; the number of variations having four 1s is 15; the number of variations having five 1s is 6; the number of variations having six 1s is 1.

The number of variations is the same as those with no 2s, one 2, three 2s, and so on. When 1 is allotted to a short/light syllable ⏑ and 2 long/heavy ones – this table is applied in metrics. This shows all 64 possible arrangements of six syllables. 1.– – – – – ––. 2.⏑ – – – – – –. 3.– ⏑ – – – ––. 63.– ⏑⏑⏑⏑⏑ 64.⏑⏑⏑⏑⏑⏑.

In Figure 5.3 the last row indicates: for the 27 variations in Table 5.2 the number of variations having no 1s is 8; the number of variations having one 1 is 12; the number of variations having two 1s is 6; the number of variations having three 1s is 1.

1									
1	1								
1	2	1							
1	3	3	1						
1	4	6	4	1					
1	5	10	10	5	1				
1	6	15	20	15	6	1			
1	7	21	35	35	21	7	1		
1	8	28	56	70	56	28	8	1	
1	9	36	84	126	126	84	36	9	1
1	10	45	120	210	252	210	120	45	10
1	11	55	165	330	462	462	330	165	55
1	12	66	220	495	792	924	792	495	220
1	13	78	286	715	1287	1716	1716	1287	715
1	14	91	364	1001	2002	3003	3432	3003	2002
1	15	105	455	1365	3003	5005	6435	6435	5005
1	16	120	560	1820	4368	8008	11440	12870	11440
1	17	136	680	2380	6188	12376	19448	24310	24310
1	18	153	816	3060	8568	18564	31824	43758	48620
1	19	171	969	3876	11628	27132	50388	75582	92378
1	20	190	1140	4845	15504	38760	77520	125970	167960

FIGURE 5.5 Application of a *meru*

3 Application of a *Meru* to the Problem

The *meru* used in metrics is $q = 2$. The *meru* in Figure 5.5 is used for the sum of the sums of a natural series.

The numbers in the second column are a natural arithmetic series. The numbers in the third column are the sums of the natural arithmetic series:

$S(17) = 153$. The numbers in the fourth column are the sums of the sums of the series: $S_2(n)$; $S_2(14) = 560$. The numbers in the fifth column are the sums of the sums of the sums of the series: $S_3(11) = 1001$. In this way $S_4(8) = 792$, $S_5(14) = 210$, and $S_6(2) = 8$, which is what Nārāyaṇa intended.

4 Calf Examples in Western Books

I have found questions on the number of calves in two modern mathematical books: *536 Puzzles & Curious Problems* by Dudeney; and *50 Mathematical Ideas You Really Need to Know* by Crilly.

Dudeney prescribes the following sequence: "'Supposing,' said my friend Farmer Hodge, 'that cow of mine to have a she calf at the age of two years, and supposing she goes on having the like every year, and supposing every one of her young to have a she calf at the age of two years, and afterwards every year likewise, and so on. Now, how many do you suppose would spring from that cow and all her descendants in the space of twenty-five years?' I understand from Hodge that we are to count from the birth of the original cow, and it is obvious that the family can produce no feminine beef or veal during the period stated."[13]

The sequence produced in this question is the Fibonacci sequence, 0, 1, 1, 2, 3, 5, 8, 13, 21, 34, 46, …368.[14] The Fibonacci sequence consists of numbers such that $a_0 = 0, a_1 = a_2 = 1$ and $a_n = a_{n-2} + a_{n-1}$. Dudeney transforms rabbits into cows and months into years in a Fibonacci sequence.[15] This question is not equivalent to Nārāyaṇa's.

In topic 11, titled "Fibonacci numbers"[16] Crilly explains the Fibonacci sequence and shows "family resemblances", where there is an intermediate year in the maturation process. His answer applies a sequence, such as 1,1,1,2,3,4,6,9, 13,19,28,41,60,88,129,189,277,406,595. The generation skips a value at 41=28+13, and 60=41+19. Therefore, the rule is: $a_{n+1} = a_n + a_{n-2}$ $(n \geq 2)$. $a_0 = 0, a_1 =$

---

13 Dudeney p. 56, p. 175. A Cow's Progeny.

14 The Fibonacci sequence is given in the *Liber Abbaci*, chapter twelve on questions of abbaci, pp. 283–284 in the edition by Boncompagni. This text is based on an incomplete manuscript. Chapter 2 part 8 is missing although it is mentioned at the beginning of the chapter. A critical edition should be published. For informations about manuscripts see Hannah p. 307. An English translation is by Sigler. For the Fibonacci sequence, see pp. 404–405.

15 Dudeney gives two formulas for addition of the terms: $a_1 + a_2 + a_3 + \cdots + a_n = a_{n+1} - 1$, and $a_{n-3} + 2a_{n-2} = a_n$.

16 Crilly pp. 44–47.

1, $a_2 = 1$. When the rest of the sequence is generated as: 277,406,595,872,1278, 1873,2745, it reaches 2745, as required by Nārāyaṇa.[17]

5 Nārāyaṇa Again

At the beginning of the chapter on *aṅkapāśa* Nārāyaṇa defines some sequences as *paṅkti*. An additive sequence (*sāmāsikāpaṅkti*) is defined with respect to the last digit q and a given sum s, when $q \leq s$. The sequence has $s + 1$ terms. First, write down 1 twice: 1,1. Then add them together, and write the result on the right side of 1,1; 1,1,2. When $q = 3$ and $s = 7$, add the three numbers, 1,1, and 2, and then write their sum, 4, after the 2; the sequence 1,1,2,4 is produced. Next, add together three ($= q$) numbers from the last number in reverse order, for instance, 1,2,4. Their sum is 7. The sequence 1,1,2,4,7 is produced. Next, add 2,4, and 7; then 1,1,2,4,7,13 is produced. Proceed in this way until the number of terms becomes one more than the given sum, 7: 1,1,2,4,7,13,24,44. The additive sequence is applied in pattern 4 to calculate the variations of all the enumerations and of those ending with each digit in the same way as music text teaches *tāla*.[18] Among 44 variations the number of variations ending with 1 is 24; the number of variations ending with 2 is 13; the number of variations ending with 3 is 7 (see Table 5.3).

When q is greater than 3, as when $q = 7$, there are no q terms after 1,1,2,4, counted from the last number, 4. In this case, add all the numbers together; and the result is 8. A sequence 1,1,2,4,8 is produced. Again, because there are no q terms, add together all the numbers available and write down the result to the right; a sequence 1,1,2,4,8,16 is produced. The additive sequence for $q = 7$ and $s = 7$ is 1,1,2,4,8,16,32,64. From this sequence we learn the number of variations: among 64 variations in Table 5.4, extension 4, the number of variations ending with 1 is 32; those ending with 2 is 16; those ending with 3 is 8 ...; those ending with 7 is 1.

---

17 It is not my intention to give mathematical proof here. Instead I calculate when $n = 19$ using Nārāyaṇa's method.

$$n = 19 - 3 \cdot 1 = 16, \quad m = 1; \quad \frac{16 \cdot 17}{1 \cdot 2} = 136 = S(16).$$
$$n = 19 - 3 \cdot 2 = 13, \quad m = 2; \quad \frac{13 \cdot 14 \cdot 15}{1 \cdot 2 \cdot 3} = 455 = S_2(13)$$
$$n = 19 - 3 \cdot 3 = 10, \quad m = 3; \quad \frac{10 \cdot 11 \cdot 12 \cdot 13}{1 \cdot 2 \cdot 3 \cdot 4} = 715 = S_3(10)$$
$$n = 19 - 3 \cdot 4 = 7, \quad m = 4; \quad \frac{7 \cdot 8 \cdot 9 \cdot 10 \cdot 11}{1 \cdot 2 \cdot 3 \cdot 4 \cdot 5} = 462 = S_4(7)$$
$$n = 19 - 3 \cdot 5 = 4, \quad m = 5; \quad \frac{4 \cdot 5 \cdot 6 \cdot 7 \cdot 8 \cdot 9}{1 \cdot 2 \cdot 3 \cdot 4 \cdot 5 \cdot 6} = 84 = S_5(4)$$
$$n = 19 - 3 \cdot 6 = 1, \quad m = 6; \quad \frac{1 \cdot 2 \cdot 3 \cdot 4 \cdot 5 \cdot 6 \cdot 7}{1 \cdot 2 \cdot 3 \cdot 4 \cdot 5 \cdot 6 \cdot 7} = 1 = S_6(1)$$

Therefore, the answer is $136 + 455 + 715 + 462 + 84 + 1 + 19 + 1 = 1873$. This number is also obtained from the sequence.

18 See Kusuba pp. 107–108.

TABLE 5.3 Extension 3

1	133	12	2122	23	2131	34	3211
2	223	13	11122	24	11131	35	12211
3	1123	14	1312	25	1321	36	21211
4	313	15	2212	26	2221	37	111211
5	1213	16	11212	27	11221	38	13111
6	2113	17	3112	28	3121	39	22111
7	11113	18	12112	29	12121	40	112111
8	232	19	21112	30	21121	41	31111
9	1132	20	111112	31	111121	42	121111
10	322	21	331	32	2311	43	211111
11	1222	22	1231	33	11311	44	1111111

For $q = 2$, the sequence becomes $1, 1, 2, 3, 5, 8, 13, 21\cdots$, which is consistent with the Fibonacci sequence.[19] Any sequence named *paṅkti* is applied to subjects of *aṅkapāśa*. Nārāyaṇa does not use the term *paṅkti* in *śreḍhīvyavahāra*. He employs this term to mean a line, on which digits are written down successively, concatenated, and taken as a number, such as *labdhaṃ dvighnaṃ paṅktyāṃ vinyasya*[20] (setting down twice the quotient at a line) in the rule for the square root of integers. In the *Līlāvatī* Bhāskara also uses the term: *paṅkyāṃ paṅktihṛte paṅkter dala*. Colebrooke translates this portion as "in a line [with the preceding double number], divide by the [number as noted in a] line, halve the [number in the] line."[21]

19 Suppose $q = 2$ and allot 1 short syllable and 2 long one. For $s = 1$, ⏑. When $s = 7$ and $q = 2$ there are 21 sequences. It also shows extensions of seven *mora* as Knuth p. 6 enumerates.

20 GKP1 p. 7.

21 Cf. Colebrooke p. 10.

TABLE 5.4 Extension 4

1	7	17	52	33	61	49	511
2	16	18	142	34	151	50	1411
3	25	19	232	35	241	51	2311
4	115	20	1132	36	1141	52	11311
5	34	21	322	37	331	53	3211
6	124	22	1222	38	1231	54	12211
7	214	23	2122	39	2131	55	21211
8	1114	24	11122	40	11131	56	111211
9	43	25	412	41	421	57	4111
10	133	26	1312	42	1321	58	13111
11	223	27	2212	43	2221	59	22111
12	1123	28	11212	44	11221	60	112111
13	313	26	3112	45	3121	61	31111
14	1213	27	12112	46	12121	62	121111
15	2113	28	21112	47	21121	63	211111
16	11113	29	111112	48	111121	64	1111111

16			
15	120		
14	105	560	
13	91	455	1820
12	78	364	1365
11	66	286	1001
10	55	220	715
9	45	165	495
8	36	120	330
7	28	84	210
6	21	56	126
5	15	35	70
4	10	20	35
3	6	10	15
2	3	4	5
1	1	1	1

FIGURE 5.6
Meru according to Nārāyaṇa

6 Concluding Remarks

It was thought that Piṅgala's rule for *lagakriyā* in the *Chandaḥsūtra* '*pare pūrṇam iti*' (the full in front) requires a *meru* construction. It is not certain what figure Piṅgala had in mind. Is it the same as Pascal's triangle? Commentaries on the text have given various figures. Piṅgala does not use the term *meru*, whereas Brahmagupta does use it at stanza 15 of chapter 20, *chandaścity-uttarādhyaya* of the *Brāhmasphuṭasiddhānta* written in 628 CE. Nobody has explained its shape. Bhaṭṭotphala (fl. 966 CE.) gives a figure in his commentary on the stanza by Varāhamihira in *Bṛhatsaṃhitā* 76.22 *pūrveṇa pūrveṇa gatena yuktaṃ sthānaṃ vināntyaṃ* ([a number] is combined with each preceding [number] that has passed except in the last place).[22] Bhaṭṭotphala explains how to construct it (see Figure 5.6) and how to use it.[23] Without the commentary it is difficult to construct the figure that Bhaṭṭotphala names *loṣṭakaprastāra*.[24] On the other hand, Nārāyaṇa shows how to construct the *meru*. The *prastāra* Nārāyaṇa intended is an extension, that is, the serial arrangement of the digits as seen in Tables 5.1–5.4. Nārāyaṇa did not create a sequence for the cow problem and used the *meru* for $q = 2$ as a mathematical tool.

Acknowledgement

I thank Dr. Akio Yasuhara and Mr. Taro Tokutake for their help.

Bibliography

Primary Sources

Bhāskara. *Līlāvatī* of Bhāskarācārya

1) with the commentary *Kriyākramakarī* of Śaṅkara and Nārāyaṇa. Ed. by K.V. Sarma. Vishveshvaranand Indological Series 66. Hoshiapur: Vishveshvaranand Vedic Research Institute, 1975
2) with Gaṇeśa's *Buddhivilāsinī* and Mahīdhara's vivaraṇa by G. Apate, Ānandāśrama Sanskrit Series 107, 2 vols. Poona 1937. An English Translation by Colebrooke, H.T. pp. 1–127.

Fibonacci. Ed. by B. Boncompagni. *Scritti de Leonard Pisano*. 2 vol., Roma, 1857–1862. An English Translation by Sigler.

Nārāyaṇa Paṇḍita. *Gaṇitakaumudī* ed. by Padmakara Dvivedi, Sarasvati Bhavana Texts 57, Benares 1936 (vol. 1) and 1942 (vol. 2).

Nīlakaṇṭha. *Tantrasaṁgraha* with commentaries *Yuktidīpikā* and *Laghuvivṛti* of

22 Kusuba and Plofker p. 51.

23 Kusuba pp. 151–161 discusses how to use *loṣṭa* (a lump of earth or clay) based on the commentary.

24 Edwards p. 170 calls the figure *meruprastāra*.

Śaṅkara Vāriyar. Cr. Ed. by K.V. Sarma. Punjab University Indological Series 10. Hoshiarpur: Vishveshvaranand Vishva Bandhu Institute of Sankrit and Indological Studies, 1977.

Secondary Sources

Colebrooke, H.T. *Algebra, with Arithmetic and Mensuration, from the Sanscrit of Brahmegupta and Bhāskara* 1817. Reprinted in India under the title: *Classics of Indian Mathematics: Algebra, with Arithmetic and Mensuration from the Sanskrit of Brahmagupta nd Bhāskara* Delhi. 2005.

Crilly, Tony *50 Mathematical Ideas You Really Need to Know* Quercus. 2007.

Dudeney, Henry Ernest *536 Puzzles & Curious Problems*, 1967. New York.

Edwards, A.W.F 'The arithmetic triangle' [Robin Wilson and John J. Watkins ed. 2013 pp. 167–180]

Hannah, John 'False position in Leonard of Pisa's *Liber Abbaci*' *Historia Mathematica* 34. 2007. pp. 306–332.

Knuth, Donald E. 'Two Thousand Years of Combinatorics' Robin Wilson and John J. Watkins 2013 pp. 3–37.

Kusuba, Takanori. *Combinatorics and Magic Squares in India: A Study of Nārāyaṇa Paṇḍita's Gaṇitakaumudī* Chap. 13–14. Ph.D dissertation. Providence, RI: Brown University. 1993.

Kusuba, Takanori and Plofker, Kim 'Indian Combinatorics' Robin Wilson and John J. Watkins ed. pp. 40–62.

Sigler, Laurence E. *Fibonacci's Liber Abaci*, New York. 2002

Wilson, Robin and John J. Watkins ed. *Combinatorics: Ancient & Modern*, Oxford University Press. 2013.

CHAPTER 6

Calculation and Correction for the Mean Position of the Planets in the *Jyotirmīmāṃsā*

Setsuro Ikeyama

1 Introduction

Jyotirmīmāṃsā (hereafter JM) is an astronomical treatise written by Nīlakaṇṭha Somayājin who flourished in South India in the 15th–16th centuries. For editing the published text, *Jyotirmīmāṃsā, Investigations on Astronomical Theories*, the editor uses only one manuscript, of which the beginning and end are no longer extant. Both the book title and the author are presumed by the editor.

Nīlakaṇṭha discusses the basis of astronomy in this small treatise. As the editor says, the author "insists that results arrived at by computation should tally with observation and that astronomical parameters and other constants should be revised periodically towards this purpose."[1]

Among the assorted contents of JM are records of two annular eclipses (1502 CE Oct. 1 and 1517 CE Jun. 19) (JM 1.9–2.11) and a unique method for deriving sines in units of 150 minutes[2] (JM 48.8–49.14), what I wish to show in this chapter is how Nīlakaṇṭha's description of the mean motions of the planets are calculated and rectified, which is the central problem and forms a large part of JM.

1.1 *Editorial Notes*

Translations from JM are:
- Indented.
- Texts in parentheses are my addition.
- Page numbers and line numbers are also in parentheses.
  Example: (3.15–4.2) means from 15th line of page 3 to 2nd line of page 4.

1 P. ix of the edition.
2 For more information, see Michio Yano, *An Indian Sine Table of 36 entries*, History of Science in South Asia 7 (2019) 42–51.

© SETSURO IKEYAMA, 2023 | DOI:10.1163/9789004536869_008

2 Nīlakaṇṭha's Basic Attitude

Indian astronomy respects its traditions. Every school has its canonical books, which no one disputes. However, Nīlakaṇṭha tried to rectify the astronomical constants not only of other schools but also of his own tradition, and he insisted that his canon, *Āryabhaṭīya* (AB) supported his attempts.

2.1 *One's Own Wisdom Is the Source of Knowledge*

Citing a verse from AB, Nīlakaṇṭha strongly criticizes people who insist that astronomical canons are not revisable because they are given by omniscient gods and are therefore free from errors.

> (2.12–21)
> If (opponents) say, "no, Brahmā who was satisfied by ascetic practices and so on taught Āryabhaṭa the number of the rotations and (the lengths of) the circumferences (of epicycles), which are the means of calculation. Some think that Āryabhaṭa organized what he was taught into ten *gītis*. Why do we need to investigate it? (Its) truthfulness is ascertained because Brahmā is omniscient and free from desire and hate." Stupid! It is not so. The graciousness of gods is just a cause of the clearness of thought. It never occurred to them that Brahmā or Āditiya himself came and taught them.
>
> He (Āryabhaṭa) mentions the same thing in the following verse:
> With the graciousness of gods the sunken excellent jewel of true knowledge is salvaged from the sea of true and false knowledge by me using the boat of my intelligence. (AB 4.49)

2.2 Āryabhaṭīya *Was Written to Investigate Astronomy*

In JM the word "investigation (*parīkṣā, parīkṣaṇa*)" means to compare the result of a calculation for an astronomical phenomenon with the observation of that phenomenon to check the astronomical constants in an astronomical canon. Nīlakaṇṭha insists that *Āryabhaṭīya* was written for such investigations.

> (5.4–7)
> The aim of the *Āryabhaṭīya* is correct because it focuses mostly on investigation and not because its constants and so on are better than those of other canons. Therefore, it is a treatise for investigations, which it emphasizes.

> (7.21–22)
> Therefore, Āryabhaṭa composed an astronomical canon for describing the method of, and a set of reasoning useful for, investigations.

2.3 *Constants of Various Canons Should Be Compared and Rectified*

In Indian astronomy all constants are given in integer numbers over a fairly long period. The most popular period is *yuga* (*caturyuga* or *mahāyuga*), which corresponds to 4,320,000 years. The sun rotates 4,320,000 times in a *yuga* according to almost all astronomical canons.

But the length of a *yuga* differs from one astronomical school to another because the each number of civil days in a *yuga* is different. Before comparing constants from different schools a common time frame must be set up and the constants adjusted to it.

However, despite comparing constants in a common time frame some remain inconsistent. Nīlakaṇṭha says that all constants are sufficiently accurate when they are created, but because errors accumulate over time, they need to be checked regularly and rectified to coincide with the results of observation.

> (6.12–16)
> The five canons should be understood as standards for a certain period. Thus the canon that corresponds to those observations should be sought for. The correspondence with observation should be known using eclipses by investigators of the time. If it is not so, that is, there are differences between the taught canons (and observations), having investigated using astronomical instruments and having known the numbers of the rotations of the planets and so on, a new canon should be produced …

3 Astronomical Constants in Question

JM mentions in many places a variety of astronomical constants listed in astronomical canons. Some are then used in the subsequent sections of JM but others are just listed.

3.1 *Longitude of* Mandocca

Mandocca (*manda-ucca*, literally 'slow high-point', hereafter MU) is the point corresponding to the apogee of our astronomy. The longitudes of MU are given in various locations in JM, all of which are listed in Table 6.1.

- *Mandocca*s and nodes of the planets in AB (JM 8.1–4)
- *Mandocca*s in KKp and Mars' *Mandocca* in KKu (JM 18.13–26)

TABLE 6.1 Longitude of MU given in AB, KK, SSS, and by Master

–	Mars	Mercury	Jupiter	Venus	Saturn
AB1.7	118°	210°	180°	90°	236°
KKp2.6	110°	220°	160°	80°	240°
KKu2.1	127°	–	–	–	–
SSS	–	–	172° 15′ 36″	–	–
Master	–	–	–	–	247°

TABLE 6.2 Rotations in a *Yuga*

	Sunrise school	Midnight school
Sun	4,320,000	=
Moon	57,753,336	=
Moon MU	488,219	=
Moon Node	232,226	=
Mars	2,296,824	=
Mercury	17,937,020	17,937,000
Jupiter	364,224	364,220
Venus	7,022,388	=
Saturn	146,564	=
Civil Days	1,577,917,500	1,577,917,800
Tithi	1,603,000,080	=
Avama	25,082,580	25,082,280

– *Mandocca* of Jupiter calculated using its rotation in SSS (JM 19.1–13)
– *Mandocca* of Venus in KK, that of Saturn given by a master (JM 19.14–19.19)
We cannot identify the master mentioned but he is apparently not Āryabhaṭa.

3.2 *Civil Days and Planet Rotations in the Sunrise-Epoch and Midnight-Epoch Schools*

In JM 8.16–18, quoting MB 7.21cd–22, there are difference in a *yuga* between the sunrise-epoch school (*audayika*) and the midnight-epoch school (*ārdharātrika*) in the number of civil days and the rotations of Mercury and Jupiter.

The sunrise-epoch school uses Āryabhaṭa's AB as its canon. *Mahābhāskarīya* (MB) of Bhāskara, *Śiṣyadhīvṛddhida* (SDV) of Lalla, *Grahacāranibandhana*

TABLE 6.3 Rotations in a *Brāhma* School's *Kalpa*

Sun, Venus, Mercury, SU (Mars, Jupiter, Saturn)	4,320,000,000
Moon	57,753,300,000
Mars	2,296,828,522
SU (Mercury)	17,936,998,984
Jupiter	364,226,455
SU (Venus)	7,022,389,492
Saturn	146,567,298
MU (Sun; Moon)	480; 488,105,858
MU (Mars; Mercury; Jupiter; Venus; Saturn)	292; 332; 855; 653; 54
Node (Moon)	232,311,168
Node (Mars; Mercury; Jupiter; Venus; Saturn)	267; 521; 63; 893; 584
Civil Days	1,577,916,450,000

(GCN) of Haridatta, and an anonymous *Grahacāranibandhanasaṅgraha* (GCNS) belong to this school. The midnight-epoch school was also founded by Āryabhaṭa, and its principal texts are the *Sūryasiddhānta* (SSL) of Lāṭadeva who is a pupil of Āryabhaṭa, and *Khaṇḍakhādyaka* (KK) of Brahmagupta.

Table 6.2 shows the rotations of the planets of both schools. They are identical, except for the number of civil days and the rotations of Mercury and Jupiter.

3.3 Brāhma *School's Rotations*

The *Brāhma* school was founded by Brahmagupta in the 7th century. Its canon is his *Brāhmasphuṭasiddhānta* (BSS), but Nīlakaṇṭha quotes constants of this school from another canon, *Siddhāntaśekhara* (SSS).

SSS 1.26–31, 41ab are quoted in JM 22.19–24.10 where rotations of the planets in a *kalpa* (= 1,000 *yugas*) are listed (see Table 6.3).

In Table 6.3, SU stands for *śīghrocca* (*śīghra-ucca*, literally, rapid high-point), which is an imaginary point to explain apparent deviations in geocentric theory. The rotations of SUs of Mercury and Venus correspond to those planet's rotations in ancient Greek theory, and the number of rotations of SUs of Mars, Jupiter, and Saturn are equal to the number of the sun's rotations.

3.4 *Mean Longitudes at the Beginning of* Kaliyuga

Kaliyuga or *kali* is a subordinate division of a *yuga*. The mean longitudes at that time are used as initial values in astronomical calculations. In JM 24.11–25.2 those values seem to be listed, but they are hard to understand and do not coincide with our calculations. There is room for further investigation.

TABLE 6.4 Circumferences of *Manda* Epicycles of the Sun and Moon

	Quadrant	East	Noon	West	Midnight
Sun	3, 4	13°20′	13°40′	14°	13°40′
	1, 2	14°	13°40′	13°20′	13°40′
Moon	3, 4	32°28′	31°36′	30°44′	31°36′
	1, 2	30°44′	31°36′	32°28′	31°36′

TABLE 6.5 Circumferences of Epicycles (for Venus odd quadrant/even quadrant)

Planet	Mars	Mercury	Jupiter	Venus	Saturn
Manda	70°	38°	33°	9°/11°	30°
Śīghra	243°40′	132°	68°	263°/258°	35°

3.5 Brāhma *School's Circumference of Epicycles*

SSS 3.19–20 and 3.35–39 are quoted in JM 26.1–27.14 and include the lengths of the circumferences of *manda* epicycles of the sun and moon (19–20) (Table 6.4), and the *manda* and *śīghra* epicycles of five other planets (35–39) (Table 6.5).

4 Various Rectifications

A large part of JM contains rectifications to the results of calculations that used the canon's original constants.

4.1 *Rectification Using* Bīja *to* AB

Bīja, which literally means a seed, is used to mean a source of correcting values in the results of astronomical calculations.

- Original constants give correct results when the canons were written. The correction values derived from *bīja* are therefore zero at that time.
- *Bīja* is the difference between the result derived from observations at a certain time and the result of the calculation using original constants for that time.

- Every correction value is calculated by the rule of three, that is, proportion. Let Y_0 be the year of the *śaka* era when a canon was written, Y be the year of the *śaka* era the observation to determine *bīja* B. Correction value b for a desired year y can be calculated as:

$$b = B \cdot \frac{y - Y_0}{Y - Y_0}.$$

b is zero when $y = Y_0$. Then increasing linearly, b becomes equal to B when $y = Y$.

Nīlakaṇṭha discussed on this subject in more than 10 pages (JM 10.10–20.21):

1. (10.10–11.2): Rectification 1. Quoted from GCNS 20b–22a and 19–20a. $Y = 679$, $Y_0 = 444$.
2. (11.3–11): Rectification 2. Quoted from an unknown source and GK. A variation of Rectification 1. $Y = 679$, $Y_0 = 444$.
3. (12.3–13): Rectification 3. Quoted from SDV 1.50–51. $Y = 679$, $Y_0 = 421$ and 420.
4. (13.1–14.15): Discussion about when AB was written. Is Y_0 444 or 421?
5. (14.16–17.5, 19.20–20.21) Discussion about why Lalla uses 420 as Y_0 in Rectification 3.

4.1.1 Rectification 1

(10.13–11.1)
Bīja of 25 is known for the moon, 114 for the (*manda-*)*ucca* of the moon, 50 for Mars, 430 for Mercury, 50 is known for Jupiter, 160[3] for Venus, 21 is said for Saturn, 96[4] is memorized for the moon node.

When a *śaka* year diminished by 444 is multiplied by a *bīja* and divided by 235, the result is (expressed in) minutes and seconds. It is positive for Mercury, Mars, and Saturn. It is negative for the moon node, the moon, the (*manda-*)*ucca* of the moon, Jupiter, and Venus.

These verses from GCNS 19–22a are also quoted by Nīlakaṇṭha in his commentary on *Āryabhaṭīya* chapter 4.[5] According to these verses, the correction value b_1 for a *śaka* year y can be calculated using *bīja* B_1:

3 Corrected according to the edition of GCNS.
4 Corrected according to the edition of GCNS.
5 NAB4 p. 149.19–150.1.

$$b_1 = B_1 \cdot \frac{y - 444}{235},$$

where 235 is the result of 679 − 444, that is, $Y_0 = 444$ and $Y = 679$.

4.1.2 Rectification 2

This is a variation of Rectification 1 that uses the same calculation but with different *bījas*.

> (11.4–10)
> Multiply a *śaka* year diminished by 444 by 09, 65, 13 and (divide) by 85, 134, 32. (The longitude of) each planet diminished by the result is (the corrected longitude of) the moon, its (*manda-*)*ucca* and its node.
> The results of (a *śaka* year) multiplied by 45, 420, 47, 153, and 20 and divided by 235 are (the correction values) for Mars and so on. They are positive for Mercury, Mars, and Saturn, negative for Jupiter and Venus. The sun is excluded.

These verses are also quoted in NAB4.[6] According to the editor of JM, they are cited from the *Mahāmārganibandhana* (MMN) of Haridatta—manuscripts of which have not yet been found.

Nīlakaṇṭha immediately quotes another half-verse, which gives different *bījas* for Mars and Saturn in Govindasvāmin's *Govindakṛti* (GK) which is also lost. Though it is not clear how it relates to the first verses, we tentatively assume that it gives variations of *bījas* in Rectification 2 for Mars and Saturn.

> (11.9)
> (A *śaka* year multiplied by) 40 minutes (and divided by 235) should be added to the mean (longitude of) Mars, 18 for Saturn.

Let B_2 be *bījas* used in Rectification 2. The correction value b_2 at year y can be calculated in the following way:

$$b_2 = B_2 \cdot \frac{y - 444}{235},$$

where $Y_0 = 444$ and $Y = 679$ is as in Rectification 1, while simplified calculation formulas are given for the moon and its MU and node:

6 Edition p. 150.3–6.

$$b_2^{moon} : 9 \cdot \frac{y - 444}{85} \approx 25 \cdot \frac{y - 444}{235},$$

$$b_2^{MU} : 65 \cdot \frac{y - 444}{134} \approx 114 \cdot \frac{y - 444}{235},$$

$$b_2^{node} : 13 \cdot \frac{y - 444}{32} \approx 95 \cdot \frac{y - 444}{235}.$$

4.1.3 Rectification 3

Yet another calculation cited from SDV.

> (12.4–11)
> Multiply a *śaka* year diminished by 420, by 25 for the moon, 114 for its (*manda-*)*ucca*, 96 for its node, 47 for Jupiter, 153 for the (*śīghra-*)*ucca* of Venus and divide by 258.[7] (The results are) subtracted. (SDV 1.50)
> (Multiply) by 48 for Mars, by 20 for Saturn, by 430[8] for (the *śīghra-ucca* of) Mercury (and divide by 258). The resulted additional numbers should be added to the minutes (of the mean longitude) of the planets or the (*śīghra-*)*ucca*s. (SDV 1.51)

Let B_3 be the *bīja*s given here. The correction value b_3 at a *śaka* year y is:

$$b_3 = B_3 \cdot \frac{y - 420}{258}.$$

Here, Y is 679, identical to that of Rectifications 1 and 2, but Y_0 is troublesome. Y_0 = 421 for deriving the divisor 258 (= 679 – 421), while Y_0 = 420 for multiplier. This question is taken up in section 4.1.5.

All formulas for calculating the *bīja* correction and the *bīja*s are shown in Tables 6.6 and 6.7.

4.1.4 When Was *Āryabhaṭīya* Composed?

As we have already seen, two different values are used as Y_0, that is, the year the *Āryabhaṭīya* was composed. It is 444 in Rectifications 1 and 2, and 421 in Rectification 3. Nīlakaṇṭha discusses it in JM 13.1–14.15 and advocates the latter.

Nīlakaṇṭha cites as the main reason that the motion of the sun's road, i.e., precessional movement, is not mentioned in AB. It means Āryabhaṭa considers precession to be zero. This is because, Nīlakaṇṭha says, according to Indian

---

7 250 in the edition of SDV and quoted verses in NAB4.
8 420 in the edition of SDV.

TABLE 6.6 Formulas for *Bīja* Rectification

$b_1 = B_1 \cdot \frac{y - 444}{235}$	GCNS
$b_2 = B_2 \cdot \frac{y - 444}{235}$	MMN?
$b_2^{moon} = 9 \cdot \frac{y - 444}{85}$	MMN?
$b_2^{MU} = 65 \cdot \frac{y - 444}{134}$	MMN?
$b_2^{node} = 13 \cdot \frac{y - 444}{32}$	MMN?
$b_3 = B_3 \cdot \frac{y - 420}{258}$	SDV

TABLE 6.7 *Bījas*

Planet	B_1	B_2	B_3
Moon	25	9 ⟨24.88⟩	25
Moon MU	114	65 ⟨113.99⟩	114
Moon Node	96	13⟨95.47⟩	96
Mars	50	45 (40)	48
Mercury	430	420	430
Jupiter	50	47	47
Venus	180	153	153
Saturn	21	20 (18)	20

⟨ ⟩: Adjusted values when divisor is 235. ( ): Values of GK.

astronomical theory, it must be zero in the 3,600th year of the *kali* era, which is the 421st year of the *śaka* era.

4.1.4.1 *Calculation of Precession in* kali *3600*

Nīlakaṇṭha deals with the theory of precession in JM 36.20–37.17, where he quotes SS 3.9–10ab:

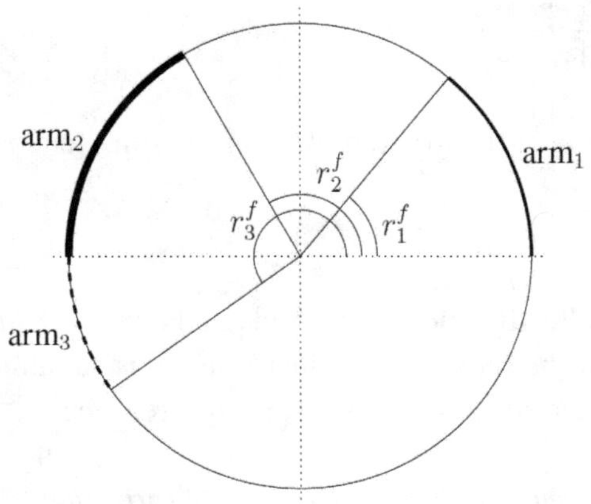

FIGURE 6.1
Arm

(37.2–4)
The ring of stars rotates eastward 600 times in a *yuga*. Whatever is the result of multiplying accumulated days by it and dividing by the number of civil days (in a *yuga*), multiply its arm by 3 and (divide) by 10. The degrees obtained are known as (the motion of) the sun's road.

Let d be accumulated days and D the number of civil days in a *yuga*, the rotation of "the ring of stars" in the accumulated days (r) can be calculated:

$$r = \frac{600d}{D}.$$

At the beginning of a year, we can calculate r using accumulated years (y) and the number of years in a *yuga* (Y) instead:

$$r = \frac{600y}{Y}.$$

In Figure 6.1, let r^f be the fractional part of r. We can get "the arm" of r using r^f. According to SS 2.10, an arm is a past arc in an odd quadrant, while it is an arc to come in an even quadrant. In Figure 6.1, arm_1, arm_2, and arm_3 are the arms of r^f_1, r^f_2, and r^f_3 respectively.

Thus the arm of r ranges from 0° to 90°. Consequently, the motion of the sun's road, which is 3/10 of the arm, changes periodically between 0° to 27°.

At the beginning of *kali*, the motion of the sun's road is set to zero. The r at the beginning of 3,600 of the *kali* era is:

$$r = \frac{600 \cdot 3,600}{4,320,000} = 0.5,$$

which means $r^f = 180°$. Therefore, the arm and the motion of the sun's road is zero.

4.1.5 Y_0 of Rectification 3

As already seen, two Y_0 are used in Rectification 3: 420 and 421. Being evident from the overall discussion, Nīlakaṇṭha seems to consider that 421 is suitable for Y_0 while it is still reasonable to use 420 as a multiplier for two reasons:

4.1.5.1 *Reason 1: Rectification 3 needs another fixed amount of correction.*

Nīlakaṇṭha produces a long argument on this subject (JM 14.16–17.5).

> (14.19–21)
> When 420 is subtracted, the "desire" always increases by 1. Because its difference is constant, the difference of the desired result is also constant. It does not increase after a long time. The desired result (always) exceeds by (a certain amount).

Generally in a proportional calculation:

$$a\,(\text{the measure}) : b\,(\text{the result}) = c\,(\text{the desire}) : d\,(\text{the desired result}),$$
$$d = \frac{bc}{a},$$

when the "desire" c increases by 1, the "desired result" d increases to d' by b/a:

$$d' = \frac{b(c+1)}{a} = \frac{bc}{a} + \frac{b}{a} = d + \frac{b}{a}.$$

This increment b/a is always constant, irrespective of the value of c.

In Rectification 3,

$$\begin{aligned} b_3 &= B_3 \cdot \frac{y - 420}{258} \\ &= B_3 \cdot \frac{y - 421}{258} + \frac{B_3}{258}, \end{aligned}$$

that is, the correction value b_3 is always greater by $B_3/258$ than is the case when 421 is subtracted.

Nīlakaṇṭha confirms this excess, especially concerning the moon, saying:

> (14.22)
> As for the moon, (the observed longitude) is less by 3 to 4 minutes than computed values even when the book was composed.

However, the excess computed for the moon is,

$$B_3^{moon}/258 = 25/258 \text{ minutes} \approx 5.8 \text{ seconds},$$

which is far less than 3 to 4 minutes.

Nīlakaṇṭha also quotes a verse from a lost work of Āryabhaṭa:[9]

> (14.24–15.1)
> It is also the desire of Āryabhaṭa that he introduces a correction for the sun, the moon, and its (*manda-*)*ucca* and node, which are causes of eclipses:
>
> Multiply the number of *yuga*s in a *manu* by 8, 1, 5, 4. The results of positive, negative, negative, and positive are (corrections) for the mean (longitudes) of the four, beginning with the sun.

In SLB 2.22[10] the whole correction procedure is quoted, beginning with this verse:

> vasvekeṣuyugaghnaṃ manuyugam arkādimadhyamacaturṇām /
> dhanam ṛṇam ṛṇam ṛṇam atha[11] kṛtiguṇitaṃ cakreśabhair labdham //
> bhaumāṅgiraśśanīnāṃ deyam ṛṇaṃ deyam abdhinandahṛte /
> sitabudhayor heyaṃ deyaṃ saptahataṃ[12] budhasyoktam //

> Multiply the number of *yugas* in a *manu* by 8, 1, 5, 4. The results of positive, negative, negative, and negative are (corrections) for the mean (longitudes) of the four, beginning with the sun.
>
> The results obtained from multiplying by 20 and dividing by 12, 11, 27 are (corrections) of positive, negative, and positive for Mars, Jupiter, and

9 The same verse is quoted in NAB4, p. 151.
10 edition p. 36.
11 JM: atha deyaṃ for ṛṇam atha.
12 JM: °hṛtaṃ.

TABLE 6.8 Correction by Āryabhaṭa

Planet	Correction
Sun	222
Moon	27.75
Moon MU	138.75
Moon Node	111
Mars	46.25
Jupiter	$50.\dot{4}\dot{5}$
Saturn	$20.\dot{5}$
Venus	138.75
Mercury	$21.58\dot{3}$

> Saturn. What are divided by 4 and 9 are subtracted from and added to Venus and Mercury. As for Mercury what is multiplied by 7 is mentioned.

In Āryabhaṭa's system, a *manu* is a long cycle that includes 72 *yugas*. According to AB 1.5, 27 3/4 *yugas* have gone by in the current *manu*. Thus, these verses gave corrections for the beginning of the current *yuga*. The correction values in seconds[13] computed according to these verses are in Table 6.8.

4.1.5.2 *Reason 2: It is easier to say.*

This reason is very simple and understandable.

> (20.18–22)
> No, subtracting from 3,600 years that elapse from the beginning of *kali* to the *śaka* era, 421 remains. It is, however, 420 here.
> Indeed. It is less just by 1. It is, however, put that way because it is easier to say.

4.2 *Rectification to SSS*

A rectification mentioned in SSS 2.91–93 is cited in JM 25.11–23. This is a long term variation of a 12,000-year cycle.[14]

13 Nīlakaṇṭha says so in JM 16.13.

14 About this rectification, see HMAI pp. 581–582 Table V.32 and SSS part II pp. xx–xxiii.

TABLE 6.9 12,000-year correction

	bīja	Correction to SSS		Max correction
Sun	3	62.1′	1°2′6″	90′
Moon	5	103.5′	1°43′30″	150′
Mars	1	20.7′	0°20′42″	30′
Mercury	62	1283.4′	21°23′24″	1860′
Jupiter	5	103.5′	1°43′30″	150′
Venus	15	310.5′	5°10′30″	45′
Saturn	4	82.8′	1°22′48″	120′
Moon MU	2	41.4′	0°41′24″	60′
Moon Node	2	41.4′	0°41′24″	60′

> (25.11–23)
> Divide an accumulated year passed in the day of *Brahmā* by 12,000. Subtract the remainder from the divisor, divide the smaller of the two by 200. The minutes corresponding to the product of 3 or 5 and the results are subtracted from the sun and moon (respectively). From Jupiter, like the moon. From the moon's (*manda-*)*ucca* and from Venus, the products of 2 and 15 are (subtracted). For the moon's node, Mars, and Saturn, the products of 2, 1, and 4 are added. For Mercury, the product of 62 is chanted. These calculations lead to the coincidence of calculation and observation.

This 12,000-year cycle completes 164,412 times from the beginning of the current *kalpa*, that is, the day of *Brahmā*, to the beginning of the current *kali*. We are currently in the 164,413th cycle.

Let us calculate the correction values of this rectification for 1039 CE, the epoch of SSS. 1039 CE is the 4,140th year from the beginning of the current *kali*. Because 12,000 – 4,140 = 7,860, "the smaller of the two", that is, 4,140 is divided by 200. The result is 20.7. Multiplying it by *bīja*s of each planet given in the verses above, we get the correction value for each, in minutes, as shown in Table 6.9:

This correction takes the maximum value at the 6,000th year of the *kali* era (2899 CE). It is not clear for what purpose this correction is made.

5 Comparing Astronomical Canons

5.1 *Comparing the Rotations of Sun and Moon in* AB *and* SS

The numbers of rotations of the sun (R_s) and moon (R_m) in a *yuga* is 4,320,000 and 57,753,336 respectively in both AB and SS. On the other hand, sidereal days in a *yuga* are, 1,582,237,500 in AB, but 1,582,237,828 in SS, 328 greater than AB. Because the number of civil days in a *yuga* (D) is the difference between the number of sidereal days and the sun's rotation, the difference is the same as between the number of civil days in AB and SS, which is 328.

A mean longitude of the moon (r_m) of accumulated days (d) can be calculated by proportion:

$$r_m = R_m \cdot \frac{d}{D},$$

where r_m for the same d is different in AB and SS because D is different.

Nīlakaṇṭha gives an example of this calculation in JM 18.9–12:

> (18.9–12)
> (For calculating the number of rotations of the sun in SS when 1,577,917,500 days have passed,) 10 *rāśi*s increased by 23 degrees should be subtracted from the (number of) rotations of the sun (in AB, that is, 4,320,000). It means a rotation does not complete (when a *yuga* of AB completes).
>
> The (number of) *tithi*s will be the same if the subtracted number from the rotations of the sun (i.e., 10 *rāśi*s 23 degrees) is added to the number of rotations of the moon.
>
> Therefore, (there is a difference in the numbers of the synodic month in a *yuga* in AB from SS), which will be 11 1/9 synodic months because a synodic month is a rotation of *tithi*.

In these quotations, a *rāśi* is 1/12 of a circle, that is, 30 degrees, and a *tithi* is 1/30 of a synodic month.

When 1 *yuga* of AB completes, that is, 1,577,917,500 civil days elapse, the sun rotates 4,320,000 times in AB, but 4,319,999.102 in SS, which is 0.898 less than that in AB, that is, 10 *rāśi*s 23 degrees. The moon rotates 57,753,336 times in AB and 57,753,324 in SS.

Because the number of synodic months is the difference between the rotations of the sun and of the moon, it is 53,433,336 in AB. While using the numbers from SS it is:

TABLE 6.10 Rotations of Sun and Moon

	Civil days	Sun's rot.	Moon's rot.	Synodic months
SS (*yuga* of SS)	1,577,917,828	4,320,000	57,753,336	53,433,336
SS (*yuga* of AB)	1,577,917,500	4,319,999.102	57,753,324	53,433,324.898
AB (*yuga* of AB)	1,577,917,500	4,320,000	57,753,336	53,433,336

$$57,753,324 - 4,319,999.102 = 53,433,324.898,$$

that is, SS is less by 11.102 ≈ 11 1/9 than AB.

In this calculation, because

$$4,319,999.102 = 4,320,000 - 10^{r\bar{a}\acute{s}i}\ 23^{\circ},$$

$$57,753,324 - (4,320,000 - 10^{r\bar{a}\acute{s}i}\ 23^{\circ}) = (57,753,324 + 10^{r\bar{a}\acute{s}i}\ 23^{\circ}) - 4,320,000,$$

that is, "if the subtracted number from the rotations of the sun is added to the number of rotations of the moon," the number of synodic months and of *tithi*, which is 30 times the number of synodic months, in an AB's *yuga* using SS constants can be calculated in a similar way. All the results are listed in Table 6.10.

5.2 *Examples of Astronomical Calculations*

Many examples appear in JM 27.15–31.18.

- Converting the mean longitude of Mercury and Jupiter from those calculated according to AB to those calculated according to SS.
- Comparing the mean longitude of the sun, the moon, its (*manda-*)*ucca* and the node, calculated according to AB and SSS.

This paper deals with these in discussions on the mean longitude of the moon.

5.2.1 Calculation Time Point

Nīlakaṇṭha sets a point in time for comparing calculation results, which is the end of the 1,683,112th day from the beginning of the current *kali*. We will call this point "Nīlakaṇṭha's epoch." Nīlakaṇṭha calls number 1,683,112 "the accumulated days of Monday" probably because that day is a Monday.

> (27.16–28.2)
> Now, all are calculated using the same accumulated days for knowing how much the difference is among the mean (longitudes) resulting from these

> astronomical canons. For convenience the calculations are made using the accumulated days of 1,683,112. What is the convenience in that case? Because (1,683,112 is) 8/7,500 of a *caturyuga* (of AB) and 8/7,500,000 of a *kalpa*.

Let us call 1/7,500 of a *yuga* of AB a "small *yuga*", which contains 210,389 days (576 years). The 5,625th small *yuga* ends at the beginning of the current *kali*. That point when eight small *yugas* end is Nīlakaṇṭha's epoch, when 1,683,112 days (4,608 years) have passed. This corresponds to the end of March 29th 1429 of the *śaka* era (1507 CE).

5.2.2 Moon of the *Brāhma* School

First, the mean longitude of the moon of the *Brāhma* school in Nīlakaṇṭha's epoch is calculated.

5.2.2.1 *First Method*

This is straightforward and the mean longitude of the moon is calculated using only constants from the *Brāhma* School.

> (30.16–19)
> As for the moon, multiply the accumulated days of Monday by the number of rotations of the moon mentioned by Śrīpati and divide by the accumulated days he mentions (i.e., 1,577,916,450,000). The resulting mean longitude beginning with *rāśis* is rectified to match calculation with observation.
>
> For that purpose, multiply 4,608, i.e., years since (the beginning of current) *kali* (upto Nīlakaṇṭha's epoch) by 5 and divide by 200. From this a negative number beginning with minutes is obtained. It is 12 seconds 55 minutes 1 degree shown beginning with seconds.

Because the mean longitude of the moon at the beginning of the current *kali* is zero, its mean longitude at Nīlakaṇṭha's epoch can be calculated using the accumulated days of Monday by proportion:

$$57,753,300,000 \cdot \frac{1,683,112}{1,577,916,450,000} \approx 61,603 \text{ rotations } 6^{r\bar{a}\acute{s}i}\ 21^{\circ}\ 57'\ 27'',$$

but the result of this calculation is not mentioned. Nīlakaṇṭha just says, "the resulted mean longitude beginning with *rāśis*".

Then the 12,000-year cycle correction explained in Section 4.2 is calculated as mentioned:

$$5 \cdot \frac{4,608}{200} = 115.2' = 1^\circ\ 55'\ 12''.$$

Applying it to the mean longitude, we get $6^{rāśi}\ 20^\circ\ 2'\ 15''$, which is the final mean longitude of the moon in the *Brāhma* School.

5.2.2.2 *Second Method*

Another calculation which gives the same result is also mentioned. This is a bit more complicated.

> (30.19–30)
> Or the difference is calculated separately between it (i.e., the mean longitude of the moon of *Brāhma* school) and (the mean longitude of) the moon of the sunrise-epoch school of Āryabhaṭa. In that case, multipliers for the rectification are applied to rotations in a *yuga* or (in a *kalpa*). When (being applied) to rotations in a *kalpa*, the multipliers for rectification (in a *yuga*) are multiplied by 1,000. (On the other hand,) because the rotation of the moon (in a *kalpa* of SSS) has no fraction (if it is divided by 1,000, it can be converted easily to that in a *yuga*). Therefore, 41 should be subtracted (if you want to convert the rotation of the moon in a *yuga* of AB to that of SSS).
>
> Therefore, when the rotations and so on (obtained from) 41 multiplied by 8 and divided by 7,500 are subtracted from the other (i.e., the mean longitude of the moon of AB, the result is) the mean (longitude of the moon) as mentioned before when the sun's rotation has completed.
>
> It is increased by the motion of the moon during 1 day 7 *nāḍī*s 12 *vināḍī*s.[15] The (numbers) obtained is the mean longitude of the moon of the sunrise-epoch school for the accumulated days of 1,683,112.
>
> What is produced from 41 rotations is 15 degrees 44 minutes 38 seconds 24 thirds. What is produced from the difference of the end of the year is 14 degrees 45 minutes 27 seconds 4 thirds.
>
> (The difference between those two) is 59 minutes 11 seconds 20 thirds. It is negative. It is indeed the difference between the two.

15 A *nāḍī* is 1/60 of a day, a *vināḍī* is 1/60 of a *nāḍī*.

The original number of rotations of the moon per *yuga* are 57,753,336 in AB while 57,753,300 (1/1,000 of numbers in a *kalpa*) in SSS. The difference is 36 rotations per *yuga*.

A 12,000-year correction is applied to SSS. As for the moon, it is negative at a rate of 5 minutes per 200 years. This corresponds to 5 rotations per *yuga*.

Therefore, the difference between AB and SSS is at a rate of 41 rotations per *yuga* as long as the 12,000-year correction increases linearly at the current rate.

The mean longitude of the moon is zero at the beginning of the current *kali* in both AB and SSS. After 8 small *yugas* (8/7,500 *yugas*), the difference between those two (Δ) can be calculated using proportion:

$$1 : 41 = \frac{8}{7,500} : \Delta,$$

$$\Delta = 41 \cdot \frac{8}{7,500} = 15^{\circ}\ 44'\ 38''\ 24'''.$$

Subtracting it from the mean longitude of the moon of AB when 8 small *yugas* of AB (1,683,112 days) are complete, produces the mean longitude of the moon of SSS rectified by the 12,000-year correction when 8 small *yugas* of SSS (1,683,110.88 days) are complete.

Note that the number of days in the 8 small *yugas* of SSS is not 1,683,112 but

$$1,577,916,450 \cdot \frac{8}{7,500} = 1,683,110.88.$$

For comparison, we need the mean longitude of the moon of SSS rectified by the 12,000-year correction at the time when 1,683,112 days elapse, that is, at Nīlakaṇṭha's epoch.

Because the difference between the two "8 small *yugas*" is 1.12 days, (i.e., 1 day 7 *nāḍīs* 12 *vināḍīs*), when we add the motion of the moon during this time (μ) to the mean longitude of the moon of SSS rectified by the 12,000-year correction when 1,683,110.88 days complete, we get the mean longitude of the moon of SSS rectified by the 12,000-year correction when 1,683,112 days complete. We can calculate μ using the constants of SSS as:

$$\mu = 57,753,300 \cdot \frac{1.12}{1,577,916,450} \approx 14^{\circ}\ 45'\ 27''\ 4'''.$$

Thus, against the mean longitude of the moon of AB when 8 small *yugas* of AB (1,683,112 days) complete, Δ is negative and μ is positive. Therefore the difference between Δ and μ, 59′ 11″ 20‴ is subtracted from the mean longitude of the moon of AB when 8 small *yugas* of AB (1,683,112 days) complete, the result is the mean longitude of the moon of SSS rectified by the 12,000-year correction when 1,683,112 days complete.

The mean longitude of the moon of AB when 8 small *yugas* of AB (1,683,112 days) complete is:

$$57,753,336 \cdot \frac{8}{7,500} = 61,603.5584 \text{ rotations}$$
$$= 61,603 \text{ rotations } 6\, r\bar{a}\acute{s}is\ 21^\circ\ 1'\ 26''\ 24'''.$$

Subtracting from it the difference between Δ and μ, 59′ 11″ 20‴, twe get he mean longitude of the moon of SSS rectified by the 12,000-year correction when 1,683,112 days complete, that is, $6^{r\bar{a}\acute{s}i}$ 20° 2′ 15″ 4‴, virtually the same as the result of the first method.

5.2.3 Moon of the Sunrise-Epoch School

As for the mean longitude of the moon of AB, the *bīja* rectification (see section 4.1) is applied.

> (30.30–31.1)
> In addition to those two, the rectification beginning with "diminished by 444" is 1 degree 44 minutes 47 seconds, which is negative.
> (The mean longitude of) the moon of Āryabhaṭa is less than the moon of Śrīpati. The difference is 45 minutes 36 seconds.

Judging from this quote, Nīlakaṇṭha calculated the correction value δ using Rectification 1:

$$\delta = 25' \cdot \frac{1,429 - 444}{235} \approx 1^\circ\ 44'\ 47''\ 14'''.$$

It is not clear why he does not use Rectification 3, which he approves. Anyway, having applied δ to the original mean longitude of the moon of AB, $6^{r\bar{a}\acute{s}i}$ 21° 1′ 26″ 24‴, we get the corrected mean longitude of the moon, $6^{r\bar{a}\acute{s}i}$ 19° 16′ 39″ 10‴, which, as previously mentioned, is less by almost 45′ 36″ than that of SSS.

6 Concluding Remarks

Nīlakaṇṭha stops his discussion about the mean longitude of the moon here. He does not make any judgements about which value is better or which is closer to observations. Since his goal is to rectify astronomical constants by observation, his discussion is insufficient. He takes a similar attitude throughout the volume.

I do not know whether that is what he intended. I hope the missing parts will be found and that this question will be settled.

Acknowledgement

This study is an outcome of the meeting of "Kyoto Seminar for the History of Science in India" held regularly in Kyoto, Japan. Members as of 2021 are: Michio Yano, Takao Hayashi, Takanori Kusuba, Taro Tokutake, and Setsuro Ikeyama.

Abbreviations of Book Titles

AB	*Āryabhaṭīya* of Āryabhaṭa
BSS	*Brāhmasphuṭasiddhānta* of Brahmagupta
GCN	*Grahacāranibandhana* of Haridatta
GCNS	an anonymous *Grahacāranibandhanasaṅgraha*
GK	*Govindakṛti* of Govindasvāmin (lost)
HMAI	David Pingree, *History of Mathematical Astronomy in India*
JM	*Jyotirmīmāṃsā* of Nīlakaṇṭha Somayājin
KK	*Khaṇḍakhādyaka* of Brahmagupta
KKP	First (*pūrva*) part of KK
KKU	Latter (*uttara*) part of KK
LB	*Laghubhāskarīya* of Bhāskara
MB	*Mahābhāskarīya* of Bhāskara
MMN	*Mahāmārganibandhana* of Haridatta (lost)
NAB4	Nīlakaṇṭha's comm. on AB chapter 4
PS	*Pañcasiddhāntikā* of Varāhamihira
SDV	*Śiṣyadhīvṛddhida* of Lalla
SLB	Śaṅkaranārāyaṇa's comm. on LB
SS	*Sūryasiddhānta*
SSL	another *Sūryasiddhānta* of Lāṭadeva summarized in PS
SSS	*Siddhāntaśekhara* of Śrīpati

Bibliography

Articles

HMAI David Pingree, *History of Mathematical Astronomy in India*, Dictionary of Scientific Biography, Supplement I, New York 1978

Editions

AB Ed. by Kripa Shankar Shukla, New Delhi, 1976

BSS Ed. by Sudhākara Dvivedin, Benares, 1902

GCN, GCNS Ed. by K.V. Sarma, Madras, 1954

JM *Jyotirmīmāṃsā, Investigations on Astronomical Theories*, ed. by K.V. Sarma, Hoshiarpur, 1977

KK, KKP, KKU Ed. with Bhaṭṭotpala's comm. by Bīna Chatterjee, 2 vols., Calcutta, 1970

LB, SLB Ed. with Śaṅkaranārāyaṇa's comm. by P.K. Nārāyaṇa Pillai, Trivandrum, 1949 (Reprinted 1974)

MB Ed. with Govindasvāmin's *Bhāṣya* and Parameśvara's supercommentary *Siddhāntadīpikā* by T.S. Kuppanna Sastri, Madras, 1957

NAB4 Ed. by K. Sāmbaśiva Śāstri, University of Travancore Sanskrit Series 185, Trivandrum, 1957

PS, SSL Ed. by O. Neugebauer and D. Pingree, København, Part I 1970, Part II 1971; Ed. by G. Thibaut and Sudhākara Dvivedī, Benares, 1889

SDV Ed. with Mallkārjuna Sūri's comm. by Bīna Chatterjee, New Delhi, 1981

SS Ed. with his own *Tattvāmṛta* comm. by Kapileśvara Śāstri, Kashi Sanskrit Series 144, Benares, 1936

SSS Ed. by Babuāji Miśra, Calcutta, Part I 1932, Part II 1947

CHAPTER 7

Integrating Logarithm Concepts into Syntheses of Sanskrit, Muslim and European Astronomy in 18th-Century Jaipur

Clemency Montelle and Kim Plofker

1 Introduction

Our present understanding of the interactions between Sanskrit and European knowledge systems in early modern India owes a great deal to the work of Christopher Minkowski. His contributions to this subject include not only seminal articles such as (Minkowski 2004a) and (Minkowski 2004b), but also his participation in a fieldwork team led by the late David Pingree in January 1998, with the gracious permission of His Highness Maharaja Sawai Bhawani Singh of Jaipur, to study and catalogue the astronomical manuscripts in Jaipur's Maharaja Man Singh II Museum. Chris produced a significant part of the manuscript cataloguing notes for the astronomical material (in addition to personally couriering a portable photocopier from New York to Jaipur—a fairly challenging mission with the technology of a quarter-century ago!—to facilitate its study). The manuscripts described in the resulting catalogue (Pingree 2003) furnish the main subject of our discussion in this modest tribute to Chris's many accomplishments in the study of Sanskrit literature and knowledge systems.

As described in, e.g., (Pingree 1999) and (Pingree 2002b), the chief catalyst of direct interactions between European astronomy and its Indian counterpart at the court of Maharaja Sawai Jai Singh II of Jaipur (r. 1700–1743) was the Maharaja's acquisition in 1730 of the *Tabulae astronomicae* compiled by the French astronomer-mathematician Philippe de La Hire. At least two decades previously the Maharaja had begun intensely pursuing and promoting the study of astronomy in both Sanskrit and Greco-Islamic traditions, eventually assembling a learned team of Hindu and Muslim scientists with extensive knowledge of Sanskrit, Arabic and Persian text corpora. The theoretical and computational differences between the two traditions[1] did not prevent the accomplishment of

1 Trigonometric orbital models in Indian and Islamic versions of geocentric astronomy, begin-

 DOI:10.1163/9789004536869_009

some brilliant syncretic work by various Jaipur scholars (Montelle, Plofker, and Van Brummelen Forthcoming).

For much of the 1720s and '30s Jai Singh's court astronomers had been engaged upon the creation of an updated *zīj* or system of astronomical tables named *Zīj-i Muḥammad Shāh* in honor of the Mughal emperor. Upon learning in 1727 of recent developments in European astronomy, the Maharaja sought to incorporate this knowledge too into his ambitious scientific project, sending to Europe a delegation headed by Father Miguel Figuereda of the Jesuit mission at Agra. The company returned to Jaipur in 1730 bringing a Portuguese astronomer named Pedro da Silva, along with various instruments and texts including de La Hire's *Tabulae*. A Latin note appended to a manuscript copy of the *Tabulae* in Jaipur in 1732 by the French physician and amateur astronomer Joseph Du Bois summarizes these events in part as follows (Mercier 1993, pp. 161–162); see also (Pingree 2003, pp. 146–148):

> Eventually God, thrice best, by ways unknown to me, led me to the service of a certain great native Ruler by name Sawai Yassang, Prince of Astronomers ... Growing daily, in the various cities of this Empire, are magnificent Observatories with great structures of his own invention and, if one may say it, certainly made by his own hands ... Then this Ruler, discovering the Almagest of Father P. [*recte* G.B.] Riccioli, saw as he had known previously, that there was a great discordance in his native tables, so that I had translated into the language of the Indians the Persian tables ordered by Shahjahan, formerly the Emperor of the whole of East India, which cost 100,000 rupees. Here he found a discordance of up to one degree. As a result a certain Father in the Society of Jesus, of the Portuguese nation, and Rector of the College in Agra of the same Empire, was sent by him (the Ruler) to seek an expert astronomer in Europe. The Father went and returned, and brought with him the tables which I have described, along with other mathematical aids, as a gift from the King of Portugal. A certain young man, educated by the Father in India, and born endowed with great ability, by name Petrus da Silva, also Portuguese, studied astronomy at Riet Clarissima with Father John Baptist Carbone, and came to the Ruler. The Ruler very happily ordered the tables to be transcribed into his script, and ordered all his astronomers to make calculations by them.

---

ning in the early and the mid-first millennium respectively, originally evolved in somewhat different directions from a shared distant ancestor in Hellenistic spherics. They re-converged to some extent when medieval Islamic mathematicians adopted the Indian innovation of the sine and associated functions.

> Now he longs for someone to go to Paris and London to drink of astronomy at the source.

Although the Maharaja's Indian scholars apparently never embarked on the proposed foreign study program, in 1734 he requested from the Jesuit mission at Chandernagore additional European astronomy expertise, which arrived in the form of two French priests, Claude Boudier and Jean François Pons. All these visitors appear to have expounded their subject to the Jaipur court scientists with the aid of de La Hire's *Tabulae*.

In this work de La Hire espoused a 'heliocentrism-agnostic' viewpoint, complaining of Kepler's Rudolphine Tables that their discrepancies with observed data 'can hardly be assigned any other cause than Kepler's own hypothesis, on which his Tables are based. Wherefore I have decided to construct my Tables not on any hypothesis but only upon observations, without regard to any System.'[2] This disavowal of Copernican assumptions was convenient for Jesuit scholars, who employed the *Tabulae* in their astronomical mission work in China (Li 2020) as well as in India. And it seems to have proved acceptable likewise to Jai Singh's Hindu and Muslim geocentrists, who have left no known record of cosmological controversy beyond the syncretic model described in (Minkowski 2004b, pp. 378–379).

But a different aspect of de La Hire's treatise may have posed more of a challenge to its Jesuit expositors and/or its Jaipur students: namely, its use of tabulated logarithms of trigonometric function values. Trigonometric tables were a standard feature of both Sanskrit and Islamic spherical astronomy, but they did not employ the computational aid of composing trigonometric with logarithmic functions that had been devised in early 17th-century Europe by, e.g., Henry Briggs and Adriaan Vlacq on the basis of Napierian logarithms. Such 'log trig' tables, which remained in widespread use until superseded by electronic calculators and computers in the late 20th century, rely on the following fundamental properties of logarithms to reduce multiplicative operations to additive ones:

$$\log(mn) = \log(m) + \log(n), \qquad \log\left(\frac{m}{n}\right) = \log(m) - \log(n),$$

for $m, n > 0$. These conversions were crucial to the complicated calculations of early modern European astronomy, constantly demanding multiplications

2 '... cujus rei vix alia afferri causa potest, quàm Kepleri ipsius hypothesis, cui innixae sunt ejus Tabulae. Quamobrem id statui Tabulas meas nulli hypothesi, sed observationibus tantummodò superstruere, nullâ cujusvis Systematis habitâ ratione' (La Hire 1727, Epistola 2).

and divisions in the spherical versions of the Law of Sines and Law of Cosines in order to solve spherical triangles.

De La Hire in the *Tabulae* neither described the basic principles of logarithm rules nor provided any general tables of logarithms, either of common numbers or of trigonometric function values. But his tables of secondary functions and his computational algorithms depend on such log trig values. One such set of log trig tables that seems to have been associated with de La Hire's procedures was published in 1651 by Adriaan Vlacq (see Appendix B.2).

The assimilation of these methods into Sanskrit at the Jaipur court, on the basis of instruction by Europeans with varying levels of mathematical expertise and linguistic fluency, was unsurprisingly a somewhat slow and difficult process. (Pingree 2002b) reconstructed the timeline of Sanskrit renderings of de La Hire's *Tabulae* as follows:

- About 1732, Jai Singh's chief astronomer or *jyotiṣarāja* Kevalarāma composed the verse *Dṛkpakṣasāraṇī* 'Tables of the Observation school', a 'clumsy effort' that mistranslated some of the Latin algorithms.
- Kevalarāma also made a *nāgarī* transcription of base-10 logarithm tables of common numbers, evidently derived from those of Vlacq: Pingree called this the *Jīvāchāyāsāriṇī* 'tables of sine and tangent' (lit. 'shadow').
- Probably in 1734, there appeared an anonymous 'Sanskrit prose version' omitting eclipse calculations but including all the tables of the original, with mathematically correct algorithms.
- Late 1734 or 1735 saw another anonymous prose version called *Phiraṅgicandracchedyopayogika* '[Concerning] the use of the lunar diagram of the Phiraṅgī (Europeans)', showing better theoretical understanding of the subject, presumably under the more competent instruction of the French Jesuit Fathers.

Our purpose in the remainder of this article is to trace this evolutionary process in the Sanskrit representations of the 'foreign' techniques, as a preliminary to a forthcoming detailed study of their different versions.

2 The Sanskrit 'Translations' of the *Tabulae* and Logarithm Tables

2.1 *The* Dṛkpakṣasāraṇī *of Kevalarāma*

This verse treatise is known or thought to be attested in diverse recensions (Pingree 2002a, 30, n. 12) in the manuscripts listed below; Pingree's anticipated critical edition from five of the manuscripts (Pingree 2002b, p. 430) never appeared in print. Somewhat surprisingly, no copies of the work appear to be preserved in the surviving materials from Jai Singh's court libraries (Pingree 2003, p. xi).

Baroda Oriental Institute 3162. (Pingree 1970–1994, A2, 63b)
Benares 35726. Sampurnanand Sanskrit University Sarasvati Bhavan, Varanasi; accession title *phiraṅgīsiddhānta*. The first chapter ends after verse 41 on f. 3v with the note *iti phiraṃgimate grahaspaṣṭīkaraṇaṃ*, 'thus the true [position] method for the planets in the European view'.
Calcutta Sanskrit College jyotiṣa 55. (Pingree 1970–1994, A3, 23b)
Poona BORI 537 of 1895/1902. (Pingree 1970–1994, A5, 54a)
Poona BORI 926 of 1886/92. (Pingree 1970–1994, A2, 63b)
RORI (Jodhpur) 37733.
RORI (Kota) 728.

2.2 *The 'Jīvāchāyāsāriṇī' or Logarithm Tables of Kevalarāma*

RORI (Jodhpur) Cat. VI 23958. See Appendix B

2.3 *The Anonymous 'Sanskrit Prose Version' of ca. 1734*

This work is preserved in the following manuscripts, along with some fragments of Sanskrit adaptations of the *Tabulae* mentioned in (Pingree 2003, pp. 144–151):

Baroda Oriental Institute 3162. (Pingree 2002b, p. 432)
Khasmohor (Jaipur) 5292, ff. 1–39v. (Pingree 2003, pp. 149–150)
Khasmohor (Jaipur) 5609, ff. 33–39v. (Pingree 2003, pp. 149–150)

2.4 *The Anonymous Prose* **Phiraṅgicandracchedyopayogika**

This work, which may be a revision or continuation of the foregoing, is attested in the following manuscripts:

Berlin or. fol. 2973? Ff. 1–4 and 1 bis–13. With accession title *sūryagrahaṇakarttavyatā* '[what is] to be done for the solar eclipse', evidently from a colophon on f. 13r *iti sūryyagrahaṇakarttavyatā*; compare the ending of the text in the Khasmohor manuscripts (Pingree 2003, p. 150), *iti ravigrahasya gaṇite iti karttavyatā* 'thus in the computation of the solar eclipse, thus [what is] to be done'.
Khasmohor (Jaipur) 5292, ff. 40–52v. (Pingree 2003, pp. 149–150)
Khasmohor (Jaipur) 5609, ff. 40–48. (Pingree 2003, pp. 149–150)
RORI (Kota) 727. (Pingree 2002a, p. 129)

The two separately foliated sections of MS. Berlin or. fol. 2973 contain two related but distinct forms (the first apparently incomplete) of what may be a separate recension of the *Phiraṅgicandracchedyopayogika* or perhaps yet another Sanskrit translation of de La Hire's *Tabulae*.

The second section ends on f. 13^v with the following post-colophon:

> *savāī japure sūryyagrahaṇagaṇikriyate || saṃ 1791 śāke 1656 vaiśākhakṛṣṇā 30 caṃdre madhyādgataghaṭyādi 6 | 47 | 30 horādi 2 | 43 sannamahammadaśahī 16 māsa jīkāda tārīṣa 30 varttamāna madhyādgataghaṭyādi 7 | 48 | 31 varttamāna sannahijarī varttamāna 3146*
>
> In Sawai Jaipur the calc[ulation] of a solar eclipse is made. Saṃ[vat] 1791, Śaka 1656, the dark [half] of Vaiśākha, [*tithi*] 30, on Monday afternoon in *ghaṭīs* etc., 6|47|30, in hours etc, 2|43. In year 16 of [the reign of] Muḥammad Shāh, month from the *zīj* [?], date (*ta'rīkh*) 30, current, afternoon in *ghaṭīs* etc., 7|48|31, current; current Hijra year 3146 [1146?].

This date, which appears unrelated to any historical solar eclipse, corresponds to Tuesday (not Monday) 1 June 1734, which did indeed fall in the sixteenth regnal year of Muḥammad Shāh and in AH 1146 (not 3146).

3 Concepts of Logarithm Use in Sanskrit Prose Recensions of de La Hire's *Tabulae*

Some of the ways in which Kevalarāma wrestled with the use of log trig functions in his *Dṛkpakṣasāraṇī* have been discussed in detail in (Pingree 2002b) and (Pingree 2002a), while his meticulous transcription of common logarithm values is described below in Appendix B. In the remainder of this work, we examine how these concepts fared in the prose successors to the *Dṛkpakṣasāraṇī*, specifically in their treatment of Praeceptum VIII of the *Tabulae* on determining the true longitude and latitude of a superior planet (La Hire 1727, pp. 14–19); see Appendix A.

In our *nāgarī* transcriptions of excerpts from these mansucripts, we have somewhat arbitrarily regularized sandhi, *daṇḍa* punctuation, and paragraphing to emphasize readability of the text. Doubtful readings have been transcribed more faithfully. Longitudes from both Latin and Sanskrit texts are represented in the standard form $x^s, y^\circ; z, w$ signifying x zodiacal signs, y degrees within a sign, z arcminutes and w arcseconds. For simplicity's sake we use the standard modern terms sin, cos and tan for trigonometric functions, even though they are not normalized to unity radius.

3.1 *Computing the Longitude of a Superior Planet in the 'Sanskrit Prose Version'*

3.1.1 The *exemplum* from de La Hire's Praeceptum VIII

De La Hire's exposition of Praeceptum VIII includes a diagram with a list of key features identified, a series of numbered passages describing the procedure (see Appendix A), and then two worked examples for Saturn and Mercury. Each of these two worked examples begins with a prose passage which sets out its date and time, corrected for the equation of time. After this follows a tabular grid with the numerical working (see figure 7.1, which we have rubricated with the numerals of the corresponding steps laid out in Appendix A).

De La Hire's worked example for determining the true longitude and latitude of a superior planet, in this case Saturn, is computed for the date September 24, 1705, $21^h\ 23'1''$. The three data points which form the crux of this example are the mean longitudes of Saturn (2^s, 2°; 24, 7), its apogee (8^s, 19°; 22, 35), and its node (3^s, 22°; 3, 25). The resulting true longitude and latitude of Saturn for this moment are found to be 62°; 25, 29 and 2°; 16, 5 respectively.

Positions, the data from table look-ups and the results of basic arithmetic are tabulated on the left-hand side along with their identification, and the numerical computations involving the logarithms are laid out on the right-hand side. The three 'rule-of-three' proportions (*proportio*) to determine the longitude are set out by listing the names of the four elements vertically, with the corresponding numerical data to the right. The computation of the true latitude of the planet occupies its own cell at the bottom right.

As quickly becomes clear, the worked example is not designed for a reader to follow as a stand-alone procedure. The data in this numerical grid are the results of computations which are not made explicit. There is no indication of the arithmetical operations or table look-ups that have produced the data. Presumably the reader is to follow this example with the more general description of the procedure close at hand.

3.1.2 The Worked Example (*udāharaṇa*) in MS Khasmohor 5292

The description of the procedure for computing the true longitude of a superior planet in the 'Sanskrit prose version' differs substantially from the Latin original. Here we focus on the version of the description given in MS Khasmohor 5292 (f. $7^{r/v}$), which along with MS Baroda 3162 (Pingree 2002b, p. 431) provided the basis for the translated passage in (Pingree 2002b, pp. 450–452). Employing Mars in its example, this description has captured the numerical working in prose, almost always making explicit the arithmetical operations required to produce the numerical data from one step to the next.

The passage in Khasmohor 5292, which does not explicitly mention a date, opens with the phrase *atha udahāraṇaṃ* 'Now, an example'. Presumably the

Exemplum.

Anno 1706. Octobris 26. horâ 10 matutinâ cum 25′ 3″ temporis apparentis Romæ ſtylo novo quærimus veram longitudinem Saturni ♄ & ejus latitudinem.

Tempore propoſito in aſtronomicum completum converſo & ad meridianum Obſervatorii Regii reducto, ac præterea ope longitudinis mediæ Solis ex apparenti facto medio habebimus 1705. annos, Septembrem complet. dies 24. horas 21. cum 23′ & 1″. ſubtraximus enim 42′ propter differentiam meridianorum Romam inter & Obſervatorium, ac prætereà 20′ 2″ pro æquatione temporis; erunt igitur ex tabulis ♄

	♄ s	°	′	″	Aphel. ♄ s	°	′	″	☊ ♄ s	°	′	″
1700 Epocha	11	21	16	1	8	29	14	41	3	21	56	29
5 anni	2	1	9	28			6	49			5	57
Sept. complet.		9	8	37			1	0				54
24 dies			48	14				5				5
21 hor.			1	45	8	29	22	35	3	22	3	25
23′				2	æquatio ☊ ♄ add.						25	51
1″				0	verus locus ☊				3	22	29	16
	2	2	24	7								

	s	°	′	″
Locus med. ♄	2	2	24	7
Aphelium ſubt.	8	29	22	35
Anom. media	5	3	1	32
Æquatio centri ſubtr.		3	8	41
Locus ♄ excentricus	1	29	15	26
II. Anomalia vera	4	29	52	51
Locus medius ☊ ♄	5	22	3	25
Argum. latit. medium	10	7	12	1
Verus locus ☊	3	22	29	16
III. Argument. lat. verum	10	6	46	10
IV. Inclinatio		2	2	55
Reductio addenda			1	37
V. Locus ♄ reduct.	1	29	16	53
Locus Terræ	1	2	31	46
VI. Angul. ad ☉ TSR	0	26	45	7
½ Ang. ad ☉		13	22	33
VII. Côpl. & ſemi-ſum ang.		76	37	27
Semi-diff. ang.		73	28	50
XI. Angul. ad Terr. STR		150	6	17
Verus locus ☉ ab æq.		212	31	46
XII. Longitudo ♄		62	25	29
vel	II	2	25	29

	Diſt. logar. S ♄	4. 95584
	Diſt. logar. ST	3. 99709
proportio	R.	VIII.
	S. C. Incli.	9. 99972
	S ♄	4. 95584
	Diſt. curt. SR	4. 95556
proportio	ST	IX.
	SR	14. 95556
	T. 45°	3. 99709
	T. V	10. 95847
		83° 43′ 15″
		45
	angulus X	38 43 15
proportio	T. 45°	X.
	T. X	9. 90404
	T. ſemiſ.	10. 62381
	T. ſemidiff.	10. 52785
		73° 28′ 50″
	Pro Latitudine	
proportio	S. TSR	Lat.
	S. STR	9. 69759
	T. Inclinat.	8. 55352
		18. 25111
		9. 65334
	T. Latitud.	8. 59777
	Latitud. ♄ Auſt.	2°. 16′. 5″

FIGURE 7.1 An excerpt from de La Hire's *Tabulae* (pp. 17–18) showing an *exemplum* for determining the true longitude and latitude of Saturn.

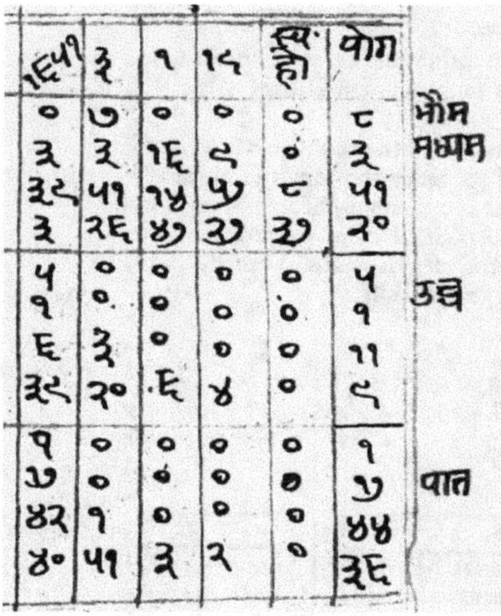

FIGURE 7.2
An excerpted table from MS Khasmohor 5292 (f. 2ᵛ) showing in its rightmost column the mean longitudes of Mars, its apogee and node. These are individually summed up from the longitudinal increments tabulated in the other columns for the year 1651, 3 additional years, month 1, day 19, and some unspecified hour, presumably the afternoon time 6:34 stated earlier.

data are specified for the date given earlier in the manuscript (f. 1ᵛ) as *saṃvat* 1789/*śaka* 1654, Vaiśākha, 14th *tithi* of the bright fortnight, Monday, 6 hours 34 minutes after noon. A small passage using tabular data (f. 2ʳ) converts this into an astronomical date which can be used directly for table look-up: namely, year 1654, month 1, day 19, 6 hours 34 minutes after noon.

A series of small tables follows, providing the mean longitudes of the sun and five planets along with their apogees and nodes, evidently for the abovementioned year, month, day, and time. The relevant table for Mars is reproduced in figure 7.2. From the final column of the table the mean longitudes of Mars, its apogee and node can be read off as $8^s\,3°$; 51, 20, as $5^s\,1°$; 10, 9,[3] and as $1^s\,17°$; 44, 36, respectively.

3.1.3 Sanskrit Text (Khasmohor 5292 f. $7^{r/v}$)

अथोदाहरणं । तत्र तावत्द्युफलसंस्कृतो मध्यमभौमः ८।३।५१।९ । अयमुच्चेन ५।१।१०।९ ऊः ३।२।४१।० । इदं भौममन्दकेन्द्रं जातं । अस्माद्भौममन्दफलं सानुपातमंशाद्यं १०।३९। १८ । मन्दकेन्द्रस्य मेषादिस्थत्वान्मध्यमभौमे ऋणं जातो मन्दस्पष्टो भौमः ७।२३।११।५१ । मन्दफलसंस्कृतमन्दकेन्द्रं २।२२।१।५१ । अथ पातोनो मन्दस्पष्टः ६।५।२७।१५ । अस्य भुजः ०।५।२७।१५ । अस्मान् मध्यशरः सानुपातोंशाद्यः ०।१०।३३ अयं पृथक् स्थापितः । अथ प्रथमफलसंस्कृतपातोनभौमादिद्विकलात्मकं शरफलं १० ऋ । अनेन संस्कृतो मन्दस्पष्टो भौमः ७।२३।११।४१ । शरफलसंस्कृतं मन्दस्पष्टमन्दकेन्द्रं २।२२।१।४१ । अथ

3 The manuscript incorrectly states this as $5^s\,1°$; 11, 9.

रविस्पष्टमन्दकेन्द्रं १०।७।५७।२८। अस्माद्रविकर्णः सानुपातः ४००४६२। अथ सषड्भसूर्यः ७।१८।५।४१। अयं भूध्रुवकः। इदमेव शीघ्रनीचं। अथ भूध्रुवकोनो मन्दस्पष्टः ०।५।६।१० । अस्यार्द्धं ०।२।३३।५। भूध्रुवकोनमन्दस्पष्टस्य षड्राल्पत्वादिदमेव शीघ्रकेन्द्रं। अस्य कोटिः २।२७।२६।५५। अस्यांशाः ८७।२६।५५। अथ स्पष्टमन्दकेन्द्रं २।२२।१५।५१। अस्माद्भौमकर्णः ४१८४७६ । अथ मध्यमशरदंशाः ८९।५० । एषां जीवा ९९९९९९ । अस्याः भौममन्दकर्णस्य योगः १४१८४७५। इदं दशलक्षहीनं शरकोटिज्यामानं रेखा संज्ञं। पुनरिदं दशलक्षयुतं रविकर्णोनं १०१८०१३। इदं प्रथमकोणछा यामानं। अस्माछायासारणी तश्चापं ५३।३३ । इदं पश्चचत्वारिंशदशहीनं ११।३३ । इदं द्वितीयसंज्ञं अस्य छाया ९३१०३९८। शीघ्रकेन्द्रकोदंशछाया ११३५१२९। छायाद्वयंयोगः २०६६१६। दशलक्षहीनः १०६६१६। अस्यछायासुचापं ७७।४२। अयं अन्तरार्द्धकोणः[4] । अयं शीघ्रकेन्द्रकोदंशेषु ८७।२६।५५ शोधितः शेषं ९।४५। अयं फलकोणो जातः। अथ शीघ्रकेन्द्रकोदंशाः अन्तरार्द्धकोणो योजिताः १६५।९। इदं स्पष्टसूर्यस्य स्पष्टभौमान्तरं जातं। इदं स्पष्टसूर्ये शोधितं शेषं ८।२।५६।३१। जातोयं स्पष्टो भौमः। इति भौमस्पष्टीकरणं ॥

Now, an example. In it, the mean [longitude of] Mars corrected for the corresponding *dyuphala* 8|3|51|9. This diminished by [the longitude of] the apogee (*ucca*), 5|1|10|9, 3|2|41|0. This is the resulting *manda*-anomaly of Mars. From that, the *manda*-equation of Mars [is found] by proportion in degrees and so on 10|39|18. Because of the *manda*-anomaly's position in [the six signs] beginning with Aries, [it, i.e., the *manda*-equation, is applied] negatively to the mean [longitude of] Mars; [this] is the resulting [longitude of] Mars, corrected for the *manda*[-equation] 7|23|11|51. The *manda*-anomaly corrected by the *manda*-equation 2|22|1|51.[5] Now, the *manda*-corrected [longitude] diminished by the node 6|5|27|15. Its [reduced] arc (*bhuja*)[6] 0|5|27|15. From that, the mean latitude [is found] by proportion in degrees and so on 0|10|33. This is to be put down separately. Now, from [the mean longitude of] Mars corrected by the first equation and diminished by [the longitude of] the node, the latitude-equation (*śara-phala*), being in seconds (*vikalās*) and so on, 10 neg., [is obtained]. Corrected by this [latitude-equation], the *manda*-corrected [longitude of] Mars 7|23|11|41. The *manda*-anomaly corresponding to the *manda*-corrected [longitude of Mars] corrected by the latitude-equation 2|22|1|41. Now, the corrected *manda*-anomaly of the sun 10|7|57|28. From that, the hypotenuse (*karṇa*) of the sun [obtained] by proportion 400462. Now, [the longitude of] the sun increased by six signs 7|18|5|41. This is

4 Text has been emended from जातोरार्द्धकोणः.

5 This last number should be 42.

6 I.e., reduced down to the first quadrant.

the fixed longitude of the earth (*bhū-dhruvaka*). This is also [known as] the *śīghra*-perigee (*nīca*). Now, the *manda*-corrected [longitude of Mars] diminished by the fixed [longitude of the] earth 0|5|6|10. Half of this 0|2|33|5. Because the *manda*-corrected [longitude of Mars] diminished by the fixed longitude of the earth is less than six [signs], the *śīghra*-anomaly [i.e., 0|5|6|10] is [taken] as it is. Its [i.e., its half] complement (*koṭi*) 2|27|26|55. Its degrees 87|26|55. Now, the corrected *manda*-anomaly 2|22|1|51.[7] From that, the hypotenuse (*karṇa*) of Mars 418476. Now, the degrees [of the complement] of the mean latitude 89|50. The 'sine' (*jīvā*) [i.e., the logarithm of the sine corresponding to this arc found in the table] of these 999999. The sum of these and of the *manda*-hypotenuse of Mars: 1418475. This, diminished by a million (*daśalakṣa*) [lit. ten hundred thousands], is the measure of the [logarithmic] cosine of the 'latitude', called the 'line' (*rekhā*). Again, this is increased by a million (*daśalakṣa*) [and] diminished by the solar hypotenuse 1018013. This is the measure of the 'shadow' (*chāyā*) [i.e., the logarithm of the tangent] of the first-angle (*prathama-koṇa*). From this, [using] the table of the 'shadow', its arc 53|33. This decreased by forty-five 11|33. This is called the 'second'-[angle] (*dvitīya*). Its 'shadow' [i.e., logarithm of its tangent] 9310398. The 'shadow' of the degrees of the complement (*koṭi*) of [half of] the *śīghra*-anomaly 1135129. The sum of the pair of 'shadows' 206616. Diminished by a million 106616. The arc [retrieved from the table in] the 'shadows' [column] of this 77|42. This is the angle of half the difference. This is subtracted from the degrees of the complement of the *śīghra*-anomaly–87|26|55—the remainder 9|45. This is the resulting angle of the equation (*phala-koṇa*). Now, the degrees of the complement of the *śīghra*-anomaly and the angle of half of the difference are joined 165|9. This is the resulting difference of true Mars and the true sun. This subtracted from the true sun; the remainder 8|2|56|31. This is the resulting true [longitude of] Mars. Indeed, the procedure for computing true Mars.

3.1.4 The Steps of the Mars Example in MS Khasmohor 5292

The worked example for Mars in MS Khasmohor 5292 omits many of the helpful intermediary steps and longer explanatory passages included in the translation in (Pingree 2002b, pp. 450–452), among them some more detailed expositions of interpolation procedures and alternative computation methods. The Khasmohor manuscript and the Pingree translation exhibit some distinct numer-

---

7 The MS has stated the incorrect value 2|22|15|51.

ical differences, most notably when citing large numbers retrieved from the logarithmic trigonometric tables. If the numbers in the translation are, as we infer, derived from MS Baroda 3162 (which we have not been able to consult), then their consistent discrepancies in, e.g., table lookups suggest interesting diversity in the ways that individual scribes understood and manipulated logarithmic quantities.

The passage begins by stating the mean longitude $\bar{\lambda}$ of Mars and its apogee longitude λ_A for the time in question, including a correction for the equation of time (*dyuphala*):

$$\begin{aligned} \bar{\lambda} &= 8^s 3^\circ; 51, 9 \\ \lambda_A &= 5^s 1^\circ; 10, 9. \end{aligned}$$

The mean *manda*-anomaly (*manda-kendra*) $\overline{{}_m\kappa}$ is computed by taking their difference:

$$\begin{aligned} \overline{{}_m\kappa} &= \bar{\lambda} - \lambda_A \\ &= 8^s 3^\circ; 51, 9 - 5^s 1^\circ; 10, 9 \\ &= 3^s 2^\circ; 41, 0. \end{aligned}$$

With this computed mean anomaly as argument, the associated *manda*-equation μ ('equation of center') is to be retrieved from a table. This table look-up is indicated somewhat obliquely by the phrase *sānupāta* 'by proportion' (or literally 'increased by a proportional part'), which refers to the process of linear interpolation from tabular entries to produce a non-tabulated value. The appropriate table for Mars is included further on in MS Khasmohor 5292 on f. $25^{r/v}$ and is a direct copy of de La Hire's table XLI on pp. 58–59 (see figure 7.3 and the following excerpt).

⋮	⋮	⋮
2	10°; 38, 45	0,48
3	10°; 39, 33	0,35
⋮	⋮	⋮

The argument value for the tabular look-up is $3^s 2^\circ; 41, 0$. Using linear interpolation between the two entries 2 and 3 for 'sign' 3 (i.e., the zodiacal sign Cancer corresponding to an anomaly between 90° and 120°) with the tabular differences in the third column produces the corresponding *manda*-equation μ:

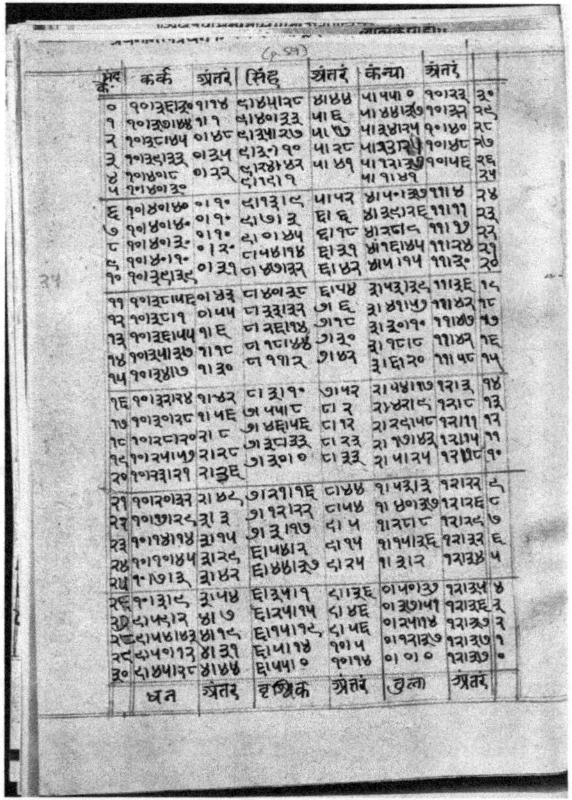

59

Tabula XLI. Æquatio centri ♂.

Subtrahe descendendo.

Ano. Med.	S. III.			Differ.		S. IV.			Differ.		S. V.			Differ.		
G.	G.	M.	S.	M.	S.	G.	M.	S.	M.	S.	G.	M.	S.	M.	S.	
0	10	36	30			9	45	28			5	55	0			30
1	10	37	44	1	14	9	40	33	4	55	5	44	37	10	23	29
2	10	38	45	1	1	9	35	27	5	6	5	34	5	10	32	28
3	10	39	33	0	48	9	30	10	5	17	5	23	25	10	40	27
4	10	40	8	0	35	9	24	42	5	28	5	12	37	10	48	26
5	10	40	30	0	22	9	19	1	5	41	5	1	41	10	56	25
6	10	40	40	0	10	9	13	9	5	52	4	50	37	11	4	24
7	10	40	40	0	0	9	7	3	6	6	4	39	26	11	11	23
8	10	40	30	0	10	9	0	45	6	18	4	28	9	11	17	22
9	10	40	10	0	20	8	54	14	6	31	.4	16	45	11	24	21
10	10	39	39	0	31	8	47	32	6	42	4	5	15	11	30	20
11	10	38	56	0	43	8	40	38	6	54	3	53	39	11	36	19
12	10	38	1	0	55	8	33	32	7	6	3	41	57	11	42	18
13	10	36	55	1	6	8	26	14	7	18	3	30	10	11	47	17
14	10	35	37	1	18	8	18	44	7	30	3	18	18	11	52	16
15	10	34	7	1	30	8	11	2	7	42	3	6	20	11	58	15
16	10	32	24	1	43	8	3	10	7	52	2	54	17	12	3	14
17	10	30	28	1	56	7	55	8	8	2	2	42	9	12	8	13
18	10	28	20	2	8	7	46	56	8	12	2	29	58	12	11	12
19	10	25	57	2	28	7	38	33	8	23	2	17	43	12	15	11
20	10	23	21	2	36	7	32	0	8	33	2	5	25	12	18	10
21	10	20	32	2	49	7	21	16	8	44	1	53	3	12	22	9
22	10	17	29	3	3	7	12	22	8	54	1	40	37	12	26	8
23	10	14	14	3	15	7	3	17	9	5	1	28	8	12	29	7
24	10	10	45	3	29	6	54	2	9	15	1	15	36	12	32	6
25	10	7	3	3	42	6	44	37	9	25	1	3	2	12	34	5
26	10	3.	9	3	54	6	35	1	9	36	0	50	27	12	35	4
27	9	59	2	4	7	6	25	15	9	46	0	37	51	12	36	3
28	9	54	43	4	19	6	15	19	9	56	0	25	14	12	37	2
29	9	50	12	4	31	6	5	14	10	5	0	12	37	12	37	1
30	9	45	28	4	44	5	55	0	10	14	0	0	0	12	37	0
	S. VIII.					S. VII.					S. VI.					G. Ano. Med.

Adde ascendendo.

FIGURE 7.3 The second half of Mars' equation table in MS Khasmohor 5292, f. 25v (left), and Tabula XLI in de La Hire, p. 59 (right)

$$\begin{aligned}\mu &= 10^\circ; 38, 45 + 0, 48 \cdot 41, 0\\ &= 10^\circ; 38, 45 + 0; 0, 32, 48\\ &= 10^\circ; 39, 17, 48.\end{aligned}$$

This is stated, without working, to be (the rounded amount) 10°; 39, 18. This equation is applied negatively (following the indicated symmetries of the tabular data) to the mean longitude of Mars to produce a *manda*-corrected longitude λ^μ and *manda*-corrected anomaly ${}_m\varkappa^\mu$, namely:

$$\begin{aligned}\lambda^\mu &= 8^s 3^\circ; 51, 9 - 10^\circ; 39, 18\\ &= 7^s 23^\circ; 11, 51\end{aligned}$$

and

$$\begin{aligned}{}_m\varkappa^\mu &= \lambda^\mu - \lambda_A\\ &= 7^s 23^\circ; 11, 51 - 5^s 1^\circ; 10, 9\\ &= 2^s 22^\circ; 1, 51 \quad \text{(should be } 2^s 22^\circ; 1, 42)\end{aligned}$$

respectively. The scribe has made an arithmetical error in the last place of the *manda*-corrected anomaly; we retain it since this erroneous value is used in subsequent steps.[8]

Next, the nodal elongation ω or argument of latitude is found. Implicit in this computation is the position of the node $\lambda_N = 1^s 17^\circ; 44, 36$, which can be found[9] earlier in the work:

$$\begin{aligned}\omega &= \lambda^\mu - \lambda_N\\ &= 7^s 23^\circ; 11, 51 - 1^s 17^\circ; 44, 36\\ &= 6^s 5^\circ; 27, 15.\end{aligned}$$

Reducing this arc to its half-circle equivalent[10] $\omega = 5^\circ; 27, 15$, the planetary 'inclination' i (or 'heliocentric latitude', $\angle PSR$; see figure 7.11 and Appendix A) is computed, again using a table. As before, the text indicates this table look-up only by the phrase *sānupāta* ('by proportion'). This refers to the Sanskrit copy of

8 Interestingly, the same error is observed in the translation (Pingree 2002b, p. 450), suggesting that it may have appeared in both MSS.

9 The nodal position is given explicitly in the translation (Pingree 2002b, p. 451), presumably on the basis of the more detailed MS Baroda 3162.

10 The table-values are tabulated for arguments 1–180° and the remaining 180 values are found using symmetry.

de La Hire's table of 'inclination' (table XLIII); Mars' table is found on p. 61 and is copied in MS Khasmohor 5292 f. 26v (see figure 7.4 and the following excerpt).

⋮	⋮
5	0°; 9, 41
6	0°; 11, 36
⋮	⋮

The inclination i corresponding to the argument[11] $\omega = 5°; 27, 15$ can be found via linear interpolation:

$$\begin{aligned} i &= 0°; 9, 41 + (0°; 11, 36 - 0°; 9, 41) \cdot 27, 15 \\ &= 0°; 9, 41 + 0; 0, 52, 13, 45 \\ &= 0°; 10, 33, 13, 45. \end{aligned}$$

The Sanskrit version simply states the rounded value $i \approx 0°; 10, 33$, without any working.

Next the 'latitude-equation' (*śara-phala*) *e*, or longitude correction due to reduction of the planet's position from its orbit to the ecliptic, is to be found. This also must be retrieved from a table (although the text makes no mention of this, not even invoking the phrase *sānupāta*). This involves the Sanskrit copy of de La Hire's table of *reductio* (table XLIV) p. 62, which appears on f. 26v (see figure 7.4). Entering the table at the position shown in the excerpt below with the elongation value $\omega = 5°; 27, 15$ as argument, the user finds (probably by inspection, given the low precision of the values) the corresponding $e = 10$ sub(tractive).

⋮	⋮
5	9
6	11
⋮	⋮

The *nāgarī* symbol ऋ (*ṛ*, abbreviating *ṛṇam* 'subtractive'), repeated twice in the top right-hand cell of the table (probably mistakenly put there instead of the top left-hand side), indicates that these corrections for the first two quadrants are negative. By symmetry, there is a *nāgarī* ध (*dh*, for *dhanaṃ* 'additive') repeated twice in the bottom right-hand side, indicating that the values are to

11 While an additional step is presented in the Latin text to correct the node, and hence find a corrected argument of latitude, this step has not been included in the Sanskrit version.

be added when dealing with arguments in the second half of the circle (quadrants 3 and 4). This is how the Sanskrit scholars appeared to have incorporated de La Hire's phrase *Reductio subtrahenda descendendo* in the table's paratext.

This latitude equation e is then applied to the *manda*-corrected longitude of Mars and its anomaly, whose new values are here denoted by $'$:

$$\lambda^{\mu'} = 7^s 23^\circ; 11, 41$$
$${}_m\varkappa^{\mu'} = 2^s 22^\circ; 1, 41.$$

The text now states that the *manda*-corrected anomaly of the sun is:[12]

$$\varkappa_S^{\mu} = 10^s 7^\circ; 57, 28.$$

From this, the hypotenuse (*karṇa*) of the sun—namely, the distance between the sun and the earth (ST in de La Hire's model)—is found. Again, the phrase *sānupāta* suggests a tabular look-up. This refers to the Sanskrit table on MS Khasmohor 5292 f. 15ᵛ corresponding to de La Hire's table XIV entitled *Solis Diametri, Motus Horarius & distantia a Terra in Logarithmis* (La Hire 1727, p. 20) (see figure 7.5). But the Sanskrit text does not specify the logarithmic nature of these values.

Of note is the use of *daṇḍas* (the vertical punctuation mark |) in the Sanskrit transcription of this table. In the first table (second column), they are used as an indicator of a successive sexagesimal place. Notably, however, in the tabulation of logarithmic values ($\log(ST)$) in the third table (fourth column), a *daṇḍa* corresponds to the decimal point used in the Latin version. This *daṇḍa* marking the decimal point is not carried over into the worked example, and the value is transcribed as a large number with six significant figures. It seems likely the scribe was not quite certain of the meaning of this notation. Entering the table with an argument of $10^s 7^\circ; 57, 28$ (and reading the values from bottom to top, as the argument falls in the second half of the circle):

<table>
<tr><td>⋮</td><td>⋮</td><td>⋮</td></tr>
<tr><td>4|00489</td><td>10</td><td></td></tr>
<tr><td>4|00424</td><td>5</td><td></td></tr>
<tr><td></td><td>0</td><td>10</td></tr>
<tr><td>⋮</td><td>⋮</td><td>⋮</td></tr>
</table>

12 Earlier in MS Khasmohor 5292 (f. 3ᵛ), a worked example (*udāharaṇa*) computes the longitude of the true sun. In the course of the example, it gives the solar apogee as $3^s 8^\circ; 38, 35$ and the solar *manda*-corrected anomaly as $10^s 7^\circ; 57, 28$, from which the true longitude of the sun can be determined: $1^s 18^\circ; 5, 31$.

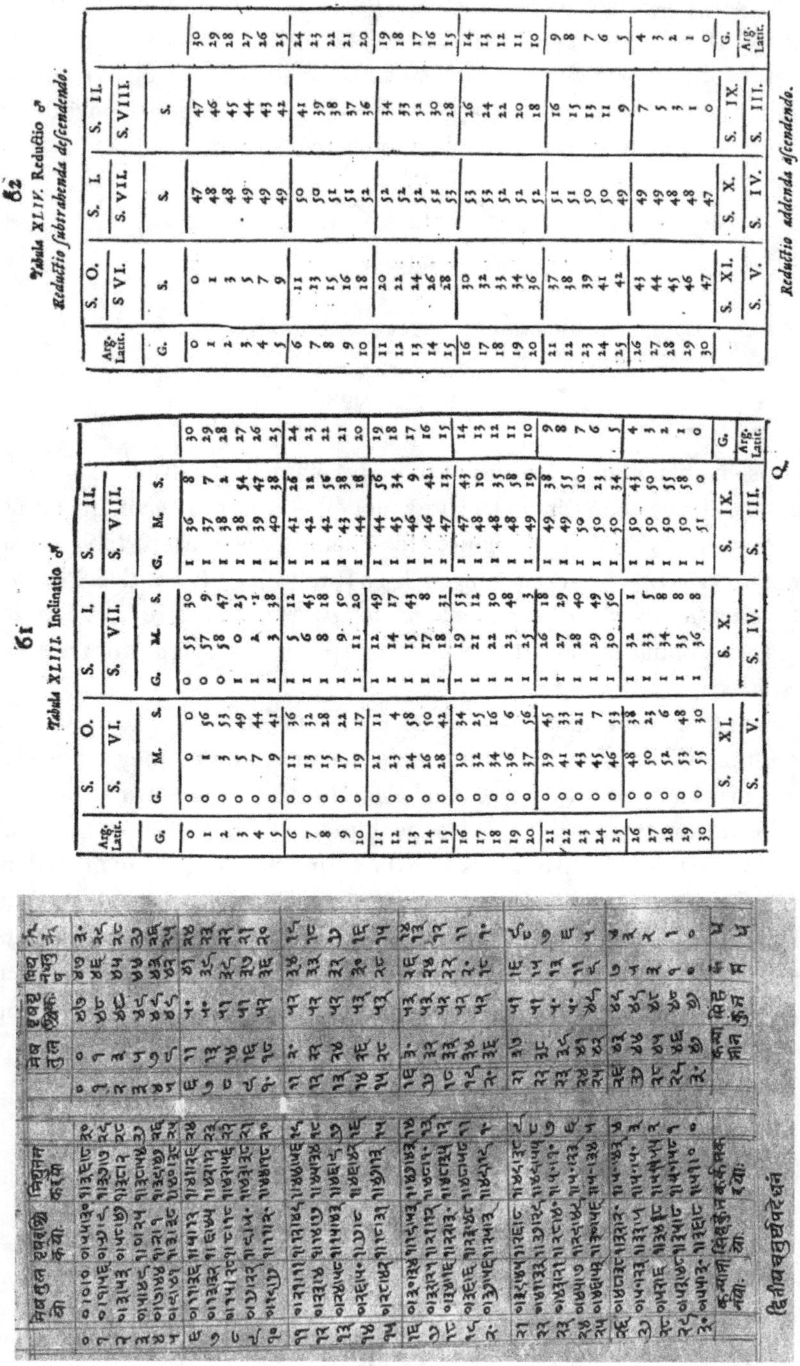

FIGURE 7.4 Tables of Mars' latitude and latitude equation in MS Khasmohor 5292, f. 26ᵛ (left), and de La Hire's corresponding tables of *Inclinatio*, p. 61 (middle) and *Reductio*, p. 62 (right).

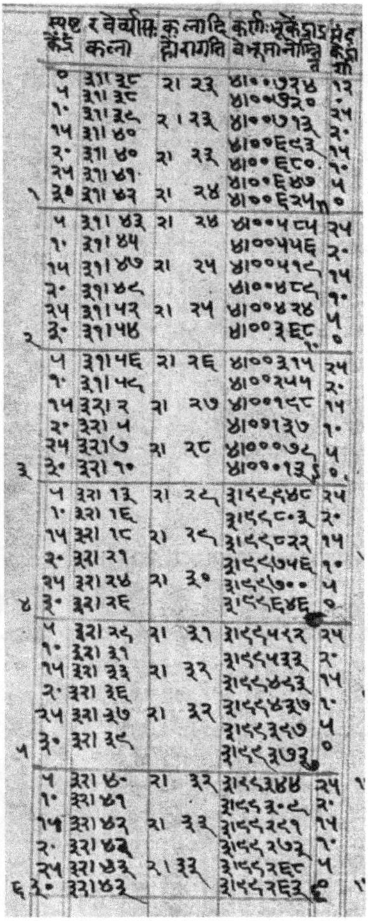

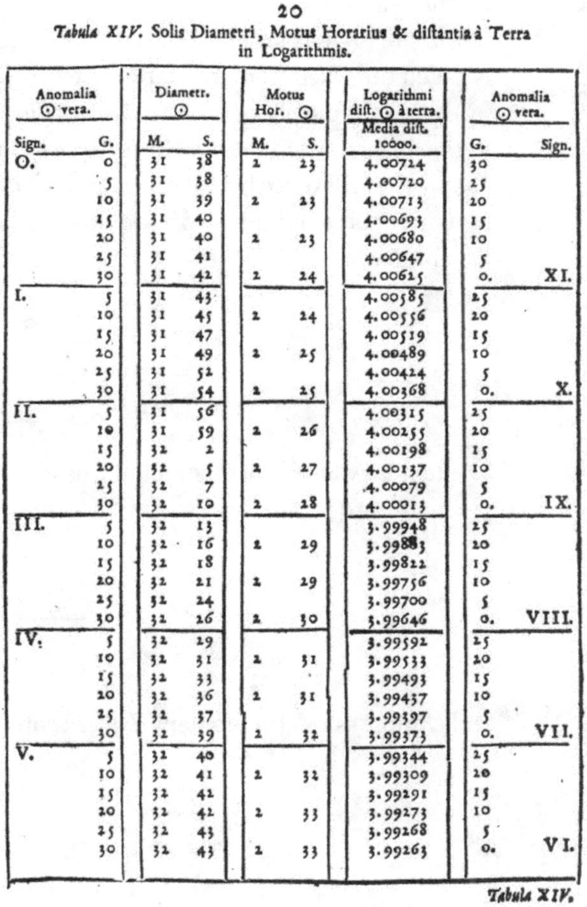

20

Tabula XIV. Solis Diametri, Motus Horarius & diſtantia à Terra in Logarithmis.

Anomalia ⊙ vera.		Diametr. ⊙		Motus Hor. ⊙		Logarithmi diſt. ⊙ à terra. Media diſt. 10000.	Anomalia ⊙ vera.	
Sign.	G.	M.	S.	M.	S.		G.	Sign.
O.	0	31	38	2	23	4.00724	30	
	5	31	38			4.00720	25	
	10	31	39	2	23	4.00713	20	
	15	31	40			4.00693	15	
	20	31	40	2	23	4.00680	10	
	25	31	41			4.00647	5	
	30	31	42	2	24	4.00625	0.	XI.
I.	5	31	43			4.00585	25	
	10	31	45	2	24	4.00556	20	
	15	31	47			4.00519	15	
	20	31	49	2	25	4.00489	10	
	25	31	52			4.00424	5	
	30	31	54	2	25	4.00368	0.	X.
II.	5	31	56			4.00315	25	
	10	31	59	2	26	4.00255	20	
	15	32	2			4.00198	15	
	20	32	5	2	27	4.00137	10	
	25	32	7			4.00079	5	
	30	32	10	2	28	4.00013	0.	IX.
III.	5	32	13			3.99948	25	
	10	32	16	2	29	3.99883	20	
	15	32	18			3.99822	15	
	20	32	21	2	29	3.99756	10	
	25	32	24			3.99700	5	
	30	32	26	2	30	3.99646	0.	VIII.
IV.	5	32	29			3.99592	25	
	10	32	31	2	31	3.99533	20	
	15	32	33			3.99493	15	
	20	32	36	2	31	3.99437	10	
	25	32	37			3.99397	5	
	30	32	39	2	32	3.99373	0.	VII.
V.	5	32	40			3.99344	25	
	10	32	41	2	32	3.99309	20	
	15	32	42			3.99291	15	
	20	32	42	2	33	3.99273	10	
	25	32	43			3.99268	5	
	30	32	43	2	33	3.99263	0.	VI.

Tabula XIV.

FIGURE 7.5 Tables of the Sun's diameter, hourly motion and distance from the earth in MS Khasmohor 5292, f. 15ᵛ (left), and the corresponding tables in de La Hire, p. 20 (right).

Linear interpolation with these tabular entries:

$$
\begin{aligned}
\text{hypotenuse} &= 4|00424 + \left(\frac{(4|00489 - 4|00424)}{10-5} \cdot (7, 57, 28 - 5)\right) \\
&= 4|00424 + \left(\frac{0|00065}{5} \cdot 2, 57, 28\right) \\
&= 4|00424 + 0|00038 \\
&= 4|00462.
\end{aligned}
$$

The text transcribes this number in the worked example as 400462.

The example now turns to the sun, giving the amount of the mean longitude of the sun (given earlier on f. 3ᵛ as $1^s 18°; 5, 41$) increased by six signs. This value is called the 'fixed (longitude) of the earth' (*bhūdhruvaka*, λ_T) and is identified by the text as the perigee of the *śīghra*-epicycle (*śīghra-nīca*). This corresponds to the angle $\angle \Upsilon ST$ in de La Hire's exposition and called *Locus Terrae* (*ex Sole*). Taking the difference between this and the *manda*-corrected longitude of Mars, remainder is called the '*śīghra*-anomaly' (*śīghra-kendra*) ${}_s\varkappa$ (this is $\angle TSR$ in de La Hire's text):

$$\begin{aligned} {}_s\varkappa &= \lambda^\mu - \lambda_T \\ &= 7^s23°; 11, 51 - 7^s18°; 5, 41 \\ &= 0^s5°; 6, 10. \end{aligned}$$

The text now instructs that half of this is to be taken:

$$\begin{aligned} \frac{{}_s\varkappa}{2} &= \frac{\lambda^\mu - \lambda_T}{2} \\ &= \frac{0^s5°; 6, 10}{2} \\ &= 0^s2°; 33, 5 \end{aligned}$$

and computes its complement (*koṭi*),[13] converting it to degrees:

$$\begin{aligned} 90° - \frac{{}_s\varkappa}{2} &= 3^s0°; 0, 0 - 0^s2°; 33, 5 \\ &= 2^s27°; 26, 55 \\ &= 87°; 26, 55 \quad \text{beginning with degrees.} \end{aligned}$$

The next step involves determining the hypotenuse (*karṇa*) of Mars. This is referred to as *distantia planetae a Sole* or log *SP* in de La Hire's text. The text is not explicit about how this is to be found, but the value is retrieved from a table using linear interpolation. This value too is a logarithmic quantity found in a Sanskrit copy of de La Hire's table entitled *Logarithmi distantiae* (table XLII) on p. 60 and is copied on MS Khasmohor 5292 f. 26ʳ (see figure 7.6). The argument used to enter the table is the corrected *manda*-anomaly ${}_m\varkappa^\mu = 2^s22°; 1, 51$. The value 418476 is recorded by the text as the associated hypotenuse (*karṇa*) of Mars.

13 Pingree translates this term as 'cosine' (Pingree 2002b, p. 451), but the quantity is actually the angular argument entered into the log-sine table in order to extract the cosine.

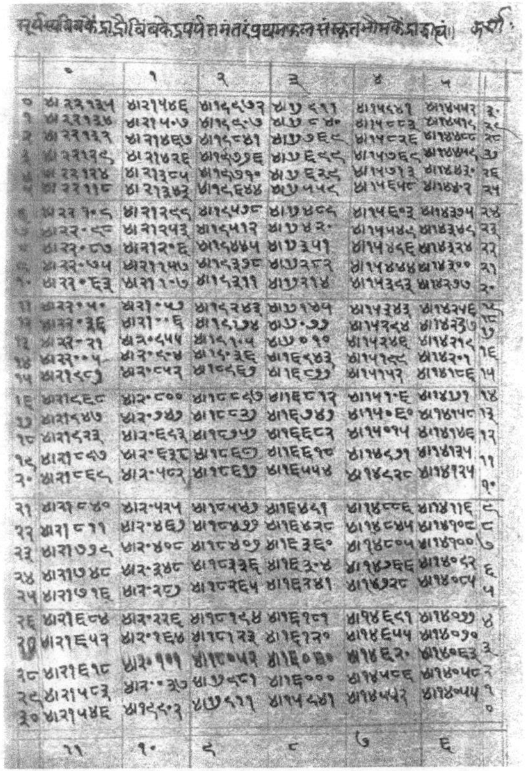

60

Tabula XLII. Logarithmi distantiæ ♂ à ☉ in singulos gradus Anomaliæ veræ vel Coæquatæ.

Ano. vera.	S. O.	S. I.	S. II.	S. III.	S. IV.	S. V.	
0	4. 22135	4. 21546	4. 19972	4. 17911	4. 15941	4. 14552	30
1	4. 22134	4. 21507	4. 19907	4. 17840	4. 15883	4. 14519	29
2	4. 22132	4. 21467	4. 19842	4. 17769	4. 15826	4. 14488	28
3	4. 22129	4. 21426	4. 19776	4. 17699	4. 15769	4. 14459	27
4	4. 22124	4. 21385	4. 19710	4. 17629	4. 15713	4. 14430	26
5	4. 22118	4. 21343	4. 19644	4. 17559	4. 15658	4. 14402	25
6	4. 22109	4. 21299	4. 19578	4. 17489	4. 15603	4. 14375	24
7	4. 22098	4. 21253	4. 19512	4. 17420	4. 15549	4. 14349	23
8	4. 22087	4. 21206	4. 19445	4. 17351	4. 15496	4. 14324	22
9	4. 22075	4. 21157	4. 19378	4. 17282	4. 15444	4. 14300	21
10	4. 22063	4. 21107	4. 19311	4. 17214	4. 15393	4. 14277	20
11	4. 22050	4. 21057	4. 19243	4. 17145	4. 15343	4. 14256	19
12	4. 22036	4. 21006	4. 19174	4. 17077	4. 15294	4. 14237	18
13	4. 22021	4. 20955	4. 19105	4. 17010	4. 15246	4. 14219	17
14	4. 22005	4. 20904	4. 19036	4. 16943	4. 15199	4. 14202	16
15	4. 21987	4. 20852	4. 18967	4. 16877	4. 15152	4. 14186	15
16	4. 21968	4. 20800	4. 18897	4. 16812	4. 15106	4. 14171	14
17	4. 21947	4. 20747	4. 18827	4. 16747	4. 15060	4. 14158	13
18	4. 21923	4. 20693	4. 18757	4. 16682	4. 15015	4. 14146	12
19	4. 21897	4. 20638	4. 18687	4. 16618	4. 14971	4. 14135	11
20	4. 21869	4. 20582	4. 18617	4. 16554	4. 14928	4. 14125	10
21	4. 21840	4. 20525	4. 18547	4. 16491	4. 14886	4. 14116	9
22	4. 21811	4. 20467	4. 18477	4. 16428	4. 14845	4. 14108	8
23	4. 21779	4. 20408	4. 18407	4. 16366	4. 14805	4. 14100	7
24	4. 21748	4. 20348	4. 18336	4. 16304	4. 14766	4. 14092	6
25	4. 21716	4. 20287	4. 18265	4. 16242	4. 14728	4. 14085	5
26	4. 21684	4. 20226	4. 18194	4. 16181	4. 14691	4. 14077	4
27	4. 21652	4. 20164	4. 18123	4. 16120	4. 14655	4. 14070	3
28	4. 21618	4. 20101	4. 18052	4. 16060	4. 14620	4. 14063	2
29	4. 21583	4. 20037	4. 17981	4. 16000	4. 14586	4. 14058	1
30	4. 21546	4. 19972	4. 17911	4. 15941	4. 14552	4. 14055	0
	S. XI.	S. X.	S. IX.	S. VIII.	S. VII.	S. VI.	Ano. vera.

Tabula XLIII.

FIGURE 7.6 Table of the distance between Mars and the sun in MS Khasmohor 5292, f. 26r (left), and de La Hire's corresponding table XLII, p. 60 (right). Again, the Sanskrit version has used *daṇḍas* in place of the decimal points in the Latin original.

The text then recalls the inclination i and states its complement

$$\begin{aligned} 90^\circ - i &= 3^s 0^\circ; 0, 0 - 0^s 0^\circ; 10, 33 \\ &= 2^s 29^\circ; 49, 27 \\ &= 89^\circ; 49, 27. \end{aligned}$$

It follows this by declaring 'the sine (*jīvā*) of this is 999999'. This term *jīvā*, usually reserved for the trigonometric sine function, in this context directs one to look up an entry in a table of logarithmic trigonometric functions; thus *jīvā* in this text here means 'the logarithm of the sine of a given arc'. In this case, one needs a set of tables, such as the *Jīvāchāyāsāriṇī* (see Appendix B), which tabulates the log-sines and log-tangents for given arcs. Entering the table with an argument arc of 89°; 49, 27 in the *jīvā* column (see figure 7.7):

$$\begin{aligned} \log\cos(0^\circ; 10, 33) &= \log\sin(90 - 0^\circ; 10, 33) \\ &= \log\sin(89^\circ; 49, 27) \\ &\approx 99999980 \end{aligned}$$

as

89	⋮	⋮
	49	99999978
	50	99999982
	⋮	⋮

using linear interpolation and rounding.

However, MS Khasmohor 5292 records this as 999999, dropping the last two digits. This may suggest the scribe was using a slightly less precise table of log-trig functions, or was purposefully including less precision than the tabular entries provided. In comparison, the translated version (Pingree 2002b, pp. 451–452) occasionally provides more precision in its log-trig quantities, implying that the scribe of MS Baroda 3162 may have had access to the more precise table. The precision of six significant figures is reminiscent of the number given by de La Hire in his *exemplum* (see, for instance, figure 7.8). The scribe of MS Khasmohor 5292 may have been influenced by this.

Now, the example turns to the three rule-of-three proportions. De La Hire invokes the first rule-of-three which computes the *distantia curtatam* ('shortened distance', *SR*) in his Praeceptum as

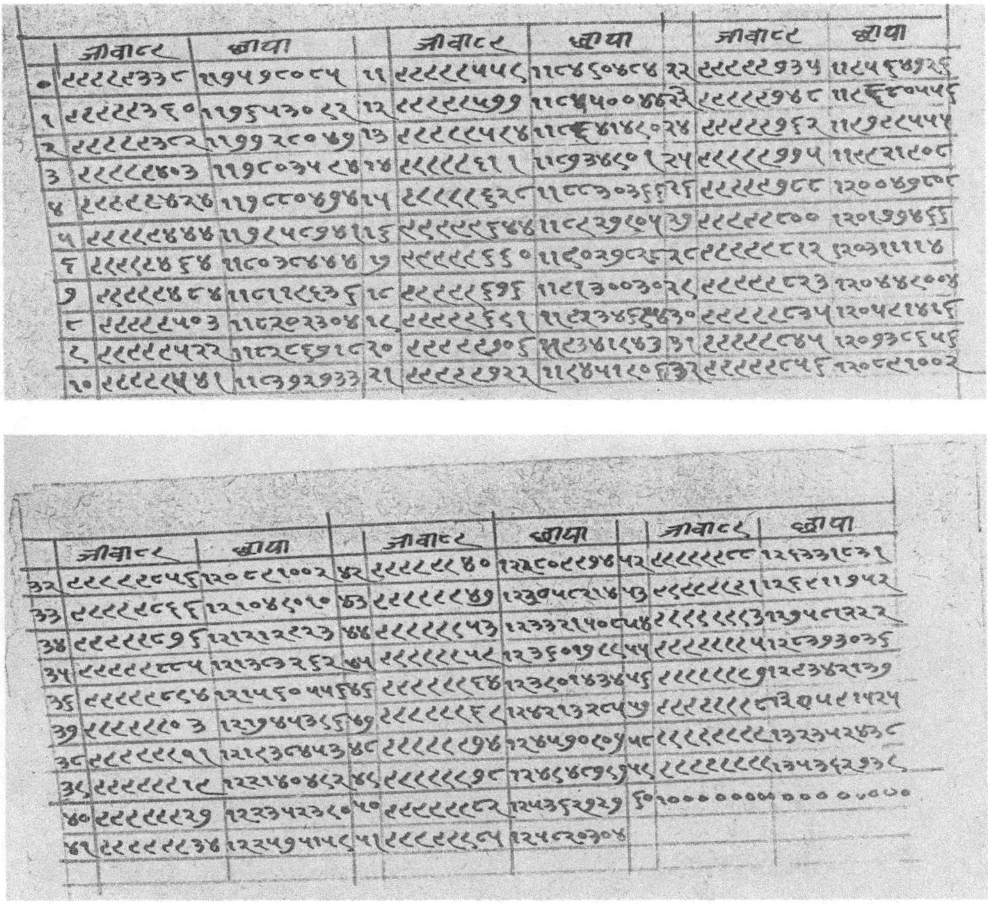

FIGURE 7.7 An excerpt from the *Jīvāchāyāsāriṇī* (RORI 23958, ff. 90ᵛ–91ʳ) showing the log sine and log tangents for degree 89 and its 60 subdivisions.

VIII. If the Radius should be to the Sine of the complement of the inclination of the planet as the planet's distance from the Sun taken from the table is to a fourth term, the last will be the distance of the reduced planet from the Sun, which they call the shortened distance.

Equivalently,

$$\frac{\text{Radius}}{\text{sine of complement of the inclination}} = \frac{\text{planet's distance from sun}}{\text{distance of reduced planet from sun}}$$

$$\frac{\text{Radius}}{\cos(\angle PSR)} = \frac{SP}{SR}.$$

De La Hire's technique for easing the arithmetical complexity of computing this rule-of-three is to take the logs of both sides, so that operations of division and multiplication can be reduced to addition and subtraction:

$$\log SR = \log \cos(\angle PSR) + \log SP - \log \text{Radius}.$$

This is set out in the *exemplum* as shown in figure 7.8:

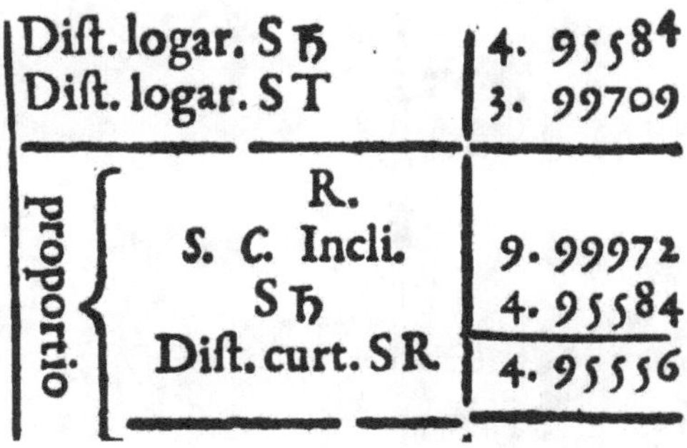

FIGURE 7.8
Shortened-distance computation from de La Hire's *exemplum* for Saturn (figure 7.1).

where *R.* is the Radius, *S.C. Incli.* is the sine of the complement (i.e., the cosine) of the inclination, and *Dist. curt. SR* is the shortest distance. These are all, albeit not explicitly indicated to be, logarithmic values. The value of the log of Radius (10.00000) is not expressed and its subtraction has been carried out silently.

In comparison, the Sanskrit version (see section 3.1) simply states:

> The sum of these and of the *manda*-hypotenuse of Mars: 1418475. This, diminished by a million is the measure of the cosine of the 'latitude', called the 'line' (*rekhā*).

suggesting the following arithmetical operations:

$$\begin{aligned} \textit{rekhā} &= \log \cos(i) + \textit{karṇa} - 1000000 \\ &= 99999 + 418476 - 1000000 \\ &= 1418475 - 1000000. \end{aligned}$$

While the Sanskrit text tells the calculator to reduce the sum by a million, this subtraction is not registered numerically. The text retains the number 1418475. Overall, this doesn't affect the computation, as in the next rule-of-three, one is to *add* a million.[14]

14 The translated version of the text (Pingree 2002b, p. 451) makes no mention of this subtraction.

The next rule-of-three proportion is expressed by de La Hire as follows:

> IX. Besides, let the distance of the earth from the Sun be to the shortened distance in the cases of the superior planets (namely Saturn, Jupiter, and Mars), but the shortened distance to the distance of the earth from the Sun in the cases of the inferior planets (namely Venus and Mercury), as the Tangent of 45 degrees is to the Tangent of a certain angle which I call V, from which, when 45 degrees have been subtracted, I call the remaining angle X.

In other words:

$$\frac{\text{distance of the earth from the Sun}}{\text{shortened distance}} = \frac{\text{tangent of 45 degrees}}{\text{tangent of V}}$$

$$\frac{ST}{SR} = \frac{\tan(45^\circ)}{\tan(V)}.$$

Again, de La Hire's technique is to take the logs of both sides, reducing the operations of division and multiplication to addition and subtraction:

$$\log\tan(V) = \log SR + \log\tan 45 - \log ST.$$

This is set out in the *exemplum* as shown in figure 7.9:

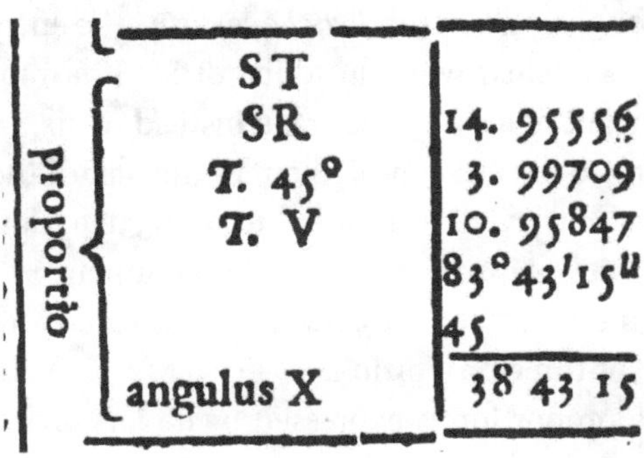

proportio	
ST	14. 95556
SR	3. 99709
T. 45°	10. 95847
T. V	83°43′15″
	45
angulus X	38 43 15

FIGURE 7.9
Earth-sun distance computation from de La Hire's *exemplum* for Saturn (figure 7.1).

In this same tabular cell, the arc V corresponding to $\log\tan(V)$ is extracted from the tables. This arc is then decreased by 45° to produce X.

The Sanskrit version states:

> Again, this is increased by a million (*daśalakṣa*) [and] diminished by the solar hypotenuse 1018013. This is the measure of the 'shadow' (*chāyā*, i.e., the logarithm of the tangent) of the 'first angle' (*prathama-koṇa*).

This 'first angle' is equivalent to *V* in de La Hire's text. In other words:

$$\begin{aligned}\log\tan(\textit{prathama-koṇa}) &= \textit{rekhā} + 1000000 - \text{solar } \textit{karṇa}\\ &= 1418475 - 400462\\ &= 1018013.\end{aligned}$$

The Sanskrit text then states the arc corresponding to this value, by indicating it is to be retrieved from the table of 'shadows' (*chāyā-sāriṇī*). This is the most explicit reference in MS Khasmohor 5292 to a table like the *Jīvāchāyā-sāriṇī*. Table look-up produces a corresponding arc of 56°; 33. The tabular value corresponding to an arc of 56°; 33 in the *Jīvāchāyāsāriṇī* (f. 58r) is 101800408. The tabular value corresponding to an arc of 56°; 34 is 101803156. The Sanskrit text has settled on the closest tabulated arc, 56°; 33, without introducing further sexagesimal places. Next, the first angle:

$$\textit{prathama-koṇa} = 56^\circ; 33$$

is decreased by 45° to produce the second arc (*dvitīya-koṇa*). This is equivalent to angle *X* in de La Hire's example.

$$\begin{aligned}\textit{dvitīya-koṇa} &= 56^\circ; 33 - 45^\circ\\ &= 11^\circ; 33.\end{aligned}$$

The 'shadow' of this arc is stated to be 9310398. In the *Jīvāchāyāsāriṇī*, the entry for argument 11°; 33 is 93103985. It is unclear why the scribe of MS Khasmohor 5292 has included seven significant figures at this stage instead of six. In addition, the 'shadow' corresponding to the arc-complement of the half of the *śīghra*-anomaly[15] (87°; 26, 55) is also retrieved. It is stated to be 1135129, notably six significant figures. In the *Jīvāchāyāsāriṇī*, the entry for argument 87°; 26 is 113484621 and the entry for argument 87°; 27 is 113412956. These values are required for the third and final rule-of-three proportion.

This third and final rule-of-three proportion is expressed by de La Hire as follows:

> x. Finally let the Tangent of 45 degrees be to the Tangent of angle *X* as the Tangent of half the sum of the unknown angles which was found in

15 MS Khasmohor 5292 fails to state that the half of this *śīghra*-anomaly is intended.

the seventh article is to the Tangent of half the difference of the same unknown angles.

In other words:

$$\frac{\text{tangent of 45 degrees}}{\text{tangent of angle} X} = \frac{\text{tangent of half the sum of the unknown angles}}{\text{tangent of half the difference of the same unknown angles}}$$
$$\frac{\tan(45^\circ)}{\tan(X)} = \frac{\tan\left(\frac{\angle STR}{2} + \frac{\angle SRT}{2}\right)}{\tan\left(\frac{\angle STR}{2} - \frac{\angle SRT}{2}\right)}.$$

Here, the unknown angles in the phrase 'half the sum of the unknown angles ('semi-sum')' and 'half the difference of the unknown angles ('semi-difference')' are $\angle STR$ and $\angle SRT$ respectively. The angle $\angle STR$ is called *angulus dictus ad Terram* ('the so-called angle at the earth') by de La Hire. De La Hire notes that half the sum of these angles was found in the seventh article, which invokes the triangle $\triangle STR$, namely

$$180 = \angle TSR + \angle STR + \angle SRT$$

so that

$$90 = \frac{\angle TSR}{2} + \frac{\angle STR}{2} + \frac{\angle SRT}{2}$$

or

$$90 - \frac{\angle TSR}{2} = \frac{\angle STR + \angle SRT}{2}.$$

This is where the term 'semi-sum' arises.

Again, de La Hire's technique is to take the logs of both sides, to reduce the operations of division and multiplication to addition and subtraction:

$$\log\tan\left(\frac{\angle STR - \angle SRT}{2}\right) = \log\tan(X) + \log\tan\left(\frac{\angle STR + \angle SRT}{2}\right) - \log\tan 45^\circ.$$

This is set out in the *exemplum* as shown in figure 7.10:

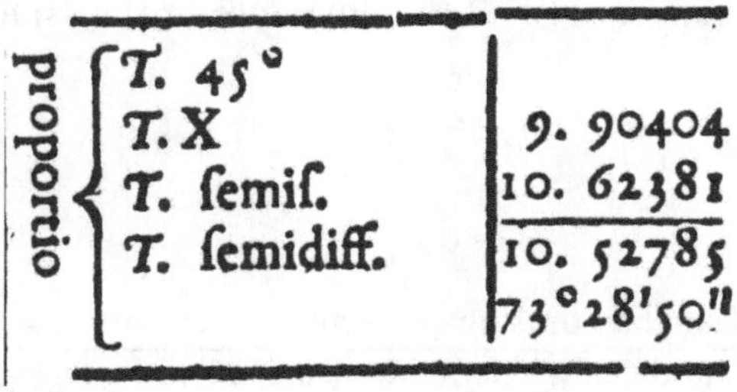

proportio

T. 45°	
T. X	9. 90404
T. ſemiſ.	10. 62381
T. ſemidiff.	10. 52785
	73°28′50″

FIGURE 7.10
Earth-angle computation from de La Hire's *exemplum* for Saturn (figure 7.1).

where *T. semis.* and *T. semidiff.* are abbreviations for the tangent of the semi-sum and semi-difference respectively.

The Sanskrit version states:

> The 'shadow' of the degrees of the complement (*koṭi*) of (half of) the *śīghra*-anomaly 1135129. The sum of the pair of 'shadows' 206616. Diminished by a million 106616.

Notably the Sanskrit version does not invoke a term equivalent to semi-sum. But, as in de La Hire's example, this rule-of-three involves this quantity. The semi-sum is equivalent to half the *śīghra*-anomaly subtracted from 90° which was computed previously:

$$
\begin{aligned}
90 - \frac{\angle TSR}{2} &= 90^\circ - \frac{{}_{s}\varkappa}{2} \\
&= 87^\circ;26,55 \\
&= \frac{\angle STR + \angle SRT}{2} \quad \text{the semi-sum in de La Hire.}
\end{aligned}
$$

There are some numerical issues in the Sanskrit version of this rule-of-three concerning the number of significant digits carried through in the computations.[16] This may indicate a certain unfamiliarity with the arithmetic of logarithms. Mathematically speaking, the correct amounts must be interpreted to be:

16 This is made even more prominent by the fact that the translation (Pingree 2002b, p. 451) includes more significant figures in these computations. It states these relevant values as 93103985 and 113512956, their sum 206616941, and the amount produced when reduced by the Radius (the text explains this as 'the terminal number is diminished by 1') as 106616941.

$$\begin{aligned}\text{log tan(‘angle of half}& \\ \text{of the difference’)} &= \text{log tan (‘second angle’)} + \text{log tan}(90 - \frac{{}_s\varkappa}{2}) - 1000000 \\ &= 1135129 + 931039 - 1000000 \\ &= 2066168 - 1000000 \\ &= 1066168.\end{aligned}$$

The text declares the arc associated with this log tan value (1066168) is 77°; 42. In the *Jīvāchāyāsāriṇī* (f. 79r), the entry for argument 77°; 42 is 106614733 and the entry for argument 77°; 43 is 106620806. The Sanskrit prose text has obviously rounded to the closest tabulated value using six significant figures. By this point, all the relevant data has been generated and the true longitude of Mars can be determined by the simple arithmetic of angles.

According to de La Hire's procedure for this final stage, the true longitude of the planet is given by angle $\angle ITR$, which can be computed via:

$$\lambda = \angle ITR = \angle ITS \pm \angle STR$$

where $\angle ITS = \lambda_S$ is the true longitude of the sun and $\angle STR$ determined via:

$$\begin{aligned}\angle STR &= \text{sum of semi-sum and semi-difference} \\ &= \frac{\angle STR + \angle SRT}{2} + \frac{\angle STR - \angle SRT}{2} \\ &= \angle STR.\end{aligned}$$

The Sanskrit text proceeds as follows. First, it finds the difference between the semi-sum and the semi-difference:

$$87^\circ; 26, 55 - 77^\circ; 42 = 9^\circ; 45.$$

It calls this the 'angle of the equation' (*phala-koṇa*).[17] MS Khasmohor 5292 proceeds no further with this quantity, although the translation (Pingree 2002b, p. 452) notes that this 'equation' can be added to the corrected longitude of Mars to produce the true longitude of Mars:

$$\begin{aligned}\lambda &= 7^s 23^\circ; 11, 41 + 9^\circ; 45 \\ &= 8^s 2^\circ; 56, 41.\end{aligned}$$

17 Glossed as 'the *śīghra* equation' in (Pingree 2002b, p. 452).

The Khasmohor 5292 and translated versions agree in computing the sum of the semi-sum and semi-difference:

$$\begin{aligned} 87^\circ; 26, 55 + 77^\circ; 42 &= 165^\circ; 9 \\ &= 5^s 15^\circ; 9. \end{aligned}$$

This is called the 'difference of true Mars and the true Sun'. The desired true longitude of Mars is then extracted by subtracting this value from the true longitude of the sun:

$$\begin{aligned} \lambda &= \text{(true sun)} - \text{(difference of true Mars and true Sun)} \\ &= 1^s 18^\circ; 5, 41 - 5^s 15^\circ; 9 \\ &= 8^s 2^\circ; 56, 41. \end{aligned}$$

(MS Khasmohor 5292 erroneously records this difference as $8^s\ 2^\circ; 56, 31$.) The scribe definitively ends the example here with the phrase *iti bhaumaspaṣṭīkaraṇam*. It does not continue on to find the true latitude of Mars.

3.2 *Mars Computations in* MS *Berlin or. fol. 2973*, 'Phiraṅgicandracchedyopayogika'

3.2.1 Superior Planet Orbit Diagram

Although we have not yet been able to compare MS Berlin or. fol. 2973 directly with the known manuscripts of the *Phiraṅgicandracchedyopayogika* listed in section 2.4, its content shares a number of features with the description of that work in (Pingree 2002b, pp. 431, 436). In particular, as shown in figure 7.11, it painstakingly replicates a slightly modified version of de La Hire's diagram in the *Tabulae* of the orbit of a superior planet (labeled *bhauma* 'Mars' as well as *pa* '*P*' in the Sanskrit version). The rayed disk of the sun is labeled *sū* for *sūrya* 'sun' (or '*S*'), the earth on its eccentric orbit is labeled *ta* for '*T*', and so forth.

The text at this point briefly sums up de La Hire's description of the diagram, beginning as follows on f. 7^r.

अथ भौमादीनं स्फुटीकरणं सच्छेद्यकं प्रोच्यते ॥

तत्र जीचादिष्टे ऽह्नि सूर्यमध्यमानयनवदेतेषां मध्यमानानीय सूर्याद्युफलं च गृहीत्वा तद्गतिमुत्थाप्य तेन संस्कृतं मध्यमादिकं कृत्वा पुनर्मन्दकेन्द्रेण मन्दफलं समानीय तेन संस्कृतं कृत्वा तदनछेद्यकेन स्फुटीकरणं विधेय ॥

तदत्र छेद्यकमुच्यते ॥

तत्र सू इति सूर्यबिम्बम् ॥ पुनरत्र सूतरनबओ क्रान्तिवृत्तधरातलम् । तथा अनपओ विमण्डलधरातलसूर्यादुपरिस्थेष्वस्ति नओ सम्पातौ स्तः । नओरेखाविमण्डलक्रान्तिवृत्तयोर्धरात-

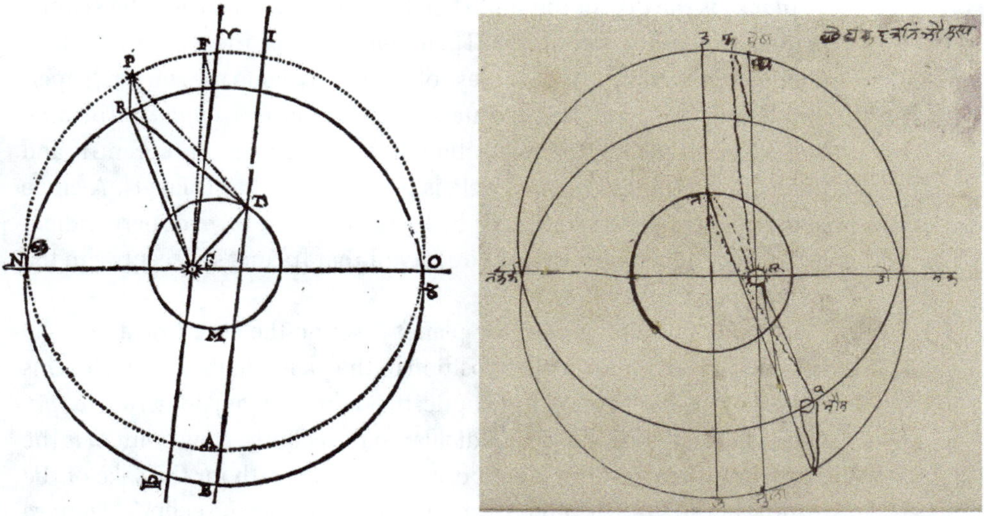

FIGURE 7.11 Left: De La Hire's diagram of superior planet orbit (La Hire 1727, p. 14). Right: Diagram of Mars orbit from f. 7^v of MS Berlin or. fol. 2973.

लयोः सम्पातरेखास्ति नरओबवृत्तं तद्वृत्तमस्ति । यद्विमण्डलपाले रानीताभिः रेखाभिः क्रान्ति-वृत्तधरातले लम्बरूपाभिः क्रान्तिवृत्तधरातले वृत्तं भवति ॥

अथ विमण्डलात्क्रान्तिवृत्तोपरिग्रहचिह्नमानीतम् । यथा पचिह्नं विमण्डले ग्रहस्थानम् । इदं क्रान्तिवृत्ते पररेखयानीतं तत् रचिह्नं यातम् । इयं पररेखा क्रान्तिवृत्ते लम्बरूपास्ति । अथ तचिह्नं भूः । तमवृत्तं भुव उपरि सूर्यकर्णभ्रमणार्थं कल्पितं क्रान्तिवृत्तधरातल एव ॥

रचिह्नं स्पष्टग्रहस्थानं क्रान्तिवृत्ते ऽस्ति ॥

Now, the correction of [the star-planets beginning with] Mars is explained with a diagram.

In that, having determined their mean [longitudes] (*madhyama*) like the determination of the mean [longitude] of the sun on the desired day from the *zīj*; and having obtained the *dyuphala* from the sun, obtained its velocity (*gati*), [and] made the mean [longitude] etc. corrected by that; [and] furthermore computed the *manda*-equation (*manda-phala*) with the *manda*-anomaly (*manda-kendra*); [and] made [the planet's longitude] corrected by that: that correction is to be established by means of a diagram [*anachedyaka*?].

Here that diagram is stated.[18]

18 Compare the first few lines with the corresponding part of the Latin original (La Hire 1727, pp. 14–15):

In it *sū* is the disk of the sun. Then here *sū-ta-ra-na-ba-o* is the ecliptic plane (*krantivṛtta-dharātala*). Then *a-na-pa-o* is the eccentric plane (*vimaṇḍala-dharātala*) in the [case of planets] beyond the sun [i.e., superior planets]. *na-o* are the two intersections [nodes] (*sampāta*). The *na-o* line (*rekhā*) is the line of intersection of the two planes, the eccentric and the ecliptic. The *na-ra-o-ba* circle is the circle of that [planet]. A circle in the ecliptic plane is [formed] by lines having the form of perpendiculars (*lamba*) drawn from the boundary of that [planet's] eccentric to the ecliptic plane.

Now the point (*cihna*) of the planet is set on the eccentric above the ecliptic, namely point *pa*, the position of the planet in the eccentric. This [is] led by line *pa-ra* onto the ecliptic; that is known [as] point *ra*. This *pa-ra* line has the form of a perpendicular to the ecliptic. Now point *ta* is the earth. The circle *ta-ma*[19] is supposed above the earth for the sake of the turning of the sun's hypotenuse, right in the plane of the ecliptic. Point *ra* is the position of the true planet (*spaṣṭa-graha*) in the ecliptic.

3.2.2 The Steps of Praeceptum VIII in Sanskrit

Skipping the remainder of the diagram labels, we turn to the Sanskrit summary of de La Hire's Praeceptum VIII algorithm, beginning on f. 7ᵛ in the manuscript. The Sanskrit rendering closely tracks the numbered steps (added in square brackets to the translation) of the Latin original; see Appendix A.

छेदकदर्शनम् ॥

अथेष्टकाले भौमस्फुटाकरणशरज्ञानं चोच्यते ॥

तत्र सूर्यमध्यमानयनवदेषां मध्यमानयनं कार्यम् । पुनः स्पष्टार्कः कार्यः । ततः सूर्यस्य स्पष्टकेन्द्रेण सूर्यकर्णो ग्राह्यः ॥

पुनः सूर्यस्पष्टवद्भौमस्य प्रथमसंस्कारः कार्यः । ततो भौमादीनां पातानयनं कार्यम् ॥ अथ शनिपातश्चेत्तदा तत्र पातफलं दत्त्वा स्पष्टः कार्यः । अन्येषां तु यथागतमेव ॥ पुनः स्पष्टं मन्दकेन्द्रं कार्यम् । तेन मन्दकेन्द्रेण स्वस्वकर्णा ग्राह्याः ।

S.	Sol	'sun'
STRNBO.	Planum Eclipticae.	'ecliptic plane'
ANPO.	Orbita Planetae, in qua	'orbit of the planet, in which …'
NO.	Linea Nodorum, vel occursus plani per orbitam cum Eclipticae plano.	'… [is] the intersection of the plane of the orbit with the plane of the ecliptic'
NROB.	Orbitae planetae reductio vel projectio in planum Eclipticae per lineas parallelas inter se & ad planum Eclipticae perpendiculares.	'reduction or projection of the orbit of the planet into the plane of the ecliptic by lines parallel to one another and perpendicular to the plane of the ecliptic'

19 Point *ma* is not marked in the Sanskrit diagram, but clearly corresponds to point *M* in its Latin counterpart.

ततः प्रथमफलसंस्कृतग्रहे तस्य पातः शोध्यः शेषं शरकेन्द्रं भवति ।

ततः शरकेन्द्रात्तिर्यक् फलं विमण्डलक्रान्तिचापान्तररूपं सारणीतो गृहीत्वा ।

यथागतं धनार्णं प्रथमफलसंस्कृमध्यमे देयं तदा क्रान्तिवृत्ते सूर्योपरिस्थदृष्ट्यनुसारेण भवति ॥

पुनः सषड्भार्कः कार्यः पुनः सषड्भार्कस्पष्टग्रहस्थानयोरन्तरं कार्यम् । अनयोर्मध्ये लघुर्महतः शोध्यः । यदि च महान् षड्भाधिकः स्याल्लघुश्च षड्भाल्पं तदा लाघावेव द्वादशराशोन्संयोज्य महान् शोध्यः । इदं तसरकोणप्रमाणं स्यात ।

तदर्धं तु तरकोणयोर्योगार्धमस्ति ॥ पुनस्तसरकोणार्धस्य भुजः कार्यः पुनरस्य कोटिः कार्या । इयं तसरत्रिभुज उपयुक्ता स्यात् ॥

तत्र त्रिज्यानिष्पत्तिः पसरकोणकोटिज्यया तथास्ति । यथा सूर्यबिम्बग्रहबिम्बयोरन्तरस्य सररेखयास्ति । एवं सररेखा ज्ञाता भविष्यति ॥

अथ सूर्यादुपरिस्थग्रहेषु सतनिष्पत्तिः सररेखया तथास्ति । अधस्थेषु च सतनिष्पत्तिः सररेखया तथास्ति । यथा पञ्चचत्वारिंशदंशछायानिष्पत्तिरिष्टकोणछाययास्ति । अत्र कोणे पञ्चचत्वारिंशदंशोन इष्टकोणो भवति ॥

पुनः पञ्चचत्वारिंशदंशछायानिष्पत्तिस्तिर्यक् फलकोणछायया तथास्ति । यथा प्रागानीतकोणयोगार्धस्य छायाया निष्पत्तिस्तयोः कोणयोरन्तरार्धछाययास्ति ।

ततस्तयोः कोणयोरन्तरार्धं विदितामिदं सूर्यादुपरिस्थग्रहेष्टयोगार्धे योज्यम् । अधस्थेषु शोध्यं शेषं सूतरकोणप्रमाणं स्यात् ॥

इदं स्पष्टार्के शोध्यं यदि सषड्भार्कोनमन्दस्फुटग्रहः षड्भाल्पो भवेत् ॥ यद्यधिकस्तदा योज्यम् । एवं स्फुटग्रहः स्यदिति भौमादिस्फुटीकृतिः ॥ ॥

अथ शरानयनम् ॥

तत्र तसरकोणज्याया निष्पत्तिः सतरकोणज्यया तथास्ति । यथा प्रागानीततिर्यक्कोणछायाया निष्पत्तिः शरकोणछाययास्ति । एवं शरज्ञानं भविष्यति । अयं शरो मेषादिषट्के शरकेन्द्र उत्तरस्तुलादौ ॥ दक्षिण इति ॥ ॥

View of the diagram.

Now the knowledge of the latitude (*śara*) [and] the correction procedure of Mars at the desired time is stated.

[Step I.] In it, their mean [longitude] determination is to be done like the mean [longitude] determination of the sun. Again, the true sun is to be made. Then with the true anomaly of the sun the hypotenuse of the sun is obtained.

[Step II.] Again, the first correction (*prathama-saṃskāra*) of Mars is to be made like the true [position] of the sun. Then the determination of the nodes of [the superior planets] beginning with Mars is to be made. Now, if the node of Saturn [is determined], then having applied the node-correction to it, the true [position] is to be made. But [the nodes] of the others [are] just as [they] come. Then, the true *manda*-anomaly is to be made. With that *manda*-anomaly, the respective hypotenuses are obtained.

[Step III.] Then from the planet corrected by the first equation (*prathama-phala*), its node is subtracted. The remainder is the latitude-argument (*śara-kendra*).

[Step IV.] Then having obtained from the table the latitude-equation (*tiryak-phala*, 'oblique equation') from the latitude-argument (? *śarakeṃdrāptiryyakphalaṃ*) in the form of difference of the arcs in the eccentric and ecliptic;

[Step V.] according as [it is] positive [or] negative, [it is] applied to the mean [position] corrected by the first equation; then [it] is in the ecliptic circle in accordance with viewing [from] a position above the sun.

[Step VI.] Then, the sun plus six signs is to be made; again, the difference of the positions of the sun plus six signs and the planet is to be made. The lesser between those two is to be subtracted from the greater. And if the greater should be more than six signs and the lesser less than six signs, then having added twelve signs to the lesser itself, the greater is to be subtracted [from it]. This should be the amount of angle *ta-sa-ra*.[20]

[Step VII.] And half of that is half the sum of the two angles *ta* and *ra*. Then, the arc (*bhuja*) of half the angle *ta-sa-ra* is to be made; then, its complement (*koṭi*) is to be made. This should be used in triangle *ta-sa-ra*.

[Step VIII.] Then the ratio (*niṣpatti*) of the Radius to the cosine of angle *pa-sa-ra* is as much as [that] of the distance [between] the disk of the sun and the disk of the planet to the line *sa-ra*. Thus the line *sa-ra* will be known.

[Step IX.] Now the ratio of line *sa-ta*, among planets standing above the sun [i.e., superior], to line *sa-ra* is as much—and among the inferior (*adaḥstha*) [planets], the ratio of *sa-ta* to line *sa-ra*[21] is as much—as is the ratio of the tangent of forty-five degrees to the tangent of the desired angle. Now when the angle is diminished by forty-five degrees, it is the desired angle.

[Step X.] Again, the ratio of the tangent of forty-five degrees to the tangent of the latitude-equation angle is as much as is the ratio of the tangent of half the sum of the previously-determined angles to the tangent of half the difference of those two angles.

[Step XI.] From that the half-difference of those two angles is known; this is added to half the sum of the desired [angles] for a planet stand-

20 The syllable *sa* is used here interchangeably with *sū*, i.e., the position of the sun *S*.

21 In the case of inferior planets the ratio of *ST* and *SR* should be inverted.

ing above the sun; in [the case of] the inferior [planets], subtracted. The remainder should be the amount of angle *sū-ta-ra*.

[Step XII.] This is subtracted from the true sun if the manda-corrected planet minus the sun plus six signs should be less than six signs; if greater, then [it is] added [to the true sun]. In this way should be the corrected planet: thus the correction of [the planets] beginning with Mars.

Now the determination of the latitude.

In it, the ratio of the sine of angle *ta-sa-ra* to the sine of angle *sa-ta-ra* is as much as is the ratio of the tangent of the previously-determined inclination angle to the tangent of the angle of latitude. In this way will be the knowledge of the latitude. This latitude, when the latitude-argument is in the six [signs] beginning with Aries, [is] north, in [those] beginning with Libra, south. Thus.

3.2.3 The Worked Example for Mars

The manuscript then immediately proceeds to a worked example for the position of Mars, for which no date is specified. Unlike the example for Mars worked in MS Khasmohor 5292 as described in section 3.1.4, the working makes no mention of logarithms or the use of log trig function values; all the proportions between trigonometric quantities are handled multiplicatively rather than additively. The following translation and the procedure layout shown in section 3.3 suggest that the calculator and/or scribe left some significant gaps towards the end of the computation process for both longitude and latitude.

अथेतत्प्रपचंस्योदाहरणं निरूप्यते ॥ ॥

तत्रोक्तवञ्चीचान्मध्यमो भौमः ८ । २३ । ४३ । १८ । उच्चं ५ । १ । १२ । २६ । पातः १ । १७ । ४५ । ५२ । द्युफलमृणं ० । १० । द्युफलसंस्कृतमध्यमो भौमः ८ । २३ । ४३ । ८ ।

उच्चोने मन्दकेन्द्रं ३ । २२ । ३० । ४२ । कोष्टकेभ्यः प्राप्तं भौमप्रथमफलमृणं १० । १५ । ४९ । अनेन संस्कृतो भौमः ८ । १३ । २७ । १८ । केन्द्रं च ३ । १२ । १४ । ५२ ।

पातोनभौमः शरकेन्द्रं ६ । २५ । ४१ । २६ ।

अतः कोष्टकेभ्यो द्वितीयफलमृणं ॥ ० । ० । ० । ४३ । तिर्यक्कोणश्च प्रथमशरसंज्ञः ० । ४८ । ६ ।

द्वितीयफलसंस्कृतभौमः ८ । १३ । २६ । ३५ । द्वितीयफलसंस्कृतकेन्द्रं ३ । १२ । १४ । ९ ।

स्पष्टो ऽर्कः १ । १२ । ३८ । ३५ । सषड्भः ७ । १२ । ३८ । ३५ । सषड्भार्कोनभौमः १ । ० । ४८ ।

अस्यार्धं ० । १५ । २४ । ० । अस्य कोट्यंशाः ७४ । ३६ । ० ।

सूर्यस्य स्पष्टमन्दकेन्द्रं १० । ३ । ५६ । सूर्यमन्दकर्णः ६० । ३३ । ७ । प्रथमशरकोणकोट्यंशाः ८९ । ११ । ५४ । एषां ज्या ५९ । ५९ । ३९ । भौमस्य स्पष्टमन्दकेन्द्राद्भौममन्दकर्णः ८८ । ५२ । ३३ । अनयोर्घातः ८८ । ५२ । १ । ५३ । ३६ । सूर्यकार्णाप्तः १ । २८ । ३ । २६

अस्य चापं छायासु ५५ । ४३ । ४८ । ३ । अत्र पञ्चचत्वारिंशदंशाः शोधिताः शेषमनयोर्घातः ४१ । १६ । ४२ ।

त्रिज्याभक्तो ऽस्य छायासु चापमंशाद्यं ३४ । ३१ । ६ ।

सषड्भार्केन भौमार्धकोणकोष्वंशेष्टुयुक्तः १०९ । ७ । ३६ । राश्याद्यं भूकोणाभिधं ३ । १९ । ७ । ३६ । स्पष्टार्के १ । १२ । ३९ । ३५ । शोधिः सषड्भार्कोनमन्दस्फुटग्रहस्य षडल्पत्वात् तं जातः स्पष्टो भौमः ९ । २३ । ३१ । ५९ । इति भौमस्फुटीकरणम् ॥ ॥

अथ भौमशरानयनम् ॥

तत्र सषड्भार्कोनप्रथमफलद्वयसंस्कृतभौमः सूर्यकोणाख्यः १० । ४८ । ० । अस्य ज्या ३० । ४३ । २१ । भूकोणाख्यः ३ । १९ । ७ । ३६ । अस्य भुजांशाः ७० । ५२ । २४ । अस्य ज्या ५६ । ४१ । १५ । प्रथमकोणः ० । ४८ । ६ । ४ । अस्य छायासु चापं जातः शरः १ । २८ । ५० । दक्षिणः ॥ इति ॥ ॥

Now, an example of that *prapacam* [Lat. *praeceptum*?] is examined.

In it, the mean [longitude of] Mars from the *zīj* as stated 8|23|43|18. The apogee (*ucca*) 5|1|12|26. The node 1|17|45|52. The *dyuphala*, negative 0|10. The mean [longitude of] Mars corrected by the *dyuphala*, 8|23|43|8.

When [this is] diminished by the apogee, the *manda*-anomaly 3|22|30|42. The first equation of Mars obtained from the tables [is] negative 10|15|49. Corrected by this, Mars 8|13|27|18. And the anomaly 3|12|14|52.

Mars diminished by the node [is] the latitude-argument 6|25|41|26.

From that, the second equation from the tables [is] negative 0|0|0|43. And the oblique angle called the 'first latitude' 0|48|6 [inclination *i*]. Mars corrected by the second equation 8|13|26|35. The anomaly corrected by the second equation 3|12|14|9.

The true sun 1|12|38|35. Plus six signs 7|12|38|35. Mars diminished by the sun plus six signs 1|0|48.

Half of that 0|15|24|0. The degrees of its complement (*koṭi*) 74|36|0.

The true *manda*-anomaly of the sun 10|3|56. The *manda*-hypotenuse of the sun 60|33|7.

The degrees of the complement of the angle of the first latitude 89|11|54. The sine of them 59|59|39.

From the true *manda*-anomaly of Mars, the *manda*-hypotenuse of Mars 88|52|33. The product of the two 88|52|1|53|36. Divided by the hypotenuse of the sun 1|28|3|26. The arc of that in the [table of] tangents 55|43|48|3. In it forty-five degrees are subtracted; the remainder [...?] the product of the two 41|16|42.

Divided by the Radius; the arc of it in the [table of] tangents beginning with degrees, 34|31|6.

The half-angle of Mars increased by [...?] the sun plus six signs 109|7|36. Beginning with signs, the so-called 'earth angle' 3|19|7|36. Subtracted from the true sun 1|12|39|35.[22] Because of the *manda*-corrected [longitude of the] planet subtracted from the sun plus six signs [being] less than six [signs], that [is] the resulting true [longitude of] Mars 9|23|31|59. Thus the correction procedure for Mars.

Now the determination of the latitude of Mars.

In that, [the longitude of] Mars corrected by the first pair of equations [and] diminished by the sun plus six signs [is] the so-called 'sun angle' 1[|]0|48|0. Its sine 30|43|21. The so-called 'earth angle' 3|19|7|36. The degrees of its [reduced] arc 70|52|24. Its sine 56|41|15. The first angle 0|48|6|4. Its arc in the [table of] tangents; [...?] resulting latitude 1|28|50, south. Thus.

3.3 *Comparison of the Mars Worked Examples in MSS Khasmohor 5292 and Berlin or. fol. 2973*

The following schematic lays out the steps of the computation procedures according to the working of the examples for Mars in the manuscripts discussed in sections 3.1.4 and 3.2.3. We have indicated in square brackets corrections of evident errors or omissions in the details of the computations.

Khasmohor 5292	Berlin or. fol. 2973	Notes
$\bar{\lambda} = 8^s3^\circ; 51, 9$ $\lambda_A = 5^s1^\circ; 10, 9$ $\overline{{}_m\kappa} = \bar{\lambda} - \lambda_A$ $= 3^s2^\circ; 41, 0$	$\bar{\lambda} = 8^s23^\circ; 43, 18 - 0^\circ; 0, 10$ $= 8^s23^\circ; 43, 8$ $\lambda_A = 5^s1^\circ; 12, 26$ $\lambda_N = 1^s17^\circ; 45, 52$ $\overline{{}_m\kappa} = \bar{\lambda} - \lambda_A$ $= 3^s22^\circ; 30, 42$ $[3^s12^\circ; 30, 42]$	Mean longitude $\bar{\lambda}$ (corrected by *dyuphala*) and apogee longitude $\overline{\lambda_A}$ used to find mean *manda*-anomaly $\overline{{}_m\kappa}$ (Step II)
$\mu = 10^\circ; 39, 18\ (-)$ $\lambda^\mu = 8^s3^\circ; 51, 9 - 10^\circ; 39, 18$ $= 7^s23^\circ; 11, 51$ ${}_m\kappa^\mu = \lambda^\mu - \lambda_A$ $= 7^s23^\circ; 11, 51 - 5^s1^\circ; 10, 9$ $= 2^s22^\circ; 1, 51$ $[2^s22^\circ; 1, 42]$	$\mu = 10^\circ; 15, 49\ (-)$ $\lambda^\mu = 8^s23^\circ; 43, 8 - 10^\circ; 15, 49$ $= 8^s13^\circ; 27, 18$ $[8^s13^\circ; 27, 17]$ ${}_m\kappa^\mu = \lambda^\mu - \lambda_A$ $= 8^s13^\circ; 27, 18 - 5^s1^\circ; 12, 26$ $= 3^s12^\circ; 14, 52$	Tabulated *manda*-equation μ corrects mean longitude and anomaly (Step II)

22 The minutes place of the previously stated true sun position 1|12|38|35 has here been changed to 39 and used in subsequent calculations, apparently without noticing the error.

(*cont.*)

Khasmohor 5292	Berlin or. fol. 2973	Notes
$\omega = \lambda^{\mu} - \lambda_N$ $= 7^s23^\circ;11,51 - 1^s17^\circ;44,36$ $= 6^s5^\circ;27,15 = 5^\circ;27,15$ $i \approx 0^\circ;10,33$ $e = 0^\circ;0,10$	$\omega = \lambda^{\mu} - \lambda_N$ $= 8^s13^\circ;27,18 - 1^s17^\circ;45,52$ $= 6^s25^\circ;41,26$ $e = 0^\circ;0,43$ $i = 0^\circ;48,6$	Nodal elongation ω is argument of inclination i; look up latitude-equation e corresponding to this i (Steps III–IV)
$\lambda^{\mu\prime} = \lambda^{\mu} - e$ $= 7^s23^\circ;11,41$ ${}_m\varkappa^{\mu\prime} = {}_m\varkappa^{\mu} - e$ $= 2^s22^\circ;1,41$	$\lambda^{\mu\prime} = \lambda^{\mu} - e$ $= 8^s13^\circ;26,35$ ${}_m\varkappa^{\mu\prime} = {}_m\varkappa^{\mu} - e$ $= 3^s12^\circ;14,9$	Use latitude-equation e to further correct longitude and anomaly (Step V)
$\varkappa_S{}^{\mu} = 10^s7^\circ;57,28$ hyp. $= 400462$ (of sun)	$\lambda_S = 1^s12^\circ;38,35$ $\lambda_S + 6^s = 7^s12^\circ;38,35$	Solar anomaly and hypotenuse (Step I)
${}_s\varkappa = \lambda^{\mu} - \lambda_T$ $= 7^s23^\circ;11,51 - 7^s18^\circ;5,41$ $= 0^s5^\circ;6,10$ $\frac{{}_s\varkappa}{2} = \frac{0^s5^\circ;6,10}{2}$ $= 0^s2^\circ;33,5$ $90^\circ - \frac{{}_s\varkappa}{2} = 87^\circ;26,55$	${}_s\varkappa = \lambda^{\mu\prime} - (\lambda_S + 6^s)$ $= 8^s13^\circ;26,35 - 7^s12^\circ;38,35$ $= 1^s0^\circ;48,0$ $\frac{{}_s\varkappa}{2} = \frac{1^s0^\circ;48,0}{2}$ $= 0^s15^\circ;24,0$ $90^\circ - \frac{{}_s\varkappa}{2} = 74^\circ;36,0$ $\varkappa_S{}^{\mu} = 10^s3^\circ;56$ hypotenuse $= 60;33,7$ (of sun)	*śīghra*-anomaly ${}_s\varkappa$ (or $\angle TSR$); its half; its complement (Step VII)
$90^\circ - i = 90^\circ - 0^\circ;10,33$ $= 89^\circ;49,27$ $\log\cos(i) = \log\sin(89^\circ;49,27)$ $= 999999$	$90^\circ - i = 90^\circ - 0;48,6$ $= 89^\circ;11,54$ $\cos(i) = \sin(90^\circ - i)$ $= 59;59,39$	Inclination i and its cos (Step VIII)
rekhā $= \log\cos(i)$ $+ \log$ *karṇa* $-$ Radius $= 999999 + 418476 - 1000000$ $= 1418475 - 1000000$	Mars hypotenuse $= 88;52,33$ $\frac{(\text{Mars hyp.})(\cos(i))}{\text{solar hyp.}} = 1,28;3,26$ $= 88;3,26$ $[= SR]$	Represents the following procedure (Step VIII): $\frac{\text{Radius}}{\cos(i)} = \frac{SP}{SR}$ $\log SR = \log\cos(\angle PSR)$ $+ \log SP$ $- \log \text{Radius}$

(*cont.*)

Khasmohor 5292	Berlin or. fol. 2973	Notes
chāyā = 1418475 − 400462 = 1018013 arctan(1018013) − 45° = 11°; 33	arctan(88; 3, 26) = 55°; 43, 48, 3 = V [?] $V - 45°$ = [?]	Represents the following procedure (Step IX): $\frac{ST}{SR} = \frac{\tan(45°)}{\tan(V)}$ $\log\tan(V) = \log SR + \log \text{Radius} - \log ST$ $X = V - 45°$
chāyā = 1135129 + 931039 − 1000000 = 1066168 arctan(1066168) = 77°; 42	$\left[\frac{\tan(X)\tan(\frac{t+r}{2})}{\tan(45°)} = \tan(\frac{t-r}{2})\right]$ $\arctan\left(\frac{41;16,42}{\text{Radius}}\right) = 34°;31,6$	Represents the following procedure, setting $t = \angle STR$ and $r = \angle SRT$ (Steps X–XI): $\frac{\tan(45°)}{\tan(X)} = \frac{\tan(\frac{t+r}{2})}{\tan(\frac{t-r}{2})}$ $\log\tan\left(\frac{t-r}{2}\right) = \log\tan(X) + \log\tan\left(\frac{t+r}{2}\right) - \log \text{Radius}$
87°; 26, 55 − 77°; 42 = 9°; 45 87°; 26, 55 + 77°; 42 = 165°; 9 = $5^s15°; 9$ = $\angle STR$ $\lambda = \lambda_S - \angle STR$ = $1^s18°; 5, 41 - 5^s15°; 9$ = $13^s18°; 5, 41 - 5^s15°; 9$ = $8^s2°; 56, 41$	[?] = $3^s19°; 7, 36$ = $\angle STR$ $\lambda = \lambda_S - \angle STR$ [= $1^s12°; 38, 35 - 3^s19°; 7, 36$] = $1^s12°; 39, 35 - 3^s19°; 7, 36$ = $13^s12°; 39, 35 - 3^s19°; 7, 36$ = $9^s23°; 31, 59$	(Steps XI–XII) $\angle STR = \frac{t+r}{2} + \frac{t-r}{2}$ $\lambda = \lambda_S - \angle STR$

(*cont.*)

Khasmohor 5292	Berlin or. fol. 2973	Notes
[none]	${}_s\varkappa = \angle TSR$ $= \lambda^{\mu\prime} - (\lambda_S + 6^s)$ $= 8^s13°; 26, 35 - 7^s12°; 38, 35$ $= 1^s0°; 48, 0$ $\sin(1^s0°; 48, 0) = 30; 43, 21$ $\sin(3^s19°; 7, 36) = \sin(70°; 52, 24)$ $= 56; 41, 15$ $i = 0°; 48, 6, 4$ [?] $\beta = 1°; 28, 50$ S	Latitude procedure: $\frac{\sin(\angle TSR)}{\sin(\angle STR)} = \frac{\tan(i)}{\tan(\beta)}$ $\beta = \arctan\left(\frac{\sin(\angle STR) \cdot \tan(i)}{\sin(\angle TSR)}\right)$

Appendix A. De La Hire's Planetary Computation Algorithms

Praeceptum VIII of the *Tabulae astronomicae* (La Hire 1727, pp. 15–17), titled 'To find the true Longitude and Latitude of superior and inferior planets at a given time', comprises twelve numbered steps for the longitude procedure and an additional section for the latitude procedure. We reproduce here Pingree's translation in (Pingree 2002b, pp. 438–439), adding in square brackets de La Hire's symbols corresponding to the various quantities.

On Longitude

I. Let the apparent proposed time be corrected and made mean, as we taught in the precept on finding the place of the Sun, and let the true place [S] of the Sun in the ecliptic be found together with the distance of the earth from the Sun [ST] from Table 14 in logarithmic numbers.

II. Let the eccentric place of the planet, which is the place of the planet seen from the Sun, be sought by the same method by which the true place of the Sun (was sought), and also the place of the planet's ascendant node, to which a correction will be applicable in the case of Saturn, and the true anomaly together with the distance of the planet from the Sun in logarithmic numbers as they are in the tables of the individual planets.

III. Let the place of the planet's node be subtracted from the eccentric place, and the remainder will be the argument of the planet's latitude.

IV. With the argument of the latitude, let there be found in their own tables the inclination [$\angle PSR$] of (each) planet and (its) reduction to the ecliptic. However, the reduction will be additive or subtractive as the apposite title shall indicate.

V. If it is additive let the reduction be added to the true eccentric place of the planet or let it be subtracted from it if it is subtractive, and the place [R] of the planet reduced to the ecliptic will be obtained, which will be the longitude of the planet from the Sun.

VI. When six signs have been added to or subtracted from the true place of the Sun, the sum or difference will be the true place of the earth seen from the Sun. Let the reduced place of the planet be subtracted from the place of the earth or the latter from the former, that is, the smaller from the larger; but if the larger should exceed the smaller by an amount greater than six signs, let twelve signs be added to the smaller before the subtraction is made; but then the smaller becomes the larger. The remainder will be the angle at the Sun [$\angle TSR$], in all cases less than six signs.

VII. Let half of the angle at the Sun be taken: its complement to a quadrant or to three signs will be half the sum of the angles [$\angle STR$ and $\angle SRT$], which (half-sum) we use in the resolution of the triangle [$\triangle TSR$] whose angles are at the Sun, at the earth, and at the planet reduced on the plane of the ecliptic.

VIII. If the Radius should be to the Sine of the complement of the inclination of the planet as the planet's distance from the Sun taken from the table is to a fourth term, the last will be the distance of the reduced planet from the Sun, which they call the shortened distance [SR].

IX. Besides, let the distance of the earth from the Sun be to the shortened distance in the cases of the superior planets (namely Saturn, Jupiter, and Mars), but the shortened distance to the distance of the earth from the Sun in the cases of the inferior planets (namely Venus and Mercury), as the Tangent of 45 degrees is to the Tangent of a certain angle which I call V, from which, when 45 degrees have been subtracted, I call the remaining angle X.

X. Finally let the Tangent of 45 degrees be to the Tangent of angle X as the Tangent of half the sum of the unknown angles which was found in the seventh article is to the Tangent of half the difference of the same unknown angles.

XI. Let this half-difference be added to the half-sum of the same angles in the cases of the superior planets, but let it be subtracted from it in the cases of the inferior planets; the angle at the earth will be obtained.

XII. But let the angle at the earth be added to the true place of the Sun if the distance of the earth from the reduced place of the planet in the order of the signs is less than a semicircle or six signs; on the contrary, if this distance is greater, the angle at the earth is subtracted; the true longitude of the planet will be obtained.

On the Latitude

Let the Sine of the angle at the Sun [$\angle TSR$] be to the Sine of the angle at the earth [$\angle STR$] as the Tangent of the planet's inclination [$\angle PSR$] is to the Tangent of its latitude [$\angle PTR$]. The latitude will be northern in the first six signs of the argument of its latitude, but southern in the last.

De La Hire goes on to discuss other aspects of latitude calculations which are less relevant to the Sanskrit versions.

Appendix B. The *Jīvāchāyāsāriṇī*

In this appendix, we describe the table text *Jīvāchāyāsāriṇī*, known to us through a single manuscript.

B.1 *Description of MS RORI 23958*

The copy of the *Jīvāchāyāsāriṇī* available to us was made from the original manuscript housed in the Rājasthān Oriental Research Institute, Jodhpur, as RORI Cat VI 23958. Black and white photocopies of this manuscript are held in the John Hay Library, Brown University, USA. The manuscript is 91 ff. and is written in *nāgarī* script on hand-made paper, with column headers in Sanskrit. In the top left-hand corner margin of f. 1^{v} is written the invocation *śrīgaṇeśāya namaḥ* ('Homage to Lord Gaṇeśa').

The manuscript contains one large continuous numerical table. The tables themselves are oriented in a landscape format, with the table argument running vertically and tabulated function values written horizontally adjacent to the argument values. The very last value on a page is repeated as the first value of the next page. On f. 91^{v}, top left-hand corner, a partial colophon is written giving the date: *śailanāgavasucaṃdra 1887 hāyane śukraśuklava-*

susaṃkhyakātithau | ('[*saṃvat*?] 1887 year, Friday, 8th *tithi* of the bright fortnight'). The date Saṃvat 1887 corresponds to 1830/31 CE. No month is specified, but in *saṃvat* 1887 (*amānta*) there were three months in which the 8th *tithi* of the bright fortnight was a Friday: Vaiśākha (30/04/1830), Bhādrapada (27/08/1830), and Maghā (21/01/1831).[23]

Nowhere in this manuscript is the name *Jīvāchāyāsāriṇī* stated. Rather it is written on the first page in *nāgarī* by the cataloguer and in transliteration in the handwriting of David Pingree. This title appears to have been created using the column headers of the numerical tables which are *jīvā* 'sine' and *chāyā* 'shadow/tangent' respectively and compounded with the Sanskrit word *sāriṇī*, 'numerical table'.

B.2 *Content of MS RORI 23958*

The *Jīvāchāyāsāriṇī* is a single numerical table spreading across 180 pages. The argument values range from 0 to 89 degrees which are further subdivided into argument values ranging from 0 to 60 minutes each. The table tabulates two separate functions that share the same argument, so that there are 10,800 distinct numerical entries.[24]

There is no additional paratext to reveal the purpose and contents of the numerical table. In order to crack this particular table, we needed to rely on mathematical analysis and the clues left by the column headers, *jīvā* and *chāyā*. *Jīvā* is a Sanskrit word that means 'sine', and *chāyā* 'shadow', and by extension 'tangent'. The fact that the entries spanned a quadrant further suggested that this numerical table had something to do with trigonometry, although the values didn't support a direct sine or tangent interpretation. The numerical entries all contain 8 significant digits. The very first numerical entry for argument value $0°0'$ was 00000000 for both the *jīvā* and *chāyā*; the very last for argument value $89°60'$ were 100000000 and 00000000 respectively (see the first entries in the table of figure 7.14 and the last in that of figure 7.7). These extremal values for the tangent in particular appeared a little unusual.

This very last value of 100,000,000 in the 'sine' column suggested a trigonometric Radius of 10^8, but as noted the entries in the table did not follow a strict sine curve. Trigonometric tables with Radii of this sort of magnitude were very common in early European traditions, and while examining European tables

23 These dates were generated using Michio Yano's *Pañcāṅga* programme http://www.cc.kyoto-su.ac.jp/yanom/pancanga/index.html.

24 There is some duplication: the 60th minute for each degree is the same as the 0th minute of the very next degree, and the last value on each page (usually) argument value 32 is repeated on the next page, presumably to help keep the folia in order.

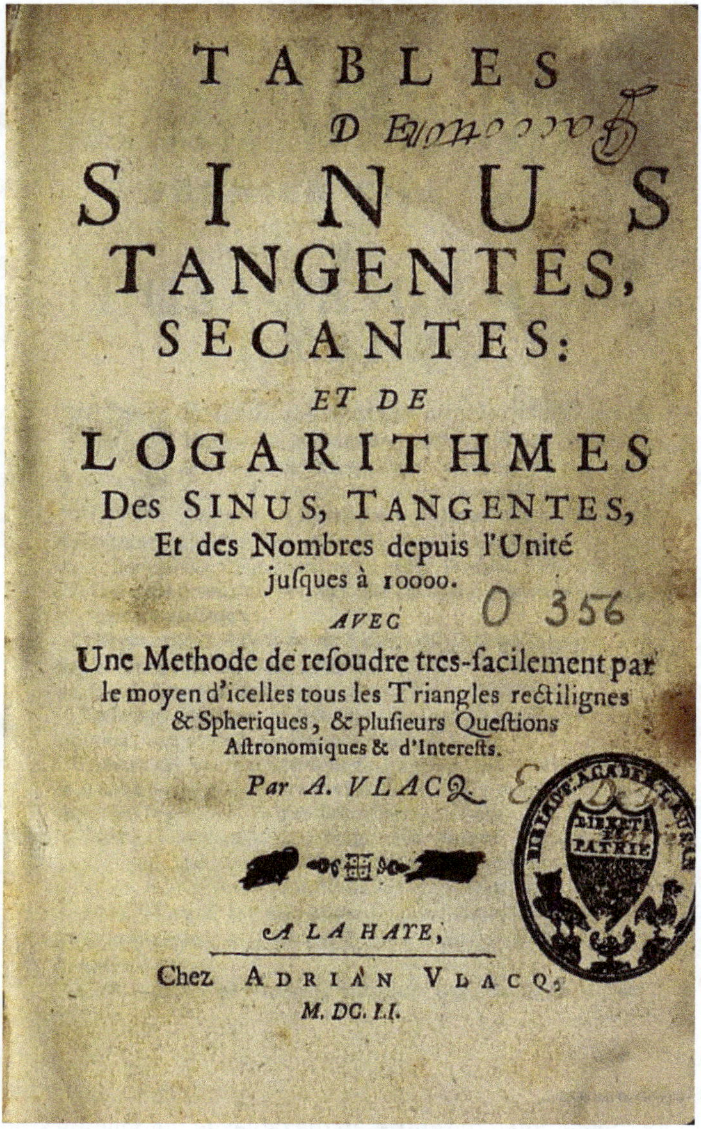

TABLES
D
SINUS
TANGENTES,
SECANTES:
ET DE
LOGARITHMES
Des SINUS, TANGENTES,
Et des Nombres depuis l'Unité
juſques à 10000.
AVEC
Une Methode de reſoudre tres-facilement par
le moyen d'icelles tous les Triangles rectilignes
& Spheriques, & pluſieurs Queſtions
Aſtronomiques & d'Intereſts.
Par A. VLACQ.

A LA HAYE,
Chez ADRIAN VLACQ,
M. DC. LI.

FIGURE 7.12 The title page of Vlacq's 1651 work on logarithms.

of this type from a similar time period and related ones, we came across tables of logarithms, many of which had similarities in structure and layout.

One in particular jumped out: Dutch mathematician Adriaan Vlacq's *Tables de sinus tangentes, secantes: et de logarithmes des sinus, tangentes, et des nombres depuis l'unité jusques à 10000* published in 1651 (see figures 7.12 and 7.13).

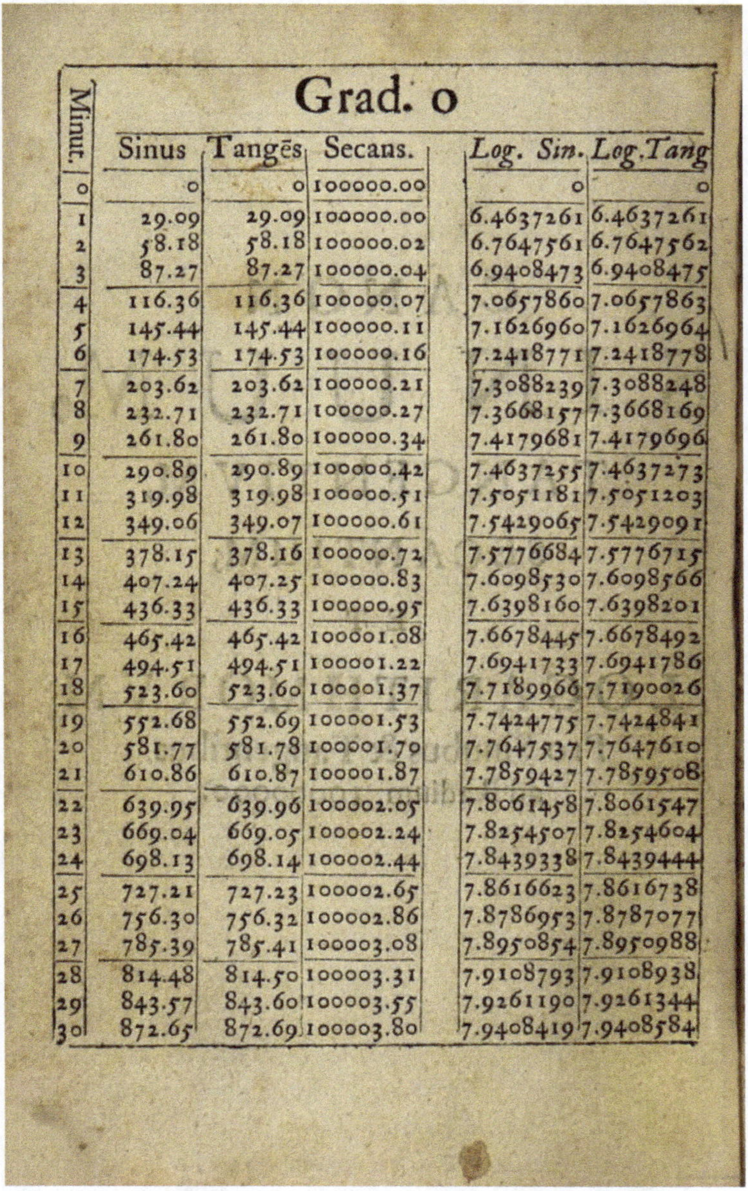

Grad. o

Minut.	Sinus	Tangēs	Secans.	Log. Sin.	Log. Tang
0	0	0	100000.00	0	0
1	29.09	29.09	100000.00	6.4637261	6.4637261
2	58.18	58.18	100000.02	6.7647561	6.7647562
3	87.27	87.27	100000.04	6.9408473	6.9408475
4	116.36	116.36	100000.07	7.0657860	7.0657863
5	145.44	145.44	100000.11	7.1626960	7.1626964
6	174.53	174.53	100000.16	7.2418771	7.2418778
7	203.62	203.62	100000.21	7.3088239	7.3088248
8	232.71	232.71	100000.27	7.3668157	7.3668169
9	261.80	261.80	100000.34	7.4179681	7.4179696
10	290.89	290.89	100000.42	7.4637255	7.4637273
11	319.98	319.98	100000.51	7.5051181	7.5051203
12	349.06	349.07	100000.61	7.5429065	7.5429091
13	378.15	378.16	100000.72	7.5776684	7.5776715
14	407.24	407.25	100000.83	7.6098530	7.6098566
15	436.33	436.33	100000.95	7.6398160	7.6398201
16	465.42	465.42	100001.08	7.6678445	7.6678492
17	494.51	494.51	100001.22	7.6941733	7.6941786
18	523.60	523.60	100001.37	7.7189966	7.7190026
19	552.68	552.69	100001.53	7.7424775	7.7424841
20	581.77	581.78	100001.70	7.7647537	7.7647610
21	610.86	610.87	100001.87	7.7859427	7.7859508
22	639.95	639.96	100002.05	7.8061458	7.8061547
23	669.04	669.05	100002.24	7.8254507	7.8254604
24	698.13	698.14	100002.44	7.8439338	7.8439444
25	727.21	727.23	100002.65	7.8616623	7.8616738
26	756.30	756.32	100002.86	7.8786953	7.8787077
27	785.39	785.41	100003.08	7.8950854	7.8950988
28	814.48	814.50	100003.31	7.9108793	7.9108938
29	843.57	843.60	100003.55	7.9261190	7.9261344
30	872.65	872.69	100003.80	7.9408419	7.9408584

FIGURE 7.13 The first page from Vlacq's 1651 work on logarithms, with the logarithms of sine and tan in the two rightmost columns.

As in the *Jīvāchāyāsāriṇī*, only the log sines and log tangents have been tabulated and to the same number of significant figures. The entries are identical. It seemed likely then, that this was a copy of this European table that had somehow made its way to India. This is, to our knowledge, the first such table

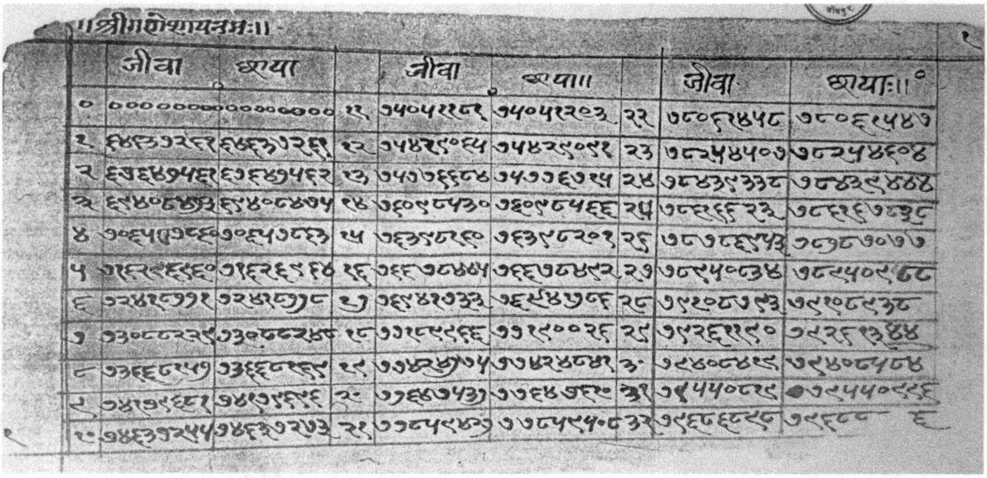

FIGURE 7.14 First page of log sine and log tangent tables from the *Jīvāchāyāsāriṇī* (RORI 23958) f. 1v.

appearing in Indian sources. As our study shows, it is clear that Sanskrit scholars were using this table, or a table very much like it, as they made sense of new techniques for computing true planetary longitudes.

References

de la Hire, Philippe (1727). *Tabulae Astronomicae Ludovici magni jussu et munificentia exaratae et in lucem editae*. 2nd edition. Paris: Montalant.

Mercier, Raymond (1993). "Account by Joseph Dubois of astronomical work under Jai Singh Sawā'ī." In: *Indian Journal of History of Science* 28, pp. 157–166.

Minkowski, Christopher (2004a). "A nineteenth century Sanskrit treatise on the revolution of the earth: Govinda Deva's *Bhūmibhramana*." In: *SCIAMVS*, pp. 199–224.

Minkowski, Christopher (2004b). "Competing Cosmologies in Early Modern Indian Astronomy." In: *Studies in the History of the Exact Sciences in Honour of David Pingree*. Ed. by Charles Burnett et al. Brill, pp. 349–385.

Montelle, Clemency, Kim Plofker, and Glen Van Brummelen (Forthcoming). *The* jyācāpa *section of the* Siddhāntakaustubha *of Jagannātha on al-Kāshī's and Ulugh Beg's iterative methods for computing the sine of one degree*.

Pingree, David (1999). "An astronomer's progress." In: *Proceedings of the American Philosophical Society* 143, pp. 73–85.

Pingree, David (2002a). "Philippe de La Hire at the Court of Jayasiṃha." In: *History of Oriental Astronomy*. Ed. by S.M.R. Ansari. Kluwer Academic Publishers, pp. 123–131.

Pingree, David (2002b). "Philippe de La Hire's Planetary Theories in Sanskrit." In: *From*

China to Paris: *2000 Years Transmission of Mathematical Ideas*. Ed. by Yvonne Dold-Samplonius et al. Stuttgart: Franz Steiner Verlag, pp. 429–453.

Pingree, David (2003). *A Descriptive Catalogue of the Sanskrit Astronomical Manuscripts preserved at the Maharaja Man Singh II Museum in Jaipur, India*. Philadelphia: American Philosophical Society.

Pingree, David (1970–1994). *Census of the Exact Sciences in Sanskrit*. 5 vols. Series A. Philadelphia: American Philosophical Society.

CHAPTER 8

Cyclical Time in Brahmanical India: Origin and Development

Johannes Bronkhorst

Classical Brahmanical texts—most notably the *Purāṇas*—present a cyclical vision of time, which Axel Michaels (2004: 300) describes as follows:

> The best-known expression of the cyclical image of the world is the doctrine of the four ages of the world (*yuga*). The calculation of time is complex and has not been handed down uniformly. ... According to [a common calculation], the course of the world consists of the change of unfolding (*sarga*, *pratisarga*) or creation (*sṛṣṭi*) and dissolution (*pralaya*), presented as day and night in the hundred-year life of the god Brahmā. Such a Brahmā day lasts from creation to the decline of an eon (*kalpa*). And every Kalpa lasts a thousand great ages of the world (*mahāyuga*), corresponding to twelve thousand god years or 4,320,000 human years, which are divided into four ages of the world (*yuga*), and each of them has a tenth long dawn and dawn and dusk, in which Brahmā (or Viṣṇu) rests. The whole thing is repeated a thousandfold, a hundred Brahmā years or 311 billion and forty million human years. When this epoch (*para*) is over, the world declines. The coarse material will again become subtle primeval material, in which the constituents are in balance, until they are shaken—either by themselves or by a divine impulse—and the cycle of the emergence and passing away of the world (*saṃsāra*) continues.
>
> According to this calculation, the current age, the Kaliyuga, began on February 18, 3102 B.C.

Vedic literature says nothing about cyclical time. Early post-Vedic literature, on the other hand, introduces the claim that the world is eternal. It does so most clearly with respect to Sanskrit, the language which, it claims, is eternal and therefore beginningless. Sanskrit, of course, is the language of the *Veda*, and the notion that the *Veda* is eternal proved attractive to the Brahmins of that period, too.[1] Indeed, this notion became fundamental to ritual Mīmāṃsā,

1 Bronkhorst forthcoming.

 | DOI:10.1163/9789004536869_010

the orthodox school of Vedic interpretation, which supported the beginninglessness of the *Veda* with the claim that every teacher of the *Veda* had had a teacher, and this without beginning. This argument remained popular with ritual Mīmāṃsā well into the second millennium, and distinguished this school from other schools of orthodox philosophy, which had meanwhile turned to a different understanding of the history of the world, the vision of cosmos and cyclical time corresponding to the description given above.

The first claim I wish to make in this article is that the cyclical vision of time came from the eastern region of northern India, from the culture of Greater Magadha. The main reasons supporting this claim are:

- This vision of cyclical time is completely absent in Vedic literature (as pointed out above); it appears only gradually in post-Vedic literature.
- It appears to be present in Buddhist and Jaina literature from the beginning.

Uncertainty about the dates at which early Buddhist and Jaina texts reached their present form tends to weaken chronological conclusions based on those texts. However, there is good reason (if further justification were necessary) to think that cyclical time was part of the culture of Greater Magadha when Buddhism and Jainism still shared space with Ajivikism.

The most important passage in the Buddhist canon that informs us about Ajivikism (i.e., about the teaching of Makkhali Gosāla, presumably its founder; see below) occurs in the *Sāmaññaphala Sutta* of the *Dīgha Nikāya*. Basham (1951: 13–14; modified) paraphrases it as follows:[2]

> There is neither cause nor basis for the sins of living beings; they become sinful without cause or basis. Neither is there cause or basis for the purity of living beings; they become pure without cause or basis. There is no deed performed either by [one's own] self (*ātman*) or by [the self of] others, no action belonging to the self (*puruṣa*), no strength, no courage, no endurance connected with the self or prowess connected with the self. All beings, all that have breath, all that are born, all that have life, are without power, strength, or virtue, but are developed by destiny, chance, and nature, and experience joy and sorrow in the six classes (of existence).
>
> There are 1,400,000 chief uterine births, 6,000 and 600; 500 *karmas*, 5 *karmas*, 3 *karmas*, a *karma*, and half a *karma*; 62 paths; 62 lesser *kalpas*; 6 classes (of human existence); 8 stages of man; 4,900 means of livelihood

2 For philological details and specifications, see Bronkhorst 2003: 160f. Graeme MacQueen (1988: 195) compares the different versions, including the Chinese versions, of this Sūtra and arrives at the conclusion "that [the Pāli version], of all the versions, preserved the most ancient state of the text". See also Meisig 1987.

> (?); 4,900 ascetics; 4,900 dwellings of *nāgas*; 2,000 faculties; 3,000 purgatories; 36 places covered with dust (?); 7 sentient births; 7 insentient births; 7 births from knots (?); 7 gods; 7 men; 7 *pisāca* (births?); 7 lakes; 7 knots (?), and 700; 7 precipices, and 700; 7 dreams, and 700; and **8,400,000 great *kalpas*** through which fool and wise alike will take their course, and make an end of sorrow. There is no question of bringing unripe *karma* to fruition, nor of exhausting *karma* already ripened, by virtuous conduct, by vows, by penance, or by chastity. That cannot be done. *Saṃsāra* is measured as with a bushel, with its joy and sorrow and its appointed end. It can neither be lessened nor increased, nor is there any excess or deficiency of it. Just as a ball of thread will, when thrown, unwind to its full length, so fool and wise alike will take their course, and make an end of sorrow.

A passage in the Śvetāmbara Jaina canon, too, informs us about the teaching of Gosāla. It occurs in the *Viyāhapannatti* (= *Bhagavatī Sūtra*) and reads:

> All those who have reached or are reaching or will reach salvation must finish in order **8,400,000 *mahākappas***, seven divine births, seven groups, seven sentient births, seven 'abandonments of transmigration' (*paüṭṭa-parihāra*), 500,000 *kammas*, and 60,000 and 600 and the three parts of *kamma*. Then, being saved, awakened, set free, and reaching *nirvāṇa* they have made or are making or will make an end of all sorrow.

The two passages have enough in common to justify the conclusion that they borrow from a common source which, we may safely assume, belongs to a time well before the composition of the *Mahābhārata*, and therefore well before the cyclical notion of time appears in the Brahmanical tradition. This common source mentioned *kalpas* that succeed each other.

This common source also uses a multiple of the number 84. This number is omnipresent in cosmological contexts in Buddhist and Jaina sources, but completely absent in Vedic literature. It appears for the first time in the *Mahābhārata*, in connection with Mount Meru, another notion that the Brahmanical tradition appears to have borrowed from Greater Magadha. In the *Bhīṣmaparvan* of the *Mahābhārata* (Mhbh 6.7.10), as well as in the *Purāṇas*, it is stated that *Meru* rises 84,000 yojanas above the earth and descends 16,000 yojanas below it. The height of Mount Meru in the *Mahābhārata* appears to be the first occurrence in Brahmanical literature of a number with cosmological significance from the group of 84 and its multiples.

With respect to early Jaina literature, Ruth Satinsky (2015: Abstract) observes:

> The number 84 and its multiples are an important concept in Jaina cosmology. Ṛṣabha and his son Bharata are held to have had lifespans of 8,400,000 pūrvas respectively; Śreyāṃsa 8,400,000 years; and Ara, 84,000 years. In addition, four out of the five Mount Merus to which the fordmakers (tīrthaṅkara) are linked rise 84,000 yojanas above the earth ... In a different context, the number 8,400,000 is for Jainas the sum total of conceivable birth-situations (yoni), in which souls may find themselves, over and over, as they circle through saṃsāra.

Satinsky further points out that "in both the Śvetāmbara and Digambara traditions, the number 84 and its multiples are omnipresent in the category of 'calculable' (*gaṇita*) (*gaṇiya*) time measures".

In Buddhism, the *Aṅguttara Nikāya* states that Mount Sineru (Meru) is 84,000 yojanas high and wide, and that it descends 84,000 yojanas beneath the sea. There are other significant occurrences of the number 84,000 in Buddhist literature. For example, there are the 84,000 *dharmaskandha*s of the Buddha—i.e., portions of the teaching relating to laws (*dharmaskandhavaśena caturaśīti-sahasravidham*)—and 84,000 stūpas containing the relics of Śākyamuni that Aśoka distributed.

All this suggests that the notion of *kalpa*—a key notion of cyclical time—was associated with a multiple of the number 84, itself a "cultural marker" of Greater Magadha.

There is one striking distinction between the cosmic cycles adopted in Brahmanism and those current in the traditions derived from Greater Magadha: the latter do not use the expression *yuga*,[3] which is yet a key term in the Brahmanical vision of time.[4] The Brahmanical tradition, on the other hand, does use the expression *kalpa*,[5] a term very common in the traditions that originated in Greater Magadha. We will consider these terminological issues below.

A small number of early Brahmanical texts present a picture of cyclical time that is rather different from the one presented above. These texts, too, speak

3 The Brahmanical *yuga*s do show up in more recent Buddhist texts; see Eltschinger 2012: 68 ff., 30 ("by the turn of the sixth century, certain Indian Buddhists ... modified their apocalyptic schemes by taking up the hitherto purely Brahmanical pattern of four ages (*yuga*)"); Bronkhorst 2016: 17 n. 45.

4 Note further that multiples of 84, though common in connection with Mount Meru etc., were not adopted in the new Brahmanical measurements of cosmic cycles; the numbers here used may have Vedic antecedents, at least in part (González-Reimann 2009: 417).

5 See the next note.

of four *yugas* that succeed each other, but these *yugas* do not now have the extraordinary lengths that later tradition came to ascribe to them. The lengths of the four *yugas* in the *Mānava Dharmaśāstra* (1.68–70), for example, are 4,000, 3,000, 2,000 and 1,000 years respectively. We find exactly the same lengths for the *yugas* in the prophecy of Mārkaṇḍeya in the *Mahābhārata* (3.186.18–23).[6] These texts contain no indication to suggest that anything but ordinary human years are meant. The emphasis on the evils of the end of the aeon (*yugānta*)—rather than on the *kaliyuga*, the last of the four—suggests that also the *Vāyu Purāṇa* (esp. 32.58–65), the *Brahmāṇḍa Purāṇa* (1.2.31) and the *Yuga Purāṇa*[7] presuppose short lengths.[8]

It appears, then, that the classical Brahmanical notion of long world-periods was derived from an earlier one of short world-periods. The mechanism that allowed this to happen is simple. Ludo Rocher (1986: 124) said the following about the lengths of the classical Purāṇic world-periods:

> One feature that sets the *yugas* apart from similar systems in other civilizations is that, in India, the world ages have been assigned specific durations. The four *yugas* extend over periods of 4000, 3000, 2000, and 1000 years. Each of these is preceded by a dawn (*saṃdhyā*) and followed by a twilight (*saṃdhyāṃśa*) equal to one tenth of the duration of the *yuga* proper. The figures for the *yugas* which appear most often in the purāṇas are, therefore, 4800, 3600, 2400, and 1200, the *caturyuga* being equal to 12,000 years. More often than not these years are said to be divine years. To convert them into human years they have to be multiplied by 360, i.e. 1,728,000 + 1,296,000 + 864,000 + 432,000 = 4,320,000.

Multiplication by 360 turned short world-periods into long world-periods, by moving from human to divine measurements.

It goes without saying that—in spite of major differences—the long Purāṇic world-periods share an important feature with the long world-periods that were current in the culture of Greater Magadha, viz. their great length. This raises another question: Why did Brahmanism initially borrow the notion of

6 Together with transitional periods, the four *yugas* add up to 12,000 years, which is a *mahāyuga*; one thousand of these great *yugas* constitute one *kalpa* (González-Reimann 2009: 417.) Note that the term *kalpa* is here used for ultra-long periods of time, even though *yuga* refers to relative short periods. Was this an attempt to maintain the notion that the duration of a *kalpa* is enormous, as it was and remained in its original milieu?

7 The *Yuga Purāṇa* is really part of a longer work called *Gārgīya-jyotiṣa* (and other names; Mitchiner 2002: 1f.). Another part of this work is studied in Zysk 2016.

8 See further below.

cyclical time without borrowing its great lengths of time? A closer look at the *Yuga Purāṇa* may suggest an answer.

There are good reasons to believe that the *Yuga Purāṇa*, or at least important parts of it, date back to a date well before the other Brahmanical texts we have considered. It describes certain historical events in the form of a prophecy. It does so in great detail, and in this respect it is quite unique in early India (Parasher 1991: 239). It mentions the Greeks (*yavana*) and the Śakas and the war and destruction these invaders bring. Most interesting from our present perspective is that it views these disasters as indicators of the approaching end of an era (*yugānta*). The invasions of the Greeks and the Śakas take place at, and signify, the end of the era. The text concludes with an indication that a new *kṛtayuga* will begin soon.

It seems clear from this text that its author really believed that the end of the era was at hand (Mitchiner 2002: 86; González-Reimann 2002: 98–99; 2009: 417). He thought it would take place soon after the invasions of the Śakas, and this is indeed the time when the text must have been composed.[9] At this time, then, at least certain Brahmins, presumably those who suffered most from the repeated invasions that struck northwestern India, thought that the end of the world—or more precisely, the end of the present world-period—was near.

The *Yuga Purāṇa* as we have it mentions the four *yuga*s. However, there are reasons to think that they originally had no place in it. In a recent article, Vincent Eltschinger (2020) questions the simultaneous appearance and parallel, organic growth of all four *yuga*s. He points out that Aśvaghoṣa's extant writings (which cannot be later than 100–150 CE) show no acquaintance with the second and third *yuga* (the *tretāyuga* and the *dvāparayuga* respectively), whereas they do know the first one (the *kṛtayuga*) and perhaps the fourth one (the *kaliyuga*). Moreover, "the early portions of the *Rāmāyaṇa* reflect no awareness of the *tretā-* and *dvāparayuga*s. They only rarely allude to the *kṛta-* and the *kaliyuga*s …, operating almost exclusively with the *yugānta*." Eltschinger concludes: "The evidence from the *Rāmāyaṇa* and the works of Aśvaghoṣa suggests that in first-century CE Ayodhyā/Sāketa (the Rāmaite center from which, according to colophons, Aśvaghoṣa was a native), cosmological and eschatological concep-

9 Second half of the first century BCE, according to Mitchiner (2002: 93). The massive eruption of Alaska's Okmok volcano in 43 BCE—one of the largest volcanic eruptions of the past 2,500 years which probably resulted in crop failures, famine and disease in many parts of the world, and which exacerbated social unrest and contributed to political realignments throughout the Mediterranean region—may also have contributed to the conviction that the end of the world was near (McConnell et al. 2020).

tions revolved around the three motifs of an 'Edenic' *kṛtayuga* ..., a cataclysmic *yugānta* ... and a still very discrete *kaliyuga/tiṣya* with unspecified mutual connections and no organic pattern of gradual degeneration." An analysis of the *Yuga Purāṇa* and the prophecy of Mārkaṇḍeya in the *Mahābhārata* brings to light that these texts contain several distinct layers. According to Eltschinger, there are

> some reasons to believe that the four-*yuga* framework of the *Yugapurāṇa* is a late addition intended to recontextualize an older, possibly end-of-the-first-century BCE account, not of the four *yuga*s, but of the *yugānta*, 'the end of the *yuga*'. The formal, compositional and doctrinal features of the *Yugapurāṇa*, *Mahābhārata* 3.186 and 3.188 point to two distinct layers: older apocalyptic accounts of the *yugānta* (first century BCE to second century CE?), and a later reframing of these prophecies made in order to align them with the new Brahmanical, essentially Vaiṣṇava ideas concerning cosmology and, in one case at least, the historical manifestation of Viṣṇu (second to fourth century CE?).

Eltschinger hypothesizes that the doctrine of the four *yuga*s developed some time between the late first century and the late second century CE, probably on the basis of an analogy with the dice game (*kṛta*, *tretā*, *dvāpara* and *kali* are the names of the four throws in the dice game).

The preceding reflections suggest that the creation of the classical Purāṇic notion of cyclical time went through several phases. There were first the catastrophes that befell Brahmanism during the final centuries preceding the Common Era. After the collapse of the Maurya Empire, around 185 BCE, the north of India suffered a succession of foreign invasions. The Indo-Greeks were among the first to extend their power on the ruins of the Maurya Empire. Indo-Scythian (or Śaka, to use the Indian term) invasions followed soon. The result was a breakdown of society, or at any rate of the Brahmanical order of society. A number of texts give expression to the Brahmanical disarray during this period. One of these is the *Yuga Purāṇa*, whose oldest parts may date back to the end of the first century BCE. The prophecy of Mārkaṇḍeya in the *Mahābhārata* does the same, but adds to the list of oppressive ruling dynasties. One of these (the Ābhīras) appears to justify the conclusion that this prophecy was written, or given its present shape, in 250 CE or later (Mitchiner 2002: 46).

At or before this date, and presumably under the influence of cyclical notions of time that existed in other traditions, Brahmanism adopted the cyclical notion of four *yuga*s. We must assume that the term *yuga* was chosen because of the pre-existing concerns with the end of the *yuga* and the notion of a *kṛtayuga* that had been around. The new cyclical notion of four *yuga*s was much more precise than *yugānta* and *kṛtayuga* had been so far. However, it was initially still based on the fear that the end of the world was near. When, presumably under the reign of the Guptas, this fear no longer looked justified, a reinterpretation—or rather: a recalculation—of the four *yuga*s put the end of the world at a safe distance. What is more, this recalculation attributed great lengths to the cycles, not just to the *kalpa*s but even to the individual *yuga*s (remember that one *kalpa* is one thousand *mahāyuga*s; see note 6). This brought these cycles closer to what must have been their source of inspiration: the time-cycles of Greater Magadha.

One text remains to be considered. It is older than the *Yuga Purāṇa*, older also than any of the other texts we have so far considered, but yet younger than the collapse of the Maurya Empire. It is Patañjali's *Mahābhāṣya*, which may have been composed (or at any rate completed) in the second century BCE at a date perhaps closer to 100 than to 150 BCE, presumably in Kashmir.[10] The *Mahābhāṣya* does not use the terms *yuga* and *kalpa* to designate time periods, but it does show awareness of long stretches of time. In its introductory chapter it tells the story of Bṛhaspati (the god of learning) who tries to teach the words of the Sanskrit language to the god Indra by enumerating them. He does not succeed in enumerating them all even after a thousand divine years (*divyaṃ varṣasahasram*).[11] Patañjali appears to situate this story in a remote past, for he contrasts it with the present, in which few live longer than a hundred years.[12] All this seems to presuppose a decline over time, not dissimilar to the opposition *kṛtayuga* vs. *kaliyuga* that presumably underlay the account in the original *Yuga Purāṇa*. Note further that this may be the earliest attestation of the notion of "divine year", so common in later Purāṇas.[13]

---

10 Bronkhorst 2016: 42–46.

11 Mahā-bh I p. 5 l. 25–27.

12 Mahā-bh I p. 5 l. 27–28: *kiṃ punar adyatve / yaḥ sarvathā ciraṃ jīvati sa varṣaśataṃ jīvati /*.

13 As pointed out by Limaye (1974: 17), the same expression *divyaṃ varṣasahasram* occurs in the *Ārṣeya Brāhmaṇa* (1.1.5; Sharma 1967: 3). This is a "secondary Brāhmaṇa" (*anubrāhmaṇa*), i.e. "a portion of a Brāhmaṇa apportioned to a Kalpa-sūtra" (Gaurinath Sastri 1982: 111; Satya Shrava 1977: 12; further Gonda 1975: 320), or a work similar to a Brāhmaṇa (*brāhmaṇasadṛśo 'yaṃ grantho 'nubrāhmaṇam*; *Kāśikā* on P. 4.2.62); it belongs "to the latest period of the Vedic literature" (Sharma 1967: 10). The expression *divyaṃ varṣasahasram* also occurs in the *Agnyādheya Brāhmaṇa* (§ 6, p. 17), a portion of an otherwise lost Kāṭhaka Brāhmaṇa (Gonda 1975: 351).

Patañjali does not say in so many words that his remarks presuppose time cycles, but it seems plausible to assume that they do. Must we conclude that his text was influenced by the notions of cyclical time that had originated in Greater Magadha?

Elsewhere I have argued at length that Buddhist ideas influenced the *Mahābhāṣya.*[14] To the arguments there presented one could add that the *Mahābhāṣya*'s introductory chapter mentions "the earth with its seven continents" (*saptadvīpā vasumatī*) and the three worlds.[15] This reminds one of Buddhist cosmology as we find it in the *Abhidharmakośa* and the earlier *Lokaprajñapti.* These describe, most in particular, seven concentric mountain ranges (*parvata*) that surround Mount Meru.[16]

In view of the Buddhist influence it is hardly reckless to assume that already Patañjali's *Mahābhāṣya* was acquainted with the cyclical notion of time, and with long periods that succeed each other. If so, its appearance in the *Yuga Purāṇa* is not without precedent in Brahmanical literature.

What happened to the belief that the world is eternal and has no beginning in time? Interestingly, the *Mahābhāṣya* is one of the earliest texts to attribute beginninglessness to the *Veda.*[17] Strictly speaking there is no contradiction between an eternal and a cyclical world. Time cycles can repeat themselves from beginningless time, and this is what most Indian thinkers ended up believing. The eternality of the *Veda* posed more difficulties, and frequently took the shape that God reintroduces the same *Veda* at the beginning of each cycle. Some went to the extent of maintaining that the *Veda*, though authorless, was non-eternal.[18]

Abbreviations

Abhidh-k-bh(P)	Vasubandhu, Abhidharmakośabhāṣya, ed. P. Pradhan, rev. 2nd ed. Aruna Haldar, Patna 1975 (TSWS 8)
BORI	Bhandarkar Oriental Research Institute, Poona
Kāśikā	Kāśikā by Vāmana & Jayāditya, ed. Aryendra Sharma, Khanderao Deshpande, D.G. Padhye, Osmania University, Hyderabad, 1969–1970, 2 vols

14 Most recently Bronkhorst 2016: 308–360.

15 Mahā-bh I p. 9 l. 20.

16 Abhidh-k-bh(P) 3.48–49, pp. 158–159. Cf. Sadakata 1997: 25 ff.: La Vallée Poussin 1914–1918: 296–297.

17 Bronkhorst forthcoming.

18 See, e.g., Łucyszyna 2020.

Mahā-bh	Patañjali, (Vyākaraṇa-)Mahābhāṣya, ed. F. Kielhorn, Bombay 1880–1885
Mhbh	Mahābhārata, crit. ed. V.S. Sukthankar et al., Poona 1933–1966 (BORI)
P.	Pāṇinian sūtra
TSWS	Tibetan Sanskrit Works Series, Patna

References

Agnyādheya Brāhmaṇa. = Sūryakānta 1981: 1–21.

Balcerowicz, Piotr (ed.) (2012): *World View and Theory in Indian Philosophy*. New Delhi: Manohar. (Warsaw Indological Studies Series, 5.)

Basham, A.L. (1951): *History and Doctrines of the Ājīvikas. A Vanished Indian Religion.* Reprint: Motilal Banarsidass, Delhi.

Brahmāṇḍa Purāṇa. Ed. J.L. Shastri. Delhi. 1973.

Bronkhorst, Johannes (2003): "Ājīvika doctrine reconsidered." *Essays in Jaina Philosophy and Religion*. Ed. Piotr Balcerowicz. Delhi: Motilal Banarsidass. (Lala Sundarlal Jain Research Series, 20.) Pp. 153–178.

Bronkhorst, Johannes (2016): *How the Brahmins Won: From Alexander to the Guptas.* Leiden—Boston: Brill. (Handbook of Oriental Studies 2/30.)

Bronkhorst, Johannes (forthcoming): "When and why did the Veda become eternal?" Proceedings of the 7th International Vedic Workshop, Dubrovnik, August 19–24, 2019.

Eltschinger, Vincent (2012): "Apocalypticism, heresy and philosophy." = Balcerowicz 2012: 29–86.

Eltschinger, Vincent (2020): "On the early history of the Brahmanical yugas." *"At the Shores of the Sky". Asian Studies for Albert Hoffstädt*. Ed. Paul W. Kroll & Jonathan A. Silk. Leiden—Boston: Brill. Pp. 38–53.

Gaurinath Sastri (1982): *A History of Vedic Literature*. Calcutta: Sanskrit Pustak Bhandar.

Gonda, Jan (1975): *Vedic Literature (Saṃhitās and Brāhmaṇas)*. Wiesbaden: Otto Harrassowitz.

González-Reimann, Luis (2002): *The Mahābhārata and the Yugas: India's Great Epic Poem and the Hindu System of World Ages*. New York.

González-Reimann, Luis (2009): "Cosmic cycles, cosmology, and cosmography." *Brill's Encyclopedia of Hinduism, 1*. Ed. Knut A. Jacobsen et al. Leiden: Brill. Pp. 411–428.

La Vallée Poussin, Louis de (1914–1918): *Vasubandhu et Yaçomitra. Troisième chapitre de l'Abhidharmakoça, Kārīkā, Bhāṣya et Vyākhyā*. Avec une analyse de la Lokaprajñapti et de la Kāraṇaprajñapti de Maudgalyāyana. London: Kegan Paul, Trench, Trubner.

Limaye, V.P. (1974): *Critical Studies on the Mahābhāṣya*. Hoshiarpur: Vishveshvaranand Vedic Research Institute.

Łucyszyna, Ołena (2020): "The Yuktidīpikā on the origin of the Vedas." *International Journal of Hindu Studies* 24, 239–256. https://doi.org/10.1007/s11407-020-09278-0.

MacQueen, Graeme (1988): *A Study of the Śrāmaṇyaphala-Sūtra*. Wiesbaden: Otto Harrassowitz. (Freiburger Beiträge zur Indologie, 21.)

Mānava Dharmaśāstra. See Olivelle 2005.

McConnell, Joseph R. et al. (2020): "Extreme climate after massive eruption of Alaska's Okmok volcano in 43 BCE and effects on the late Roman Republic and Ptolemaic Kingdom." *Proceedings of the National Academy of Sciences of the United States of America* (PNAS) 117(27), 15443–15449. https://doi.org/10.1073/pnas.2002722117.

Meisig, Konrad (1987): *Das Śrāmaṇyaphala-Sūtra*. Synoptische Übersetzung und Glossar der chinesischen Fassungen verglichen mit dem Sanskrit und Pāli. Wiesbaden: Otto Harrassowitz. (Freiburger Beiträge zur Indologie, 19.)

Michaels, Axel (2004): *Hinduism: Past and Present*. Princeton & Oxford: Princeton University Press.

Mitchiner, John E. (2002): *The Yuga Purāṇa*. Critically edited, with an English translation and a detailed introduction. Second revised edition. Kolkata.

Olivelle, Patrick (2005): *Manu's Code of Law. A critical edition and translation of the Mānava-Dharmaśāstra*. Oxford University Press.

Parasher, Aloka (1991): *Mlecchas in Early India: A Study in Attitudes towards Outsiders upto AD 600*. Delhi.

Rocher, Ludo (1986): *The Purāṇas*. Wiesbaden. (A History of Indian Literature, II/3.)

Sadakata, Akira (1997): *Buddhist Cosmology. Philosophy and origins*. Tokyo: Kōsei Publishing.

Satinsky, Ruth (2015): "What the lifespans of Ṛṣabha, Bharata, Śreyāṃsa, and Ara can tell us about the history of the concept of Mount Meru." *International Journal of Jaina Studies* (Online) 11(1), 1–24.

Satya Shrava (1977): *A Comprehensive History of Vedic Literature: Brāhmaṇa and Āraṇyaka Works*. New Delhi: Pranava Prakashan.

Sharma, Bellikoth Ramachandra (ed.) (1967): *Ārṣeya Brāhmaṇa, with Vedārthaprakāśa of Sāyaṇa*. Tirupati: Kendriya Sanskrit Vidyapeetha.

Sūryakānta (ed.) (1981): *Kāṭhaka-Saṃkalana*. New Delhi: Meharchand Lachhmandas Publications.

Vāyu Purāṇa. Vāyu Mahāpurāṇa. An exhaustive introduction, Sanskrit text, English translation, scholarly notes and index of verses. Vol. I (chapters 1–64). By Sudarshan Kumar Sharma. Delhi. 2008.

Yuga Purāṇa. See Mitchiner 2002.

Zysk, Kenneth G. (2016): *The Indian System of Human Marks*. Leiden: Brill. (Sir Henry Wellcome Asian Series, 15.)

CHAPTER 9

Caraka on the Human Being: A New Translation of *Carakasaṃhitā*, Śārīrasthāna, Chapter 1

Dominik Wujastyk

1 Introduction

This translation is offered to Chris Minkowski in friendship and appreciation of his interest in, and contributions to the history of science in India. The translation presented here is based on the collation of manuscripts of the Sanskrit text of this part of the *Carakasaṃhitā* prepared by the "Philosophy and Medicine in Early Classical India" (PMECI) project at the University of Vienna led by Karin Preisendanz.[1] This part of the text was prepared by myself and other project participants, especially Philipp Maas.[2] It is to be hoped that the text of the Sanskrit collation, on which this paper is based, will in due course be formulated as a critical edition. It is also to be hoped that publication will follow. The translation and annotation presented here are entirely my own.[3]

Modern studies on the human body in Ayurveda and the history of Indian medicine began over forty years ago.[4] The present paper will not recapitulate those studies, but presents a clearer, modern English translation that is more philologically-oriented than previous translations.

1 Preisendanz et al. 2002–.

2 Project participants since 2000 have included Karin Preisendanz (PI), Ernst Prets, Philipp Maas, Christina Pecchia, Vitus Angermeier and myself as well as the assistants who are listed on the project website.

3 This paper may be read as a companion to Wujastyk 2021.

4 Amongst important studies are: Dissanayake 1993; Flood 1993; Staal 1993; Thatte 1994; Holdrege 1998; Zarrilli 1998; Filliozat 1999; Kuriyama 1999; Leslie 1999; Bouillier and Tarabout 2002; Wujastyk 2002; Zimmermann 2002; Alter 2004; Ciurtin 2004; Lyssenko 2004; Preisendanz 2005; White 2006; Maas 2008; Wujastyk 2008; Michaels and Wulf 2009; Wujastyk 2009; Zimmermann 2009; Kitada 2012; Doniger 2014; G. Flood 2015; Preisendanz 2015; Wujastyk 2015; Cerulli 2016; Robertson 2017. Meulenbeld's *A History of Indian Medical Literature* offers a birds-eye view of the contents of this chapter, with substantive notes for further reading (HIML: IA.38–IB.76–77).

 | DOI:10.1163/9789004536869_011

2 The Manuscript Base

In his path-breaking article of 2010, Philipp Maas showed that the textual data presented by 52 available manuscripts of the *Carakasaṃhitā* that included the Vimānasthāna section could be represented for scholarly purposes by a carefully-evaluated sub-group of ten manuscripts.[5] Maas concluded that,

> In the case of the present transmission of CS Vi 8, ten manuscripts can be identified that were apparently not strongly contaminated. The most parsimonious cladogram … [of] these witnesses has a CI of ca. 0.89 … [that] increases to a CI of ca. 0.99 when all non-substantial convergent variants … are excluded.[6] *I take this very high degree of consistency as justifying the hypothesis that this cladogram mirrors the history of the transmission of the text passage under investigation as faithfully as possible.*[7]

The italic emphasis is added by me. What Maas has shown is that 42 manuscripts can be eliminated from further consideration. The transmission of this part of the *Carakasaṃhitā* is 99 % accurately and completely represented by the ten manuscripts in his figure 8 (p. 95), namely: $\mathrm{P1}_s$, Ch_d, $\mathrm{Jp1}_d$, A_d, $\mathrm{Ap1}_d$, $\mathrm{B5}_d$, $\mathrm{B1}_d$, $\mathrm{Ba1}_d$, $\mathrm{P4}_d$ and Km_d.[8] The transmission is bifurcated at the highest level of the stemma into just two branches, K and E, that correspond approximately to manuscripts transmissions in the west, including Kashmir and in the east, including north India and Bengal. In almost every case of difference, the vulgate edition and Cakrapāṇidatta's commentary both (Bo) read with the E branch.

During the period 2011–2012, Philipp Maas and Dominik Wujastyk collated four of the manuscripts containing this chapter, Śārīrasthāna 1: $\mathrm{Jp1}_d$, A_d, $\mathrm{Ap1}_d$ and Ch_d.[9] We worked under the assumption that the stemmatic results of Maas (2010) are applicable to other parts of the *Carakasaṃhitā*, an assumption that should at some future time be tested. Under this assumption, the four manuscripts so far collated represent the K branch of the stemma rather

5 Maas 2010: 94–95.

6 The Consistency Index (CI) is a statistical measure of "how many variants are in conflict with the structure of this stemma." (Maas 2010: 71). A CI of 1.0 indicates a perfect congruence between variants upon the stemma structure, while a lower figure indicates more complexity and contamination in the stemma structure.

7 Maas 2010, 95–96.

8 Subscript "d" indicates Devanāgarī; "s," Śāradā. Within the project, we jocularly refer to these MSS as "The Golden Ten."

9 MS Jaipur MSMSM 2608, MS Alwar RORI 2498, MS Alipur BLII 5285 and MS Chandigarh DAV 2315 respectively.

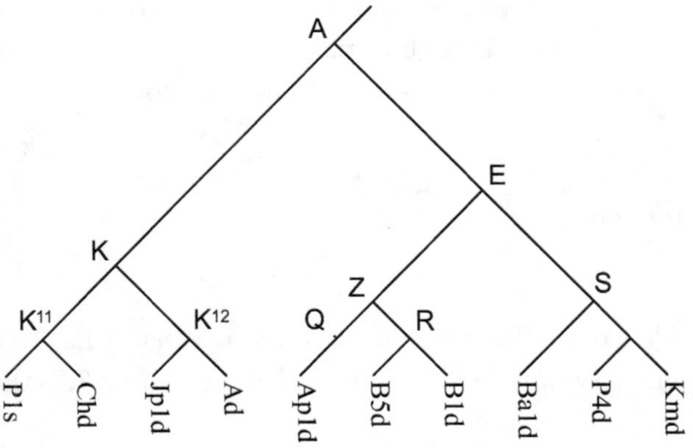

FIGURE 9.1 Maas 2010: page 95, figure 8

fully (75 %) and the E branch only weakly (17 %). Thus, the translation below is somewhat weighted towards the western and Kashmiri transmission of the text. After a discussion of the textual and historical problems surrounding the so-called Kaśmīrapāṭha of the *Carakasaṃhitā*, Meulenbeld concluded that,

> This survey of the available material proves unequivocally that the Kāśmīrapāṭha is not identical with Dṛḍhabala's text at all, being a version of the *Carakasaṃhitā* that was current in Kaśmīr and comparable with the versions of the Saindhavas, Gauḍīyas, etc.[10]

Translation of *Carakasaṃhitā, Śārīrasthāna* 1

1 "Now I shall explain the chapter about the body that begins, 'how many parts does a man have (*katidhāpuruṣīya*)?'"

2 said Lord Ātreya.

[How many parts are there in a person?]

3 [Agniveśa asked:]

"Into how many parts is a person divided, Wise One, according to the division of constituents (*dhātu*)? How is that person the cause? What is the person's origin?

10 HIML, 1A 136–139.

4 "Is he the one who knows or the one who does not know? Is he shown to be permanent or impermanent? What is original matter (*prakṛti*), what are the alterations? And what is the characteristic mark (*liṅga*) of a person?"

[The nature of the self]

5 [Ātreya answered:]

"Those with self-knowledge say that the self does not act, that it has free will (*svatantra*), is compelling (*vaśin*), omnipresent and all-pervading. It is the embodied knower (*kṣetrajña*), the witness."

[Problems with the above view]

6 [Agniveśa asked:]

"Without action, it has activity, O Lord. How is that possible? And if it has free will (*svatantra*), why is it sometimes born into unfavourable wombs?

7 "If it is compelling (*vaśin*), why is it forcefully assailed by unpleasant conditions? And, since it is omnipresent, why does it not experience all sensations (*vedanā*)?

8ab "It is all-pervading (*vibhu*), so why does it not see things that are behind mountains and walls?

8cd–9 "A doubt exists: which comes first, the embodied knower (*kṣetrajña*) or the body (*kṣetra*) itself? For without a body to be known in the first place, it is impossible to have an embodied knower. And if the body is there first, the knower of the body would not be eternal.

10 "What is it the witness of, since there is no other agent than it? And how can something that is without any disease (*vikāra*) have a symptom (*viśeṣa*) caused by pain (*vedanā*)?

[Questions about pain]

11 "Moreover, Lord, which pain (*vedanā*) that the patient has, out of the three, does a physician treat: past, present, or future?

12 "The future pain has not yet arisen; the past one will not come back. Nor does a present one have continuity (*sthāna*). So that leads to uncertainty (*saṃśaya*).

13 "What is said to be the cause of pains? What is their basis (*adhiṣṭhāna*)? Where do all these pains find a complete end (*nivṛtti*)?

[Question about the self]

14 "What are the marks (*liṅga*) by which the soul (*bhūtātman*) is discerned (*upa√labh*), that knows all, that renounces all,[11] that is without attachment to anything, single, peaceful?"

[Ātreya Punarvasu answered]

15 Punarvasu, the best of the wise, inwardly peaceful, listened to what Agniveśa said, and then explained everything in the proper manner.

[The constituents of the human being]

16 "The human being (*puruṣa*) is traditionally said to consist of the constituents (*dhātu*) starting from ether (*kha*), with consciousness (*cetanā*) as the sixth. There is also a tradition that the technical term '*puruṣa*' applies to the constituent of consciousness alone.

17 "Furthermore, it is traditionally said to consist of twenty-four parts according to the division of the constituents (*dhātu*): mind (*manas*), the ten sense organs, the objects (*artha*), and matter (*prakṛti*), that has eight constituents.

[*The mind*]

18–19 "The defining feature (*lakṣaṇa*) of mind is the existence or[12] non-existence of knowledge. For, it being present, it functions (*vṛt-*) because of a contiguity (*sannikarṣa*) of the self, the senses and their objects.[13] Because of disjunction

---

11 The Kashmir recension reads "who lays down all (*sarvavinnyāsī* (*sic*))."

12 Reading वा with the MSS against च in the vulgate (Bo).

13 This much-cited statement is almost the same as *Vaiśeṣikasūtra* 3.2.1 (VS_2: 28). Comba (1987) offered a seminal study of the passage and summarized earlier scholarship. Preisendanz (1989) developed the analysis in the context of the theory of vision. Further references were cited by Meulenbeld (HIML: 1B.76, n. 6). The phrasing of this passage in the vulgate 1941 edition (Bo) is problematic because of the न "not" that inappropriately negates the sense of the statement in the first line. All the manuscripts collated read not सति ह्यात्मेन्द्रियार्थानां सन्निकर्षे न वर्तते "… where there does not exist a contiguity" but instead सति ह्यात्मेन्द्रियार्थानां सन्निकर्षेण वर्तते, "… by means of contiguity." Editions C1 (327) and B_{1911} (1.659) agree with the MSS; editions C6 (367) and C8 (2.1763) agree with the vulgate.

(*vaivṛttya*) from the mind, no knowledge exists;[14] and from contiguity (*sānnidhya*) it does exist. Mind is traditionally said to have two qualities, minuteness (*aṇutva*) and unity (*ekatva*).

20 “Whatever the mind can know, be it what is thought about, what is considered, what is to be investigated, meditated upon or willed, all that is technically termed ‘the object (*artha*).’

21 “The proper action of the mind is limited to the control of the senses and to investigation and considering. Beyond that, the intelligence (*buddhi*) engages.

22 “Because a sense object (*indriyārtha*) is grasped mentally by means of the sense organ. It is subsequently considered by the mind with respect to its merit and demerit.[15]

[*The organs of sense*]

23 “The intelligence (*buddhi*), whose essence is certainty, that arises about that sense object (*viṣaya*) determines how to speak or act in accordance with intelligence.[16]

24 “Intelligence functions by means of the five sense organs, joined increasing one at a time amongst the elements like space, and inferred through action.[17]

25–26 “The five organs of action are the hands, the feet, the rectum, the nether parts (*upastha*), and the organ of the tongue.[18] The feet are used for walking, the rectum and nether parts are for evacuation,[19] the hands are for grasping and holding, the tongue is an organ of speech. Speech that is true is light (*jyotis*); speech that is untrue is darkness (*tamas*).

27 “The great entities (*mahābhūta*) are space, wind, fire, water and earth. Their qualities are sound, touch, form, taste and smell.

---

14 I have assumed an avagraha before ज्ञान, thus reading, before sandhi, वैवृत्त्यात् मनसस् अज्ञान “… no knowledge exists.” Since this is not marked in the MSS, the reading cannot be decided stemmatically. The editions C1; C6; B_{1911}; Bo read ज्ञन, problematically. The commentator Cakrapāṇidatta is of no help here since he explains the meaning of the verses without revealing his readings.

15 Read यथा with the K MSS, against ऽथवा.

16 Read तथा with the K MSS, against तया. In this account, commentators like Cakrapāṇidatta worry that *ahaṃkāra* is not part of the *Carakasaṃhitā*'s scheme of the psyche (Bo: 288b). They argue that it is implied in the functioning of *buddhi*, but this is not true to what the *Carakasaṃhitā* actually says.

17 This verse again presents a view that contrasts with classical Sāṅkhya, which has the senses evolving from *ahaṅkāra* (Bo, 289a, some discussion by Sharma 1981–1994: 3.352).

18 The K branch MSS read “organ of the tongue (*jihvendriya*)” against “organ of speech (*vāgindriya*).”

19 The K branch MSS read an explicit dual here, *pāyūpasthau*.

28 "Amongst these, there is one quality in the first;[20] later ones each increase by a quality. Amongst qualities, the former one has the former qualities in order, it is said.

29 "The perceived characteristic of earth is hardness, of water, liquidity, of wind, motility and of fire, heat. The characteristic of space is lack of resistance (*apratighāta*).

30 "Each one of these characteristics is the range of function (*gocara*) of the sense of touch,[21] because touch and its opposite can be cognized by the sense of touch.[22]

31 "The qualities that are displayed (*nirdiṣṭa*) are a signal (*chihna*) of the qualificands (*guṇin*) within the body. The 'ranges of function (*gocara*),' the 'sense objects (*viṣaya*),' and the 'qualities (*guṇa*)' are understood to mean the objects (*artha*) such as sound.

32 "The perception (*buddhi*) that a person has by reliance on a particular sense gets to be designated (*nirdeśa*) by that sense; and one had by reliance on the mind (*manas*) is designated 'mental.'

33 "Perceptions (*buddhi*) are traditionally said to be plural because of the differences arising from goals (*kārya*), sense organs (*indriya*) and sense objects (*artha*).

"Each different one arises from the proximity (*sannikarṣa*) of the self, the sense organs, the mind and the sense objects (*artha*).

34 "A perception (*buddhi*) is seen to arise from contact (*saṃyoga*), just as a sound is seen to arise from a finger hitting the base of the thumb or from a nail plucking the string of a veena.[23]

---

20 The K branch reads "one quality in the first (*eko guṇaḥ pūrve*)" against "the first has one quality (*ekaguṇaḥ pūrvaḥ*)."

21 All MSS read "the sphere of the sense of touch (*sparśanendriyagocaraḥ*)", a masculine nominative, in apposition to "characteristic (*lakṣaṇam*)," that can only agree with touch (*sparśaḥ*).

22 Cakrapāṇidatta explains that whatever one of the senses may grasp, the absence of that can also be grasped by the selfsame sense. A residual problem about how space, that does not have the quality of touch, can be known through the sense of touch is unconvincingly skirted. Cakrapāṇidatta tries to integrate the *Carakasaṃhitā*'s scheme with the classical Sāṅkhya theories of the *tanmātras*, *mahābhūtas*, that are not mentioned here. The *Carakasaṃhitā* does not know the theory of *tanmātras*, and mentions the *mahābhūtas* only infrequently (and in a compound with *-vikāra* seven out of the eight times).

23 At the time of the composition of the *Carakasaṃhitā*, the *vīṇā* was probably a lute, having evolved from an earlier harp-like instrument, as seen in Gupta and earlier coins and sculptures (Rowell 1992: 115 *et passim*).

35 "One should know the combination (*yoga*) of perception, senses, mind and the sense objects (*artha*) each with the other.[24] For a person (*puruṣa*) is defined as this congeries (*rāśi*) known as 'the twenty four items'.[25]

36 "For someone who is joined to passion and darkness, this combination (*saṃyoga*) is without end. By doing away with those two, by increasing purity, it is brought to an end.

37 "On this is based karma and the outcome, and knowledge. On this is based delusion, happiness, pain, life, death, selfhood (*svatā*).

38 "Someone who truly understands this understands dissolution and creation, continuity, therapy and everything that can be known.

[*The person*]

39 "If there were no person (*puruṣa*), there would be no light, no darkness, no truth or untruth, no Vedas, no good or bad karma (*karma*). There would be no agent nor any knower (*veditā*).[26]

40 "There would be no support (*āśraya*),[27] no happiness, no suffering, no going, no staying, no speech. There would be no knowledge, no scholarship (*śāstra*), no birth and no death.

41 "If there were no person (*puruṣa*), there would be no bondage and no liberation. The person (*puruṣa*) is the cause: therefore it is so declared by those who know causes.

42 "If the self (*ātman*) were not the cause, then light etc.[28] would have no reason to be (*hetu*). Knowledge about these things could not come into existence, nor could they have any point (*prayojana*).

43 "'The pot is produced out of clay, a stick and a wheel without a potter.[29] The house made with mud, grass and sticks has no builder.'

---

24 This translates the the K recension, which reads बुद्धीन्द्रियार्थमनोर्थानं विद्याद्योगं परस्परम्।. This is quite different from the vulgate as published (Bo, 290) and as read by Cakrapāṇidatta. The latter reading involves the introduction of a difficult discussion concerning "the beyond (*para*)," which Cakrapāṇi glosses as "the unmanifest (*avyakta*)" and manages to work into a classical Sāṅkhya interpretation.

25 The K recension reads चतुर्विंशक इत्येष ..., giving a slightly different sense than the vulgate.

26 The K recension reads "knower (*veditṛ*)" against the E reading of "experiencer (*ca bhoktā*)." Neither branch reads the vulgate "perceiver (*boddhā*)" (Bo, 291).

27 Cakrapāṇidatta explains this as the body (*śarīra*), the basis for all experiences that the self has.

28 Ācārya (Bo, 291, n. 3) notes a variant reading "space, etc. (*khādayaḥ*)" from an unknown manuscript. This fits better with the narrative of this chapter (but *lectio difficilior potior*?), but it is not attested in the MSS so far collated for K and E.

29 The manuscript record is uneven, but the best reading seems to be *mṛddaṇḍacakraprakṛtaṃ* against *kṛtaṃ mṛddaṇḍacakraiś ca*, "and made by means of clay, a stick and a wheel" in the vulgate Ācārya (Bo, 291, n. 3).

44 "Anyone who would say this would ignorantly say that the body, having come into being by means of instruments, has no creator, placing himself beyond reason or authoritative tradition (*āgama*).

45 "All valid means of knowledge lead one to understand that the person (*puruṣa*) is the cause. So do all the authoritative traditions (*āgama*) from which what should be authoritatively known is correctly known (*pra√mā*).

46 "Different new things that are not identical to what they came from arise in succession (*pāramparya*). Because there is similarity of form, people refer to the new things as the being the exact same things.

[*The Buddhist view*]

47 "Some people hold the view that a living being (*sattva*) is defined as a collection of these without a lord (*nirīśa*), as entities (*bhāva*), that the person (*puṃs*) is not the agent or the experiencer.[30]

48 "Amongst them, those who do not teach the self arrive at the view that the same new entities (*bhāva*) enjoy the result of what was done by different entities.[31]

[*The person is permanent*]

49 "An agent may be seen to have a multiplicity of different instruments, yet he is the same agent.[32] For the agent connected with the instruments is the cause of all actions.

50 "When passing away, the time of entities is quicker than the blink of an eye. There is no returning of what has been broken. What has been done does not accrue to someone else.[33]

51 "This is the opinion of those who know the truth. Because of this, it follows that he who is termed 'the person (*puruṣa*),' who is eternal, is the cause with respect to creatures' experience of their actions.

---

30 Cakrapāṇidatta makes explicit that the "some people" of these verses are the Buddhists and in particular that the verses express the doctrine of momentariness (*kṣaṇabhaṅga*). It is not obvious what the *Carakasaṃhitā* means by the polysemic word *bhāva* in this and subsequent passages; Cakrapāṇidatta ignores the term.

31 The syntax of this verse is difficult and there are no helpful variant readings in the collated MSS. The problem the author is characterizing is that if entities exist a discontinuous series, there is a problem in assigning identity and responsibility to later instances of the entity since it is not ontologically the same entity.

32 Cakrapāṇidatta clarifies that "instruments" refers to "mutable bodies" and the agent is the self.

33 This is the counter-argument to the Buddhist position on momentariness. If change is constant and instant, there can be no continuity of karma.

52 "A sense of identity (*ahaṅkāra*), consequences (*phala*), karma, the passage to another body, memory: given that there exists a cause, this exists without the physical body.

53 "The supreme self (*paramātman*) has no source (*prabhava*) because it is without any beginning. The person (*puruṣa*) defined as a conglomeration, is created out of delusion, craving, hatred and karma.

54 "The self is the knower.[34] Its knowledge proceeds out of the conjunction (*yoga*) with the instruments. But it does not proceed if the instruments have impurities or if there is no conjunction.

55 "For someone who is looking, when the mirror is tarnished or the water is turbulent there is no reflection, it is the same for truth when the mind (*cetas*) is damaged.[35]

56 "The mind (*manas*), reason (*buddhi*), and the organs of knowing (*buddhi*) and of action. The agent's action, sensation (*vedanā*) and reason (*buddhi*) arise out of conjunction.

57 "The embodied self (*bhūtātman*) does not initiate action or reap the result by itself. Everything happens because of conjunction. Without it, there is nothing.

58 "Because no existing entity (*bhāva*) exists alone, nor can it be without a cause. An existing entity cannot avoid the fact of existing because it changes rapidly.[36]

59 "The person (*puruṣa*) is permanent, what arises from a cause while is the opposite. A thing that exists with no cause is permanent; something that arises from a cause is deemed otherwise.[37]

60 "That permanence cannot be grasped from existence (*bhāva*) by any means.[38] That unmanifest entity beyond thought is manifested, it is different from an existing entity.

---

34 The K recension reads "the knower of the self (*ātmajñaḥ*)." This can just about be construed ("One with self-knowledge develops his knowledge out of the conjunction with the instruments") but it is slightly awkward.

35 The *Carakasaṃhitā* normally does not distinguish sharply between *cetas* and *manas*.

36 The readings of the vulgate make this passage almost unintelligible. The K recension reads शीघ्रत्वात्स्वभावत्वं भावो न व्यतिवर्तते, which is translated here.

37 The second half of this verse appears to be a commentary on the first half.

38 There is a missing *anusvāra* in the printed vulgate (Bo, 293); its presence is confirmed by Cakrapāṇi's reading. The K recension reads "because of being permanent (*nityatvān*)" against the vulgate's permanence (*nityatvaṃ*). It is hard to construe "from existence (*bhāvād*)" convincingly without perhaps adding an idea like "different from," as *anyathā* does in the next line.

61 "The self (*ātman*) is unmanifest and the embodied knower (*kṣetrajña*). It is eternal, present everywhere, unchanging. The manifest is anything different from that. In what follows, the two will be discussed.

[The manifest and the unmanifest]

62 "That which is grasped by the senses is manifest and sensory. The unmanifest is, once again, other than this. It is beyond the senses, to be understood through signs.[39]

63 "The origin of all beings (*bhūtaprakṛti*) is explained as being space etc., the intelligence, the unmanifest, and eighth of all the sense of self. And there are sixteen modifications (*vikāra*).[40]

64 "The modifications are technically defined as follows. The intelligence and the senses make five. There are also five organs of action. The five objects (*artha*) together with the mind.

65 "Thus the embodied (*kṣetra*) has been explained as everything apart from the unmanifest. The unmanifest is the embodied knower (*kṣetrajña*) of the embodied. The sages know this.

66 "The intelligence (*buddhi*) is born out of the unmanifest.[41] 'I am' is thought (*man-*) by the intelligence. Then space etc., arise in sequence out of the sense of self.

67 "Out of that one with all its parts fully formed is born.[42] The person is said to have arisen.

"And at the time of dissolution, the person is separated from the desired entities once again.[43]

68 "He moves from the unmanifest to being manifest; and again from manifested to being unmanifest. Subject to passion and darkness, he rotates like a wheel.

---

39 These signs are listed from verse 70 onwards.

40 This is not the standard Sāṅkhya account.

41 The K recension, reading *avyaktā*, would mean "intelligence is born as the unmanifest." There are other odd readings in the K recension of this verse that appear to be grammatical errors.

42 Cakrapāṇidatta appears to have interpreted this as referring to the original creation, whereas it seems clear that the verse is talking about the physical formation of a person. This is a perennial hermeneutic problem with Sāṅkhya thought.

43 Cakrapāṇidatta understands the "entities" to be intelligence and the other items described earlier. This verse could be understood, against Cakrapāṇidatta, as "separated from desired rebirths."

69 "Those who have an ultimate attachment to duality, who have the sense of self as their highest value, they have birth and death. But not those who are different from that.

[The signs of the self]

70–72 "The signs of the supreme self are: breathing in and breathing out, things like blinking, living, the movement of the mind,[44] moving from one sense to another, motivation and persistence,[45] dreaming about going to another place, comprehending death, comprehension by the left eye what is seen by the right,[46] desire, aversion, happiness pain, effort, consciousness, will, intelligence, memory, and the sense of self.

73 "Wise sages have said that because these signs belong to a being that is alive, not dead, they are the signs of the self.

74 "When it has departed, the body is an empty house, without consciousness. All that is left is the five elements, and so they say, 'it has become five'.[47]

[Consciousness and mind]

75 "The mind (*manas*) is full of action but unconscious. It is the supreme that is the conscious one. Actions are assigned to that omnipresent one being joined with the mind.

76 "The self is called the agent because it has consciousness. Not so the mind, even though it is active, because it has no consciousness.

77 "By means of the breaths, a living being leads itself by means of itself into all the wombs according to its own wish. There is no other controller of this than he.[48]

---

44 Cakrapāṇidatta charmingly exemplifies this, saying, "travelling mentally to Patna."

45 Cakrapāṇidatta specifies that these are mental acts.

46 The K recension reads "going out (*apagama*)" for "comprehension (*avagama*)," thus, cryptically, "the departure by the left eye of what is seen by the right eye."

47 The K recension reads, "when that body has departed (*śarīre hi gate tasmin*), there is an empty house that is unconscious," which might just about make sense, but seems more obvious that it is the departure of the self that is the circumstance of the predicate.

48 This cryptic verse has caused trouble since Cakrapāṇidatta, who is evasive about how it should be construed. The nominative "all (*sarva*)" seems to hang, and is even more confusingly read as "mind (*manas*)" in the K recension.

78 "The controller (*vaśin*) performs an action, and having done it, that is the result that he consumes. It is the controller who integrates the mind; it is the controller who renounces everything.[49]

79 "The embodied self, even though omnipresent knows only its very own sensations and senses. That is why the self does not know all the sensations that reside in all subjects.

80 "Thus, because he is great and omnipresent, he is everywhere. By concentrating the mind, the self is able to see hidden things.

81 "One should know that what has a permanent connection by means of the mind, with its consequences from the actions of the body, is present in every womb, as well as present in one womb.

82 "The self has no beginning; the continuity of embodiment (*kṣetra*) has no beginning. So, because these two have no beginning one does not ask 'which came first?'[50]

83 "They say that the witness is the one that knows, not the one that does not know. That is why tradition calls the self the witness. Amongst all creatures, all beings (*bhāva*) have the self as a witness.

84 "The unique embodied self can never be understood by characteristics. There does not exist a distinguishing mark that belongs to that one thing that cannot be understood.

85 "The embodied person (*saṃyogapuruṣa*) does have the required distinguishing mark, formed out of pain (*vedanā*). Where a pain is definite, there is a distinguishing mark formed from it.

[Past and future]

86 "The logic of, 'the physician treats pains belonging to all three times,' that some people declare, is a logic that ought to be considered.

87–89 "'The headache is back; the fever is back; that severe cough is back and the vomiting has come back again too.' Well-known expressions like this give rise to the idea that the what is past returns. 'The time for these past pains has comes back again.' The medicine that is used that relates to the time of that pain is called the soothing of past pains.

49 Cakrapāṇidatta is exercised to distinguish between the self's independence and control that have been described, insisting that the self is internally guided by its own will, and not motivated by likes and dislikes.

50 Cakrapāṇidatta worries that if the created world too is beginningless, it might also be like the self. He argues that there are different kinds of continuity. The beginninglessness of the self is unique and inherent, whereas that of the world can be divided into parts.

90 "'Those waters that formerly destroyed the corn must not come again.'[51] Just as a dyke is constructed, so a remedy is applied in the case of the body (*āśraya*).

91 "The treatment that is applied after observing the premonitory signs of diseases that are in the course of appearing, destroys future pain.

92 "By means of the therapy that causes well-being, the continuous symptom of pains is extinguished and well-being is established.

93 "Equal things do not become unequal and unequal things do not become equal. The appropriate elements in the body always come into being for a reason.[52]

94 "The physician keeps this reasoning before him and 'destroys the pain that belongs to three times.'[53]

[Pain and renunciation]

"But the perfect therapy is said to be the one free from dishonesty (*upadhā*).[54]

95 "For dishonesty is the greatest cause that leads to pain and a painful body. The renunciation of all dishonesty is what wards off all pain.

96 "As a silkworm (*kośakāra*) accumulates threads that produces its destruction, in the same way an ignorant person who is always ill accumulates craving (*tṛṣṇā*) from sense objects (*artha*).

97 "But a person who knows the sense objects to be like fire, realizes that and turns away. Because of not starting, not having contact, he does not come close to that pain.

---

51 The grammar of this sentence is archaic.

52 The term "element (*dhātu*)" is here used as a synonym for the term "humour (*doṣa*);" for the importance of this shifting terminology see the discussion by Scharfe (1999).

53 There is an odd disjunction in the flow of ideas at this point.

54 Reading with K's "therapy is said (*uktā*)" against the vulgates "it is said (*uktaṃ*)." The word "pretence (*upadhā*)" is glossed by Cakrapāṇidatta as "craving, thirst (*tṛṣṇā*)," picking the word up from verse 96 below. But the word more commonly means "trick, fraud, imposture." Cakrapāṇidatta reads this as being about subjective moral responsibility for disease rather than as an attack on quackery. Cf. Dasgupta's remarks on this passage (Dasgupta 1952–1961: v. 2, 414–415), where he insightfully suggested the word *upādāna*—a word that in Buddhist contexts is commonly glossed as craving (*tṛṣṇā*)—would suit the context better. One might conjecture a reading such as नैष्ठिक्योपदानम्विना उपादानम्परो हेतुः, but unfortunately the manuscripts collated do not offer any evidence at all in this direction.

[The abrogation of wisdom]

98 "The causes of pain should be understood to be the deterioration (*vibhraṃśa*) of intelligence (*dhī*), will-power (*dhṛti*) and memory (*smṛti*), the arrival of time and deeds, and the onset of unconducive sense objects.[55]

99 "The deterioration of intelligence (*buddhi*) is known to be an adherence to irregularity (*viṣama*) about what is permanent and what is fleeting, what is wholesome and what is not. For the intelligence sees straight (*sama*).[56]

100 "The deterioration of will-power (*dhṛiti*) makes it impossible to restrain a mind that inclines towards the objects of sense from the unwholesome object. For restraint is the nature of will-power.

101 "Someone whose memory has deteriorated is someone whose self is wrapped in passion and delusion and whose memory of the knowledge of truth has been lost. What ought to be remembered resides in memory.[57]

102 "That impure act that someone with a deteriorated intelligence, will-power and memory performs should be known as an abrogation of wisdom (*prajñāparādha*). It inflames all the humours.

103–108 "The abrogation of wisdom causes illness. It is described by the wise as an action arising out of passion and darkness such as the following: forcing out movements or suppressing an urge,[58] indulging in violent activities and over-indulging in sex with women, procrastination or starting actions at the wrong time, forsaking good manners and behaviour, being disrespectful during worship, pursuing one's own goals even though they are known to be bad for one, pursuing ideas that lead to complete madness, wandering about at the wrong time or place, friendship with those who behave reprehensibly, ignoring the virtuous behaviour described in the chapter '*indriyopakrama*',[59] envy, pride, fear,

55 These are standard etiological categories, discussed in *sūtrasthāna* 11 (see, e.g., Wujastyk 2003: 28–31; 2017) and elsewhere.

56 Here and in the next two verses, the author is unpacking the terms *dhī, dhṛti* and *smṛti* that he introduced in verse 98. In the *Carakasaṃhitā*, these three key terms are the defining constituents of *prajñā*, wisdom, and the abrogation of wisdom (*prajñāparādha*) is the *Carakasaṃhitā*'s final cause of all disease (Dasgupta 1952–1961: v. 2, 415–423, 321; Wujastyk 2017: 44).

57 On the meaning of memory (*smṛti*) and its discussion in this context, see Wujastyk 2012.

58 Unusual vocabulary, but referring to the well-known ayurvedic concept of not suppressing natural urges described in *Carakasaṃhitā sūtrasthāna* 7 and elsewhere. See Wujastyk 2003, 7, 15–17, *et passim*.

59 This is an internal reference to chapter 8 of the *sūtrasthāna* of the *Carakasaṃhitā*, where codes of good conduct are described.

anger, greed, delusion, intoxication and confusion, or distressing action stemming from these, or distressing acts on the body.

109 "One should understand that the abrogation of wisdom is improper understanding (*viṣamavijñāna*) by the intelligence and improper procedure. It is a mental thing.

[*Diseases implicating time*]

110 "The development of diseases over time has been described in the chapter '*vyādhisaṃgraha*',[60] such as the accumulation, inflammation and soothing of bile and the other humours.

111 "So have the causes of disease characterized as wrong, excessive or inadequate, those at the ends of seasons, and those based on right and wrong times for food being digested, eaten or badly digested.[61]

112 "And those diseases that are specific to the early, mid and late times of the day and to the three watches of the night are called 'born from time (*kālaja*).'

113 "Diseases that develop on alternate days, or that take hold every two days, or every three or four days, have their own time for they grow strong at their own time.

114 "These and any of the many other 'born of time' diseases yet to come should be treated by someone who understands strength and time.

[*Inherent ailments*]

115 "Diseases that arise from old age and death, through the passage of time, are seen as inherent (*svābhāvika*). Something inherent cannot be counteracted.

116 "The word 'fate (*daiva*)' designates an act (*karma*) that is connected with a former embodiment. That too can cause diseases that are acquired through time.

117 "One cannot avoid experiencing the consequence of any major act. Diseases that arise from karma destroy therapy. They are soothed after it is gone.

---

60 Chapter 17 of the *sūtrasthāna* of the *Carakasaṃhitā*.

61 These are all standard categories of etiology that are elaborated on in other parts of the *Carakasaṃhitā*. The vulgate records an variant from an unknown manuscript but it does not help the slightly confused text at this point (Bo, 297). The word प्रजीर्ण in K and the vulgate has been problematic since at least the time of Cakrapāṇidatta, who glosses it with *vidagdha* "scorched." It is a hapax in medical literature and interpretations are commonly derived recursively from this very passage. The E recension records प्रकीर्ण "scattered," which is not a normal descriptor for digestion, but rather a sign of discomfort with the difficult reading.

[*Ailments of excess, deficiency and error*]

118–127 "The threefold contact with disagreeable (*asātmya*) objects that inflames the humours is as follows:[62]

"They become dumb (*jaḍa*) by hearing excessively fierce sounds, and by not hearing any sound at at all or by hearing too diminished sounds. Erroneous contact is described as the contact of hearing with sounds that are rough, frightening, censured, unlovely or that point to evil.

"Not touching tactile objects, or touching them too much or too little, is said, in short, to damage the sense of touch. Untimely contact of the sense of touch with oily, hot or cold things belonging to ghosts, poison and winds is called inappropriate contact (*mithyāyoga*).

"Gazing at shining shapes is destructive because of gazing too much. Gazing at things that are too small does too, as does too not gazing anywhere. Gazing at repulsive, frightening, disgusting, distant and excessively close items and gazing at forms that are of the nature of darkness is called inappropriate contact (*mithyāyoga*).

"Excessive consumption, no consumption, by the standards of personal affinity (*okasātmya*) etc., the abnormal consumption of flavours, or consuming too little, is harmful.[63]

"Habitually smelling smells that are too mild and too sharp, and not smelling anything at all, are destructive of the sense of smell. The contact of the nose with smells that are stinking, ghostly, poisonous or hateful, and with those that are unseasonable, is called inappropriate contact (*mithyāyoga*).

"Something that does not become part of oneself (*ātmatā*) should be known as disagreeable (*asātmya*).[64]

128 "The wise call a disease that arises from an erroneous, excessive or deficient contact with sound etc., 'an illness relating to sense perception.'

129 "These are traditionally said to be the causes of pains that are disagreeable (*asātmya*).[65] The one cause of happiness is being equable (*sama*). It is hard to achieve equable contact.

---

62 This is taken from the last verse in this group. The author now presents a version of one of the standard etiological models that is laid out at greater length in *sūtrasthāna* 11 (see Wujastyk 2003, 28–30; 2017; 44; Dasgupta 1952–1961: v. 2, 320–321). Having presented issues relating to hearing, he goes on to discuss touch, sight, taste (consumption) and smell. It is curious to see that neither the author nor Cakrapāṇidatta seem to notice the difference between inability to hear and inability to speak.

63 I read ओकसात्म्य-, a well-known concept (Wujastyk 2017: 41), with the E recension and the vulgate, but against the odd, meaningless एकसात्म्य- of K.

64 An etymology for the expression *asātymya*, "disagreeable, unwholesome, without personal affinity." On this concept, see Wujastyk, 2017: 41–43.

65 Reading असात्म्यानाम् with the K recension against अशान्तानाम् "not pacified" of the E recen-

130 "The causes of happiness and pain are neither the senses nor the sense objects. But the cause of happiness and pain is observed to be the four kinds of connection (*yoga*).[66]

131 "The senses exist, sense objects exist: there is no contact and there is no disease, no happiness.[67] Therefore, it is the fourfold connection (*yoga*) that is the cause.

132 "There is no happiness or pain without the self, the senses, the mind, the intelligence, a sense object and karma. This is to be explained in a manner so that it can be understood.

133 "Touch, which is tactile contact between the organ of touch and the touching, is in fact mental. It is twofold, and generates sensations of happiness and pain.

134 "Craving, which consists of desire and aversion, is set in motion because of happiness and pain. And craving, once again, is said to be the cause of happiness and pain.

135 "Craving grasps at entities that are known as the foundations of pain.[68] One without any grasping cannot be touched; without being touched one cannot know sensations.[69]

136 "The basis for sensations is the mind and the body with its senses. The exceptions are the hair on the head, the hair on the body, the fingernails, food, excretions, liquids and qualities.[70]

[The path to liberation]

137 "In yoga and liberation, all pains stop recurring. In liberation, the cessation is complete; yoga promotes liberation.[71]

138–139 "Happiness and pain result from the close proximity of the self, the senses, the mind, and their objects. By not commencing, while the mind is

---

sion and the vulgate, which records the K reading as a variant in both the main text and Cakrapāṇi's commentary (Bo, 299a, footnote 1–2).

66 Given that the author will shortly embark on a treatise on yoga as a spiritual path, it is hard not to see here a deliberate ambiguity in the use of terms.

67 Reading with the K recension, योगो नास्ति न चास्ति रुक्, which is hardly different but stylistically more pleasant.

68 The terminology and thought here are clearly Buddhist.

69 Reading *nānupādāno* with the K recension.

70 Cakrapāṇidatta glosses liquids as "urine," and qualities as "sound, etc."

71 Fuller discussion of the verses from here to the end of the chapter was given in Wujastyk 2012.

steady and focussed on the self, that pair stops and an embodied person becomes powerful. Sages who understand yoga know that that is yoga.[72]

140–141 "Entering someone's mind, knowledge of objects, free action, vision, hearing, recollection, beauty, and invisibility at will: these are called the eightfold lordly power that yogins have. All that arises from meditation (*samādhāna*) with a purified mind.[73]

142 "Liberation comes from the absence of passion (*rajas*) and lethargy (*tamas*), due to the disappearance of potent karma. The disjunction from all karma is called non-rebirth.[74]

143–144 "Constant attendance on good people, and the avoidance of those who are not good; fasting and performing vows as well as each of the separate restraints (*niyama*); steady concentration (*dhāraṇā*), knowledge (*vijñāna*) of the teachings about dharma, a penchant for solitude, an aversion to the objects of sense, striving towards liberation, and supreme willpower.

145–146 "Not initiating actions, destruction of past deeds, unworldliness,[75] freedom from ego, seeing the danger in contact (*saṃyoga*),[76] concentration (*samādhāna*) of mind and intellect, critical analysis of philosophical reality. All this develops from from abiding in the memory of reality.[77]

147 "Recollection (*smṛti*) arises by means of attendance on good people, etc., through to willpower.[78] After having recollected the essential nature of states of being, recollecting, one is released from suffering (*duḥkha*).

148–149 "The eight causes that are said to bring recollection about are:

1 perception of the cause (*nimitta-grahaṇa*),
2 perception of the form (*rūpa-grahaṇa*),
3 similarity (*sādṛśya*),
4 contrast (*saviparyaya*),
5 attachment to goodness (*sattvānubandha*),[79]

72 Reading *tad yogam* with K for *tam yogam* in E and the vulgate.

73 I have not adopted the K reading कर्मतः for छन्दतः.

74 Reading कर्मसंयोगैर् "conjunctions with karma" with K, against सर्वसंयोगैर् "conjunctions with all things" of E, the vulgate and Cakrapāṇidatta (Bo, 300b).

75 "Unworldliness (*naiṣkramya*)" follows Cakrapāṇidatta's gloss, "'a tendency to escape' means a desire to leave samsara" (Bo, 300b). Every scribe seems to have had a different difficulty with this word.

76 Cakrapāṇidatta glosess this specifically as "the contact of the self to the body etc." (Bo, 300b).

77 The expression तत्त्वस्मृतेरुपस्थानात् "abiding in the memory of reality" paraphrases the Buddhist meditation term, स्मृत्युपस्थान.

78 I.e., the verses 143–144 above.

79 Cakrapāṇidatta is unusually reticent about how *sattva* should be understood; he blusters about how the mind should be devoted to remembering what should be remembered.

6 practice (*abhyāsa*),[80]
7 the yoga of knowledge (*jñānayoga*),[81] and
8 what is heard again (*punaḥ śruta*).[82]

"Recollection is said to come from recollecting what has been seen, heard and experienced.[83]

150–151 "This is the one path of liberation, having the strength of recollecting the truth (*tattva*), revealed by liberated people; those who have travelled by it have not returned again.[84]

"Yogins call this the way of yoga. Those followers of Sāṅkhya who have reckoned (*saṅkhyāta*) the dharmas and those who are liberated call it the way of liberation.[85]

152–153 "Everything that has a cause is pain, dependent, and impermanent. For that is not the created by the self (*ātmakṛtaka*). And personal attachment (*svatā*) arises with regard to that,[86] as long as the true realization has not arisen by which the knower, having known "I am not this, this is not mine," transcends everything.

154–155 "When this final renunciation exists, all pains (*vedanā*) and their causes, with consciousness (*saṃjñā*), knowledge (*jñāna*) and understanding (*vijñāna*), stop completely.[87]

"After that, the corporeal self (*bhūtātman*) that has become Brahman is not perceived. Having departed from all conditions, not a sign can be found. But an ignorant person cannot know the state (*gati*) of those who understand Brahman."[88]

There is a verse about this:

80 Similarly here, Cakrapāṇidatta says only that one gets good at whatever one practices.

81 The K recension reads ज्ञानरूप "the form of knowledge."

82 Again, Cakrapāṇidatta says only that if you hear something once you forget it; you have to repeat things to remember them.

83 Cakrapāṇidatta cites a variant reading स्मरणं स्मृतिरुच्यते "recollection is said to be recollecting," rather than "from recollection." This reading is reported in a manuscript of the E recension.

84 Cakrapāṇidatta glosses "recollecting the truth" as "strength (*bala*)," because, he says, people on the path to liberation need strength. This interpretation suggests again that he is more of a physician than a yogin.

85 This verse is missing in the manuscripts of both recensions. It is included here because it is in the vulgate.

86 This verse is hard to understand but no readings offer any clarifications to the text.

87 The Buddhist terminology here is again unmistakable: see, for example, Buddhaghosa's account of the process of experience described by Dasgupta 1952–1961, v. 1, 96–97.

88 Reading गति with all the manuscripts, against the vulgate's ज्ञान (Bo, 301).

> Those who see the truth have resolved the twenty-three highest questions relating to the person (*puruṣa*) in this chapter "Into how many parts is a person divided?"

Here ends the first chapter of the section about the body called "About the body, into how many parts is a person divided" in the Tantra created by Agniveśa and edited by Caraka.

Acknowledgements

This study is partly based on work done during the period 2010–2014 as part of the project "Philosophy and Medicine in Early Classical India" (PMECI) at the University of Vienna. The project is led by Prof. Dr Karin Preisendanz, and was funded by the Austrian Science Fund (http://www.fwf.ac.at/, grant no. P23330-G15). The project's website offers rich information about the history and aims of the project, and the reader is referred to that site for further details, including a bibliography of publications from 2000 onwards that have stemmed from the project.[89]

Index of Manuscripts

The numbers after the colon refer to pages in this document.

Abbreviations

B_{1911} Śivaśarman (1911–1931) (ed.), चरकसंहिता श्रीमन्महर्षिप्रवरचरकप्रणीता ... रामप्रसादवैद्योपाध्यायविरचिता प्रसदनी—भाषाटीकासहिता ... शिवशर्मणा संशोधिता = *The Carakasaṃhitā by Caraka with a Bhāṣā commentary by Rāmaprasāda Vaidyopādhyāya of Patiala, edited by Śivaśarman of Lahore* (Mumbai: Gaṅgāviṣṇu Śrīkṛṣṇadāsa at the Lakṣmīveṅkaṭeśvara Steam Press), URL http://indianmanuscripts.com/charaka-samhita-book-1, accessed 21/05/2021.

89 Preisendanz et al., 2002.

Bo Ācārya, Yādavaśarma Trivikrama, ed. महर्षिणा पुनर्वसुनोपदिष्टा, तच्छिष्येणाग्निवेशेन प्रणीता, चरकदृढबलाभ्यां प्रतिसंस्कृता चरकसंहिता, श्रीचक्रपाणिदत्तविरचितया आयुर्वेददीपिकाव्याख्यया संवलिता. (3rd edn., Mumbayyāṃ: Nirnaya Sagara Press), ARK: ark:/13960/t48q2f20n, accessed 01/01/2018.

C1 Bhaṭṭācārya, Jīvānanda Vidyāsāgara (1877) (ed.), चरकसंहिता, सूत्र-निदान-विमान-शारीरेन्द्रिय-चिकित्सित-कल्प-सिद्धिस्थानात्मका. आयुर्वेदीय भगवता आत्रेयेण पुनर्वसुना उपदिष्टा अग्निवेश-नामधेयेन तत्शिष्येण विरचिता चरकाभिधेन ऋषिणा प्रतिसंस्कृता, वि. ए. उपाधिधारिणा श्रीजीवानन्द विद्यासागर भट्टाचार्य्येण बहूनि आदर्शपुस्तकानि समालोच्य संस्कृता प्रकाशिता च = *Charakasanhita* (Kolkata: Sarasvati Press), ARK: ark:/13960/t0sq9gf44.

C6 Sen, Debendra Nath, and Sen, Upendra Nath (1897) (eds.), चरक-संहिता = *Charaka Samhita by Mahamuni Agnibesha* (Calcutta: Dhanvantari Machine Press), URL: http://n2t.net/ark:/13960/t2n59qh2r.

C8 Senagupta, Narendranātha, and Balāicandra Senagupta (1927–1933), (eds.), चरक-संहिता. महामुनिना भगवताग्निवेशेन प्रणीता महर्षिचरकेण प्रतिसंस्कृता. चरकचतुराननश्रीमच्चक्रपाणिदत्तप्रणीतया आयुर्वेददीपिकाख्यटीकया महामहोपाध्याय-श्रीगङ्गाधरकविरत्नकविराजविरचितया जल्पकल्पतरुसमाख्यया टीकया च समलङ्कृता. कविराज श्रीनरेन्द्रनाथ सेनगुप्तेन कविराज श्रीबलाइचन्द्र सेनगुप्तेन च सम्पादिता संशोधिता प्रकाशिता च (1st edn., Kalikātānagaryyāṃ: Dīnanātha at the Dhanvantari Steam Machine Press), URL: https://archive.org/search.php?query=gangadhara%20charak, accessed 20/01/2020.

HIML Meulenbeld, Gerrit Jan (1999–2002), *A History of Indian Medical Literature*, 5 vols. (Groningen: E. Forsten) ISBN: 9069801248.

VS_2 Jambūvijaya, Muni (1982) (ed.), चन्द्रानन्दविरचितया प्राचीनया वृत्त्या समलङ्कृतं कणादप्रणीतं वैशेषिकसूत्रम् = *Vaiśeṣikasūtra of Kaṇāda, with the Commentary of Candrānanda.* (Gaekwad's Oriental Series; 2nd edn., Baroda: Oriental Institute), ARK: ark:/13960/t54g0dx5q.

References

Alter, Joseph S. (2004), *Yoga in Modern India: The Body between Science and Philosophy* (Princeton, NJ: Princeton University Press), ISBN: 0-691-11873-6. DOI: 10.2307/j.ctv1h9djhv.

Bouillier, Véronique, and Tarabout, Gilles (2002) (eds.), *Images du corps dans le monde hindou* (Collection Monde Indien, Sciences sociales, 15–20 siècle; Paris: CNRS éditions), ISBN: 2-271-06060-5.

Cerulli, Anthony (2016), 'Body, Self, and Embodiment in the Sanskrit Classics of Āyurveda', in Barbara A. Holdrege and Karen Pechilis (eds.), *Refiguring the Body. Embodiment in South Asian Religions* (SUNY Press), ISBN: 9781438463162. DOI: 10.17613/m6590s.

Ciurtin, Eugen (2004) (ed.), *Du Corps Humain, Au Carrefour, De Plusieurs Savoirs En*

Inde. Mélanges Offerts À Arion Roşu Par Ses Collègues Et Ses Amis À L'occasion De Son 80e Anniversaire / the Human Body, at the Crossroads of Multiple Indian Ways of Knowing. Papers Presented to Arion Roşu by His Colleagues and Friends on the Occasion of His Eightieth Birthday (Studia Asiatica: International Journal for Asian Studies, IV–V; Bucharest: University of Bucharest).

Comba, Antonella (1987), 'Carakasaṃhitā, Śārīrasthāna I and Vaiśeṣika Philosophy', in Gerrit Jan Meulenbeld and Dominik Wujastyk (eds.), *Studies on Indian Medical History* (Groningen: Egbert Forsten), chap. 3, 43–61.

Dasgupta, Surendranath (1952–1961), *A History of Indian Philosophy*, 5 vols. (Cambridge: Cambridge University Press), ISBN: 0-521-04778-1, URL: https://tinyurl.com/ratwdof, accessed 26/01/2018; Reprint of 1922 edition.

Dissanayake, Wimal (1993), 'Self and Body in Theravada Buddhism. A Tropological Analysis of the Dhammapada', in *Self as Body in Asian Theory and Practice* (New York: SUNY), chap. 6, 123–145, ISBN: 0-7914-1080-3.

Doniger, Wendy (2014), 'Medical and Mythical Constructions of the Body in Sanskrit Texts 1', in *On Hinduism*. DOI: 10.1093/acprof:oso/9780199360079.003.0006, accessed 16/08/2017.

Filliozat, Pierre-Sylvain (1999), 'Homologies Du Monde, De La Parole Et De L'homme Dans Les Religions De L'inde (des Veda Aux Tantra)', in *Ressembler au monde: nouveaux documents sur la théorie du macro-microcosme dans l'antiquité orientale*, chap. 1, 11–40.

Flood, Gavin D. (2015), 'The Body', in *Oxford Bibliogaphies Online* (Oxford University Press). DOI: 10.1093/obo/9780195399318-0154.

Flood, Gavin D. (1993), *Body and Cosmology in Kashmir Śaivism* (San Francisco: Mellen Research University Press), ISBN: 0773499741.

Holdrege, Barbara A. (1998), 'Body Connections: Hindu Discourses of the Body and the Study of Religion', *International Journal of Hindu Studies*, 2/3: 341–386. DOI: 10.1007/s11407-998-0002-8.

Kitada, Makoto (2012), *The Body of the Musician. an Annotated Translation and Study of the Piṇḍotpatti-prakaraṇa of Śārnṅgadeva's Saṅgītaratnākara* (Bonn etc.: Peter Lang), ISBN: 9783034303194. DOI: http://dx.doi.org/10.3726/978-3-0351-0417-2.

Kuriyama, Shigehisa (1999). *The Expressiveness of the Body and the Divergence of Greek and Chinese Medicine* (New York: Zone Books), ISBN: 0-942299-88-4.

Leslie, Julia (1999), 'The Implications of the Physical Body: Health, Suffering, and Karma', in Roy Porter and John Hinnells (eds.), *Religion, Health, and Suffering* (London: Kegan Paul International), 23–45.

Lyssenko, Viktoria (2004), 'The Human Body Composition in Statics and Dynamics: Āyurveda and the Philosophical Schools of Vaiśeṣika and Sāṃkhya', *Journal of Indian Philosophy*, 32: 31–56. DOI: 10.1023/b:indi.0000014006.14364.9b.

Maas, Philipp André (2008), 'The Concepts of the Human Body and Disease in Classical Yoga and Āyurveda', *Wiener Zeitschrift für die Kunde Südasiens = Vienna Journal of South Asian Studies* 51: 125–162. DOI: 10.1553/wzkslis123.

Maas, Philipp André (2010), 'Computer Aided Stemmatics—the Case of Fifty-two Text Versions of Carakasaṃhitā Vimānasthāna 8.67–157', *Wiener Zeitschrift für die Kunde Südasiens* 52–53: 63–119; Special issue edited by Jürgen Hanneder and Philipp A. Maas, *Text Genealogy, Textual Criticism and Editorial Technique*.

Michaels, Axel, and Wulf, Christoph (2009) (eds.), *The Body in India: Ritual, Transgression, Performativity* (Paragrana: Internationale Zeitschrift für Historische Anthropologie, 1; British Library; Oxford: Akademie Verlag), ISBN: 3050046015.

Preisendanz, Karin (1989), 'On Ātmendriyamanorathasannikarṣa and the Nyāya-vaiśeṣika Theory of Vision', *Berliner Indologische Studien*, 4/3: 141–213.

Preisendanz, Karin, et al. (2002–), 'Philosophy and Medicine in Early Classical India', URL: http://www.istb.univie.ac.at/caraka/, accessed 11/08/2016.

Preisendanz, Karin (2005), 'Soul, Body and Person in Ancient India' in Hans-Dieter Klein (ed.), *Der Begriff der Seele in der Philosophiegeschichte* (Würzburg: Königshausen & Neumann), 119–175, ISBN: 3826027965, URL: https://jainelibrary.org/book-detail/?srno=269591, accessed 20/05/2021.

Preisendanz, Karin (2015), 'Between Affirmation and Rejection. Attitudes Towards the Body in Ancient South Asia from the Ṛgveda to Early Classical Medicine', in Gert Melville and Carlos Ruta (eds.), *Thinking the Body as a Basis, Provocation, and Burden of Life* (Berlin: de Gruyter Oldenbourg), 113–144, ISBN: 9783110407389. DOI: 10.1515/9783110407389-010.

Robertson, Matthew I. (2017), 'The Identity of Person and World in Caraka Saṃhitā 4.5', *Journal of Indian Philosophy*, 45/5 (Aug.): 837–861. DOI: 10.1007/s10781-017-9325-3.

Rowell, Lewis (1992), *Music and Musical Thought in Early India* (Chicago: University of Chicago Press), ISBN: 978-0-226-73034-9. DOI: 10.7208/chicago/9780226730349.001.0001.

Scharfe, Hartmut (1999), 'The Doctrine of the Three Humors in Traditional Indian Medicine and the Alleged Antiquity of Tamil Siddha Medicine', *Journal of the American Oriental Society*, 119/4: 609–629. DOI: 10.2307/604837.

Sen, Debendra Nath, and Sen, Upendra Nath (1897) (eds.), चरक-संहिता = *Charaka Samhita by Mahamuni Agnibesha* (Calcutta: Dhanvantari Machine Press), URL: http://n2t.net/ark:/13960/t2n59qh2r.

Sharma, Priya Vrat (1981–1994), *Caraka-saṃhitā: Agniveśa's Treatise Refined and Annoted by Caraka and Redacted by Dṛḍhabala (text with English Translation)* (The Jaikrishnadas Ayurveda Series, 36; 1st edn., Varanasi, Delhi: Chaukhambha Orientalia), ISBN: 8176370126; 8th edition, 2003.

Staal, J. Frits (1993), 'Indian Bodies', in *Self As Body in Asian Theory and Practice* (New York: SUNY), chap. 4, 59–102, ISBN: 0-7914-1080-3, URL: https://goo.gl/jXdHg4, accessed 26/01/2018.

Thatte, Dinkar Govind (1994), *Śārīrasthānam: Suśruṭa* (sic) *Saṃhitā. Section on the Study of the Human Body* (Varanasi: Chaukhambha Orientalia); Foreword by Priya Vrat Sharma.

White, David (2006), ''Open' and 'Closed' Models of the Human Body in Indian Medical and Yogic Traditions', *Asian Medicine: Tradition and Modernity*, 2/1: 1–13. DOI: 10.1163/157342106777996466.

Wujastyk, Dagmar (2015), 'On Perfecting the Body. Rasāyana in Sanskrit Medical Literature', *AION: Annali dell'Università degli Studi di Napoli "L'Orientale": Elisir Mercuriale e Immortalità. Capitoli per una Storia dell'Alchimia nell'Antica Eurasia. A cura di Giacomella Orofino, Amneris Roselli e Antonella Sannino*, XXXVII: 55–77. ISSN: 1128–7217, URL: https://www.libraweb.net/articoli.php?chiave=201509901&rivista=99, accessed 16/08/2017.

Wujastyk, Dominik (2002), 'Interpréter l'image du corps humain dans l'Inde prémoderne' in Véronique Bouillier and Gilles Tarabout (eds.), *Les représentations du corps dans le monde hindou* (Collection Monde Indien, Sciences sociales, 15–20 siècle; Paris: CNRS éditions), chap. 2, 71–99, plates I–V, ISBN: 2-271-06060-5. DOI: 10.4000/books.editionscnrs.9310.

Wujastyk, Dominik (2003), *The Roots of Ayurveda: Selections from Sanskrit Medical Writings* (Penguin Classics; 3rd edn., London, New York, etc.: Penguin Group), ISBN: 0-140-44824-1.

Wujastyk, Dominik (2008), 'A Body of Knowledge: The Wellcome Ayurvedic Anatomical Man and His Sanskrit Context', *Asian Medicine: Tradition & Modernity*, 4/1: 201–248. DOI: 10.1163/157342109X423793.

Wujastyk, Dominik (2009), 'Interpreting the Image of the Human Body in Premodern India', *International Journal of Hindu Studies*, 13/2: 189–228. DOI: 10.1007/s11407-009-9077-0.

Wujastyk, Dominik (2012), 'The Path to Liberation through Yogic Mindfulness in Early Ayurveda' in David G. White (ed.), *Yoga in Practice* (Princeton: Princeton University Press), chap. 1, 31–42, URL: https://academia.edu/3216968.

Wujastyk, Dominik (2017), 'Models of Disease in Ayurvedic Medicine', in Mark Jackson (ed.), *The Routledge History of Disease* (Abingdon: Routledge), chap. 3, 38–53, ISBN: 9780415720014. DOI: 10.4324/9781315543420-3, URL: https://www.academia.edu/25517629/, accessed 25/01/2018.

Wujastyk, Dominik (2021), 'A New Translation of Carakasaṃhitā, Vimānasthāna, Chapter 1, Based on the Vienna Critical Edition' in Toke Lindegaard Knudsen, Jacob Schmidt-Madsen and Sara Speyer (eds.), *Body and Cosmos. Studies in Early Indian Medical and Astral Sciences in Honor of Kenneth G. Zysk* (Leiden, Boston: Brill), chap. 6, 77–109. DOI: 10.1163/9789004438224_007.

Zarrilli, Phillip B. (1998), *When the Body Becomes All Eyes: Paradigms, Discourses and Practices of Power in Kalarippayattu, a South Indian Martial Art* (New Delhi: Oxford University Press), ISBN: 0195639405.

Zimmermann, Francis (2002), 'Ce Qu'un Hindou Dit à Son Corps—La Réécriture Des Représentations Traditionnelles', in Véronique Bouillier and Gilles Tarabout (eds.), *Images du corps dans le monde hindou* (Paris: CNRS éditions), 49–69.

Zimmermann, Francis (2009), 'A Hindu to His Body: The Reinscription of Traditional Representations' in *The Body in India: Ritual, Transgression, Performativity* (Paragrana: Internationale Zeitschrift für Historische Anthropologie 1. British Library; Oxford: Akademie Verlag), chap. 2, ISBN: 3050046015. DOI: 10.1524/para.2009.0002.

Glossary

abhyāsa
  practice 248
abrogation of wisdom
  prajñāparādha 243
act
  karma 244
adhiṣṭhāna
  basis 232
āgama
  authoritative tradition 237
  authoritative traditions 237
ahaṅkāra
  sense of identity 238
all
  sarva 240
all-pervading
  vibhu 232
aṇutva
  minuteness 234
apagama
  going out 240
apratighāta
  lack of resistance 235
artha
  object 234
  objects 233, 235, 239
  sense objects 235, 236, 242
asātmya
  disagreeable 245
āśraya
  body 242
  support 236
ātmajñaḥ
  the knower of the self 238
ātmakṛtaka
  created by the self 248
ātman
  self 236, 239
ātmatā
  oneself 245
attachment to goodness
  sattvānubandha 247
authoritative tradition
  āgama 237
authoritative traditions
  āgama 237
avagama
  comprehension 240
avyakta
  unmanifest 236
bala
  strength 248
basis
  adhiṣṭhāna 232
because of being permanent
  nityatvān 238
beings
  bhāva 241
bhāva
  beings 241
  entities 237
  existence 238
  existing entity 238
bhāvād
  from existence 238
bhūtaprakṛti
  origin of all beings 239
bhūtātman
  corporeal self 248
  embodied self 238
  the soul 233
boddhā

PART 3

Statecraft and Jurisprudence

CHAPTER 10

To Kill or Not to Kill: The Hermeneutics of the Ethical Axiom *ahiṃsā*

Patrick Olivelle

The ethical axiom of *ahiṃsā*—non-killing, non-injury, non-violence—seems, at first sight, simple and straightforward. It has been and continues to be a central pillar of the moral philosophies of most Indian religions. When one looks a bit closer, however, complexity and ambiguity begin to emerge. What sort of killing/injury/violence does the concept comprehend? At what external entity is it directed: at humans, at all animals, at all living beings? Is injury/violence envisioned solely as an external act or does it comprehend also internal dispositions or violent intentions? As Hans-Peter Schmidt in his "The Origin of *ahiṃsā*" notes: "In spite of its great importance for the religious attitude of the Indians, the history of the idea of ahiṃsā has rarely being investigated into, and the handbooks of Indian religions devote little space to it" (1968: 625). This paper will not attempt to trace the origins of *ahiṃsā* or even the history of this ethical principle. Instead, I will focus on how the Brahmanical tradition of hermeneutics (Mīmāṃsā and Dharmaśāstra) have engaged intellectually with the many theological problems and contradictions that *ahiṃsā* raises, both within Brahmanical theology and in its interactions with outsiders, such as Buddhists, Jains, and other ascetic traditions.

Some inherent inconsistencies of the Brahmanical position that both advocated *ahiṃsā* and defended killing animals within the Vedic sacrificial context are exposed in an interesting conversation between Kapila, the putative founder of the Sāṃkhya system and the paradigmatic ascetic, and a cow being taken to slaughter for a divine guest, in the *Go-Kapila-Saṃvāda* of the *Mahābhārata* (12.260–262). The seer Syūmaraśmi enters the cow and engages Kapila in a dialogue about both the appropriateness of killing animals and the relative superiority of the ascetic and the householder. An even more poignant confrontation is found in the medieval text, *Dharmāraṇya Purāṇa*, between Jain monks and Brahmins in the court of a Jain king. Veena Das (2013: 17) summarizes the debate:

> Jain monks accuse the Brahmins of indulging in violence and urge the king to withdraw patronage from the Brahmins. The Brahmins retort that

 | DOI:10.1163/9789004536869_012

> it is hypocritical to assume that anyone can live without violence for, in order to live, it is necessary to eat, and that act means that we are bound to inflict violence, whether on animals or plants. They argue further that violence is much more expansive than physical harm and who, they ask, is free of harboring anger and jealousy, which are also forms of violence.

Brahmanical moral theologians, heirs to the Vedic sacrificial religion, found themselves in a quandary resulting from a value system inherited from Buddhism and other śramaṇic traditions and anchored in the principle of *ahiṃsā*, not killing or injuring living beings. *Ahiṃsā* had become a central ethical principle at least by the time of Aśoka, who includes it in all his definitions of dharma and is uncompromising with regard to it. It became a central ethical principle not only in the ascetic religions of Buddhism and Jainism, but also in the mainstream of the Brahmanical tradition represented by the Dharmaśāstras.[1] Given the constraints of space, let me give you just one eulogy of *ahiṃsā* from the *Mahabhārata* (13.117.37–38):

> *ahiṃsā paramo dharmas tathāhiṃsā paro damaḥ* |
> *ahiṃsā paramaṃ dānam ahiṃsā paramaṃ tapaḥ* ||
> *ahiṃsā paramo yajñas tathāhiṃsā paraṃ balam* |
> *ahiṃsā paramaṃ mitram ahiṃsā paramaṃ sukham* |
> *ahiṃsā paramaṃ satyam ahiṃsā paramaṃ śrutam* ||

> Ahiṃsā is the highest dharma; and ahiṃsā is the highest restraint. Ahiṃsā is the highest gift; ahiṃsā is the highest austerity.
>
> Ahiṃsā is the highest sacrifice; and ahiṃsā is highest strength. Ahiṃsā is the highest friend; ahiṃsā is the highest bliss. Ahiṃsā is the highest truth; ahiṃsā is the highest Vedic command (*or* highest learning).

In spite of such eulogies, the ethical principle of *ahiṃsā* ran into a problem within the Brahmanical tradition—it was in conflict with entrenched religious and ritual practices and social and political customs, in other words, against other central prescriptions of dharma. Many Vedic sacrifices required the killing and the ritual offering of one or more animals. The most effective *śrāddhas*, the offerings to deceased ancestors, required the flesh of a variety of animals, birds, and fish.[2] The welcome feast for an honored guest (*madhu-*

1 See, for example, MDh 10.63; 11.222; 12.83; YDh 1.122; 3.213; ViDh 2.16; BDh 1.8.2; 3.1.27; 3.10.13; VaDh 4.4; 30.8.

2 See MDh 3.267–272.

parka), as we saw in the dialogue between Kapila and the cow, involved the killing and cooking of a cow, because of which such a guest is given the epithet *goghna*, "cow-killer". How can the new ethic and the old practices, both claiming to be dharma, be reconciled? What course should someone follow when ethical principles and ritual rules are in conflict? This is where the well-honed hermeneutics of Mīmāṃsā came to the rescue, or at least attempted to resolve the conflict.

1 *Ahiṃsā* in Mīmāṃsā

Among the numerous maxims called *nyāyas* used as hermeneutical tools for resolving conflicts between injunctions, I will feature three. The first is enunciated in the *Gautama Dharmasūtra* (GDh 1.4): *tulyabalavirodhe vikalpaḥ*, "When two injunctions of equal force contradict each other, there arises an option"—that is, one may follow either of the injunctions. But, not so fast: the Mīmāṃsakas never liked to leave anything that has to do with dharma to the free choice of an actor. For the most part, then, choice is eliminated by showing that these conflicts are only apparent. So, there is the maxim called *utsargāpavādanyāya*, stating: *utsargād apavādo balīyān*, "An exception has greater force than a general rule." Thus, if two rules contradict each other, it may well be that one of them is a general injunction while the other is more specific and is an exception to the general rule. In that case, one must follow the exception in spite of the general rule. Further, there is another interpretive rule called *viṣayavyavasthā*, that is, two rules that appear to contradict each other may in fact have different people, or different times or circumstances in mind. So, there really is no conflict at all; simply our inability to understand the scope of each rule. And the list of such interpretive strategies can go on. But back to *ahiṃsā*.

Both the origin of *ahiṃsā* and its relationship to Brahmanical dharma have been dealt with by several scholars, most notably by Wilhelm Halbfass.[3] I do not intend to plough that ground again here. Instead, I want to bring into the discussion some medieval discussions of the topic that introduce several new and innovative aspects.

Let me, however, present a quick survey of how Brahmanical scholars dealt with the conflict between *ahiṃsā* and various kinds of killing sanctioned by Brahmanical dharma and Vedic injunctions. In general, these discussions focus

3 For the origin of *ahiṃsā*, see Alsdorf 1962, where he posits that the source of *ahiṃsā* is not the Veda but non-Āryan sources. Hans-Peter Schmidt (1968) responds to Alsdorf and seeks to find the roots of *ahiṃsā* within the Vedic ritual itself. For the Mīmāṃsā discussions of *ahiṃsā* and ritual killing, see Halbfass 1983.

on two issues: killing in ritual settings and killing in the context of a conflict. These two broad categories in which *hiṃsā* or killing is permitted are set forth by Viśvarūpa, the ninth century commentator of the *Yājñavalkya Dharmaśāstra* (1.8), which mentions *ahiṃsā* in a list of six religious acts constituting dharma. Viśvarūpa gives two exceptions to the prohibition of killing: *ahiṃsā yajñādyātatāyivyatirekeṇa*, "not killing, with the exception of a sacrifice etc. and an assailant." All ritual exceptions appear to be included in *yajñādi*, and all situations of conflict, in *ātatāyin*.

It is the ritual killings authorized by the Veda and the Dharmaśāstras that principally occupy the minds of Brahmanical authors. Manu (5.41) gives three categories of ritual where killing animals is justified:

> *madhuparke ca yajñe ca pitṛdaivatakarmaṇi* |
> *atraiva paśavo hiṃsyā nānyatrety abravīn manuḥ* ||
>
> The honey-mixture, a sacrifice, an offering to gods or ancestors—at no other occasion than these, Manu has declared, may animals be killed.

The first is a hospitality rite for especially honored guests; the second is Vedic sacrifice; and the third is any rite in honor of gods or ancestors, especially *śrāddhas*. Both in the Dharmaśāstras and especially in Mīmāṃsā, however, the discussion is focused on Vedic sacrifices. Manu (5.39) states clearly and boldly that "in sacrifice killing is not killing" (*yajñe vadho 'vadhaḥ*). And again at 5.44, Manu says that killing sanctioned by the Veda (*yā vedavihitā hiṃsā*) should be recognized as *ahiṃsā* (*ahiṃsām eva tāṃ vidyāt*). Manu, however, is conflicted. He is attached to the ethical value of *ahiṃsā* and speaks against meat eating because it necessarily involves the killing of animals (5.33–57). He is also cognizant of the fact that a householder's life involves *hiṃsā*, saying that a householder has five slaughter houses and to expiate sins committed through them he has to perform the five great sacrifices (3.68–81).

Kumārila, the great Mīmāṃsaka, on the other hand, has none of the qualms exhibited by Manu. For him, *ahiṃsā*, much like any other moral imperative, has any validity only if it is the subject of a Vedic injunction, *vidhi*, or a prohibition, *niṣedha*. As Halbfass (1983: 8) observes, "Ahiṃsā is not a rule which in itself alone would be 'rationally' or 'morally' self-evident; it is itself valid only insofar as it is scripturally enjoined." Morality as dharma, therefore, has no validity unless it is the subject of a Vedic injunction. There cannot be a "rational" or "humanly envisioned" moral philosophy. Given that the killing of animals in Vedic sacrifices is the subject of a *vidhi*, it is legitimate, and the principle of *ahiṃsā* cannot be raised against it as an objection.

It is important in this context to focus a bit on the nature of dharma according to Mīmāṃsā and, to a somewhat lesser degree, Dharmaśāstra. For these traditions, dharma is not a nebulous ethical philosophy discovered through reason, emotion, or through some yogic power; not the ethics of an Aśoka, or even of a Buddha or a Jina. The definition of dharma is given at the very beginning of the *Pūrva Mīmāṃsāsūtra* (1.1.2): "dharma is an *artha*, something beneficial, contained in a Vedic injunction," *codanālakṣaṇo 'rtho dharmaḥ*. Dharma is not a philosophy or an ethic; it is the sum total of Vedic and, derivatively, Dharmaśāstric injunctions. So, when a particular act, for example, killing a goat, is said to be adharmic, it must refer not to some abstract idea of what is moral and immoral but to an actual prohibition, *niṣedha*, of such killing. So, when the proverbial injunction, *agnīṣomīyaṃ paśum ālabeta*, "the animal dedicated to Agni and Soma should be killed in sacrifice," no question can be asked about its dharmic nature, because dharma is what the Veda enjoins.

The Dharmaśāstra tradition, and even scholars from other traditions such as Advaita and Nyāya, did not follow this strictly atomistic view of dharma. When we speak of *rājadharma, strīdharma*, or *jātidharma*, we are talking about a broader set of customary practices obligatory on individuals belonging to these groups. Further, Dharmaśāstras frequently list important virtues as constituting the essence of dharma. Manu, for example, gives ten virtues such as forbearance, mastering the organs, and truthfulness as the ten defining characteristics of dharma, *daśakaṃ dharmalakṣaṇam* (6.92). Yājñavalkya (1.6) presents gift giving, *dāna*, as the distinguishing characteristic of dharma, *dharmalakṣaṇam*. And again, he (1.8) lists several practices but calls knowledge of the self the highest dharma: "Sacrifice, normative practice, self-control, abstaining from injury (*ahiṃsā*), gift giving, and Vedic recitation—among these activities, however, this is the highest dharma: to perceive the self by means of yoga."

Within this expanded view of dharma, *ahiṃsā* comes to be viewed both as a general prohibition against killing living beings and as the inner virtue underpinning non-violent behavior. Texts present the Vedic injunction relating to *ahiṃsā*: *na hiṃsyāt sarvā bhūtāni*, "one should not kill any being," an injunction, however, that cannot be traced in any extant Vedic text. This, in Mīmāṃsā classification, is an *utpattividhi*, an originative injunction (*utsarga*) of broad applicability. The Mīmāṃsā argument presents ritual killing, such as that contained in the injunction: *agnīṣōmīyaṃ paśum ālabeta*, "one should offer in sacrifice an animal dedicated to Agni and Soma," as an exception (*apavāda*) to that general prohibition, according to the aforementioned maxim: *utsargād apavādo balīyān*, "an exception has greater force than a general rule."

The common Mīmāṃsā distinction between *kratvartha* and *puruṣārtha* is also significant for the hermeneutics of *ahiṃsā*. The first, *kratvartha*, relates to acts that are performed for the sake of the ritual—so that a ritual is performed properly with all its subsidiary elements—and are only secondarily beneficial to the performer of the ritual. The second, *puruṣārtha*, on the other hand, are acts undertaken for the sake of the performer of the ritual. Halbfass (1983: 7) notes that in Prabhākara's view the *hiṃsā* that is prohibited relates to a *puruṣārtha* killing, such as those done in a rite of sorcery or, for that matter, killing an animal for food. When *hiṃsā* is *kratvartha*, done because the rite requires it—such as killing an *agnīsomīya* animal—then it is not prohibited. I think this distinction, although not used extensively in discussions of *ahiṃsā*, is significant in the Indian hermeneutical arguments regarding this topic.

One reason that *kratvartha* killing is not considered *ahiṃsā* is that it does not entail any kind of desire (*kāma*) or passion (*rāga*). The latter is prominent in the arguments presented by the 9th century legal scholar Medhātithi. In responding to an opponent who claims: *hiṃsāyāṃ lokavedayor na viśeṣaḥ* ("With respect to killing, there is no difference between the worldly and the Vedic"; on MDh 2.10), Medhātithi says: *rāgalakṣaṇā laukikī hiṃsā, vidhilakṣaṇālaukikī hiṃsā* ("Worldly killing is characterized by passion, while non-worldly killing is characterized by injunctions": on MDh 2.10).

2 *Ahiṃsā* in the *Vacanamālā*

So far we have dealt with the strictly Mīmāṃsā arguments regarding ritual killing as the major exception to the general prohibition of killing in the context of *ahiṃsā*. There are, however, other areas of human and social activities where killing and violence are viewed by authors of Dharmaśāstric texts as justified. Within that context, I want now to bring into this discussion of *ahiṃsā* a significant late medieval text called *Vacanamālā*. It is a thus far unpublished sub-commentary on Viśvarūpa's commentary, *Bālakrīḍā*, on the *Yājñavalkya Dharmaśāstra*. The *Vacanamālā*, whose author is unknown but is probably from Kerala, is extant only in one manuscript in Trivandrum, along with a Devanāgarī copy of it made at the beginning of the 20th century.[4] On Viśvarūpa's brief gloss *ahiṃsā yajñādyātatāyivyatirekeṇa* (on YDh 1.8), which we have already looked at, the *Vacanamālā's* commentary runs to three and a half of manuscript pages. The first part deals with the term *yajñādi* focused

4 University of Kerala Oriental Manuscripts Library, Trivandrum. Ms. No. T 555 A and B. This transcription in Devanāgarī was transcribed probably 1896. The original manuscript written in Malayalam script is 19082. The New Catalogus Catalogurum identifies the author of the Vacanamālā as Rāmaśaurya, but provides no source for this information. But now see Olivelle 2022.

on killing within ritual contexts. Here I will focus on the second part dealing with *ātatāyin*, which contains some interesting and innovative exceptions to the general injunction of *ahiṃsā*, along with interpretive rules justifying those exceptions.

We must remember that *ātatāyin* is a term already used by early writers with the narrow meaning of an assailant with the intent to kill. Manu, for example, says: "When an *ātatāyin* attacks—whether he is a guru, a child, an old person, or a learned Brahmin—one may surely kill him without hesitation" (*guruṃ vā bālavṛddhau vā brāhmaṇaṃ vā bahuśrutam* | *ātatāyinam āyāntaṃ hanyād evāvicārayan* || *MDh* 8.350; see also *BDh* 1.18.12–13). Vasiṣṭha expands the category of *ātatāyin* beyond an immediate assailant: "An arsonist, a poisoner, a man brandishing a weapon, a robber, and a man who seizes one's land or abducts one's wife—these six are called *ātatāyin*" (*agnido garadaś caiva śastrapāṇir dhanāpahaḥ* | *kṣetradāraharaś caiva ṣaḍ ete ātatāyinaḥ* || *VaDh* 3.16).

The *Vacana*, however, expands the category even further to include other kinds of activities that involve killing, saying: *ātatāyiśabdaś cāhavādīnām api grāhakaḥ*, the term *ātatāyin* covers also military challenge *et cetera*. The *et cetera* (*ādi*) here is meant to include at least two kinds of killing: war and hunting. That killing in war is justified is a doctrine found in early texts. Indeed, the refusal to turn back and flee but to face the enemy and either kill or be killed is the essence of the Kṣatriyadharma. Manu equates being killed in battle to a Vedic sacrifice (5.98), and Gautama (GDh 10.17) says that there is no sin in doing *hiṃsā* in battle. The difference here is that the killing is not confined to a particular set of circumstances, such as a ritual or an assailant rushing to kill someone, but becomes part of the code of conduct of a particular group, in this case the King and, more broadly, Kṣatriyas. So, in an interesting way, *hiṃsā* is not simply an exception, *apavāda*, but actually becomes the *svadharma* of a group.

The second kind of killing contained in that *et cetera, ādi*, is hunting. Now, this common human activity can be a means whereby hunters obtaining meat, often sold in markets. It is also the paradigmatic pastime of royalty. This aspect of hunting has, as far as I know, not been theorized in the context of ethics in ancient Indian literature. That hunting and fishing are unethical or adharmic is stated frequently, but the conditions under which they may become ethical and legitimate have not been subjected to systematic scrutiny. The *Vacana* undertakes this task.

The author begins by stating that there are authoritative statements about the obligatory nature of hunting and similar activities for a king, just as there are about warfare: *evaṃ mṛgayādikartavyatāpratipādakāny api vacanāny udāhartavyāni*. He first cites a verse from the *Bhāgavata Purāṇa* (10.52.62)

kṣātradharmasthito jantūn avadhīr mṛgayādibhiḥ |
samāhitas tat tapasā jahy aghaṃ madupāśrayaḥ ||

Established in the Kṣatriya dharma, you have killed animals through hunting and the like. Being self-possessed, expel that sin through austerity, resorting to me.

The *Vacana* comments that the initial term *kṣātradharmasthitaḥ* indicates that, as *hiṃsā* is eliminated when killing is done for the purpose of a rite (*kratvarthahiṃsānirāsavat*), so here the text aims to praise the knowledge of the self (*ātmajñānapraśaṃsārtham*), probably meaning the knowledge of what someone should do when he is Kṣatriya. The *Vacana*'s use of the term *kratvartha* here is significant. It may well be simply a synonym of *yajñārtha*, done for the purpose of a sacrifice; but it could also have the technical meaning of an act done for the sake of the rite as opposed to the performer of the rite, the *puruṣārtha*, an issue that I have already discussed.

The author then cites a passage from Harṣa's *Naiṣadhyacarita* (2.10) to demonstrate that there is no sin attached to hunting:

abalasvakulāśino jhaṣān nijanīḍadrumapīḍinaḥ khagān |
anavadyatṛṇārdino mṛgān mṛgayāghāya na bhūbhujāṃ ghnatām ||

Fish eat the weaklings of their own kind; birds damage the trees on which they build their nests; deer cause pain to the innocent grass—kings who kill them by hunting do not incur a sin.

His final passage in defense of hunting comes from the *Anuśāsanaparvan* of the *Mahābhārata* (although he ascribes it to the *Śāntiparvan*).

mṛgayātrāṃ tu vakṣyāmi śruṇu tāṃ dharmacāriṇi |
mṛgān hatvā mahīpālo yathā pāpair na lipyate ||
nirmānuṣām imāṃ sarve mṛgā icchanti medinīm |
bhakṣayanti ca sasyāni śāsitavyā nṛpeṇa te ||
duṣṭānāṃ śāsanaṃ dharmaḥ śiṣṭānāṃ paripālanam |
kartavyaṃ bhūmipālena nityaṃ kāryeṣu cārjavam ||
svargaṃ mṛgāś ca gacchanti svayaṃ nṛpatinā hatāḥ || MBh 14, App. 15.1271–1277

I will speak about the hunting expedition. Listen to it, O follower of dharma—as to how sin does not taint a king when he kills wild animals.

> Wild animals yearn for this earth to be devoid of humans, and they eat crops. They are to be disciplined by the king.
>
> Dharma consists of disciplining the wicked and looking after those who are disciplined. It should be always carried out by a king, as also rectitude in his activities.
>
> The wild animals themselves, moreover, go to heaven when they are killed by the king.

There is a significant difference in *Vacana*'s defense of killing as compared to earlier ones by Kumārila and other Mīmāṃsakas. The older defense is based on a purely hermeneutical argument, namely that the generic prohibition of killing, which is an *utsarga*, has to be set aside when there is a specific injunction to kill, which is an *apavāda*. In that context, dharma is defined as a series of injunctions and prohibitions, following the Mīmāṃsā definition of dharma as *codanālakṣaṇa*. The *Vacana*, on the other hand, bases its defense of killing on a broader conception of dharma as *svadharma*. If the dharma of a particular class of individuals, in this case kings, requires killing, then that killing is dharmic. Such killing happens, for example, in the execution of criminals and the pursuit of war. Somewhat innovatively, the *Vacana* extends this argument to hunting, which is generally viewed as a king's pastime and sometimes even as a royal vice, but which in the eyes of the *Vacana* is obligatory as part of his dharma. Just as execution and war eliminates criminals and adversaries from society, so, according to the *Vacana*, hunting eliminates or reduces the number of animals, who are intent on destroying crops, and who would prefer an earth devoid of humans.

3 *Ahiṃsā* and *svadharma*

The line of reasoning pursued by the author of the *Vacanamālā* opens up whole new areas of killing, of *hiṃsā*, to being considered morally right and dharmic. So, for example, the *jātidharma* of a fishermen's caste involves killing fish; that their *svadharma*. Does that make fishing for people of fishing castes a dharmic act, where their *hiṃsā*, just as Vedic *hiṃsā*, is actually *ahiṃsā*? In fact, Manu (10.48) seems to back such a conclusion when he assigns the occupations of fishing to Niṣādas, hunting to Medas, Andras, Cuñcus, and Madgus, and trapping animals living in burrows to Kṣattṛs, Ugras, and Pulkasas.

This precise issue is taken up by Medhātithi in his commentary on MDh 10.63. Here five virtues, including *ahiṃsā*, are given as the gist of dharma for the four *varṇas*.

> *ahiṃsā satyam asteyaṃ śaucam indriyanigrahaḥ* |
> *etaṃ sāmāsikaṃ dharmaṃ cāturvarṇye 'bravīn manuḥ* || 10.63 ||
>
> *Āhiṃsā*, truthfulness, refraining from stealing, purification, and mastering the organs—this, Manu has declared, is the gist of dharma for the four *varṇas*.

An opponent (Pūrvapakṣa) makes the apparently logical point that when Manu assigns (in 10. 48) killing as the occupation or livelihood of various mixed castes, such *hiṃsā* should be dharmic. This takes us back to the opening verses of Manu (1.2), where the sages ask him to teach the dharmas of all social classes (*sarvavarṇānām*) and also of those born in between (*antaraprabhavānām*), that is mixed castes. The latter can be *anuloma* (husband's class higher than wife's: hypergamy) or the opposite: *pratiloma*. The origins and the occupations of these mixed castes are the focus of Chapter 10 of Manu. The jobs assigned by Manu to the *pratiloma* castes at MDh 10.48 entail *hiṃsā*. So, the Pūrvapakṣa argues, *ahiṃsā* cannot be the dharma of these castes, given that their dharma, i.e., the jobs (*vṛtti*) assigned to them, entails *hiṃsā*. That, I think, is the gist of Medhātithi's argument presented in the form of a rhetorical question:

(Pūrva) *yady ahiṃsā pratilomānāṃ dharmaḥ syāt, kathaṃ tarhi tad uktam: "matsyaghāto niṣādānām" (MDh 10.48), "bilaukovadhabandhanam" (MDh 10.49) kṣatrādīnām, "araṇyapaśuhiṃsanam" (MDh 10.48)* |

If *ahiṃsā* is the dharma of *pratiloma* people, how then is it said: "killing fish to the Niṣādas," "trapping and killing animals living in burrows" to Kṣattṛs etc., "hunting wild animal" (to Medas etc.)?

Then follow two arguments by people (here called Pūrva-A and Pūrva-B), who are overall sympathetic to Pūrvapakṣa's view, yet wants to salvage the proposition that *ahiṃsā* is dharma also for all, including the *pratiloma* castes:

(Pūrva-A) *evaṃ ca kecid āhuḥ: ye jīvikāhetutayā vadhyatvenoktās tato 'nyatrāhiṃsā* |

And in like manner, some say: *ahiṃsā* applies to areas other than those covered by statements authorizing killing for the sake of livelihood.

(Pūrva-B) *anye manyante: ahiṃsā teṣām* [= pratilomānām] *abhyudayasādhanahetutvena dharmaḥ, na tu pratiṣedho hiṃsāyāḥ saṃmata iti, yathā "na māṃsabhakṣaṇe doṣaḥ"* (MDh 5.56) *itivat* |

Others opine: for them [= *pratiloma* people] *ahiṃsā* is dharma insofar as it the cause of generating spiritual welfare, but no prohibition of *hiṃsā* is intended. This is similar to the statement: "There is no fault in eating meat."

Next, Pūrva returns to reject the views of A and B. Note that the statement here replicates the beginning of Pūrva, followed by a similar rhetorical question beginning with "*katham tarhi*":

(Pūrva) *yady ahiṃsā dharmas teṣām* [= pratilomānām] *kathaṃ tarhi tair jīvitavyam* | *yathākutaścid avagamāt, teṣām anupapannāhiṃsā* |[5]

If *ahiṃsā* was their [= pratiloma people] dharma, how then can they obtain a livelihood? So, because no source at all make us aware of [such a means of livelihood], *ahiṃsā* is inappropriate for them [= pratiloma people].

Then Pūrva answers a possible objection (*iti cet*), and wraps up his argument by showing that the livelihoods presented in the sources, such as Manu's 10th chapter, are not available or authorized to *pratiloma* people. Here, I take the sandhi to be *yathākutaścid* (= *akutaścid*), because in the way I read the text it makes better sense, and because I have never seen references to authoritative texts implied by *avagama* ascribed to a *kutascid*. So, a negative meaning makes better sense: "never seen this anywhere".[6] I also take the phrase beginning with *yathā* as referring to what follows, mostly because a reason is inappropriate after a rhetorical question. Here *yathā* seems to introduce the argument given after the question, which is an implicit answer to the rhetorical question. The

5 I have here emended the text to insert *ahiṃsā* after *anupapannā*. Here it appears that scribes have missed the double use of *hiṃsā*. In the manuscripts, which do not use white spaces between words, the text would have read: *anupapannāhiṃsāhiṃsānivṛttir*. The insertion provides a feminine noun required by *anupapannā*.

6 One can also take *avagama* as a reference to obtaining a livelihood, although the term most frequently refers to knowledge obtained through some authoritative source. Ganganath Jha, in his translation of Medhātithi, takes it in the first sense as referring to means of livelihood: "Other sources of income being not available to them."

readings here are somewhat corrupt, as I have noted in the footnote. But the overall meaning is clear enough.

> *hiṃsānivṛttir dharmāyeti cet, tadā ko jīvanopāyaḥ | anyāś ca vṛttayaḥ prativarṇaniyatāḥ | adhyāpanādayas tāvad atyantāsaṃbhavād aprāptāḥ | kṛṣyādayo 'pi vaiśyaniyatāḥ | śūdrasya sevā, anyā vā sādhāraṇī vṛttir asti |*
>
> If you say desisting from *hiṃsā* is for the sake of dharma, then what is their means of livelihood? And other occupations are restricted to specific *varṇas*. Now, teaching and so forth are not viable, because they are totally inappropriate. Agriculture and so forth are also restricted to Vaiśyas. For Śūdras there is service or some other generic occupation.

Here we have a closely argued case for why *ahiṃsā* should not be considered part of a universal dharma, given that the dharma of some social groups requires them to engage in killing. This is, of course, not Medhātithi's own view. His response to this argument is rather lame. He says that the means of livelihood for these social classes will be presented in Manu 10.116. This verse, however, only gives sources of income in times of adversity (*āpad*). The arguments of these opponents (*pūrvapakṣas*) are instructive, and they are picked up centuries later by the author of the *Vacanamālā*.

Can the *svadharma* of various castes, then, be an exception, *apavāda*, to the general injunction of *ahiṃsā*? For both the Pūrvapakṣa of Medhātithi and the author of the *Vacanamālā*, the answer appears to be yes. Whether it is persons belonging to *pratiloma* castes who must earn a living by hunting and killing animals, or kings whose pastime is the chase, these activities define their *svadharma*. Manu's (5.39) saying: *yajñe vadho 'vadhaḥ*, "in sacrifice killing is not killing", can be extended along the lines of these later arguments to read: *svadharme vadho 'vadhaḥ*, "within the dharma specific to one, killing is not killing."

Abbreviations

BDh Baudhāyana Dharmasūtra. In Olivelle 2000.
GDh Gautama Dharmasūtra. In Olivelle 2000.
MBh Mahābhārata
MDh Mānava Dharmaśāstra. In Olivelle 2005.
VaDh Vasiṣṭha Dharmasūtra. In Olivelle 2000.
ViDh Viṣṇu Dharmaśāstra. In Olivelle 2009.
YDh Yājñavalkya Dharmaśāstra. In Olivelle 2019.

References

Alsdorf, Ludwig. 1962. *Beiträge zur Geschichte von Vegetarismus und Rinderverehrung in Indiens*. Wiesbaden.

Das, Veena. 2013. "Violence and Nonviolence at the Heart of Hindu Ethics." In *The Oxford Handbook of Religion and Violence*, ed. Mark Juergensmeyer, Margo Kitts, and Michael Jerryson. New York: Oxford University Press, pp. 15–40.

Halbfass, Wilhelm. 1983. "Kumārila on Ahiṃsā and Dharma." In Wilhelm Halbfass, *Studies in Kumārila and Śaṅkara*. Studien zur Indologie und Iranistik, Monographie 9. Reinbek: Dr. Inge Wezler Verlag für Orientalistische Fachpublikationen.

Olivelle, Patrick. 2000. *The Dharmasūtras of Āpastamba, Gautama, Baudhāyana, and Vasiṣṭha*. Sanskrit editions and annotated translations. Delhi: Motilal Banarsidass.

Olivelle, Patrick. 2005. *Manu's Code of Law: A Critical Edition and Translation of the Mānava-Dharmaśāstra*. New York: Oxford University Press.

Olivelle, Patrick. 2009. *Viṣṇu's Code of Law: A Critical Edition and Translation of the Vaiṣṇava-Dharmaśāstra*. Harvard Oriental Series, No. 73. Cambridge, Mass.: Harvard University Press.

Olivelle, Patrick. 2019. *Yājñavalkya: A Treatise on Dharma* (Critical Edition and Translation). Murty Classical Library of India 20. Cambridge, MA: Harvard University Press.

Olivelle, Patrick. 2022. *Bālakrīḍāvyākhyā Vacanamālā: A Sub-Commentary on the Yājñavalkya Dharmaśāstra*. Delhi: Motilal Banarsidass Publications.

Schmidt, Hans-Peter. 1968. "The Origin of *ahiṃsā*." In *Mélanges d'Indianisme à la mémoire de Louis Renou*. Paris, pp. 625–655.

CHAPTER 11

Ascetics as Royal Advisors in the *Mahābhārata*

Valters Negribs

Early treatments of the science of statecraft are found not only in the *Arthaśāstra*, but also in the relevant parts of the *Mahābhārata*, the Dharmasūtras, and the *Mānavadharmaśāstra*. A recurrent topic in this literature is that of the qualities required of royal advisors and ministers, the supposed experts in statecraft. This paper will discuss several passages from the *Mahābhārata* in which ascetic qualities are required of royal advisors or, going a step further, renouncers are depicted as becoming royal ministers. It is difficult to say anything certain about the social background of experts in statecraft relying only on such prescriptive accounts or on early narratives in the Sanskrit epics or the Buddhist Jātakas. Nevertheless, this paper will argue that the different views that these texts express about the ideal royal advisor hint at social tensions and competition for patronage during the time of their composition.[1]

By "ascetic" qualities I mean virtues that are usually cultivated by people who have given up social and family life in order to seek liberation from rebirth. Such renouncers are known in Sanskrit as *śramaṇas*, *bhikṣus*, *saṃnyāsins* as well as by other names. It is not possible to say that any virtue belongs exclusively to these renouncers. We know that some *śramaṇa* groups identified desire (*kāma*), greed (*lobha*) or craving (*tṛṣṇā*) as the driving force of continuous rebirth in *saṃsāra*. In the Buddhist tradition, for example, "destruction of craving" (*tṛṣṇākṣaya*) is a synonym of *nirvāṇa*. There is no doubt that many people who did not belong to the renunciate traditions restrained their desires. Freedom from greed or craving can therefore be called an "ascetic" virtue in a broad sense—it is a form of self-cultivation which does not require one to adopt a lifestyle of a *śramaṇa*. There are other virtues that have much stronger links with the *śramaṇa* traditions such as, for example, being indifferent to pleasure

1 The dating of the *Mahābhārata* and the *Śāntiparvan* in particular has been discussed by Fitzgerald (2004 and 2006). According to his estimates, parts of the *Mahābhārata* may be as late as 350 CE (2006:271), while elsewhere he claims that he has "no doubt that some substantial part of the *nīti* advice of Laws for Kings [= *Rājadharmaparvan*] and Law in Times of Distress [= *Āpaddharmaparvan*] comes from earlier times, perhaps some of it even from the late Vedic era (2004: footnote 200 on page 129)." I will not attempt to situate any of the *Mahābhārata* passages more precisely within this rather wide time frame.

 | DOI:10.1163/9789004536869_013

and pain. We will see below that some *Mahābhārata* passages demand of the ideal ministers such more unusual virtues.

The passages presented below will show a spectrum of ascetic demands made on royal advisors. I will begin by looking at the *Arthaśāstra*, which probably contains the earliest surviving textual evidence that deals with the appointment of advisors and which presents a rather practical "job description" for the *amātya* (minister) and *purohita* (chaplain or counsellor) or *mantripurohita* (advisor-chaplain). The main part of this paper will discuss several passages from the *Mahābhārata* in which ascetic qualities are required of royal advisors or, going a step further, renouncers are depicted as becoming royal ministers. The "ascetic" qualities range from rather general virtues, such as being without greed (*alubdha*), to more specific virtues that elsewhere in the *Mahābhārata* are usually associated with renouncers.

All translations are my own unless stated otherwise.

1 Royal Advisors in the *Arthaśāstra*

These are the qualities required of an ideal minister according to the *Arthaśāstra*, as translated by Patrick Olivelle:

> Native of the country; of noble birth; easy to manage; trained in the crafts; insightful; intelligent; with a keen memory; skilled; articulate; bold; quick-witted; possessing energy and might; able to endure hardships; honest; friendly; firmly loyal; endowed with character, strength, health, and spirit; free of obstinacy and fickleness; amiable; and not a fomenter of enmities—these are the exemplary qualities of a minister.
>
> OLIVELLE's 2013 translation of *Arthaśāstra* 1.9.3[2]

These are the requirements of the *purohita*, royal chaplain:

> He should appoint as Chaplain a man who comes from a very distinguished family and has an equally distinguished character, who is thor-

2 It should be noted that this statement is preceded by another section on the appointment of ministers in 1.8. McClish has argued that 1.8. was inserted by a later editor who may have misunderstood the original colophon in which the word *mantripurohita* probably referred to a single person ("advisor-councillor") rather than to *purohita* and *mantrin* (2019:65). In any case, 1.8. exhibits a similarly pragmatic attitude, concluding that "capability" (*sāmarthya*) is the most important consideration when appointing an advisor.

> oughly trained in the Veda together with the limbs, in divine omens, and in government, and who could counteract divine and human adversities through Atharvan means. He should follow him as a pupil his teacher, a son his father, and a servant his master.
>
> OLIVELLE's 2013 translation of *Arthaśāstra* 1.9.9

The requirements of both the minister and the *purohita* dwell on the education and useful personal qualities, and there is nothing specifically ascetic about them. In the *Arthaśāstra* only the king is required to have an ascetic mastery of himself, as seen in the "Conquest of the Senses" (*indriyajaya*) section in 1.6. The situation is quite different in the *Mahābhārata*.

2 Royal Advisors in the Sanskrit Epics

James Fitzgerald has identified two groups of discourses in the *Śāntiparvan* of the *Mahābhārata* that deal with the selection of royal advisors: *adhyāyas* 80–86 and 116–119 (Fitzgerald 2004). These materials include straightforward instructions about the appointment of ministers as well as several narratives that explore related topics.

For the most part, the non-narrative discourses do not depart from the *Arthaśāstra* in a major way, although there is a shift in the vocabulary used. Echoing *Arthaśāstra*'s requirements of the *purohita*, most of them also emphasise noble birth (*kula*[3]) and character or conduct (*śīla* or, less frequently, *vinaya*) as requirements for royal advisors. The educational requirement is in the *Mahābhārata* usually found under the adjective *bahuśruta*—literally "he who has heard much." These three qualities dominate the descriptions of the advisor in the *Śāntiparvan* of the *Mahābhārata* as well as other passages where royal advisors are mentioned.[4]

---

3 In addition to the programmatic statements about the importance of nobility (*kula*) of the advisor, *Mahābhārata* 12.117 presents a story about a dog who was transformed into ever stronger creatures by a Brahmin sage, but who in the end wanted to devour the Brahmin itself. The story makes it clear that this was due to his ungrateful nature as a dog (12.117.39). This narrative promotes a conservative agenda, according to which the species of animals, representing the social rank of humans, determines one's conduct.

4 *Kulīna* appears with *śīlasaṃpanna* in 12.81.21. *Kulīna* and *bahuśruta* in 12.84.5 and 12.87.17. *Abhijāta* and *bahuśruta* in 12.84.25. *Kulaja* in 12.116.15. *Bahuśruta* and *kule jāta* in 12.87.27. In 2.5.29 the ideal *purohita* is described as "*vinayasaṃpannaḥ kulaputro bahuśrutaḥ*." The elder Brahmins who are experts in astrology and predict the Pāṇḍava victory to Dhṛtarāṣṭra are said to be "*bahuśrutāḥ śīlavantaḥ kulīnāḥ* (5.27.92)."

Given the centrality of the concept of *śruti* (Vedic literature) and oral transmission of Vedic literature in the Brahmanical tradition, one could expect the term *bahuśruta* to be a standard term for a learned person in the Brahmanical tradition, but this is not the case. The adjective *bahuśruta* hardly appears in such early Brahmanical dharma texts as the *dharmasūtras* or the *Mānavadharmaśāstra*,[5] and in these texts it never qualifies royal advisors. Similar remarks can be made about *śīla*. The word *śīla* does appear a number of times in the Dharmasūtras and the *Mānavadharmaśāstra*, but usually not in the sense of "virtuous conduct."[6] Meanwhile, both *bahuśruta* and *śīla* are extremely frequent terms in the Buddhist tradition, and often appear side by side in the Buddhist literature.

It might be a mere coincidence that the *Mahābhārata* talks about royal advisors using the same terms that the Buddhist tradition uses to talk about learned and virtuous monks. Yet it seems plausible that it points to a mutual interaction and possible competition for patronage. It has been argued that the emerging Brahmanical tradition that was responsible for the composition of the early Brahmanical *dharma* literature, labelled "neo-Brahmanism" by some scholars (e.g. McGovern 2019), borrowed such key terms as *gṛhastha* (householder) or even *dharma*, in its classical meaning, from Middle-Indic discourse, especially as seen in Aśokan inscriptions (Jamison 2019; Olivelle 2004). Fitzgerald has argued that much of the *Śāntiparvan* should be read as a Brahmanical response to the non-special status of Brahmins in the Maurya empire and under the Maurya emperor Aśoka in particular (2004). Given that it is in the *Mahābhārata* that the pair "*bahuśruta*" and "*śīlavān*" enter Brahmanical discourse on royal advisors, it is at the very least plausible and consistent with recent scholarship to assume that the choice of terms might have been influenced by their prestige in the Buddhist tradition, which is believed to have fared favourably with the Maurya emperor Aśoka. While terms like *bahuśruta* and *śīla* are shared by the Buddhist tradition and *Mahābhārata*'s descriptions of royal advisors, Buddhism and other *śramaṇa* traditions do not usually place equal emphasis on nobility, being born in a *kula*. Given that many passages in the *Śāntiparvan* of the *Mahābhārata* present a specifically Brahmanical ideology, it is not surprising that these Brahmin authors and editors would place an emphasis on a good birth. It is possible that some non-Brahmins, such as *kṣatriyas*, would also have belonged to the *kulaja* category. We will see below that

---

5 In these texts *bahuśruta* appears between zero (in the *Āpastamba Dharmasūtra*) and three (in the *Gautama Dharmasūtra*) times.

6 The typical usage of *śīla* in the early Brahmanical *dharma* literature is exemplified by such compounds as *snānaśīla*—"he who has the habit (*śīla*) of bathing [regularly]."

some more ascetically-oriented passages in the *Śāntiparvan* either ignore this requirement (e.g. 12.152) or even present a low-born jackal becoming a minister in a fable.

In addition to dwelling on the learning, birth (*kula*) and conduct (*śīla*) of royal advisors, the main *Śāntiparvan*'s sections on the selection of royal advisors begin introducing ascetic qualities. It is claimed that an ideal advisor who "has conquered himself" (*jitātmā*) "would not give up *dharma* because of desire, fear, anger or greed (*na ca kāmād bhayāt krodhāl lobhād vā dharmam utsṛjet*; 12.84.12)." In another description of the advisor (*saciva*), he is said to be patient (*kṣānta*), disciplined (*dānta*), with senses conquered (*jitendriya*), free from greed (*alubdha*), and satisfied with what he [rightfully] obtains (*labdhasaṃtuṣṭa*) (12.118.8). Another verse in the same discourse again juxtaposes more obvious qualifications with ascetic ones, by claiming that "he should know the essence of the lore of elephants as well as be devoid of egoity (*hastiśikṣāsu tattvajñam ahaṃkāravivarjitam*; 12.118.12)." Elsewhere we read that "the king whose companions treat pleasure and pain equally (…) enjoys the fruit of kingship (*samaduḥkhasukhā yasya sahāyāḥ* (…) *sa rājyaphalam aśnute*; 12.116.17)." The qualities mentioned in the last two examples, namely, being *samaduḥkhasukha* and without *ahaṃkāra* are commonly associated with renouncers.

Similar language is used to talk about royal advisors outside the specific *Śāntiparvan*'s passages. In the *Rāmāyaṇa* Rāma expresses his hope to Bharata that during his absence Bharata's learned ministers perceive *artha* without passion (*arthaṃ virāgāḥ paśyanti tavāmātyā bahuśrutāḥ*; 2.94.49). In the *Mahābhārata* Kṛṣṇa is said to be particularly fit to advise Yudhiṣṭhira because he has gone beyond desire and anger (*kāmakrodhau vyatītya*; 2.12.40).

In the main narrative of the *Mahābhārata*, the issue of royal advisors becomes especially important in book 5. There king Dhṛtarāṣṭra is criticised for not heeding the advice of his main minister Vidura who is also described as very learned (*bahuśruta*) and of good conduct (*śīlavān*; 5.26.12). Notably nothing is said about good birth. Vidura is just a half-brother to the king since he was born from a *śūdra* mother, a member of the lowest class (*varṇa*) in the Brahmanical worldview. While ignoring Vidura, king Dhṛtarāṣṭra has succumbed to the advice of his own son Duryodhana, who in in contrast to Vidura is described as "under the sway of anger" (*manyuvaśānuga*) and "of lustful nature" (*kāmātman*; 5.26.13). The consequences of following the wrong counsel, rooted in the vice of Duryodhana, prove catastrophic for Dhṛtarāṣṭra's lineage.

3 *Artha* and *dharma*

A significant difference between the *Arthaśāstra* passages quoted above and the epic materials on advisors is that the latter present the maintenance of *dharma* as paramount. This may signal a shift in the paradigm of statecraft more generally. McClish has argued that an earlier version of the *Arthaśāstra*, which he provisionally calls *Dandanīti*, "was produced within a cultural context distinct from that of both the Dharmaśāstras and the final recension of the *Mahābhārata*, (2019:210)." McClish then suggests that "we can typify the development of *rājadharma* within the Dharmaśāstras as the gradual appropriation of an independent statecraft tradition, an appropriation culminating in the *Mānava Dharmaśāstra*. There, instruction on statecraft is systematically intertwined with the political theology of *varṇadharma* drawn from Brāhmaṇical theology (2019:216)."

In the *Mahābhārata* and other sources *artha* and *dharma* are frequently mentioned together as the two areas of expertise required of royal advisors or even of kings themselves. For example, in *Mahābhārata* 12.81.23 it is recommended that in king's "house should dwell a minister (*gṛhe vased amātyaḥ*)" who "should know the best counsel and the nature of *artha* and *dharma* (*vidyāt paraṃ mantraṃ prakṛtiṃ cārthadharmayoḥ*)."

It should not be surprising that coupling the science of statecraft (*artha*) with competency in *dharma* has consequences for the qualities required of the advisor. All the Dharmasūtras make it clear that the experts on *dharma* are, at the very least, free from greed or desire. Thus, some form of self-cultivation and virtue is required in addition to textual learning.[7] According to *Āpastamba*, "He should model his conduct after that which is unanimously approved in all regions by Āryas who have been properly trained, who are elderly and self-possessed, and who are neither greedy nor deceitful (*Āpastamba* 1.20.8=*Āpastamba* 2.29.14; trans. by Olivelle)." *Gautama* says that "In matters that are unclear, one should follow what is endorsed by a minimum of ten persons who are cultured, skilled in reasoning, and free from greed (*Gautama* 28.40, trans. by Olivelle)." According to *Baudhāyana*, "Now, cultured people are those who are free from envy and pride, possess just a jarful of grain, and are free from covet-

7 The Dharmasūtra statements quoted in this paragraph can be compared to this verse definition of *śiṣṭa*, which emphasises learning while being silent about desirelessness or other virtues: "Those Brahmins, in whose families the study of the Veda together with its supplements is hereditary, and who are able to adduce as proofs express vedic texts, should be recognized as the cultured (*Vasiṣṭha* 6.43; trans. Olivelle)".

ousness, hypocrisy, arrogance, greed, folly, and anger (1.1.5; trans. by Olivelle)." *Vasiṣṭha* is more laconic: "Now, a cultured man is free from desires. (*Vasiṣṭha*, 1.6)." All of these statements concern the people who are called *śiṣṭa* ("cultured" in Olivelle's translation) and whose exemplary conduct or judgement is one of the sources of *dharma*, alongside authoritative texts such as the Vedas.

These statements about expertise on *dharma* were most probably pronounced in a context of possible competition with other traditions of *dharma*, some of which were represented largely by wandering ascetics or monastics who made similar claims about overcoming their desires and greed. The Brahmanical literature of the Vedic period, including the earliest Upaniṣads, does not place a similar emphasis on dharma or freedom from desire as central to Brahmanism.

The *Mahābhārata*'s main discussions of royal advisors can also be read as indicating competition for being able to advise the king. They go beyond requiring just an intellectual expertise in the relevant textual traditions and require more ascetic virtues and/or emphasise high birth. The remaining parts of this paper will look at some passages from the *Mahābhārata* and a Buddhist Jātaka that portray renouncers as royal advisors. While there is plenty of evidence in the Buddhist canonical sources that the Buddhists aspired to extend their teachings on *dharma* to royal audiences, one hardly finds similar claims about an aspiration to advise kings on matters of *artha*. A stanza from the *Therīgāthā* says that *śramaṇas* (Pali *samaṇa*) are dear to the speaker because these *bahuśruta* (Pali *bahussuta*) people "teach *artha* and *dharma* (*attham dhammañ ca desenti*; 279)." In the *Nāradakassapajātaka* the king is recommended to visit a *śramaṇa* or a *brāhmaṇa*, who is *bahuśruta* and who knows *artha* and *dharma* (verse 12). While the *śramaṇa* visited in the *Nāradakassapajātaka* is certainly not a Buddhist, these two verses suggest that *śramaṇas*, too, can be advisors not just on *dharma* but also on *artha*.

In the following sections I turn to three passages from the *Śāntiparvan* of the *Mahābhārata* that criticise bad advisors and suggest that renouncers might do a better job.

4 Ascetics as Royal Advisors in *Mahābhārata* 12.152

Mahābhārata 12.152 is another textual unit which describes the people that a king should consult, but without calling them ministers or advisors specifically. Adam Bowles, who calls it "The chapter on greed", has already noted that in this text "attributes of the learned brāhman appear to be merging with those of the *yogin* (2007:339)."

At a first glance "The chapter on greed" does not seem to be about royal advisors. It opens with Yudhiṣṭhira's question about the foundation (*adhiṣṭhāna*) of bad deeds (*pāpa*), and Bhīṣma immediately states that it is "greed alone" (*eko lobho*; 12.152.2). Such speculations about the fundamental role of *lobha* are already indicative of a more ascetic context.

From verse 15 onwards the text takes a turn and engages in polemics against people who appear learned (*bahuśruta*), but lack in virtue. Although these "learned people have memorised extensive treatises (*sumahānty api śāstrāṇi dhārayanti bahuśrutāḥ*)" they are "of small wisdom" (*alpabuddhayaḥ*), "devoted to hatred and anger" (*dveṣakrodhaprasaktāḥ*) as well as "acting outside the conduct of the virtuous (*śiṣṭācārabahiṣkṛtāḥ*; 12.152.15–16)."

After more criticism in a similar vein, Yudhiṣṭhira is instructed to "be aware that these unvirtuous people are constantly full of greed (*etān aśiṣṭān budhyasva nityaṃ lobhasamanvitān*; 12.152.19)" and to rather ask questions to the learned people of pure vows that are described in the remaining part of the discourse (*śiṣṭāṃs tu paripṛcchethā yān vakṣyāmi śucivratān*; 12.152.20). The description of their ascetic qualities, however, is much closer to descriptions of renouncers or even liberated sages.[8] They do not delight in gain (*na hṛṣyanti lābheṣu*) nor are they agitated by loss (*nālābheṣu vyathanti ca*), they see all things as equal (*samadarśinaḥ*). They do not entertain the concepts of "mine" or "self" (*nirmamā nirahaṃkārāḥ*). Gain and loss, pleasure and pain, the liked and the disliked, death and life are equal for them (*lābhālābhau sukhaduḥkhe ca* (…) *priyāpriye maraṇaṃ jīvitaṃ ca samāni*). Their selves are awakened (*buddhātman*) (12.152.30–31). The virtues of compassion (*dayā*; 12.152.22) and non-violence (*ahiṃsā*; 12.152.26) are also mentioned. While it may be wrong to claim that any social or religious group is the sole "owner" of such virtues and attitudes, one nevertheless observes that they are found predominantly in descriptions of renouncers.

Bowles has made some useful observations regarding this passage. First he quotes the description of the *śiṣṭa*, the learned or disciplined people that are authoritative in matters of *dharma*, from the *Baudhāyanadharmasūtra* 1.1.5. This includes several of the terms found in the *Mahābhārata* passage, such as *nirmama* and *nirahaṃkāra*. The *Āpastamba* and *Gautama* Dharmasūtras, that are earlier than the *Baudhāyana* according to the estimates of Olivelle (1999), do not use such strongly ascetic vocabulary in their descriptions of the *śiṣṭa*, although we have seen above that all Dharmasūtras emphasise absence of greed or desire. Thus, there appears to be a parallel development in the related

8 See, for example, the description of a sage (*muni*) in *Mahābhārata* 12.269.

disciplines of *dharma* and *artha* in so far as both disciplines come up with increasingly ascetic requirements of their ideal experts.

Bowles further notes: "However, if this ĀDhP text differs from these traditional accounts of the *śiṣṭa*, it is to the degree that it emphasises the *śiṣṭa*'s 'yogic' traits, and a corresponding downplaying of the *śiṣṭa*'s relationship to scripture (2007:337)." Indeed, he rightly notes that in this *Mahābhārata* passage "there is no reference to the sacred texts of the Veda or the *smṛti*. This is in striking contrast to the *dharma* texts already cited, which almost invariably emphasise the connection of the 'learned' to the Veda (2007:338)." Nevertheless, Bowles is still careful to say that it "would be going too far to say that scripture is dismissed in this text, apart from the ironic *bahuśruta*, 'very learned' (2007:338)."

The "Chapter on greed" remains an ambiguous text. It recommends that the king Yudhiṣṭhira consults people who possess a range of qualities that are elsewhere emblematic of renouncers such as complete indifference and freedom from the conceptions of "mine" and "self". If we take such adjectives as *buddhātman* literally, these are perhaps awakened sages. Yet, it is not explicitly stated whether this really means that such experts should be recruited from groups of wandering renouncers. It is also unclear how much significance should be given to the things *not* mentioned in the text, namely, Vedic literature or the performance of Vedic rites, as well as the requirement that advisors are of noble birth. There is some possibility that the "Chapter on greed" represents the perspective of a particularly ascetically-oriented Brahmanical or even non-Brahmanical community.

5 The Brahmin Sage Kālakavṛkṣīya Becomes a Minister

There are two narratives in the *Śāntiparvan* that depict ascetics becoming ministers.

The first of these is found in *adhyāya* 83 in one of the *Mahābhārata*'s textual clusters that deal with the selection of ministers. The context for telling the story is the announcement of the problem that some ministers may plunder king's treasury. In this regard Bhīṣma relates a tale about the sage (*muni*) Kālakavṛkṣīya. This sage found out that the minister and other servants had been taking money from the king's treasury and informed the king, offering his own services: "A minister such as myself provides assistance of wisdom in times of distress (*dadāty asmadvidho 'mātyo buddhisāhāyyam āpadi*; 12.83.32)." The reference to *āpad*, times of distress, may suggest that it is unconventional for such an ascetic to become a minister. That there is a problem with the ascetic becoming a minister is further emphasised by the ensuing discussion, during which the sage voices concerns about his personal safety due to the unpredictable and violent nature of the kings as well as the malice of other ministers, who at that point have already killed his crow. Nevertheless, in the end the ascetic is satisfied with the personality of the king, and the king for his part invites the sage to live in his house and promises that there will be no room for those who dislike him.

This narrative may point to a real conflict or tension between settled ministers and such ascetic sages. The latter derive authority from their religious observances and in the present context even claim supernatural knowledge in the form "eyesight" (*cakṣus*) which has been made "far-sighted by the practice of austerities (*tapodīrgha*; 37)." In contrast to corrupt ministers, such sages can claim to have unhindered concern for the welfare of the king:

> We Brahmins, however, are of mild punishment and compassionate. I desire the welfare of you and others as I desire my own welfare.
>
> *vayaṃ tu brāhmaṇā nāma mṛdudaṇḍāḥ kṛpālavaḥ*
> *svasti cecchāmi bhavataḥ pareṣāṃ ca yathātmanaḥ* (12.83.60)

6 The Story of Jackal Who Becomes a Minister

The final *Mahābhārata* passage to be discussed is the story of a jackal who becomes a minister in *Mahābhārata* 12.112. This is a complex story that explores several themes. It leaves the impression of being related to the Buddhist and Jaina stories that illustrate the workings of karma and spiritual progress over several lifetimes. Thus, the main character obtains the low rebirth of a jackal after being a wicked king. The jackal remembers his previous wicked life and as a consequence develops supreme *nirveda*—indifference or disgust (*saṃsmṛtya pūrvajātiṃ sa nirvedaṃ paramaṃ gataḥ*; 12.112.5)—which is typically a mental state preceding renunciation. The jackal then adopts an ascetic lifestyle, abstaining from meat (*na bhakṣayati māṃsāni*), abstaining from harming any beings (*ahiṃsraḥ sarvabhūteṣu*), and holding firm vows (*sudṛḍhavrataḥ*; 12.112.5–6).

The other jackals attempt to convince him to adopt their impure livelihood, but the jackal claims that "nobility (*kula*) is really made by virtuous conduct (*śīlataḥ kriyate kulam*; 12.112.12)." This contrasts with the descriptions of ideal ministers in other parts in the *Mahābhārata* and the *Arthaśāstra*, where *śīla* and *kula* appear as a pair of desirable qualities.

It is therefore difficult to take this text simply as a part of a homogeneous Brahminical ideology of the *Śāntiparvan*. There is no doubt that the jackal, who dwells in the cremation ground (*śmaśāna*) like some *śramaṇa* ascetics, represents low social status. Moreover, his rhetoric seems to be directed against Brahmanical sages when he claims that "[living in] a hermitage does not [necessarily] indicate *dharma* (*nāśramo dharmalakṣaṇam*; 12.112.13)." Unsurprisingly, there is no mention of Vedic literature or sacrifice. This contrasts to the story about the Brahmanical sage Kālakavṛkṣīya discussed above.

The jackal's qualities are then recognised by the king who makes him a minister:

> The tiger [king] Khyātavikrama considered that he was a pure pandita, honoured him like himself, and himself established [the jackal] as a minister.
>
> *taṃ śuciṃ paṇḍitaṃ matvā śārdūlaḥ khyātavikramaḥ*
> *kṛtvātmasadṛśāṃ pūjāṃ sācivye 'vardhayat svayam* (12.112.17)

The jackal praises the kings effort to secure companions (*sahāya*) that are "pure" (*śuci*) and skilled in *dharma* and *artha* (*dharmārthakuśala*; 12.112.21), revealing that the selection of ministers is a key theme of the narrative.

The jackal highlights how his ascetic qualities in particular are beneficial to his role as a minister: "I do not desire pleasures, enjoyments or the sovereignty that you have (*na kāmaye sukhān bhogān aiśvaryaṃ vā tvadāśrayam*; 12.112.25)." This is because of his contentment (*saṃtoṣa*), which quality is often associated with ascetics or yogins. At one point the jackal is claimed to know *nītiśāstra*, the science of statecraft (12.112.69), but it is unclear how he could have obtained this knowledge.

In this story we again find the tension between the sage and the previous servants of the king, as the jackal says:

> My virtue will not be compatible with your former servants. Themselves in the habit of causing trouble, they will set you against me.
>
> *na yokṣyati hi me śīlaṃ tava bhṛtyaiḥ purātanaiḥ*
> *te tvāṃ vibhedayiṣyanti duḥkhaśīlā madantare* (12.112.26)

The problem with these "learned men" (*paṇḍitas*) is that they are greedy and they make living be deceit (12.112.59). They are interested in appropriating the wealth of others, but will be restrained by the jackal when he becomes a minister (12.112.42). As a result, they attempt to get rid of the jackal, but it is beyond the scope of this paper to explore that aspect of the narrative further.

The jackal's speech explores another tension—between the ascetic and the social reality of advising the king. He again claims to be already satisfied with what he has (*svenāsmi saṃtuṣṭo*; 12.112.29) and contrasts the troublesome life of the servants of the king with life in the wilderness (*vanacaryā*), which he claims to be free from attachment (*niḥsaṅgā*), fear (*nirbhayā*) and restraint of movement (*niravagrahā*; 12.112.30).

Furthermore,

> The hearts of those who are satisfied in the wilderness subsisting on roots and fruit do not know the fear of those who are summoned by the king.
>
> *nṛpeṇāhūyamānasya yat tiṣṭhati bhayaṃ hṛdi*
> *na tat tiṣṭhati tuṣṭānāṃ vane mūlaphalāśinām* (12.112.31)

Despite these complaints, the jackal agrees to become an advisor on special terms—he should never hold council with the other advisors and will only advise the king alone and in secret (12.112.36–37). This appears to be an attempt at creating a special model of relationship between a king and an ascetic advisor.

The story of the jackal stands apart from the other *Mahābhārata* episodes discussed above, even if it addresses some of the same terms and issues. It offers the strongest claim of an ascetic being a suitable royal minister. Yet, in contrast to the story of Kālakavṛkṣīya, this challenge does not seem to come from the Brahminical tradition.

7 Ascetic Advisor in the *Mahābodhijātaka*

The Pali collection of Jātaka stories includes several narratives with a character who is a minister. Almost all of these depict the *bodhisattva* in a previous birth as a wise advisor to the king. Such narratives do not depart significantly from what could be labelled as the "mainstream" view of ministers found in the didactic parts of the *Mahābhārata*. The Jātaka ministers are experts in *dharma* and *artha*, receive good education, and often inherit the post. The only Jātaka narrative which joins the more subversive voices of the *Mahābhārata* is the *Mahābodhijātaka*. In this story the *bodhisattva* becomes a renouncer (*paribbājaka*) dwelling in the Himālaya mountains, but ends up visiting the royal court in search for alms. Like in the *Mahābhārata* story about the jackal just discussed, the *bodhisattva* discovers that the court is corrupt and dangerous. A special arrangement is made so that a leaf hut (*paṇṇasāla*) is built in the royal garden for the ascetic. While living at the side of the king the *bodhisattva* has a chance to demonstrate the value of his ascetic virtues. His indifference is contrasted to the injustice of the five main advisors of the king who take bribes when deciding court cases. Here again, the story emphasises that the arrangement is uncommon—the ascetic at first refuses to become a judge on a regular basis but eventually agrees to a special arrangement that allows him to remain an ascetic living in the hut.

8 Conclusion

This paper presented some passages that deal with the appointment of ministers in the *Mahābhārata*. A comparison with the *Arthaśāstra* reveals that the materials from the *Mahābhārata* introduces a range of additional ascetic qualities that are required of royal advisors, ranging from more general calls from self-control to unambiguous claims that real renouncers might be the best advisors. A somewhat similar process in the *dharmasūtra* literature has already been theorised by Timothy Lubin. He has argued that "Brahmanical theorists" formalized "a ritual and disciplinary (*vrata*-based) program to define *gṛhastha* status as an *āśrama* alongside the celibate and homeless professions (2019:105)." Lubin situates this development as a response to the policies of emperor Aśokan, who is well-known for his preference of non-Brahmanical groups. Thus, the competitors and the instigators of the increasing asceticism are the *śramaṇa* ascetics. Noting a parallel between the Brahmanical and *śramaṇa* traditions, Lubin notes "a Brahmanical practice of venerating and feeding Brahmin virtuosi in a manner parallel to laymen giving alms to Buddhist, Jainist, or Ājīvika mendicants (2019:105)."

I would argue that the emerging emphasis on the ascetic qualities of royal advisors can be explained by presuming that the authority of the experts in statecraft was challenged by renouncer groups. These include the *śramaṇa* movements such as the Buddhists and the Jainas as well as renunciate strands within the Brahmanical tradition. The qualities usually attributed to sages, such as equanimity and lack of greed, would imbue them with charisma and authority that those who are only learned may be lacking. Several of the passages discussed above are of a polemical nature and lament the harm done to king by greedy ministers.

While, for example, the Buddhist literature has little evidence of Buddhist monks aiming to become advisors specifically on matters of *artha*, the story of the jackal in *Mahābhārata* 12.112 and the *Mahābodhijātaka* provide at least two specimens of narratives that depict an apparently non-Brahmanical renouncer as a superior advisor. This text challenges the claim that royal advisors need to be born in a distinguished family (*kula*), which is found in most of the passages discussed above, but is notably absent from the more ascetically oriented "Chapter on greed" (*Mahābhārata* 12.152). The same requirement is also not met by Vidura, the chief advisor of king Dhṛtarāṣṭra.

Bibliography

Bowles, Adam. *Dharma, disorder, and the political in ancient India: the Āpaddharmaparvan of the Mahābhārata*. Leiden: Brill, 2007.

Fitzgerald, James L. *The Mahābhārata: 11, The Book of the Women; 12, The Book of Peace, Part One*. Chicago: University of Chicago Press, 2007.

Jamison, Stephanie W. "The Term Gṛhastha and the (Pre)history of the Householder." In *Gṛhastha: The Householder in Ancient Indian Religious Culture*, edited by Patrick Olivelle, 3–19. Oxford: Oxford University Press, 2019.

Lubin, T. "The Late Appearance of the Gṛhastha in the Vedic Domestic Ritual Codes as a Married Religious Professional." In *Gṛhastha: The Householder in Ancient Indian Religious Culture*, edited by Patrick Olivelle, 95–106. Oxford: Oxford University Press, 2019.

McClish, Mark. *The history of the Arthaśāstra: Sovereignty and sacred law in ancient India*. Cambridge: Cambridge University Press, 2019.

McGovern, Nathan. *The snake and the mongoose: The emergence of identity in early Indian religion*. New York: Oxford University Press, 2019.

Olivelle, Patrick. *King, governance, and law in ancient India: Kauṭilya's Arthaśāstra*. New York: Oxford University Press, 2013.

Olivelle, Patrick. *The Dharmasūtras of Āpastamba, Gautama, Baudhāyana, and Vasiṣṭha*. Oxford: Oxford University Press, 1999.

Olivelle, Patrick. "The semantic history of dharma: The middle and late Vedic periods." *Journal of Indian Philosophy*, 2004, 492–511.

Sukthankar, Vishnu S., ed. *Mahābhārata: for the first time critically edited*. 19 vols. Poona: Bhandarkar Oriental Research Institute, 1927–1959. With the cooperation of Shrimant Balasaheb Pant Pratinidhi [and Others].

CHAPTER 12

Real Settlement and Constructed Wilderness? *Grāma* and *araṇya* Revisited

Frederick M. Smith

> From one viewpoint, the forest is a cluttered environment, the antithesis of open space. Distant views are nonexistent. A farmer has to cut down trees to create space for his farmstead and fields. Yet once the farm is established it becomes an ordered world of meaning—a place—and beyond it is the forest and space. The forest, no less than the bare plain, is a trackless region of possibility. Trees that clutter up space from one viewpoint are, from another, the means by which a special awareness of space is created, for the trees stand one behind the other as far as the eyes can see, and they encourage the mind to extrapolate to infinity. The open plain, however large, comes visibly to an end at the horizon. The forest, although it may be small, appears boundless to one lost in its midst.
>
> YI-FU TUAN 1977: 56

1 Introduction

In this essay, I would like to nuance and expand, and to some extent repurpose, the observations of Charles Malamoud on *grāma* and *araṇya*, village and forest, from 1976, and update the cultural applications and changing parameters of these terms in various historical and discursive spaces in India.[1] Malamoud

1 This essay had its origins in a conference paper a decade ago. I am grateful to William Sax and Jan Weinhold for inviting me to Heidelberg for a small, intense, and delightful symposium, "Rituals of Possession: Dissociation as Therapy," 3–5 November 2011. Among the luminaries in attendance was Etzel Cardeña of the University of Lund, whose work on dissociation has influenced my thinking on possession, and whose paper at this conference I cite below. I

 | DOI:10.1163/9789004536869_014

principally addressed the distinctions indicated by these words in first millennium BCE vedic ritual texts, even if he also included other references and descriptions. The contrast between settlement or village and forest or wilderness was, he rightly understood, as psychological and sociological as it was strictly environmental. Malamoud notes "that the two zones of forest and village are generally distinguished from one another, less on the basis of their material features than on that of the religious and social significance attributed to each of them" (1996: 75).[2] He contrasts the differences between *grāma* (village) and *araṇya* (forest): each has its own spaces; its own domesticated and non-domesticated animals, reified in ritual ordering, including choice of sacrificial animals; their orderly predictability and unpredictable wildness; their mechanisms of peacefulness opposing the unbridled threat of violence; and the householder stage of life (*gṛhasthāśrama*) contrasted with retirement and renunciation (*vānaprasthāśrama* and *saṃnyāsāśrama*).[3] The last of these distinctions has never been absolute; its boundaries were always hazy (like the others), as "the figures of the *gṛhastha* and the *saṃnyāsin* are not strictly antithetical: The forest lies on the village's horizon and is, in a certain sense, integrated into village life" (Malamoud 1998: 85). Dharma, Malamoud demonstrates, does persist in the *araṇya*; it is not strictly a habitational or domestic phenomenon marked by village or urban social structure. Rather, the dharma of the forest, signifying the unknown, is subject to other forces, as the examples discussed below demonstrate.

The shifting sensibilities of *grāma* and *araṇya* require an attunement to geography, practice, agency, ethics, and space. The last of these is particularly important, not least because it is not the same as geography. Space can be both internal and external, and possesses topographical characteristics based on social, political, and psychological factors that are not necessarily pegged to physical geography. It is the relationship between the internal conceptualizations of *grāma* and *araṇya* that are crucial in an individual's understanding of his or her place in the world. It is this relationship, linked with geography on the ground, that encourages liminal spaces to arise where the disjuncts and conflicts between *grāma* and *araṇya* can be resolved. It is these liminal spaces

---

thank the attendees at that symposium for feedback on the much earlier, very different, version of this paper, and Alberta Ferrario for a critical reading of a much more recent draft.

2 Zimmermann 1987 describes the ecology of early Ayurveda as a relationship between forest and savannah, thus somewhat expanding Malamoud's terminological boundaries.

3 Olivelle 1990 (2011), 1993: 66–67, cogently discusses the last point. The most thorough examination of the early Indian perspective is Sprockhoff 1981, 1984, 1991. The most comprehensive etymological study of *araṇya* is Visigalli 2019.

that I will examine and explore here. In short, the cases discussed here illustrate that internal constructions of space condition notions of external space.

In these examples, internal space is organized by cultural or civilizational parameters that encourage the erection of buffer zones between *grāma* and *araṇya*. These parameters foreground psychological mechanisms that structure notions of safety and community (rather, the safety of community) to counter the threat of unpredictability and wilderness (rather, the unpredictability of wilderness). In this accounting, the *araṇya* may be expressed and rendered visible through possession, a feature in most of these case studies. The primary psychological mechanism involved in possession is dissociation, a distancing and buffering phenomenon that directly contributes to the reation of internal and external liminal space. Thus, dissociation must be briefly explained. It will be impossible to discuss agency, ethics, and practice as deeply as necessary, because many more examples of different dispositions on the broader distinctions and relationship between *grāma* and *araṇya* would be necessary than can be considered here.

I will present four case studies, with the caveat that, like most of the examples provided by Malamoud, these do not actually juxtapose the terms *grāma* and *araṇya* in order to highlight their differences or opposition. The first two will be drawn from the *Mahābhārata*, and the final two from my ethnographic work in modern India. The first will not require comment on possession or dissociation, but the latter three will.

- The first is an idealized description of the sacrificial arena used for Yudhiṣṭhira's horse sacrifice (*aśvamedha*) in the fourteenth book of the *Mahābhārata*. It describes the construction of a temporary liminal space that brings together elements from domesticated and non-domesticated terrain.
- The second is the climactic and vividly described meeting of Yudhiṣṭhira and his half-brother Vidura from the fifteenth book of the *Mahābhārata* as the latter dies as a renunciate in the forest. It discloses unusual behavior that internalizes dharma, agency, and ethical practice. It shows how deity (or spirit) possession serves as a medium for hosting the opposing characteristics of both *grāma* and the *araṇya*, how it provides an individual with an interior platform for negotiating the two.
- The third example is drawn from the performance of a Pāṇḍav *līlā* in a remote Himalayan village in 2013. It was a ritual of inheritance that closely resembles a practice found in the Upaniṣads more than two and a half millennia earlier, with no known intermediary texts or descriptions.
- The fourth example is based on research in two Indian healing centers, Balaji, in Rajasthan, and Chottanikkara, in Kerala, in which the forces of order and disorder, *grāma* and the *araṇya*, appear in direct conflict. The complex

mechanisms of possession and counter-possession[4] at these healing centers illustrate the power of dissociation in explicitly liminal spaces as a healing modality to reinstate balance and normality.[5]

Finally, I will offer a few conclusions on the recognition or creation of geographical, social, and psychological borderlands or buffer zones used to enhance the connections between domesticity and wilderness, between the settlement and the forest, between stability and chaos, between the assumption of single agency that undergirded the codices of the *varṇāśramadharma* that became the portents of the civilized *grāma* and the multivocal agency that characterize these case studies, and on the power of internal spaces to both domesticate and alienate external spaces.

2 A Note on Dissociation

Etzel Cardeña, one of the leading authorities on dissociation, articulates a working definition of it that is relevant for the study of possession: "Dissociation more generally should be considered a particular form of psychobiological functioning rather than something necessarily pathological. For instance, a number of dissociative alterations in consciousness and identity can be part of ritual or religious activities, are not necessarily associated with pathology, and can even foster creative endeavors as in the case of some instances of automatic writing or drawing."[6] To this we must add oracular speech and other visionary or intuitive bridging activity, such as generation of a state of aware-

4 Intriguingly, Michael Lambek uses the Freudian terms "transference and counter-transference" for this phenomenon (1993: 373ff.). I cannot go deeply into that here, but these terms are surely resonant for what is seen in certain forms of South Asian possession, and even for the primary example described below from the Mahābhārata. See also Prasko, et al. 2010; Lambek 2018: 183.

5 In this article, I have limited the main topics to those I have actively researched. Many other areas of concentration are possible however, including studies of human-animal interactions in Indian history and mythology, discussion of forest monasticism (e.g., Boucher 2008), and elements of the history of ecological practices in India (e.g., Chapple & Tucker 2000, Haberman 1994).

6 Taken from Cardeña's talk at the Heidelberg symposium, "Rituals of Possession: Dissociation as Therapy," 3–5 Nov 2011. See Cardeña 1994 for a good discussion of early definitions of dissociation. It is impossible and inadvisable to enter into the thicket of nuanced definitions and close but not identical conditions (e.g., hypnosis, amnesia, multiple personality disorder) offered of this term since its first serious theorization by Pierre Janet in 1889 (for an assessment, see Howell 2005: 50–64). The Heidelberg conference, consistent with my intentions here, focused on nonpathological and (often) ritual possession.

ness or focus that allows the resident of a settled community to access and absorb the "spirit" of the wilderness, and vice versa. The images of domesticity and wilderness, day and night, control and chaos, comfort and discomfort, and bliss and terror should help us understand that formations of possession might be rooted in formations of the natural world, specifically in the *grāma* and *araṇya* as Malamoud has interpreted them.

I propose that practices and processes of dissociation, as experienced in both the modern world and its classical antecedents, are not just analogous to the separation of wilderness and city but are necessary for and indicative of that separation. Dissociation in this context is the response to the perceived distinction between wilderness and settlement. This highlights environmental and psychological tensions inherent in civilizational processes, whether they involve adjustment to slow non-radical change in the environments of the *grāma* and the *araṇya* such as parks, managed agricultural or forest lands in order to create buffer zones that facilitate gradual acculturation, or whether they are adjustments to rapid change, including sudden immersion into zones demarcated by unpredictability and danger, including wild animals, unknown populations, and military incursions.[7] "Normal" dissociation, positive or oracular possession, and negative disease producing possession,[8] will be discussed here. Whether the *grāma* is populated territory, or whether it is familiar cultural or ideological, which is to say urbanized, territory within the mind, the result is an ambiguous isolation, or at least distancing, from one's surroundings due to unfamiliarity or selective focus. One is neither here nor there; dissociation in this sense becomes a positive leap into and appropriation of a liminal space in which joint occupation by both the *grāma* and *araṇya* is possible without the trauma of division, conflict, or disintegration. We will see this in the examples below, as internal and external barriers or buffer zones are breached. Where we

---

7 One can argue that a zookeeper is a perfect liminal occupation, or, taken to further extremes, anyone who owns and tries to manage a large number of more or less domesticated pets, such as cats, dogs, or parrots.

8 For "normal dissociation," a category within which positive or oracular possession must fall, see Cardeña 2011; Dalenberg & Paulson 2011; Sax 1991, 2002; Smith 2006: 62–78. The field of "dissociation studies" has increasingly focused on studies of reactions to trauma such as abuse or violence, multiple personality disorder, and other largely negative psychological challenges. What we consider here represents common dissociative experiences such as absorption, the redirecting or selective allocation of attention on one aspect of an experience. Such dissociative states engage a singular focus that confers on them a visage of enhanced awareness and spontaneity. They become more effective and manageable when they are planned and rehearsed, as oracular of positive possession inevitably is when practiced by shamans and healers. The neural pathways through which this occurs are not yet fully understood.

would expect to see pushback from the *araṇya* into the *grāma* and attenuation of communication, more often than not the outcome is a more nuanced appreciation of the regime of wilderness within the forces of the settlement. In short, we are addressing agents of civilizational crossroads and countercurrents that have adopted strategies of dissociation as a means to breach their boundaries and become integrative.

3 Yudhiṣṭhira's *aśvamedha*

After the cataclysmically destructive war between the Kauravas and the Pāṇḍavas was settled in favor of the Pāṇḍavas, it became the duty of the Pāṇḍava ruler, the eldest brother Yudhiṣṭhira, to perform a horse sacrifice (*aśvamedha*) to solidify and demonstrate his sovereignty. This is described, albeit briefly, in the *Āśvamedhika-parvan*, the fourteenth book of the *Mahābhārata*. Explication of this episode does not require discussion of dissociation except to say that it shares territory with positive oracular possession inasmuch as it was a conscious action with a "fated" or predicted denouement. Rather, its value is that it recognizes and negotiates the tensions between *grāma* and *araṇya* that seek resolution through dissociative intervention.

Much of the *aśvamedha*, both in the Vedic texts and the Sanskrit epics, is expended in extravagant preliminaries, including the conquest of surrounding principalities into which the carefully selected horse is allowed to roam freely for a year. Any ruler or rival potentate who threatens to capture the horse is immediately met by a well-respected and fearless warrior, in this case Arjuna, accompanied by hundreds of well-armed troops. The meanderings through India of the horse selected for Yudhiṣṭhira's *aśvamedha* and the battles of conquest that follow are recounted at length in the *Āśvamedhika-parvan* (chapters 72–86). The "conquest of the directions" (*digvijaya*), of which Arjuna's march through the subcontinent is an early example, is a recurring image in classical Indian literature. It is an act, although often symbolic, of absorbing wild, unruly, untamed, and usually largely unknown lands, replete with unforeseen challenges, into the dominant culture, even if "civilized" urban kingdoms are also captured predictably and ruthlessly.[9] It is, however, largely the absorption of little understood *araṇya*s of almost entirely unknown power into the culture, polity, and safety net of a recognizable power, the settled culture of a known and hegemonic *grāma*. In spite of this vivid action, the main ritual events of

9 Amaladass 2001 sees the *digvijaya* as a search for identity through border crossing.

the *aśvamedha* are by no means anticlimactic, even if they appear to be little more than a lengthy footnote in the *Āśvamedhika-parvan* (chapters 87–91).[10]

Among the most colorfully written chapters in the *Āśvamedhika-parvan* are two that depict the sumptuous and bountiful arrangements for the sacrifice. These chapters describe an entire city constructed at the site of the sacrifice to house the direct participants, the substantial support staff, and the invited guests, including princes from far-flung lands and hundreds of thousands of highly educated brahmans. Even if this settlement, imagined as a semi-rural bucolic idyll, was intended to be short-lived, for the duration of the sacrifice its opulence was depicted as unbounded.

> At the commencement of the sacrifice, many eloquent disputants articulated their logic, wishing to defeat each other. Joy of the Kurus, the kings observed that this sacrifice, organized by Bhīma, adhering to the highest standards, as if it were for the king of the gods himself. They saw its golden archways, its many couches, seats, and luxury items ornamented with gems. These princes witnessed water and storage pots, eating vessels, and saucepans with handles, lids, and covers, none of them not forged from gold. They saw sacrificial posts carved from wood and ornamented with gold, as described in the learned texts. These wonderfully beautiful posts were arranged in the prescribed way and at the proper time. Lord, these kings saw many animals born on land and in water, all brought together there. They saw cows and buffalos, including females and those put out to pasture, aquatic animals, predatory animals, birds, those born from embryos, egg-born creatures, sweat-born creatures, those produced from digging the earth, and animals from the mountains, shores, and forests. Seeing the entire sacrificial area decked out gaily with animals, herds of cattle, and grain, the princes entered into a state of complete amazement. A rich setting of fine, delicate food was prepared for the brahmans and commoners alike. After a hundred thousand brahmans had completed their meal, drums were beaten repeatedly, like thunder in the clouds. This

10 Dumont 1927, 1948; Koskikallio 1993, 1995, 1999; Teshima 2008, Takahashi 2021a, 2021b and much more. The horse sacrifice has long fascinated both the South Asian Sanskritic tradition and the scholarly world, for better and/or worse. The reason the ritual details here are brief appears to be because the epic narrator(s) did not want to bog themselves down in endless ritual prescriptions, which would be disruptive to the narrative and were, in any event, readily available to them at that time. In addition, the ambiguous ethics of the MBh, which achieve a denouement in the story of the golden mongoose immediately after this, at the end of the *Āśvamedhika-parvan* (MBh 14.92–96), bespeak the reasons for the attenuation of the ritual details here.

> thundering was a constant feature day after day. In this way the sacrifice of the wise Dharmarāja proceeded. The people saw gifts of food comparable to mountains, O king, rivulets of yogurt and lakes of melted ghee. O king, at the great sacrifice of that king, the entirety of Jambūdvīpa, including its many townships, was seen gathered in one place. People from thousands of community backgrounds came, bringing wealth, O bull of the Bharatas. Kings decked out in garlands and bright, well-polished earrings served the foremost of the twice-born by the hundreds and thousands. The men who were their attendants also gave these brahmans a variety of food and drink that was worthy of being consumed by these kings.
>
> *Mahābhārata* 14.87.1–16[11]

This idealized opulence, in which nature itself was transformed by the promise of the sacrifice (or by the confirmation of Yudhiṣṭhira's kingship), is followed by a brief description of the perfection of the performance itself and passing references to a few of its most noteworthy rites.[12] This grandeur and magnificence, however, is soon belied by a detailed narration of a miraculous event at the close of the sacrifice: a massive half gold mongoose with a voice booming through the heavens, akin to that of Indra, the king of the gods, enters the sacrificial arena and condemns the performance itself for its wastefulness, narrating a counter example of frugality, virtue, and soteriological attainment. It was the polar opposite of the *aśvamedha* in the realm of dharma.

Briefly, the story is as follows: The mongoose claims the stage and urges his high status listeners to hear the story of a brahman whose close-knit family had taken a vow of living only on the gleanings from cultivated fields (*uñchavṛtti*).[13] Once upon a time there was a severe famine, and no food could be found anywhere. In spite of the heat, dryness, and hunger, the brahman was always observant of his rituals. One evening, as the family was about to eat its ounce of coarsely milled gleaned barley grain, a stranger appears at the door. After initial greetings and exchanges of identification and pedigree, the father of the family, including himself, his wife, his son, and his pregnant daughter-in-law, offers the

---

11 Translation by the author, to appear in Smith, Forthcoming. Note the similarity in spirit of this passage with the passage Malamoud translates from Kālidāsa's *Kumārasaṃbhava* (Malamoud 1996: 89–90). There can be little doubt that the spirit of such idyllic passages from the Sanskrit epics influenced imaginings like Kālidāsa's, which occur throughout post-epic Indian literature.

12 For example, the occasion in which the chief wife of the patron (who must, of course, be a victorious king) should engage in sexual relations with the horse just after it was ritually slaughtered (*Āśvamedhika-parvan* 91.1–6; cf. Jamison 1996, 65–72).

13 For an extended discussion of this, see Hiltebeitel 2016.

guest his own share of coarse barley grain as food. The guest gratefully eats it, but is not satisfied. So, the brahman's wife, also half-starved and possessing an ascetic nature, offers him her share of the evening repast of coarse grain. After a good deal of discussion, the guest accepts this grain. Still, he is not satisfied.[14] The son of the brahman then offers the guest his share. After further discussion the guest accepts this grain as well. But he remains unsatisfied. Finally, the pregnant daughter-in-law offers her share of the grain. After internal family debate on the value of sons and grandsons, and of the father-in-law's guilt over taking her grain to feed the guest, they finally agree to offer it to the guest. At last, the guest is satisfied. He praises the virtue of the brahman, particularly noting that his acts of hospitality and selflessness were salvific for his ancestors and descendants, and reveals himself to be Dharma incarnate. He then delivers a lecture on the virtue of generosity and the overcoming of negative emotions, noting many examples of generosity, purity, selflessness, and sincerity. He announces to this poor righteous family that an aerial chariot is waiting outside to take them all, fully embodied, to the heavenly world. The mongoose then informs his audience that half his body turned to gold as a result of consuming dust and remnants of this family's gleanings, and that he had come to this great sacrifice in order to gather its remnants, thus turning the remainder of its body into gold. But this failed, proving the inadequacy of the *aśvamedha*. The moral of the story, as told by Vaiśaṃpāyana, the epic's principal narrator, is that actions or gifts not fully in accord with dharma are destined to fail, but that those whose actions are virtuous and borne of asceticism, performed with material legitimately earned, ascend to heaven.

The juxtaposition of these episodes highlights the tension between the ideology of the village or urban areas and the ideology of the forest, between the state and the individual, between order and disorder (Angot 2009), between householder expansion and ascetic contraction, between safety and danger, and between the demands of verdant prosperity and arid decline. The unquestioned direction in most of the world's recorded cultures since the Neolithic period is for cultivated land to encroach on and absorb forest land, for the domesticated to absorb the wild. This is no less the case in ancient and classical India (Malamoud 1996),[15] nor should this dynamic be unfamiliar to those who follow the travails of globalization in the modern world, in which the local

14 This is a variant on an old story from the Vedas; see Jamison 1996: 21–26, who relates at length a similar story from the earliest Sanskrit prose works available, the *Maitrāyaṇī Saṃhitā* (4.8.1) and the *Kāṭhaka Saṃhitā* (30.1).

15 See Zimmermann 1987 for an account of the ecology of early Ayurveda as a relationship between forest and savannah.

and the global press against each other (albeit with an eerie reversal, in which the global is viewed as expansive and attractive and the local as inhibiting, isolating, and contracting).

In the description of the cultivated opulence of Yudhiṣṭhira's sacrificial arena, we see first a magnificent city constructed on pastureland that probably served as a buffer zone between the domestic and the wild. The wealth poured into the sacrificial arena on this site facilitated the presence, and ultimate offering, of animals from all realms—domestic, wild, and aquatic.[16] On the other hand, the drought-afflicted family of the poor righteous brahman survived on the other side of the buffer zone between domestic and wild. Like birds, they subsisted on grain left on harvested fields, the drought had turned the land to the functional equivalent of the wilderness, and the brahman and his family had become the functional equivalents of renunciates, whose spiritual freedom is depicted as proportional to their exposure to (and preparation for) the dangers of the wilderness. In the final analysis, this story depicts the challenges of living with both a surfeit of responsibility for domestication and both the gifts and randomness of nature. They converge in a liminal zone of encroaching domestication and encroaching wildness. Which, however, represents a superior moral state?

In the story, the poor family, surviving on gleaning alone, ascends bodily to the heavenly world for their unflagging dedication to generosity and hospitality. However, at the end of the epic, farther along that we have recorded here, the Pāṇḍavas and their common wife, Draupadī, approach the heavenly world. One by one they expire due to the weight of their personal faults, their entrance into heaven limited to their celestial bodies. But there is one exception to this, namely Yudhiṣṭhira, the final member of the clan to reach the gates of heaven. Due to his adherence to dharma, indeed his manifestation of it, he is allowed to enter heaven in his physical body, like the poor starving brahman family. This, it appears, represents a standoff between the forces of domesticity and asceticism, the conventional sense of the security of the settlement and the chaos of nature.[17]

16 The texts note that animals from the wild that are to be sacrificed are first tied to the sacrificial posts marked for them, around which they are herded, but are then released back into the wild. The only animals actually slaughtered, then, are the domestic ones.

17 The most complete explication of the much neglected final *parvan* of the MBh, including its inexplicable tensions, is Shalom 2017.

4 The Death of Vidura in the *Mahābhārata*

The *Mahābhārata* episode that might be the most revealing for our present purposes is the story of the death of Vidura, who merges into the figure of his half-brother Yudhiṣṭhira at the moment of his passing. This story, perhaps more than the others here, highlights Malamoud's key insight about the tensions between the village and the forest.[18] This distinction found throughout the *Mahābhārata* demonstrates that the boundary between the two was both problematic and porous. Its penetrability, hence ambiguity, is found throughout the *Mahābhārata*, from the burning of the house of lac (*lakṣagṛha*, MBh 1.147), to the disrobing of Draupadī and its aftermath in the exile of the Pāṇḍavas to the forest, to the war itself, to the episode of Kṛṣṇa's meeting with Uttaṅka (MBh 14:52–57),[19] to the final trek up Svargārohaṇa mountain, where the Pāṇḍavas and Draupadī close out their earthly existences. But few better examples are found than the tale of the death of Vidura.

Vidura was the wisest of the wise, Dharma incarnate, but through the usual machinations of Mahābhārata ambiguity, he was born the son of the great sage Vyāsa, who impregnated a low caste if still celestial servant girl. As a result, Dharma itself became ambiguous, mixed, and inconsistent. Ironically (if theory is to be prioritized over practice), the offspring of this *varṇasaṅkara* or mixing of castes (*Bhagavadgītā* 1.42) turned out to be unalloyed in his dharma and wisdom. Certainly his comportment of dharma was much more straightforward than that of his curiously naïve and flawed half-brother Yudhiṣṭhira, the king of the Pāṇḍavas, born of the union of Vyāsa (through Pāṇḍu) and Kuntī, the head of the family that eventually won the devastating war against their cousins, the Kauravas. Yudhiṣṭhira, known for his righteousness and thoughtfulness, as well as for his haplessness and his ruinous error of judgment that led to the war, was regarded as the twin of Vidura, and was regularly called Dharmarāja, the King of Dharma. After many exhausting battles and stirring adventures, to cut to the quick, the war long over, Yudhiṣṭhira legitimized his power through a massive horse sacrifice that was, as we just saw in the episode of the mongoose, palimpsested with ambiguity.

Finally, this messy resolution achieved, it was time for the heroes and wise men of their generation to die. In accordance with established, if idealized, codes, the elders of the two clans ended their worldly engagement, setting aside

---

18 I did not address this episode in my earlier study of possession in the MBh, Smith 2006 ch. 6.

19 Also cf. MBh 1.3, the "Pauṣya-*parvan*" (Magnone 2012).

their belongings and left for the forest to lead lives of soul-searching renunciation. Among those who left for the forest were the aged and blind senior member of the clan, Dhṛtarāṣṭra, and his self-sacrificing wife Gāndhārī. They were accompanied to a forest hermitage by Yudhiṣṭhira and a squadron of troops, to see that they were safely settled. As he was about to take his leave of Dhṛtarāṣṭra, Yudhiṣṭhira inquired about Vidura, and was told that he too had begun his final renunciation and was now practicing extreme austerities, living on air alone.

> Even as Dhṛtarāṣṭra said this, however, the former chamberlain was sighted far off, naked and thin, hair matted, his mouth filled with stones, his body smeared with dirt and covered with dust from the forest. He looked towards the hermitage, saw all the people gathered there, and swiftly turned away. When King Yudhiṣṭhira was informed of this, he ran all alone in pursuit of him, following Vidura as he plunged deeper into the dreadful forest, sometimes in view, sometimes not. And as he ran after him with all his might, the king cried out, 'Vidura, O Vidura, I am your much-loved king Yudhiṣṭhira!' Then Vidura, foremost of the wise, halted in a solitary spot in the depths of the forest and leant against a tree. King Yudhiṣṭhria of mighty wisdom recognized Vidura of mighty wisdom from his look, though he was almost completely wasted away. Standing before him, the king said 'I am Yudhiṣṭhira' in his hearing. In answer the wise Vidura gestured with his hand; then gazing unblinking at the king and in deep meditation, he entered into him, fusing his own sight with his sight, his own limbs with his limbs, his own breath with his breath, his own senses with his senses. Thus Vidura, seeming to blaze with fiery energy, used his yogic power to enter the body of King Yudhiṣṭhira lord of *dharma*. As for his own body, it remained as it was, leaning with staring eyes against the tree. The king saw that it was now lifeless; he also sensed that he himself was now many times more powerful than before. Pāṇḍu's learned son of mighty ardour, the lord of *dharma*, recalled his own entire former history, lord of the peoples, and the way of Yoga, as Vyāsa had described.
>
> MBh 15.33. 17–29, trans. JOHN D. SMITH, 2009: 743–744[20]

This story has many of the elements of possession not just as it appears in the *Mahābhārata*, which, among other things, can justly be described as the hand-

20 15.33.25–26b: *viveśa viduro dhīmān gātrair gātrāṇi caiva ha | prāṇān prāṇeṣu ca dadhad indriyāṇīndriyeṣu ca || sa yogabalam āsthāya viveśa nṛpates tanum |*

book of possession in Indian literature, but can extend well to other layers of Indian history and doctrine (Smith 2006: 255–265). The present story from the *Āśramavāsika-parvan*, consistent with the ethics and Promethean mission of Yudhiṣṭhira, presents an uneasy transition, or even a rupture, between settlement and forest, between the comparative safety of domestic life in the world and the comparatively dangerous life of the renunciant in the forest.

In the end they merge, in part because the forest renunciant, emblematic of the *araṇya*, is absorbed into the body of the recognizable domesticated paragon of the *grāma*; it is the manifest destiny of the *grāma* to press into and absorb the *araṇya*. The dissociation that characterizes both Vidura's final state and the division of Dharma into half-brothers is healed, made complete again, by the king's ability to absorb Vidura's spirit. Yudhiṣṭhira is possessed by the wise forest renunciant Vidura, who embraces his dark side, his mixed caste birth, with unflinching rectitude and resolve. This also adds post-Vedic weight for J.C. Heesterman's proposed solution of the "conundrum" of the king's authority: Although the king is a member of the *grāma*, he is able to derive his authority only from outside of this community, from the *araṇya*, "to recuperate the strength that has left him" (Heesterman 1985: 123). Even if Yudhiṣṭhira and Vidura are equally descended from Dharma, the fulfillment of Yudhiṣṭhira's authority over both *grāma* and *araṇya* emerges only when the untamed intercaste outsider is fully absorbed into the body of the king. Both are Dharma incarnate, representing the perfection of habitation and wilderness. Vidura overcomes the dangers of the forest while Yudhiṣṭhira overcomes the equally immense challenges of life in the world. Their merging, or re-integration, is the final outcome of cultural and geographical division, of singular identity that is blown apart by the civilizational processes.[21]

21 Another Garhwali ritual that inscribes the *grāma* onto the *araṇya*, which so far as I know is found only in a few villages, with no literary precedent even as it is performed by highly literate people, is known locally simply as *ban yāt*, "forest pilgrimage" (*vana yātrā*). In this ritual the local experts construct an image of the "sacrificial man" or *yajñapuruṣa* outside the village. Then, for seven to nine days they host a bridegroom's procession (*barāt*) to a wedding in which the groom is married to the image of a goddess many times every day. After 360 such marriages, the *yajñapuruṣa* is buried in a pit. This is, to them, a union of village and wilderness. Some of the symbolism is clear, but cannot be further unpacked here. I am grateful to D.R. Purohit of Srinagar, Garhwal, for this information.

5 A Rite of Inheritance in the Himalaya

In densely populated modern India, much of what was long ago regarded as *araṇya*, together with its inherent dangers (e.g., leopards prowling at night, at least in the Himalayas), is now associated with small and remote villages and settlements, even if they are comparatively close to each other. As Malamoud notes, the difference between *grāma* and *araṇya* should be recognized as social and cultural, not necessarily physical, a settled bounded town contrasted with a deep untamed jungle. In the present case, echoes of an ancient, seemingly non-vedic, practice that entered the canonical precincts of two Upaniṣads resonate as the voice of the *araṇya*, preserved in the liminal space or buffer zone between *grāma* and *araṇya*, namely a Pāṇḍav *līlā*, which is itself a meeting ground of the classical *Mahābhārata* and the shifting songs of modern rural Himalayan culture.

This and the next case study will require greater context in order to understand how it fits into an otherwise carefully etched Indological essay. I moved my base in India to Kumaon, the name for eastern districts of Uttarakhand, in 2006 in order to begin my translation and study of the last five *parvans* of the *Mahābhārata* for the University of Chicago series (now to be published by Primus Books in Delhi), commenced half a century ago by J.A.B. van Buitenen. I did this in order to experience the *Mahābhārata* in *Mahābhārata* country. Soon enough, I realized that nearly every village, temple, place name, and clan had forged a connection with the epic in its Sanskrit and regional forms. More often then not, these connections referenced the Sanskrit *Mahābhārata*, in place names such as Garud, Karnaprayag, Markandey, in Uttarakhand, and too many place names to count in Himachal Pradesh, such as Haḍimbā (Halperin 2019).[22] However, the epic knowledge that was transmitted was of regional and local variants of the epic, radically altered from the Sanskrit version(s). What I envisioned then was deep mapping and understanding of the reinscribed *Mahābhārata* in the Himalaya.[23]

---

22 Halperin 2019 speaks of the Mahābhāratization of Haḍimbā Devī (Skt. Hiḍimbā) in Himachal Pradesh. Halperin sees Haḍimbā as a figure from the wilderness who has relocated to the domestic arena.

23 I now understand that it would take an army of scholars and ethnographers working decades, on an "epic" scale, to complete this project, which, in any case, dealing with an epic in motion, meaning that the project can never actually be completed. Among the questions that would need to be addressed are: How does today's *Mahābhārata* topography in Uttarakhand and Himachal Pradesh map onto what we find in the classical text(s)? What can we learn about its transmission and the cultural and religious factors that brought this about?

A number of scholars have studied the range and distribution of regional *Mahābhāratas*, including, most prominently, Alf Hiltebeitel (1988, 1991, 2001), who described the connected but gradually shifting geographies of many of the regional *Mahābhāratas*, based on the characters in these epics and their narrative associations with local cultic forms.[24] His seminal work will surely be nuanced by further fieldwork, especially in Rajasthan (which has strongly influenced the Himalayan epics),[25] so that perhaps after another generation or two of scholarship we can better delineate the relationship between local and regional *Mahābhārata* geography and the classical epic. In the Himalayas, this work has principally been undertaken by Claus Peter Zoller, who has published a linguistic study and translation of the only Himalayan *Mahābhārata* to appear so far, in the now nearly extinct Baṅgāṇī language of northwestern Uttarkashi district, in Uttarakhand (Zoller 2014). Zoller has described vertical and horizontal relationships between folk and classical traditions. Thematic comparisons of folk and classical epics represent its verticality. Its horizontal dimensions include the comparison of stratigraphically equal folk traditions, in this case other regional Himalayan *Mahābhāratas*. "In the case of Himalayan oral epics, in particular," Zoller writes, "it is not sufficient merely to compare them with other Himalayan oral epics." Rather, one has to see them in the wider context of at least northern South Asia. This fact remains to be generally acknowledged because, when scholars in the twentieth century (and earlier) attempted to connect specific folk traditions with other, older traditions, "they frequently tended to exaggerate relationships with the classical traditions" (2001: 83). Hiltebeitel would agree with this assessment, as does D.R. Purohit of Garhwal University, Srinagar, the preeminent folklorist of Uttarakhand, who has commissioned and has in his personal library a Garhwali language *Mahābhārata* from Agastyamuni, north of Rudraprayag, along the Mandakani river, which is quite different, linguistically and culturally, from the Baṅgāṇī *Mahābhārata*.[26]

The primary culturally recognizable modes of transmission of the Himalayan *Mahābhāratas*, patently rural, abiding in the *araṇya*, are through performance and possession. Performance provides both liminality and communi-

---

24 Hiltebeitel took his inspiration to a great extent from Madeleine Biardeau, whose work in Andhra Pradesh preceded his own in Tamilnadu (Biardeau 1989; also see Hawley and Pillai 2021, and https://ecsas2021.univie.ac.at/panels/iu8uk/), indicative of renewed attention to the exceptional variations in the *Mahābhārata*.

25 See the groundbreaking work of Gold 1992, Miller 1994, Schomer 1990, John D. Smith 1991, and Vaudeville 1965, among others.

26 I am presently involved in a project with Professor Purohit and William Sax or Heidelberg University to produce an edition, translation, and study of this unique text.

tas, to employ easily accessible terms from ritual studies (Turner 1967), while possession is a special form of barely domesticated liminality that draws the performer into personal identification with *Mahābhārata* characters, local divinities, and other psycho-spiritual forces, all serving as magnetic centers through whom the local people see and identify their deep history. Individuals who are possessed by *Mahābhārata* characters such as Bhīma, Arjuna, Karṇa, Draupadī, Kuntī, Gāndhārī, Duryodhana, Abhimanyu, Babhruvāhana (often called Nāgārjuna), and others, are generally viewed favorably, as reflections or embodiments of the culture and history that mirrors and backgrounds them.[27] I have written at length about possession as the most dispersed and trusted religious and spiritual expression in India, and it is no less here than anywhere else (Smith 2006). Men generally serve as oracles for deities on the move, which are tied on to *dolī*s or palanquins (lit. "swings") often carried long distances, up to fifty or a hundred kilometers. These oracles act as mediums for people along pilgrimage routes and at gatherings of deities, which are frequent in the Himalaya. Both men and women fall into trance or possession states during folk *Mahābhārata* performances called Pāṇḍav *līlā* (Sax 2002) and epic recitational performances such as *jāgar*s or "awakenings," long nocturnal music sessions by local shamans designed to awaken the deity in the audience.[28] It's all part of the *Mahābhārata*, so far as they are concerned; indeed, much of these performance descriptions and librettos are redacted in the local texts. The pilgrimage routes sometimes traverse only nearby villages, occasionally embrace larger networks of villages, or extend to large festivals. In their view, these are replications of the pilgrimages in deep antiquity by *Mahābhārata* characters. It is this mapping that is particularly intriguing, as it ought to be for the next generation of ethnographers and fieldworkers.[29] Still, most of the pilgrimages are comparat-

27 On multivocal agency, a potent if contentious topic within anthropology, see Sax 2006, and his application of it in Sax 2002, 2008. As it applies to the study of Indian history, see Inden 1990. A good example of its application in Buddhist studies is Adamek 2012.

28 See Jassal 2014, 2017; Negi 2019; Sax 2008; Vidal 1988; Zoller 2014. On the musical practices, see Fiol 2017. Amateur videos of snippets of different kinds of *jāgar* are mounting on YouTube. An unlimited amount of research remains to be done on this in both Garhwal and Kumaon.

29 At the moment, this is the domain of D.R. Purohit of Garhwal University, Srinagar (Garhwal), and a scant few ethnographers, some of them non-Indian. It is likely that as the younger generation of Garhwalis continues to leave their mountain villages, women's education continues to advance, and local musical and performance modes become more infrequent and even disappear, the pilgrimages themselves will be jeopardized and the traditions will slowly disintegrate, leaving a few hardy members of the community to revive them, rather like what has happened with the Vedic rituals in Maharashtra and Kerala. On this, see Smith 2001; Lubin 2001a, 2002b; Larios 2017. On Kerala, see Staal 1983, Witzel 2011,

ively small in scale; the large ones at Mandi and Kullu in Himachal Pradesh are the exception to the rule.[30]

During a Pāṇḍav *līlā* in the village of Raithal in December 2013, the man who had played the role of Arjuna for forty years, declining in health and strength, decided to pass his role on to his son. As a result of regular possession and self-identification as Arjuna, the latter came to "occupy" a greater part of his own personal identity. Arjuna, at least in this context, was a possession in both senses of the word: he "owned" Arjuna as personal property and identified him as an integral aspect of his own personhood. He performed the rite of transference near the end of the two-week long Pāṇḍav *līlā*.[31] The act of passing the torch to his son in this remote Himalayan village was a demonstration of the cultural importance and impact of the deep mapping of practice, text (whether written or oral), and geography.

The actor playing Arjuna, old, wizened, bent over, with an unusually heavy *jahnoi* or sacred thread that characters in the Pāṇḍav *līlā* in this village wear. He mounted the small brick stucco podium, perhaps two meters square and about one-meter high. It proved to be just large enough for certain dramatic ritual actions to be performed. The actor disrobed down to his *chāddī* or underwear, which indicated he felt the possession of Arjuna coming on. Then, with help, he mounted the stage, upon which two copper water vessels were placed. He quickly assumed the visage of possession, took up for the purpose of divination a handful of rice or barley grains from a medium-sized bowl, and tossed the grains vertically in the air with the palm of his right hand. This form of divination is a regular feature in such performances, including Pāṇḍav *līlā*s, in

---

Nowicka 2017, and many others. The filmmaker N.V. Ramachandran of Palakkad has in his personal archives a number of his own professionally filmed and produced video records of Nambudiri *śrauta* yajñas dating back to the 1980s.

30 In the last two decades, these festivals have seen exponential growth in attendance, analogous to the growth of Kumbhamelās, On the Kullu *melā*, see Berti 2001, 2006; Diserens 1995; Luchesi 2006; Vidal 1988. Only one article has appeared on the Mandi *melā* (Conzelmann 2005). I have gathered a great deal of information on Mandi in two trips to the *melā*. The Kullu *melā* takes place after Dussehra, in the fall, lasting about a week. The Mandi *melā* is a week-long event after Mahāśivarātrī, in February or March every year. Other valuable and relevant studies are Bindi 2009, 2013; Halperin 2019a, 2019b; Sutherland 1998. Berti, Conzelmann, Sax & Basu (2015), Sutherland 1998 and others (less directly) address the relationship between deities, possession, Mahābhārata related discourse, and local polity.

31 I am deeply indebted to Dr. D.R. Purohit of Srinagar for bringing my colleague from the Theater Department at the University of Iowa, Dare Clubb, to this Pāṇḍav līlā. Dare's camera was at the ready, and when he showed me this stunning set of photos, my jaw dropped in amazement. He had no idea what he was photographing, but I remain greatly indebted to him for it.

the towns along the Bhagirathi. It can occur dozens of times in every performance, by men and women alike, as a form of monitoring the gods' approval of their actions and impelling them forwards. Occasionally the actions are altered depending on the number of grains that are counted on the palm. The number of grains that land on the hand must be from one to seven. If it's more than that then it's done over again. The number is significant. Every participant and every member of the village appears to know what this means, although I have found that the numerical significance is different in different villages. It was not clear how many grains landed on Arjuna's right palm. In any case, he took note of it, as the participants always do, regardless of the degree to which they are entranced.

He then whirled around to his right, took up one of the large copper vessels, which was filled with milk. It could easily hold five liters. Clearly the vessels had been placed there in preparation for this rite. He poured it over his head in an act of self-consecration or self-divinization. This rite of self-*abhiṣeka* served the function and converted him fully into Arjuna, the son of Indra. He was a mess, it appeared, with milk dripping and flowing over his body and *chāddī*. He was otherwise naked. And it was cold. Everyone else was bundled up tightly. He then took up the second vessel, which was filled with water, and poured this over his head. This washed most of the milk off his body, but of course could not have been warming. However, it did not appear to affect him, which the audience instinctively believed to be integral to his state of possession. He dismounted the stage. His family members, including his wife, his son and son's wife, and a few other male kinsmen, were standing at hand in preparation for what would follow. They huddled up in a tight circle. The father embraced his son, with the others crowded around. They did not stand in place, but rotated slowly, clockwise (the opposite direction from rotating during possession), some of them taking a short dance step that is almost always used in ritually induced possession states. The father's embrace was so tight that it took half an hour to unlock his hands; indeed, others present took it upon themselves to loosen Arjuna's vicelike grip on his son. In this way the persona, identity, and essence of Arjuna was passed on to his son, who in the Pāṇḍav *līlā* had been playing the role of Babrik, the son of Arjuna, Babhruvāhana in the Sanskrit epic.

This is strikingly close to practices of linking the senses of the active and passive participants in the possession noted in the Upaniṣads and *Mahābhārata* long ago (Smith 2006: 216–218). It remains a source of wonderment that such practices persist today. I cannot claim that what I saw in Raithal was adopted (or adapted) from the *Bṛhadāraṇyaka* or *Kauṣītaki* Upaniṣads, or a latent memory of the death of Vidura as described in the *Āśramavāsika-parvan* of the *Mahābhārata*. Certainly, this was not the case. But it's a practice that is so close that it may be justifiable to conclude (rightly or wrongly) that in some fashion, through an untraced (and likely untraceable) mechanism, it has persisted as part of Himalayan ritual.

The *Bṛhadāraṇyaka* is the oldest of the Upaniṣads, possibly dating to the sixth or seventh century BCE.[32] BĀU 1.5.17 describes the following scene: Immediately before his expected death, a man is instructed to transfer his essence to his son.[33] "You are the *brahman*! You are the sacrifice! you are the world!" he tells his son. "I am the *brahman*! I am the sacrifice! I am the world!" replies his son. "Thus, he enters his son with his vital breaths (*prāṇaiḥ saha*). It is only through a son that a man finds a secure footing in this world. Thereupon, these divine and immortal vital functions enter him. From the earth and fire divine speech enters him. Divine speech is that which makes whatever one says happen. From the sky and the sun the divine mind enters him. The divine mind is that which makes a person always happy and never sorrowful. From the waters and the moon the divine breath enters him. The divine breath is that which never falters or fails, whether it is moving or at rest" (trans. Olivelle 1998: 55–57). The son thus is conferred the gifts of true speech, happiness, and unhindered, balanced breath.

The Kauṣītaki Upaniṣad (2.15) is a slightly later and more detailed text: The father lies down dressed in a fresh garment, with ritual fire kindled and a pot of water near at hand. The son lies down on top of the father, touching his father's body parts with his own corresponding parts; otherwise, they sit facing each other. Not only is the lifebreath transferred, but in a similar formulaic rhythm of announcement and response, so are speech, sight, hearing, taste, actions, pleasures and pains, bliss, delight, and procreative capacity, movements, mind, and intelligence. If the father is too unwell to speak much, he may limit the transfer to the *prāṇa*s alone. The son then turns around toward his right, walks toward the east, and his father wishes him well, saying: "May glory, the luster of sacred knowledge, and fame attend you!" The son glances over his left shoulder, hiding his face with his hand or the hem or his garment, and replies, "May you gain heavenly worlds and realize your desires!" (*svargāṁl lokān kāmān āpnuhi*). Interestingly, the *KṣU* ends its account with the statement: "If the father recov-

32 Olivelle (1998: 12) is cautious on all but relative chronology. He brings the dates of the earliest Upaniṣads forward by about a century based on renewed dating of the Buddha (Beckert 1984).

33 We know nothing about the identity of the man, whether he was a sage, a teacher, or an ordinary man. But the Upaniṣad frames the story in a manner that suggests that this is a known and prescribed practice. Nevertheless, this framing could as well be a rhetorical device to mask or canonize what was in fact little known or idiosyncratic.

ers his health, he should either live under the authority of his son or live as a wandering ascetic. But if he happens to die, they should perform the appropriate final rites for him" (trans. Olivelle 1998: 345–347).[34] Thus the power and rights are passed on to the son through this rite actualizing inheritance. The similarity, and possible connection, with the transfer of identity in the Raithal Pāṇḍav *līlā* should be obvious.

The connections between *grāma* and *araṇya* are here also fairly transparent. In village India, possession is more regularly a feature of local religion than in large urban areas. The movement of local residents out of their village environment, through marriage, education, employment opportunities, or desire to relocate for other reasons, is increasingly frequent today. For regular returnees from diasporas within India or abroad (I have met people who have returned to their remote villages for biennial or triennial winter Pāṇḍav *līlā* performances from Delhi, Bangalore, Mumbai, London, Paris, and Dubai), the capacity for the deep internal engagement and culturally regulated dissociation required for possession is markedly lacking. They are almost uniformly content to enjoy the role of spectator and the communitas derived from that. Their long exposure to the *grāma* has caused them to lose some of the "natural" talents of the *araṇya*, even if their periodic return allows them to recover a part of themselves that became lost in the *grāma*.

6 Balaji and Chottanikkara Revisited

The concepts of *grāma* and *araṇya* as applied here also contribute to diagnoses and therapies found at the great possession healing centers of Balaji in the small town of Mehndipur, in Rajasthan, a few kilometers off the main highway, about halfway between Jaipur and Agra, and at the possession healing center at the large Bhagavathy temple in Chottanikkara, in central Kerala, an hour east of Kochi. Comparatively little research has been conducted at either of these important sites.[35] Balaji has become the largest and best known of a disconnected if similar group of Hindu (and Jain[36]) temples in North India that specialize

34 For all of this plus additional comments, see Smith 2006: 216–218.

35 On Bālājī, see Pakaslahti 1998, 2006; Dwyer 1998, 1998, 203, Kakar, an early, although important, study based on only two afternoons of research (personal communication), and at that before it became the leading boom town for psychological healing after 1980, For Chottanikkara, see F. Smith 2004.

36 See the well-known study of the Padampura Svetambara Jain temple near Jaipur is Humphrey and Laidlaw 1994.

in spirit healing. Five to ten thousand pilgrims come daily, nearly all of them in family groups in which (at least) one member is believed to be possessed by a wayward spirit.[37] Balaji and Chottanikkara have become astonishingly popular and visible in the last few decades, as increasing numbers of people from all castes, classes, religions, and educational and economic levels, come in search of spirit and spiritual healing.[38] In Rajasthan, Bālājī is the child form of Hanumān, who presides over a "court" (*adālat*) in which, employing the special language of the old Mughal court system, punishment (*sazā, daṇḍ*) is meted out to possessing spirits (*bhūt*) by the messengers (*dūt*) of the foot soldiers (*phauj*) of Bālājī, through an "appearance in court" (*peśī*), during which both the afflicted individual(s) and the healer will experience possession (Pakaslahti 1995, Smith 2006: 110–114).

The operational assumption is that all spirit or spiritual healing must perforce be family therapy; if one person in a family is afflicted, others are as well, as if the perceived afflictions of the *araṇya* are a constant threat to the *grāma*. Nearly all of the families come after treatment through other available medical options has failed. Balaji is, thus, a last resort healing center.[39] A number of methods of healing are available, through complex temple ritual, temple sponsored healers and exorcists, and from a large number of healers (*bhagat*s or *ojhā*s). They typically come from towns and villages in Rajasthan, Madhya Pradesh, Uttar Pradesh, Haryana, and Delhi and stay for about a week every month in room or apartment rentals to treat their clientele. This essay is not the place to examine the healing modalities, except to say the number is much larger than one can first imagine, and that there is no central registry or organization that catalogues them.[40] All healing is local, even if it is outsourced to

37 Never have I or other researchers heard the term "evil" (in any language) used for a possessing spirit. Always they are wayward, looking for a home after an accidental death (*akālamṛtyu*), a neglected or in improperly performed funeral rite, or the spirit of a deceased family member.

38 This is not the place to examine the reasons for this, except to say that as colonialism slips deeper into the recesses of history formerly colonized people are rediscovering and reanimating their roots. In this case and in the present context, it's safe to say that the local culture of wildness, of the *araṇya*, is becoming more prominent in the lives of many in India, regardless of class, caste, or educational status.

39 Sax's statement with respect to the Himalayas can be extended unproblematically to the Hindi heartland: "When one has an illness that cannot be diagnosed or cured, one typically visits an oracle to find out the cause of the problem, and only later seeks out a healer" (2014: 831).

40 For one example, see Aaltonen 2000. Antti Pakaslahti was behind this effort, which featured the healers he had been documenting for many years.

Balaji.[41] Although there is certainly communication between healers from different villages, there is no attempt by any of them to extend their practices beyond their own highly localized lineages.[42]

These villages, with their socialized residents, represent the *grāma*, and are the counterpart to the *araṇya* as Malamoud employs it, while the Balaji temple itself functions as the counterpart of the *grāma*. The temple *ritual* is complex and extended, while the private clinics employ ritual that can be comparatively simple. The objective is to enlist Bālājī to help the possession manifest, this becoming visible and negotiable for the healer to enact a cure, or at least a remission of the negative forces that afflict the individual and the family. The physical symptoms of this kind of possession may also be regarded as belonging to the *araṇya*, undomesticated, wild, and threatening. These include loosening of the hair among women, counterclockwise gyrating or spinning, screaming, self-flagellation, and much more.[43] One of the reasons that it is effective is because Balaji and the many clinics in the town are safe spaces for this kind of exotic self-expression. It is sanctioned behavior at Balaji. Balaji is thus a buffer zone, where it is safe and comfortable to express oneself through behaviors sanctioned in both the socially complex settlement and the forest. Thus, the psychodynamic therapies, along with the psychosocial expectations, have turned Balaji into a liminal space that embraces both *grāma* and *araṇya*. The *bhūts* that are resocialized and transported from the realm of malevolent spirits into Hanumān's army of benevolent spirits by his *dūts* are most often revealed through an interactive dialogue between healer and spirit, the end result of the *peśī* or the session itself, to be family members or others within one's village who have died bad deaths or not received proper funeral ceremonies. In all cases, in my experience, the villain is a spirit of the dead, the darkest creature of the *araṇya*.

---

41 On localization, see Smith 2006: 73; Shulman 1981: 55, esp. ch. 2, "The Phenomenon of Localization" (pp. 40–89).

42 That said, it would be incorrect to say that *bhagat*s and *ojhās* are uneducated or unlearned, even if they are outside what has become the medical mainstream. They are part of respected and widespread lineages and have always apprenticed with others. This is not the place to examine what appears to be an alternative medical "system."

43 Interestingly, counterclockwise motion (*vāmāvartakrama*) is mentioned in a Abhinavagupta's *Tantrāloka* 16.42, in a description of the actions to be undertaken in the absorption of a sacrificial animal's *prāṇas* in order to remove the impurity of the sacrifice while purifying the heart of the ritualist (*Tantrāloka* 16.36–45; Abhinavagupta 1933: 15–16). The process described by Abhinavagupta also has deep resonances in the MBh and Upaniṣadic passages discussed above. To discuss this fascinating passage here, however, would take us in directions well beyond the scope of this essay, even if it bears directly on the relationship between *grāma* and *araṇya*.

A related psychodynamic process occurs at the distant Bhagavathy or goddess temple in Chottanikkara. This temple is in many respects characteristic of grand South India temples: elephants, business center, large concentric enclosures, secondary temples within the main enclosure. Most of the devotees come to worship Bhagavathy, although not for healing. But those who do come for healing suffer more expressly from psychological or mental disturbances. It is not a last resort healing center, at least not primarily. The possession that generates the healing occurs on Tuesday and Friday nights, special times for worshipping the goddess, when offerings given by devotees are made by temple priests at a shrine perhaps twenty-five meters below the main temple hall, next to a large cistern of red colored water. This is to remind the devotee of blood. Many of those who come fall into trance states and exhibit the same physical symptoms that may be seen in North India, including the counterclockwise whirling, screaming, and so on. Two large trees are nearby, one of which is hardly a few meters from the site of these offerings. It is thoroughly embedded with nails and other sharp objects, including railway spikes, which, so it is said, people drive into the tree with their foreheads in order to chase away invasive spirits. The other tree is a bit further away, but well within the immediate presence of this shrine, on which dolls and other representative objects are nailed into the tree. These, it is said, resonate with the spirits and cause them to be eliminated from the presence of the afflicted individual.

The Bhagavathy temple at Chottanikkara is much more hybrid and domesticated than Balaji. It caters to regular goddess worshippers as well as those with problems uniquely treated there. Nevertheless, the trees densely decorated with nails, spikes, dolls, articles of clothing, and so on, are constant reminders of the culture of possession and healing that has placed Chottanikkara on the map of important South Indian temples. Both have integrated the wildness of the wilderness, the raw unpredictable features of the natural world, into their otherwise architecturally sanctioned ground plans. In both cases, the *araṇya* has taken refuge within individual devotees and clients. It is no longer "out there" but is within. The abode of fear and tension ghettoized within individual bodies and minds might be considered modern if we did not already see this abundantly recorded in the *Mahābhārata* and early Ayurvedic delineations of lists of invasive entities (Smith 2006, chapters 6 and 12, respectively). The behaviors within both of these temple environments more or less seamlessly and without opposition mimic the natural world in the civilizational process that seeks to alter it. The isolation of forest from town, of unpredictable wilderness from domesticated manicured space, is encouraged to come together in both of these healing centers.

The irruption between the forest and the city has led to isolation and failure in communication, to unfriendly encroachment, to barriers and fears from both sides. What has proven most healing, then, is sanction of the wild within the cultivated, the reintroduction of the dark, wild, undomesticated jungle of the individual's personality or self, alongside the refined, public, constrained, constantly amended and filtered personality, through a process of mediation by a healer (at Balaji) of an institutional understanding (at Chottanikkara) that in one way or another understands this irruption and is able to relieve fear and guilt from the process of drawing these sides together. In this way, the positive and destructive spirits can come together as a single salutary force. This is also why witnesses are required, why both temples recognize the value of family presence. One can think about the value of role playing, mimesis, and other psychodynamic therapies that draw together different aspects of the personality. But the healing at these two temple sites in India must be important in realization of any of these for the way they integrate the undomesticated wilderness with predictable domesticated habitational spaces.[44] My argument, that possession, whatever it may be, and the behaviors that distinguish it, are expressions of the natural world, the *araṇya*, which forces the individual and family to become alienated from their socialized and regulated settlements, and that the Balaji and Bhagavathy temples, like the Pāṇḍav *līlā* stage, the arena for Yudhiṣṭhira's *aśvamedha*, and the known forest for Dhṛtarāṣṭra and Gāndhārī's *vanaprastha*, are buffer zones for resolving the crises that arise from the confrontation of *grāma* and *araṇya*.

7 Conclusions

The relationship between settlement and wilderness is notably vexed, and is often negotiated through psychomental strategies, including those used to counter dissociation induced possession. The first example, a discussion of Yudhiṣṭhira's construction of his *aśvamedha* city and the poor brahman's vow to live on gleaning alone, two episodes that are deliberately juxtaposed at the end of the Āśvamedhika-*parvan* of the *Maihābhārata*, revealed that wealth, power,

44 I observed the display of possession, perhaps even possession itself, as a mimetic process. On Tuesday and Friday evenings women experienced in possession were training younger women and teenagers in the "art" of performing possession. On mimetic possession, see Freeman 1993, 1998. One can say that for certain socially disadvantaged people in these areas (women, underclasses) it has become necessary to balance the inevitable excesses of the oppression of the *grāma* with an injection of elements of the *araṇya*.

and position within the *grāma* dictated the terms through which resources from the *araṇya* were incorporated, and conversely how poverty, small scale unity of purpose, and generosity, within a buffer zone between *grāma* and *araṇya*, enabled a family to maintain its independence, even during a drought. Both intentions were fulfilled: Yudhiṣṭhira's reign was validated by successfully performing the *aśvamedha* and the family of the poor brahman ascended to the heavenly world as a result of its unwavering dedication to their vow.

Nevertheless, the latter was adjudged superior by the mongoose, a denizen of the *araṇya*, whose body turned fully into gold through consuming its leavings after it failed to do so when consuming the leavings of the horse sacrifice. In spite of this possible bias towards the *araṇya*, the virtue of the poor brahman was as much part of the *grāma* as was the *aśvamedha*. In both cases, codified and cultivated dharma was invoked and maintained. The Kuru-Pāñcāla domain was maintained by the *aśvamedha*. Similarly, the buffer zone on which the poor family gleaned was agricultural land, presumably replanted the following year, enabling the forest beyond the scanty perishable hut of the poor brahman to continue on as before, minus whatever had been taken from it by Yudhiṣṭhira for his *aśvamedha*. Thus, the isolation and social and psychological distancing of the *araṇya* from the *grāma* was unaffected. Whatever dissociation may be read into this due to the relative absence of communication between the *grāma* and the *araṇya* was, at least to some extent, resolved by the mysterious creature from the *araṇya* that upheld the actions of the poor brahman, who was as close to a denizen of the forest as was possible under the circumstances.[45] The brahman and his family took willing and fearless risks, embracing at least some of the forces of the wilderness and their potential outcomes.

The second case presented here, the death of Vidura and the absorption of his persona into the body and mind of Yudhiṣṭhira, is a clear example of dissociation between the settlement and the forest, and their resolution through internal acceptance of the powerless, disadvantaged, moribund, half-brother in the wilderness into the persona of his powerful, royal, twin. This could only occur through the mysterious transmission of Vidura's senses and internal essence into the body of Yudhiṣṭhira, who immediately understood and felt the infusion. This act of possession strengthened Yudhiṣṭhira's control over the domain by internalizing the forces of his half-brother in Dharma, themselves

45 It is perhaps no accident that the fourth Pāṇḍava brother is named Nakula, mongoose. Among his outstanding qualities were horsemanship and knowledge of medicine. Both of these abide in the intermediate zone between the settlement and forest, while participating in both.

given their full potency by the isolation and austerity that could only have come to him in the forest.

The third case studied here, the surprising rite of transference of an alternate identity as Arjuna by a Pāṇḍav *līlā* actor in the Himalayas to his son through an act of (dis)possession, illustrates that the *Mahābhārata* in its present form in Garhwal is not simply a text or story cycle but a template through which the largely educated residents of these towns and villages live their lives. That the actor had habituated himself to "becoming" Arjuna, to becoming possessed by Arjuna, through a learned and repeated process of self-consecration, demonstrates that through a practice of dissociation, he was able to bring Arjuna to life, at least for the duration of the performance. Not only was he possessed by Arjuna, but he actually "owned" Arjuna, at least in that village, meaning that Arjuna was property subject to be transferred as inheritance. Even if the entire town knew the stories perfectly well, they gathered together for these periodic performances because the actors brought to life many of the key characters in the epic, including Arjuna, Draupadī, Kuntī, Bhīma, Karṇa, and Kṛṣṇa. It is not only the characters that are brought to life, however; it is the merging of the present with the historical presence of the *Mahābhārata* that is resolved through the act of dissociation induced possession. In the end, it is not division and loss that are paramount in this possession, but revitalizing the ground beneath the feet of the community, bringing greater depth and purpose to their lives.[46]

The final case discussed here is of loss and gain of identity at Balaji and Chottanikkara. The reason these centers are not simply thriving, but are increasingly visible as healing options for nearly everyone, is because the unpredictable wildness of the forest has become sanctioned within the confines of these otherwise orthodox and domesticated Hindu temple complexes. Clearly, the scope of the activity at the temples is much broader than at other Hindu temples in admitting, sanctioning, and positively addressing uncontrolled and deeply rooted personal expressions of dissociation, depression, and anxiety. The dissociation that manifests in the clinical situations at these healing centers is identified by clients and healers alike as possession, and is often addressed, at least at Balaji, with counter-possession, under the watchful eye of Balaji, that exorcises the invasive spirit, lifts veil of the depression, or relieves the trauma of anxiety. The healing process involves bringing together the unpredictable, untamed, uncharted and dissociational power of the deity in charge (Bālājī or

46 This process is also recognizable in Krishna *līlā*s in Vrindavan and elsewhere in Braj, and Ram *līlā*s throughout North India. The literature on these *līlā*s is large, well-known, and need not be repeated here.

Bhagavathy in these cases), which is to say the force of the wilderness within the captured and domesticated deity, with alien and uncharted forces within the client, who desperately requires this mediation or mitigation in order to live peaceably within the confines of their own domesticity. The drama between *bhūt* and *dūt* drags on endlessly, because they, like the *devas* and *asuras* in Vedic India, battle eternally to a standstill.

This is an old story, as Stephanie Jamison notes in her translation of *Ṛgveda* 10.146, a hymn of the late second millennium BCE dedicated to Araṇyānī, the "Lady of the Wilderness." Jamison comments: "It is dedicated to the spirit of the *araṇya*, the wild place, and the contrast between the wild place and the settled one, an opposition that runs throughout Indian literature, dominates the hymn. Despite the ostensible dedication, the poet tries as much as possible to evade the wilderness and conjure up the safety of the village" (Jamison in Jamison & Brereton 2014: 1631 [vol. 3]). As we have seen here, this wilderness, the *araṇya*, is as much an internally configured forest as an external one. That the forest is interiorized should not come as a surprise. Interiorization of the external world is an old practice in India, dating back (at least) to the early centuries of the first millennium BCE in the *Jaiminīya Brahmaṇa*, in which the five sacred fires were interiorized as the five breaths (*prāṇa*). "Interiorization refers to the process by which a ritual that is performed manually as a set of physical actions comes to be internalized as a mentally performed ritual" (Payne 2016: 31).[47] The dimensions of the interiorized forest are here extended to (a) the conceptualization of activities possible in the forest and the capacities generated from it (Vidura and Yudhiṣṭhira), (b) hierarchical polities distinguished by ethnicity or language (*ĀCVP*), (c) the spatial compression of multiple characters and histories within individuals and communities in the present moment (the rite of inheritance in Raithal), and (d) the diagnosis of possession by extra-human presences from unknown or speculative realms (Balaji and Chottanikkara).[48]

47 For interiorization in the *Jaiminīya Brāhmaṇa*, see Bodewitz 1973: 110–123. Zysk 1993 further elucidates the relationship of fires and *prāṇas*. The beginning of the *Bṛhadāraṇyaka Upaniṣad* provides a road map for interiorization in cosmos—the *araṇya*—in terms of its coordinates as the sacrificial horse—the *grāma*. The piecemeal drawing together of the external world with the internal world is well developed in yoga; See Eliade 1954: 111–114, and many others more recently.

48 For example, *Suśrutasaṃhitā* 6.60.10 (Smith 2006: 491) states that a person may be possessed by a *gandharva* or celestial musician if one's "character is joyous, frequents sandy river banks and the interiors of forests, is well-behaved, fond of singing, perfumes and garlands, laughs while dancing, is agreeable, and is of few words."

The focus here has been on the internal coordinates of *grāma* and *araṇya*, on their contrasts, tensions, and resolutions. But it should not be forgotten that these terms indicate, in their primary designations, physical environments.[49] As denizens of an ever-modernizing *grāma*, it is increasingly easy to forget that there is a world outside that is not always answerable to our needs, that is not as invariable and renewable as we would like. For hundreds of years, perhaps thousands, one of modernity's inexorable trends has been a slow transition indoors, away from the forest, as the habitat and habitus of towns and cities encroach onto forest and wilderness. It is common knowledge in India (like almost everywhere else) that deforestation has dramatically increased over the centuries, that forest cover has decreased as towns, highways, and urban connectivity have replaced it.[50] Consistent with this is the fact that attention to the external world has shrunk factorially. A 2002 study showed that 93% of our time is spent indoors (87%) or in cars (6%). This leaves 7% of our time outdoors (Klepais, et al. 2001). We wonder how much of that 7% is spent inattentive to our surroundings or under the glare of artificial lighting? Homo sapiens as a species has almost entirely lost touch with the world outside the increasingly technologized *grāma*. Whether it is the Kuru-Pāñcāla realm of Yudhiṣṭhira or the modern world, the *grāma* has progressively swallowed up the *araṇya*. The cascading media and technologically driven inwardness that characterizes most of the educated segments of our society has accepted, even if grudgingly, that the inner *grāma* of safety is best brought about by systematic destruction of and disregard for the external *araṇya*. Liminal spaces or buffer zones such as well-tended forests or sanctioned dissociative experience, including certain forms of possession, helped mediate the physical and emotional space between *grāma* and *araṇya* in India centuries and millennia ago, as they continue to do today.

---

49 A term of recent currency for this is "wildland-urban interface" (Theobald & Romme 2007). Although this has been used primarily to determine fire risk and urban encroachment, it has much broader implications that may be applicable to *grāma/araṇya* or other classical distinctions. This cannot be explored further here, except to say that the distinctions addressed are settled urban areas, residential forest, and deeper forest land, the same divisions addressed in this essay. The problem that anyone alive today will have in conceiving the *grāma/araṇya* relationship in ancient and classical India is that in the twenty-first century, and surely well before this, there cannot be any place in sub-Himalayan India (or in the US or most of Europe, and doubtless in much of the rest of the world), outside of strictly policed national parks, that is more than a few kilometers from a road. The environmental and psychological impact of carving up the land with ever more intricate and trafficked networks can scarcely be imagined.

50 See Kumari, et al., 2019 on recent deforestation in India. Data measurement differs according to method. See: https://ourworldindata.org/grapher/share-global-deforestation. In a 2015 survey, India loses 668,400 hectares per annum of forest loss.

Acknowledgements

I would like to take this opportunity to thank our celebrant, Christopher Minkowski, for nearly half a century of friendship, for his unfailing effusiveness, clarity, sagaciousness, and good humor, and for many pleasant experiences, including accompanying with me to several Vedic *śrauta* rituals in the early 1980s in Maharashtra, and blindfolding and leading me up a disorienting circular staircase in Cambridge, Mass., many years ago, into the deceptively spacious, and to my eye medieval, epicenter of good humor, the Harvard Lampoon. Should we expect less of our *jubilaris*? May his *vānaprasthya* be spent, at least in part, in the *vana*. May it be long and happy!

References

Aaltonen, Jouko. 2000. *Kusum*. London: Royal Anthropological Institute. 69 minutes. https://search.alexanderstreet.com/preview/work/bibliographic_entity%7Cvideo_work%7C1879379

Abhinavagupta. 1933. *Tantrāloka of Abhinavagupta*, with the Commentary by Rājānaka Jayaratha, vol. 10. Ed. Madhusūdan Kaul Shāstrī. Kashmir Series of Texts and Studies, vol. 52. Srinagar: Mahārāja of Jammu and Kashmir; Bombay: Nirṇaya Sāgara Press.

Adamek, Wendi. 2012. The Agency of Relations at Baoshan. *Literature & Aesthetics* 22(2): 50–71.

Amaladass, Anand. 2001. Conquest of the Quarters (*digvijaya*): The Quest for Identity. In Anand Amaladass & Rosario Rocha (eds.), *Crossing the Borders. Essays in Honour of Francis X. D'Sa on the occasion of his 65th Birthday*, pp. 151–167. Chennai: Satya Nilayam Publications.

Angot, Michel. 2009. Land and Location: Errant Gods, Erring Asuras and the Land of Men: Place and Space in Vedic Literature. In Daniela Berti & Gilles Tarabout (eds.), *Territory, Soil and Society in South Asia*, pp. 99–142. New Delhi: Manohar.

Beckert, Heinz. 1984. Date of the Buddha Reconsidered. *Indologica Taurenensia* 10. 29–36.

Berti, Daniela. 2001. *La parole de dieux. Rituels de possession en Himalaya indien*. Paris: CNRS.

Berti, Daniela. 2006. Ritual Kingship, Divine Bureaucracy, and Electoral Politics in Kullu. *European Bulletin of Himalayan Research* 29–30:39– 61.

Biardeau, Madeleine. 1989. *Histoires de poteaux: Variations védiques autour de la déesse Hindoue*. Paris: Ecole française d'Extrême-Orient. [Madeleine Biardeau. *Stories about Posts: Vedic Variations around the Hindu Goddess*. Trans. by Alf Hiltebeitel,

Marie-Louise Reiniche, and James Walker. Ed. by Alf Hiltebeitel and Marie-Louise Reiniche. Chicago: University of Chicago Press, 2004.]

Bindi, Serena. 2009. *La* fabrique de l'événement. Paysage de possession, de soin et de production de sens dans la vallée du Baghirati (Garhwal, Inde du Nord). Thèse Ethnologie, Paris / Sienne, EHESS / Università degli Studi di Siena.

Bindi, Serena. 2013. Patrimonialiser la culture: la mise en scène des rituels de possession en Uttarakhand. *Anthropoloigie & Santé* 6: 2–16.

Bodewitz, Hendrik W. 1973. *Jaiminīya Brāhmaṇa I, 1–65: Translation and Commentary, with a Study of Agnihotra and Prāṇāgnihotra.* Leiden: E.J. Brill.

Boucher, Daniel. 2008. *Bodhisattvas of the Forest and the Formation of the Mahāyāna. A Study and Translation of the Rāṣṭrapālaparipṛcchā*-sūtra. Honolulu: University of Hawai'i Press.

Bourguignon, Erika. 1976. *Possession.* San Francisco: Chandler & Sharp.

Braude, Stephen E. 1995. *First Person Plural: Multiple Personality and the Philosophy of Mind.* Lanham MD: Rowman & Littlefield.

Buswell, Robert & Donald Lopez. 2013. Donald. *The Princeton Dictionary of Buddhism.* Princeton: Princeton University Press.

Cardeña, Etzel. 1994. The domain of dissociation. In S.J. Lynn & J.W. Rhue (eds.), *Dissociation: Clinical and Theoretical Aspects*, pp. 5–31. New York, NY: Guilford Press.

Cardeña, Etzel, Marjolein van Duijl, Lupita A. Weiner, Devin B. Terhune. 2011. Possession/Trance Phenomena. In Paul F. Dell & John A. O'Neil (eds.), *Dissociation and the Dissociative Disorders: DSM-V and Beyond*, pp. 146–154. New York: Routledge.

Chapple, Christopher Key and Mary Evelyn Tucker (eds.). 2000. *Hinduism and Ecology: The Intersection of Earth, Sky, and Water*, 169– 289. Cambridge, MA: Harvard University Press.

Conzelmann, Elizabeth. 2006. A Royal Ritual of Mandi State. *European Bulletin of Himalayan Research* 29–30: 14–38.

Dalenberg, Constance J. & Kelsey Paulson. 2011. The case for the study of "normal" dissociation. In Paul F. Dell & John A. O'Neil (eds.), *Dissociation and the Dissociative Disorders: DSM-V and =*

Diserens, Hélène. 1995–1996. Images et symboles des déesses de la haute vallée de Kullu. *Bulletin d'études indiennes* 13–14: 91–115.

Dumont, Paul-Émile. 1927. *L'Aśvamedha, description du sacrifice solennel du cheval dans le culte védique d'apres les texts du Yajurveda blanc.* Paris: Paul Guenther.

Dumont, Paul-Émile. 1948. "The Horse-Sacrifice in the Taittirīya-Brāhmaṇa: The eighth and ninth prapāṭhakas of the third kāṇḍa of the Taittirīya-Brāhmaṇa with translation." *Proceedings of the American Philosophical Society* 92.6: 447–503.

Dwyer, Graham. 1998. The Phenomenology of Supernatural Malaise: Attribution, Vulnerability and Patterns of Affliction at a Hindu Pilgrimage Center in Rajasthan. *Social Analysis* 42.2:3–23.

Dwyer, Graham. 1999. Healing and the Transformation of Self in Exorcism at a Hindu Shrine in Rajasthan. *Social Analysis* 43.2:108–137.

Dwyer, Graham. 2003. *The Divine and the Demonic: Supernatural Affliction and Its Treatment in North India*. London and New York: RoutledgeCurzon.

Eliade, Mircea. 1958 (1954). *Yoga: Immortality and Freedom*. Trans. by Willard R. Trask. London: Routledge & Kegan Paul.

Fiol, Stephen. 2017. *Recasting Folk in the Himalayas Indian Music, Media, and Social Mobility*. Champaign-Urbana: University of Illinois Press.

Freeman, J. Richardson. 1993. Performing Posession: Ritual and Consciousness in the Teyyam Complex of Northern Kerala. In Heidrun Brückner, Lothar Lutze, and Aditya Malik (eds.), *Flags of Flame: Studies in South Asian Folk Culture*, pp. 109–138. New Delhi: Manohar.

Freeman, J. Richardson [J.R. Freeman]. 1998. Formalized Possession Among the Tantris and Teyyams of Malabar. *South Asia Research* 18.1:73–98.

Gold, Ann Grodzins. 1992. *A Carnival of Parting. The Tales of King Bharaṛhari and King Gopi Chand as Sung by and Told by Madhu Natisar Nath of Ghatiyali*. Delhi: Munshiram Manoharlal.

Hawley, Nell Shapiro & Sohini Sarah Pillai. 2021. *Many Mahābhāratas*. Albany: State University of New York Press; Delhi: Primus Books.

Hiltebeitel, Alf. 1988. *The Cult of Draupadī. Vol. 1: Mythologies: From Gingee to Kurukṣetra*. Chicago: University of Chicago Press.

Hiltebeitel, Alf. 1991. *The Cult of Draupadī. Vol. 2: On Hindu Ritual and the Goddess*. Chicago: University of Chicago Press.

Hiltebeitel, Alf. 1999. *Rethinking India's Oral and Classical Epics. Draupadī among the Rajputs, Muslims and Dalits*. Chicago: University of Chicago Press.

Hiltebeitel, Alf. 2016. *Nonviolence in the Mahābhārata: Śiva's summa on Ṛṣidharma and the gleaners of Kurukṣetra*. London & New York: Routledge.

Haberman, David. 1994. *Journey Through the Twelve Forests. An Enconter with Krishna*. New York: Oxford University Press.

Halperin, Ehud. 2019a. Is the Goddess Haḍimbā Tantric? Negotiating Power in a Western Himalayan Sacrificial Arena. *International Journal of Hindu Studies* 23.2: 195–212.

Halperin, Ehud. 2019b. *The Many Faces of a Himalayan Goddess. Haḍimbā, Her Devotees, and Religion in Rapid Change*. Oxford: Oxford University Press.

Heesterman, Jan. 1985. *The Inner Conflict of Tradition. Essays in Indian Ritual, Kingship, and Society*. Chicago: University of Chicago Press.

Howell, Elizabeth F. 2005. *The Dissociative Mind*. New York: Routledge.

Humphrey, Caroline, and James Laidlaw. 1994. *The Archetypal Actions of Ritual: A Theory of Ritual Illustrated by the Jain Rite of Worship*. Oxford: Clarendon Press.

Inden, Ronald. 1990. *Imagining India*. Oxford: Basil Blackwell.

Jamison, Stephanie. 1996. *Sacrificed Wife/Sacrificer's Wife: Women, Ritual, and Hospitality in Ancient India*. New York, Oxford: Oxford University Press.

Janet, Pierre. 1889. *L'automatisme psychologique, essai de psychologie expérimentale sur les forms inférieures de activité humaine*. Paris: Félix Alcan.

Jassal, Aftab Singh. 2014. In Search or Nagaraja: Narrative, Place-Making, and Narrative Embodiment in Garhwal. Ph.D. dissertation, Emory University.

Jassal, Aftab Singh. 2017. Divine Politicking. A Rhetorical Approach to Deity Possession in the Himalayas. *Religions* 26.1: 1–18.

Kakar, Sudhir. 1982. *Shamans, Mystics and Doctors: A Psychological Inquiry into India and its Healing Traditions*. Chicago: University of Chicago Press.

Klepeis, Neil E., et al. 2001. National Human Activity Pattern Survey (NHAPS). *Journal of Exposure Analysis and Environmental Epidemiology* 11: 231–252.

Koskikallio, Petteri. 1993. "Jaiminibhārata and aśvamedha." *Wiener Zeitschrift für die Kunde Sudasiens* 36 (Supplementalband): 111–119.

Koskikallio, Petteri. 1995. "Epic Descriptions of the Horse Sacrifice." In *International Conference on Sanskrit and Related Studies, September 23–26, 1993*. Jagiellonian University, Institute of Oriental Philology. Cracow: The Enigma Press, pp. 165–177.

Koskikallio, Petteri. 1999. "The Horse Sacrifice in the Pātālakhaṇḍa of the Padmapurāṇa." In Mary Brockington and Peter Schreiner (eds.), *Composing a Tradition: Concepts, Techniques and Relationships. Proceedings of the First Dubrovnik International Conference on the Sanskrit epics and Purāṇas, August 1997*, pp. 227–243. Zagreb: Croatian Academy of Fine Arts.

Kumari, Rima, et al. 2019. Deforestation in India: Consequences and Sustainable Solutions. *Deforestation Around the World*. New Delhi: IntechOpen. https://www.intechopen.com/profiles/279806/rima-kumari. DOI: 10.5772/intechopen.85804

Lambek, Michael. 1993. *Knowledge and practice in Mayotte: Local Discourses of Islam, Sorcery, and Spirit Possession*. Toronto: University of Toronto Press.

Lambek, Michael. 2018. *Island in the Stream: An Ethnographic History of the Mayotte*. Toronto. University of Toronto Press.

Larios, Borayin. 2017. *Embodying the Vedas. Traditional Vedic Schools in Contemporary Maharashtra*. Warsaw/Berlin: DeGruyter Opel, Ltd.

Latour, Bruno. 2005. *Reassembling the Social: An Introduction to Actor-Network-Theory*. Oxford: Oxford University Press.

Lubin, Timothy. 2001a. Science, Patriotism, and Mother Veda. Ritual Activism in Maharashtra. *International Journal of Hindu Studies* 5: 297–321.

Lubin, Timothy. 2001b. Veda on Parade: Revivalist Ritual as Spectacle. *Journal of the American Academy of Religion* 69: 377–408.

Luchesi, Brigitte. 2002. Haḍimbā Devī—Hinduistische Volksgottin Mit Uberregionaler Bedeutung. In Manfred Hutter, Karl Hoheisel, Wassilios Klein, and Ulrich Vollmer (eds.), *Hairesis: Festschrift Für Karl Hoheisel Zum 65. Geburtstag*, pp. 327– 340. Münster: Aschendorff.

Luchesi, Brigitte. 2006. Fighting Enemies and Protecting Territory: Deities as Local Rulers in Kullu, Himachal Pradesh. *European Bulletin of Himalayan Research* 29–30: 62– 81.

Lutgendorf, Philip. 2000. City, Forest, and Cosmos: Ecological Perspectives from the Sanskrit Epics. In Chapple and Tucker (eds.), *Hinduism and Ecology*, pp. 169– 289.

Magnone, Paolo. 2012. Uttaṅka's Quest. In John Brockington & Petteri Koskikallio (eds.), *Battle, Bards and Brāhmins. Papers of the XIII World Sanskrit Conference.* Vol. 2, pp. 101–128 [Edinburgh, 10–15 July 2006]. Delhi: Motilal Banarsidass.

Malamoud, Charles. 1976 (1998). Village et forêt dans l'idéologie de l'Inde brâhmanique. *Archives Européenes de Sociologie / European Journal of Sociology* 17(1), 3–20. [Reprint in Charles Malamoud, *Cuire le monde: Rite et pensée dans l'Inde ancienne*, 93–114. Paris: Editions Ia Decouverte, 1989. Charles Malamoud, Village and Forest in the Ideology of Brahmanic India, *Cooking the world: Ritual and Thought in Ancient India*, (trans. David White), pp. 74–91. Delhi: Oxford University Press, 1996.].

Miller, Joseph C., Jr. 1994. The twenty-four Brothers and Lord Devnarayan: The story and performance of a folk epic of Rajasthan, India. Ph.D. dissertation, University of Pennsylvania.

Moore, Donald, Jake Kosek, and Anand Pandian (eds.). 2003. *Race, Nature, and the Politics of Difference.* Durham, N.C.: Duke University Press.

Negi, Shachi. 2019. Understanding an Indigenous Indian Himalayan Folklore: Jagar as Emotion Discourse. *3rd International Conference on Humanities and Social Science, Berlin. 29–31 August 2019.* https://www.dpublication.com/wp-content/uploads/2019/08/28-128.pdf

Nowicka, Olga. 2017. Vedic Ritualism and Advaita Vedānta Monastic Institutions in Kerala. *Studia Religiologica* 50(2): 163–171. doi:10.4467/20844077SR.17.010.7341.

Olivelle, Patrick. 1990 (2011). Village vs. Wilderness: Ascetic Ideals and the Hindu World. In *Monastic Life in the Christian and Hindu Traditions*, ed. Austin B. Creel and Vasudha Narayanan, pp. 125—160. Lewiston NY: Edwin Mellen. [Reprint: 2011. In *Ascetics and Brahmins. Studies in Ideologies and Institutions*, pp. 43–62. London, New York, Delhi: Anthem.]

Olivelle, Patrick. 1993. *The Āśrama System. The History and Hermeneutics of a Religious Institution.* New York, Oxford: Oxford University Press.

Olivelle, Patrick. 1998. *The Early Upaniṣads: Annotated Text and Translation.* New York, Oxford: Oxford University Press.

Ortner, Sherry B. 1997. Thick resistance: Death and the cultural construction of agency in Himalayan mountaineering. *Representations* 59: 135–162.

Pakaslahti, Antti. 1998. "Family-Centered Treatment of Mental Health Problems at the Balaji Temple in Rajasthan." In Asko Parpola and Sirpa Tenhunen (eds.), *Changing Patterns of Family and Kinship in South Asia. Studia Orientalia* 84:129–166.

Pakaslahti, Antti. 2005. "Traditional Healers as Culturally Accepted/Sanctioned Mental Health Practitioners." In Savita Malhotra (eds.), *Mental Disorders in Children and Adolescents: Need and Strategies for Intervention*, pp. 215–229. Delhi, Bangalore: CBS.

Payne, Richard K. 2016. Introduction. In Richard K. Payne and Michael Witzel (eds.), *Homa Variations: The Study of Ritual Change Across the Longue Durée*. New York: Oxford University Press.

Prasko, Jan, et al. 2010. Transference and Countertransference in Cognitive Behavioral Therapy. *Biomedical Papers of the Medical Faculty of the University of Olomouc, Czech Republic* 154(3): 189–197.

Sax, William S. 1991. *Mountain Goddess: Gender and Politics in a Himalayan Pilgrimage*. New York: Oxford University Press.

Sax, William S. 2002. *Dancing the Self. Personhood, Performance and the Pāṇḍav Līlā of Garhwal*. New York: Oxford University Press.

Sax, William S. 2006. Agency. In Jens Kreinath, Jan Snoek, and Michael Stausberg (eds.), *Theorizing Ritual, vol. 1: Issues, Topics, Approaches, Concepts*, pp. 473–481. Leiden: Brill.

Sax, William S. 2009. *God of Justice: Ritual Healing and Social Justice in the Central Himalayas*. New York: Oxford University Press.

Sax, William S. 2014. Ritual healing and mental health in India. *Transcultural* Psychiatry 51(6): 829–849.

Sax, William & Helene Basu. 2015. *The Law of Possession. Ritual, Healing, and the Secular State*. Oxford: Oxford University Press.

Schomer, Karine. 1990. The "Ālha" Epic in Contemporary Performance. *The World of Music* 32.2: 58–80.

Shalom, Naama. 2017. *Re-ending the Mahābhārata: The Rejection of Dharma in the Sanskrit Epic*. Albany: State University of New York Press.

Shulman, David. 1981. Tamil Temple Myths: *Sacrifice and Divine Marriage in the South Indian Śaiva Tradition*. Princeton: Princeton University Press.

Smith, John D. 1991. *The Epic of Pābūjī: A Study, Transcription and Translation*. Cambridge: Cambridge University Press

Smith, John D. 2009. The Mahābhārata. Middlesex, UK: Penguin

Smith, Frederick M. 2001. The Recent History of Vedic Ritual in Maharashtra. *Studia Orientalia* 94: 443–463.

Smith, Frederick M. 2004. Possession as an Intermedial Event. In Hans Breder and Klaus-Peter Busse (eds.), *Intermedia: Threshold States*, pp. 122–130. Dortmund: Universität Dortmund Press.

Smith, Frederick M. 2006. *The Self Possessed. Deity and Spirit Possession in South Asian Literature and Civilization*. New York. Columbia University Press.

Smith, Frederick M. Forthcoming. *The Mahābhārata: 14. The Book of the Sacrifice of the Horse. The Mahābhārata: 15–18. Final Renunciation and Ascendance. Introduction, Translation, and Annotations*. Delhi: Primus Books.

Sprockhoff. Joachem. Friedrich. 1981. 1984, 1991. Āraṇyaka und Vānaprastha in der vedischen Literatur. Neue Erwägungen zu einer alten Legende und ihren Problemen I, II, III. *Wiener Zeitschrift für die Kunde Südasiens und Archiv für Indische Philosophie* 25: 19–90; 28: 5–43; 35: 5–46.

Staal, Frits. 1983. *Agni: The Vedic Ritual of the Fire Altar*, vols. 1–2. Berkeley, Asian Humanities Press.

Sutherland, Peter. 1998. Travelling gods and government by deity: An ethnohistory of power, representation and agency in west Himalayan polity. Ph.D. diss., Oxford University.

Takahashi Kenji. 2021a. Yudhiṣṭhira's Contested *aśvamedha*. Handout, 231st Annual Meeting of the American Oriental Society.

Takahashi Kenji. 2021b. The Dharma of Gleaners as Expounded in the Umāmaheśvarasaṃvāda of the Śivadharma Corpus. Studies on the Śivadharma and the Mahābhārata 2. In Florinda De Simini & Csaba Kiss (eds), *Śivadharmāmṛta. Essays on the Śivadharma and its Network*, pp. 255–283, 312–338. The Śivadharma Project. Studies on the History of Śaivism II. Napoli: Unior Press.

Teshima, Hideki. 2008. *Die Entwicklung des vorbereitenden Rituals im Aśvamedha ausgehend von der Darstellung im Vādhūla-Sūtra*. Berlin: Logos.

Theobald, David M. & William H. Romme. 2007. Expansion of the US wildland–urban interface. *Landscape and Urban Planning* 83: 340–354.

Tuan, Yi-Fu. 1977. *Space and Place. The Perspective of Experience*. Minneapolis: University of Minnesota Press.

Turner, Victor. 1969. The Ritual Process. Ithaca: Cornell University Press. *Process*. Ithaca: Cornell University Press.

Vaudeville, Charlotte. 1965. *Bārahmāsā: les chansons des douze mois dans les litteratures indo-aryennes*. Publications de l'Institut français d'indologie, no. 28. Pondichéry: Institut Français d'Indologie.

Vidal, Denis. 1988. *Le culte des divinités locales dans une région de l'Himachal Padesh*. Paris: Editions de l'Ostrom.

Visigalli, Paolo. 2019. Charting 'Wilderness' (*araṇya*) in Brahmanical and Buddhist Texts. *Indo-Iranian Journal* 62: 162–180.

Witzel, Michael. 2011. Report on the Atirātra ritual held in Kerala in April 2011. https://mittalsouthasiainstitute.harvard.edu/wp-content/uploads/2012/08/KERALA-report-Witzel.pdf

Zimmermann, Francis. 1987. *The Jungle and the Aroma of Meats. An Ecological Theme in Hindu Medicine*. Berkeley. University of California Press.

Zoller, Claus Peter. 2001. On the relationship between the folk and classical tradition in South Asia. *European Bulletin of Himalayan Research* 20–21: 77–104.

Zoller, Claus Peter. 2014. *Die Paṇduaṇ: Ein mündliches Mahābhārata-Epos aus dem Garhwal-Himalaya*. Wiesbaden: Harrassowitz.

Zysk, Kenneth A. 1993. The Science of Respiration and the Doctrine of the Bodily Winds in Ancient India. *Journal of the American Oriental Society* 113.2: 198–213.

CHAPTER 13

Forest People in the *Kauṭilīya Arthaśāstra*

Thomas R. Trautmann

The place in Indian civilization of what the texts call forest people is difficult to determine. All the written sources that survive from the ancient period come from the *paṇḍita* culture and the religious specialists of Buddhism and Jainism, in which forest people figure but little. The *Kauṭilīya Arthaśāstra* (hereafter the *Arthaśāstra*, or *KA*) is an exception, in which forest people are mentioned frequently. Upinder Singh, in her magnificent study of political violence in ancient India, has devoted an entire chapter ("The Wilderness") to the forest and forest people, synthesizing an extensive corpus of available evidence including that of the *Arthashastra*, resulting in a complex and internally conflicted picture. Her work builds upon earlier works, among which those of Aloka Parasher Sen and Romila Thapar stand out, and the reader on India's environmental history edited by Mahesh Rangarajan and K. Sivaramakrishnan, works that address the question on a more or less extensive basis.[1] I believe *intensive* study of the *Arthashastra*, taken by itself and in relation to its own book history, has still more to tell us on what proves to be a knotty matter. It affords the opportunity to closely read the entirety of references in a single text, and see what we can make of the patterns that emerge, as an ensemble. All the passages take the point of view of kingship, of course, viewing forest people from without, and need to be understood with that in mind.[2]

A good place to start is in the middle. Book 6, chapter 1 of the *Arthaśāstra* is, as Olivelle says,[3] a link between the two halves of the text, joining the internal administration of the kingdom in the first half with foreign affairs and war in the second half. It states the ideal qualities of the seven elements of the state, namely, lord, minister, countryside, fort, treasury, army, ally. As to countryside, it is an agrarian paradise, free, for example, of stones or mud. Its human population is mostly cultivators. Notably, the ideal countryside lacks forest people entirely (*aṭavī-hīna*).[4]

---

1 Singh 2017; Parasher-Sen 2012; Thapar 1971, 2012; Rangarajan and Sivaramakrishnan 2012.

2 I am grateful to Madhav Deshpande, Patrick Olivelle and Mark McClish for their comments on aspects of this essay.

3 Olivelle 2013:10.

4 *KA* 6.1.8. It comes at the end of a long compound of undesirables: mud, stones, brackish soil, rugged land, criminals, gangs, vicious animals, wild animals, forest people.

 | DOI:10.1163/9789004536869_015

And yet that same ideal territory *should* have forests, for elephants and forest products. Forests, then, but without the forest people, not to mention wild animals—except for elephants, the largest of them all.

Already we see a strain concerning forests and the people who inhabit them, in the thinking about the kingdom. And yet in spite of that ideal zero, forest people are ubiquitous in the *Arthaśāstra*, and appear again and again in the thinking about the king and his countryside. The text as a whole is not given to ideal-type thinking, but to the contrary it models kinds of reasoning which assume a king who wishes to expand his realm (*vijigīṣu*) but who suffers certain disadvantages that he must work around.

In this I will first discuss the concept of forest people and then examine the references to them, grouping them into three categories: forest people as servants of the king in connection with forests supplying forest produce and elephant forests; as bands of military troops; and the forest chieftain as a player in interstate politics, especially as it concerns the succession to kingship. I will then examine the development of this corpus of references in the book history of the *Arthaśāstra* as elucidated in the excellent recent analysis of Mark McClish, ending with some thoughts about the many and important roles of forest people in Indian civilization, and how they may be recovered from texts they did not compose.

1 The Name "Forest People" (aṭavī, āṭavika, vanacara)

Olivelle, in the notes to his fine translation of the *Arthaśāstra*, says, "The term *aṭavī* is often used metonymically with reference to tribal people living in remote forest areas". *Aṭavī* in the primary meaning of "forest" occurs only twice, in a compound denoting "regions separated by forests",[5] while it appears abundantly in the meaning of forest people. We also find the derivative form *āṭavika*, in the sense of forest chieftain, but the two expressions are often interchangeable.

The word refers to tribal people. But what it expresses is not an organization into *tribes*, however that may be defined, but an identification with a *landscape*, the forest, specifically, and not some other landscape in which tribal people might dwell such as the desert, the hills or the marshland. There are also a few instances of *vanacara*, "forest-dweller",[6] which confirm the tendency in

5 *Aṭavīvyavahata*, describing a fort: KA 7.6.8, 12.5.11.

6 Olivelle 2013: 496, n. on KA 2.2.4.

a different word of similar meaning; in one of them *vanacara* comprehends two distinct kinds of forest dwellers, renouncers and forest people. Leaving aside other kinds of people who dwell in the forest, the identification of tribal people with the forest is insisted upon as a distinguishing character, and we want to understand what that means. To that end I will use the expressions "forest people" or "foresters" instead of "tribal people" in this essay.

I believe this naming by means of an environmental zone has to do with a fundamental contrast between landscapes of the agrarian village and the forest, to which Francis Zimmerman draws our attention in his tracing of an ecological theme in the medical literature of ancient India, called Āyurveda. Thus in the classification of animals we find two great categories, those of the village (*grāmya*) opposed to those of the forest (*āraṇya*). Animals of the village are man, horse, ox, sheep, goat, pig; those of the forest are lion, tiger, boar, buffalo, elephant, bear, monkey. In the medical texts they fall in the contrasting landscapes called *jāṅgala*, originally meaning the dry, grassy landscape with thorny plants, as opposed to the *anūpa* landscape that is humid and afforested. It is the first of these that is the ideal landscape for humans and the agrarian villages that sustain them and their farm animals, implying that the wetter landscape and its wild animals is not suitable for settlement by agricultural people and their flocks and herds.[7] And yet kings and kingdoms have need of forests, for forest products and elephants at a minimum.

In this perspective the settled agrarian village is normative and supplies the point of view from which people of the forest are seen occupying a landscape that is the opposite of the ideal for humans. Thus the otherness of forest people is given in their very name. But otherness is not the last word, for difference among the products of landscapes tends also in the opposite direction, toward relations of exchange. In respect of the kingship, this would be the need of the king for forest products and wild elephants and therefore, to some extent, also for the practical knowledge of forest people. It is because of the complexity exchange introduces that the relation between kings and forest people becomes fruitful, unstable, unpredictable, and interesting.

2 Forest People and Forest Products

We will now examine mentions of forest people in Book 2, called *Adhyakṣapracāra*, or Duties of Overseers, which concerns the production of goods and the administration of the kingdom.

7 Zimmerman 1982:214 *et passim.*

Book 2 opens with "the settlement of the countryside" in which the king is to establish agrarian villages by settling farmers, supplying seed and oxen, and abating taxation for a number of years, in this way promoting an agricultural surplus to support the cities and expand the tax base of the kingdom in what is its largest productive sector. The firstness of the farming village is strongly marked by putting it in the very first chapter (*KA* 2.1); and in the next chapter it is consolidated by provision for the disposal of *non-agricultural* land, lying *between* villages (2.2), drawing attention to the secondness to agriculture of all other forms of production. These secondary units to which the land between villages is dedicated include pasture for farm animals; forests for religious renouncers, pleasure forests, produce forests, elephant forests; mines; and the roads for trade which connect the village with the city.

Some few references to forest people in Book 2 chime in with themes we will meet in later sections of this paper: efforts to aid a region of the kingdom devastated by an enemy king or forest people; the requirement that the Overseer of Infantry know the qualities of the various types of troops including forest troops; tracking and reporting, by homing pigeons or smoke signals, the movements of enemies and forest people, on the part of the Overseer of Passes.[8] These three passages are possibly late additions, as we shall show in the concluding section.

In the main, forest people appear in Book 2 in connection with produce forests and elephant forests.

As to the first we have the following suggestive statement:

> He should also establish a forest for each product classified as forest produce, as well as factories attached to the produce forests and forest people living in the produce forests.[9]

This presumes that forest people are already living in the forests, when the king designates part of it as a produce forest, and that they are brought into the royal service by the proffer of a wage. The forest products are specified further along: timber, reeds, vines, fibrous plants, material for rope, leaves, flowers, medicinal products (bulbs, roots, fruits), poisons (plants and insects), animal skins and body parts, common metals other than silver and gold, vessels of clay and fiber (pots and baskets), charcoal, husks and ashes; enclosures for animals; enclosures for firewood and grass.[10]

---

8 *KA* 2.1.36, a chapter-ending verse; 2.33.8; 2.34.11.

9 *KA* 2.2.5.

10 *KA* 2.27.4–16.

The same presumption of resident forest people appears to be in play with the elephant forests in the next *sūtra*:[11]

> At the frontier, he should establish an elephant forest guarded by forest people.

A few lines later, forest people (under the locution *vanacara*) are among those who keep track of the wild elephants of the elephant forest, for the making of a written record, with oversight of the elephant-forest wardens.[12]

These are the few explicit references to forest people in Book 2. Yet if we follow the elephants we may expand upon the significance of forest people in the kingdom, which are implicit in other passages of the same book.

We could wish the *sūtras* on this matter were not so cryptic. The *Arthaśāstra* gives us lists of other servants in the elephant establishment, and the superior ranks of royal servants as a class, as well as their pay, but they do not specify which of them, if any, are forest people. We have reason to believe, from ancient sources and from present-day ones, that that elephant drivers (mahouts) are often forest people, and that they formed lineages by which elephant handling skills were transmitted to sons and nephews, lineages of forest people become royal servants. One would like to know whether forest people served in the more exalted position, at a higher level of pay, of elephant trainers. Possibly so. We can surmise that the diagnosis and treatment of common ills was a part of their working knowledge, but the elephant physician had book-learning and probably high-caste status; it is unlikely he was from the forest people, though probably some of his treatments were.[13]

One additional point on elephants: the doctrine of the eight elephant forests appears in a pair of verses at the end of the second chapter of Book 2, on non-agricultural land, which not only indicate the location of the forests but grade them for the quality of their elephants.[14] As chapter-ending verses we infer that they are later additions (explained below, section 5 on book history); and indeed they tell a somewhat different story from the discussion of elephant forests which precede them. They imply not that every kingdom will have an elephant forest for its own supply, but rather that there are superior and inferior sources of supply in different regions of India, and an international trade through which they may be acquired. Another passage says that

11 *KA* 2.5.6.

12 *KA* 2.2.10.

13 See discussion at Trautmann 2015:164–169: The elephant staff.

14 *KA* 2.2.15–16, discussion and map at Trautmann 2015:12–14.

the southern route gives access to elephants, unlike the northern route, as one of its special advantages.[15] In all these matters forest people are likely to have been essential components as, to a degree, they continue to be.

As such, forest people played a critical role in the production of war elephants. We catch the measure of this in the political life of Indian civilization, when the *Arthaśāstra* states that "the extermination of an enemy's army *depends principally* on elephants."[16] Megasthenes, from Mauryan times, confirms the sentiment, and echoes of it in days of the Sultans of Delhi and the Mughals, from the twelfth and the sixteenth centuries. It was a maxim of Indian military wisdom for more than two millennia.[17]

Forest people played crucial roles in the production of war elephants, for the very good reason that the status of the wild elephant as the largest land animal of the current era, combined with the sluggish rate of reproduction and the late age at which they became sufficiently mature for war (twenty years, according to the *Arthaśāstra*[18]), made it prohibitively expensive for humans to breed them and feed them from birth. For these reasons elephants were the most marginal of domestic animals, being, in effect, domesticated from the wild one by one through the three millennia of the Indian war elephant.

Forest people played another crucial role which we may infer from the *Arthaśāstra* when it says that they are to be guardians of the elephant forest. For the *Arthaśāstra* also says that the king is to protect wild elephants by punishing, by death, those who kill them; and to reward with money those who bring in the ivory of elephants dying naturally.[19] How successful was this royal protection of the wild elephant we may judge by the fact that the countries of South and Southeast Asia in which wild Asian elephants still may be found, all of them had until recent times the institution of the war elephant, which was invented in India in the late Vedic period, very likely through a collaboration of forest people with Indian kingdoms.[20]

In these few lines of the *Arthaśāstra*, I suggest, we are given an intimate picture of forest people. Here at the very beginning of Book 2 the best land for cultivation is assigned to farming villages which the king creates, and in the land in between the villages there are forests in which forest people already

15 *KA* 7.12.22–24.

16 *KA* 7.11.16, emphasis added.

17 Trautmann 2021 (Megasthenes); Digby 1971 (Delhi Sultanate); Abul Fazl's *Ain-i Akbari* cited in Trautmann 2015:173 (Mughals).

18 *KA* 2.31.9.

19 *KA* 2.2.8–9.

20 Trautmann 2015:2, populations of Asian elephants by nation (Sukumar).

dwell, who are brought into the production of forest products and the guardianship of wild elephants as a military store from which suitable elephants may be captured and trained. In both cases the skills of forest people are essential to the kingdom and they work for the king at a wage. These passages paint a picture of forest people which is not at all menacing, and positively essential to the forest sector of royal production.

3 Forest Troops in the Military

Forest people appear frequently in the second half of the *Arthaśāstra*, among military forces (*bala*) making up the king's army and that of the enemy, or self-contained formations plundering the remote provinces of the kingdom from afar.

Ancient armies, and large armies of any period, were composites made up of levies from different sources. Knowing the strengths and weaknesses of each, and forming up the battlefield array and designating units to be held in reserve based upon comparative judgements about reliability, are essential skills of generalship. Accordingly, the *Arthaśāstra* devotes a good deal of attention to analyzing the different kinds of troops. There are six kinds of troops: hereditary, hired (i.e. mercenaries), corporate (professional fighters formed into a corporate group such as a guild), troops supplied by the ally, troops supplied by the enemy (as a result of a treaty), and forest troops. These are the terms of comparison in the various hypothetical scenarios of diplomacy and war which the *Arthaśāstra* poses and then resolves. The scenarios are often intricate and highly particular, all of them abstract and not drawn from historical instances. For our purposes what matters is the relative valuation of forest troops.

The long and the short of it is, that the hereditary troops are the most reliable, and forest troops, at the far end of the list of six, are the least. The six kinds of troops are listed in order of best to worst, with the hereditary troops (*maula*, from *mūla*, "root") at the beginning and forest troops (*aṭavī*) at the end. One chapter, taking up the topic of "proper times for deploying troops" of these six kinds, makes this abundantly clear, and states that in the list of six each previous kind is superior. The interests of hereditary troops align with those of the king, while forest troops are the opposite case, and for them prizes of war constitute the leading motivation.[21]

---

21 *KA* 9.2.1–12 (times); 9.2.13 (each previous kind superior); 9.2.14, 19 (interests).

In reliability forest troops are often paired with troops supplied by the enemy, the second-to-last in the list of six kinds of troops. They are sometimes in series with treasonable troops (*duṣya*), which speaks for itself.

In legal matters (Book 3) the holder of a deposit is not liable for its loss if a fort or a frontier province is plundered by an enemy king or a forest chief (*paracakrāṭavika*); and things robbed by an enemy king or forest chief (*paracakrāṭavī*) should be recovered by the king and restored to their owners.[22] What is clearly conveyed is the propensity of this pair to plunder the remote regions of the kingdom. Enemy troops are slightly superior: "Troops supplied by an enemy, commanded by Āryas, are better than forest troops" (9.2.18), but both pose the danger of a snake.[23]

If reliability were the only measure, forest troops would hold no appeal at all for the king, and forest people would not appear in the *Arthaśāstra* in a military capacity. But it is not the only measure. Among the others are special functions that on certain occasions bring forest troops to center stage. The obvious one is when the army passes through a forest and needs the special familiarity of forest troops to find the best way. A splendid illustration of this is found in a relief at Angkor, in Cambodia, when troops of Siamese forest people are shown in the van, scouting the way through the forest for the army of king Sūryavarman II.[24] Another function of forest troops is to counter the forest troops of the enemy army. A passage on the proper times for deploying forest troops gives a succinct overview that covers the ground especially well: "When they are needed to guide the way; when they are suitable for the enemy terrain; when they are best at countering the enemy's mode of fighting; or when the enemy has mostly forest troops (let a *bilva* fruit be smashed by a *bilva* fruit); or when a small incursion has to be defeated."[25]

Others concern the less-than-ideal kingdom facing a situation in which its shortcomings are put to the test and a way around them is urgently needed. The inventiveness with which scenarios are imagined and prescribed for in the *Arthaśāstra* in its finished form is limitless and acute. We cannot hope to encompass all of them, and it would not advance the analysis to do so. We need only say that forest troops appear, at the extreme, as the most expendible of troops, used to hold premier hereditary troops in reserve, or as the resort of the desperate, deprived of all better kinds of troops.

22 *KA* 3.12.2; 3.16.25. In these passages taken together *aṭavī* and *āṭavika* are interchangeable.

23 *KA* 9.2.18,19.

24 Poncar and Maxwell 2006: 157.

25 *KA* 9.2.8.

4 The Political Universe and the Place of Forest People in It

As I read it the political universe of the *Arthaśāstra* has three parts. First and best is king (*rājan*) and kingship (*rājya*), to whose interests and point of view the text is dedicated.

The second part is occupied by the republics (*saṅgha*), with a collective sovereignty. The *Arthaśāstra* devotes the whole of a single book, Book 11, to the *saṅghas* and says many interesting things about them. We could wish it had said much more, though we do have other sources on the *saṅghas*, including Buddhist and Jaina canonical works, the surviving histories of Alexander in Greek and Latin, and the fragments of the *Indica* of Megasthenes, ambassador to Candragupta Maurya. In the *Arthaśāstra* the republics are of two kinds, which answer reasonably well to the democracies and aristocracies or oligarchies of Greek political thought. With kingdoms and the two kinds of saṅghas we have the rule of the one, the few and the many.[26]

The names of historical republics are given as instances of the two kinds. Those among whom farmer-warriors constitute the decision-making body include the Kambojas, the Surāṣṭras, the Kṣatriyas and the Śreṇis; those who "make their living by the title of raja" include the Licchivikas, the Vṛjjikas, the Mallakas, the Kukuras, the Kurus and the Pāñcālas—the first a kind of democracy, the second an aristocracy or oligarchy of a warrior class, presumably landowner agriculturalists.[27] The aristocracy having multiple *rājas*, kingship is distinguished as the rule of the single king (*eka-rāja*).

The *Arthaśāstra* treats the *saṅghas* as the best of allies and the most formidable of foes because of their solidarity and proceeds to give the king advice about how to undermine that solidarity by sowing dissention among them.[28] But the *saṅghas* do not appear outside of Book 11, and they are notably absent from the chapters on diplomacy and warfare in the second half of the treatise, where we expect to find them.

The third part of the political universe, the forest people, are represented differently from the republics. They are ubiquitous in the text, passages in which they figure appearing in each of Books 1–13. They are not named, being treated as a generic category, and not as particular historical communities with attributes of their own. This differs from their treatment in the *Mahābhārata* in which we find, for example, a list of people renowned as fighters from atop elephants which includes peoples of kingdoms and forest people: Prācyas

26 Trautmann, forthcoming: "Democracy in ancient India".

27 *KA* 11.1.4–5.

28 *KA* 11.1.1–3.

(Easterners), Dākṣiṇātyas (Southerners), Aṅgas, Vaṅgas, Puṇḍras, Māgadhas, Tāmraliptakas, Mekalas, Kośalas, Madras, Daśārṇas, Niṣadas and Kaliṅgas.[29] Furthermore, in the *Arthaśāstra* forest people occupy a specific place among the kinds of military forces as previously mentioned, in the series, hereditary troops (*maula*), hired troops (*bhṛtaka*), corporate troops (*śreṇī*), troops supplied by the ally (*mitra*), troops supplied the enemy (*amitra*) and forest troops (*aṭavī*). They are one among the kinds of units into which military forces are analyzed—a list in which republics do not appear.

What is the nature of the polities of forest people? The issue is neither directly asked nor answered in the text. But we can infer an answer of sorts from one of the set-piece debates which are a notable feature of the text, one in which "the teachers" and Kauṭilya argue which is worse, highwaymen or forest people.[30]

The teachers say,

> Between highwaymen and forest people, highwaymen lurk under the cover of night, cause physical harm, are ever present, rob hundreds of thousands, and incite chiefs to revolt, whereas, remaining far away and operating in frontier forests, forest people go about openly and in plain sight, and they plunder a single region.

Here it is a question of criminals versus non-criminals, according to a trope that criminals act under cover of night, while the violence of warriors is ennobled by taking place in the light of day.

But Kauṭilya says,

> Highwaymen rob those who are heedless, are few in number and sluggish, and are easy to recognize and apprehend, whereas, living in their own region and being numerous and brave, forest tribes fight in the open, plunder and destroy regions, and behave like kings (*rājasadharmāṇa*).

This comes to a grudging semi-recognition that forest peoples are organized into polities that have the attributes of kingdoms and republics: a region of their own, numbers, military force, king-likeness. In short, forest people form something approximating a state, having greater scale, their own territory and other characteristics of proper polities, while highwaymen, lacking these qualities, do not constitute states.

29 *Mahabharata* 8.171.1–4.

30 *KA* 8.4.41–43.

On the other hand, there is a series of passages that focus upon the chief of the forest people, which in the reckoning of his political significance is not half-hearted, it is sky-high. The first of these passages is astonishing.

The extreme of such a scenario is presented as a fiction of regime change formulated by a secret agent to test the loyalty of a minister. The agent argues that the king is *adharmika* and needs to be replaced by a dharmic person: "a pretender from his family, a prince in disfavor, a member of the royal household, the man who is the sole support of the kingdom, a neighboring lord, a forest chieftain or a man who has risen to power. Everyone likes this idea. What about you?".[31] It is surprising to find the forest chieftain in that list of thinkable alternative rulers, and especially in the context of a king characterized as lacking in righteousness. We are forced to conclude that a forest chieftain could become a king, not of his own people, as an *āṭavika-rāja*, but in succession to a king ruling agrarian villages and cities.

In subsequent passages the set of persons among whom the forest chieftain is situated in this passage is boiled down to a compound of four, joining the neighbor king (*sāmanta*), the forest chief (*āṭavika*), pretenders from the king's family (*tatkulīna*) and princes in disfavor (*aparuddha*). The compound recurs frequently in the *Arthaśāstra*. This boiling-down into a compound used repeatedly suggests that all the later occurrences are simplifications of the test-of-loyalty passage.

Let us examine the contexts of the other mentions of the forest chieftain, whether in this compound of four or otherwise. The passages appear in books 1, 7, 9 and 12, eleven cases in all.

The first of them, from Book 1, concerns keeping a watch, with spies, over the wealthy men of the countryside who live on its grain, animals and money, and help the king with these in times of prosperity and adversity, suppressing a rebellious relative or province, or repel an enemy or forest chief: if they are disaffected he should employ division (*bheda*), dividing them against one another, and from the neighbor, forest chief, pretender from his family, and princes in disfavor (*KA* 1.13.15–18). I should add that the tribal chief himself is an object of spying from the outset, reinforcing our sense of his high significance in problems of the succession to the kingship.

A second passage from Book 1 concerns advice to the prince in disfavor. A prince who is in disfavor with his father is counselled to take refuge in the forest or with a neighboring king who is righteous (*dhārmika*) and truthful (*satyavāk*), gather a treasury and troops, making a marriage alliance with eminent warri-

31 *KA* 1.10.3.

ors, and alliances (evidently without marriage) with forest chiefs, and win over seducible factions (1.18.5–7). Here the forest chief appears in an appealing light, being the near equivalent of a neighboring king who has attributes of the ideal Hindu king, for a prince in need.

Passages from Books 7, 9 and 12 concern ways of dealing with the Gang of Four (as we may call the fourfold compound), or instigating it to act against one's enemies in the Circle of Kings. Forest chiefs figure prominently in inter-state war and politics, similar to the royal participants.

Thus forest chiefs, doubtless because they lead formidable military forces, are players in the political dynamics within the royal family and are sources of disturbance to the royal succession. It is an effect, I suggest, of the internal stresses upon the royal family that is produced by unitary, hereditary kingship passing to only one son, in contrast, for example, to the inheritance of property, which is divided among all the sons. The *Arthaśāstra* makes this explicit and historical by naming kings who were killed in their own bedchamber by close relatives: Bhadrasena by a brother, Kārūṣa by a son, and the kings of Kāśi, Vairantya, Sauvīra, Jālūtha and Vidūratha by their queens, using poison, administered in food, or smeared on an anklet, a girdle-jewel or a mirror, or a weapon concealed in her braids.[32]

It shows, in connection with that internal dynamic, that the forest chief is a continuing source of disruption to the royal succession along with the neighbor, the royal pretender and the prince in disfavor.

In sum, forest people appear frequently in the *Arthaśāstra*, in three distinct ways. As individuals, they appear as servants of the king at a wage, in connection with the production of forest products from the produce forest, and the monitoring and protection of wild elephants in the elephant forest. As military troops they are allies and components of the army, at the lowest rung in the scale of reliability among the component forces that make up the army, but valued for special functions in which they are skilled such as scouting the way through forests, and for countering the forest troops of the enemy. Distant forest troops may be enemies that despoil the remoter provinces of the kingdom. Finally, forest chiefs are players of great importance in the royal succession, such as a refuge for a prince in disfavor with his father the king. Should we understand these three strikingly different valuations simply as effects of their different functions (economic, military, political)? Or should we look to the book history of the *Arthaśāstra* for an explanation?

32 *KA* 1.20.15–16.

5 Forest People in the History of the Text

Some time ago I wrote a doctoral dissertation, and out of it a book, in which I gave evidence that the structure of the *Arthaśāstra* showed evidence of having been divided into chapters (*adhyāya*) with concluding verses some time after its original composition, whose *sūtras* were grouped into topics (*prakaraṇa*); and that statistical analysis of the longer books (2, 3, 7) suggested multiple authors.[33]

More recently Mark McClish wrote a doctoral dissertation, and out of it a book, which consolidated, revised, improved upon and expanded these findings to reach a comprehensive book history for the *Arthaśāstra*. His treatment of these issues have been incorporated, with modifications, into the translation of the *Arthaśāstra* by Patrick Olivelle.[34]

An explicit case of recension of a received text is the *Caraka Saṃhitā*, cited by McClish, a recension, by Caraka, of an earlier text by Agniveśa called *Agniveśatantra*, and subsequently Caraka's text was itself the object of recension by Dṛḍhabala, who added seventeen chapters.[35]

In the excellent analysis of McClish we have a book history of the *Arthaśāstra* that provides a solid foundation upon which we may hope to determine whether the picture of forest people we have drawn from the text is unitary, and if not what its own history may be.

Clearly not all of the citations concerning forest people belong to the original text of the *Arthaśāstra*. Citations come from each of books 1–13, while books 5, 8, 11 and 12 are judged to belong to the recension, that is, are later additions to the text. Two citations come from chapter-concluding verses, belonging therefore to the recension. This is sufficient to show that some of the material on forest people clearly is later.

Other of the citations are probably early, belonging to the original, pre-recension text. This early material will especially include what occurs in Book 2, concerning produce forests and elephant forests.

Turning to another aspect of the book history, the title of second book is *Adhyakṣapracāra*, "The Duties of Overseers". It is cited by this name in the *Kāmasūtra* of Vātsyāyana.[36] This may indicate that Book 2 of the *Arthaśāstra* was once a self-contained work; and, indeed, there is reason to think that such a

33 Trautmann 1968 and 1971

34 McClish 2009 and 2019; Olivelle 2013:6–31 ("Composition").

35 McClish 2019:22–23, citing the work of Philip A. Maas.

36 *Kāmasūtra* 1.2.10, discussed in Trautmann 1971:173.

self-contained treatise was the core around which the *Arthaśāstra* was formed. This is consistent with what the *Arthaśāstra* says about its own composition:

> This singular Arthaśāstra has been composed for the most part by drawing together the Arthaśāstras composed by former teachers for gaining and administering the earth. (1.1.1)

Now, that does not mean that the entirety of Book 2 in the *Arthaśāstra* as we have it is from the original work, without additions; some content was added at the time of the recension, at a minimum all or at least most of the chapter-concluding verses for certain, and less certainly, material which expands the original text. Three examples of the latter have previously been mentioned: one indicating royal relief of a region devastated by an enemy king or forest people (*paracakrāṭavī*), a verse and a compound used several times elsewhere; a second saying the Overseer of Infantry should have knowledge of types of troops, including forest troops, a stock list widely used in other parts of the text, part of an entire topic which has evidence of lateness;[37] and the third concerning tracking and reporting movements of enemies and forest people, toward the end of the book, a position in which expansion material is readily made, and its pairing of enemies and forest people, which is found frequently in other parts of the *Arthaśāstra*.

By contrast, the discussion of produce forests and elephant forests are situated among the first two chapters of Book 2, concerning the settlement of the countryside with agrarian villages and the allocation of the non-agricultural land between agrarian villages to other functions, the two of which form a natural pair; laying out a vision for the division of the landscape into different zones, by which the spheres of overseers are denominated. They must surely have been original to the hypothesized self-contained prior treatise called *Adhyakṣapracāra*.

That original vision takes in forest people, then, and implies they are already living in the forests, among them those which the king dedicates to the various kinds of forest produce, and also the one designated as the elephant forest. They would have been coopted, drawing them into the king's establishment for forest products and the tracking and guarding of wild elephants. They figure in this as individuals and not as quasi-states.

Viewing the references to the military forces and the chiefs of forest people, in what we identify as later additions to the *Arthaśāstra*, they have a rather different framing.

---

37 Trautmann 2021:347 fn. 26, citing McClish, personal communication.

It is one in which forest people are suppliers of one kind of troops among six kinds, of which forest troops are of the most unreliable, often paired with the second-to-last enemy troops; as despoilers, with enemy troops, from across the border of the outer regions of the kingdom; and so forth—but indispensable for their knowledge of forests and the forest troops of the opposing army.

It seems to me that the recurring compound discussed above, which joins the neighbor king, forest chief, pretenders from the king's family and princes in disfavor as causing or assisting disturbers of the succession, likely to be a later addition, expresses a later, stronger vision of the place of forest people, one in which they play, with the others, a prominent role as a resource for needy kings or princes in disfavor, or as a danger to the succession of the settled agrarian kingdom.

In both contexts they occupy a place in a world of kingdoms and republics, as quasi-states, acting like kings, but by implication illegitimately so in the debate about highwaymen versus forest people. But this does not exactly agree with the very high estimation of the role of the forest chief as a potential substitute for an *adharmika* king, on the one hand, and, at the other extreme, the description of the ideal kingdom as being forest-people-free.

6 Discussion

This has been an attempt to locate the place, or, as the evidence forces us to say, places, of forest people in the *Arthaśāstra* and the world it describes. It has quite different parts to it, and is somewhat incoherent, to say the least. Some of the incoherences of the text as we have it have been caused by the recension of the original text, and have served to lead us to its book history. Some of them are incoherences of substance as distinct from incoherences of structure, when new matter on forest people have been added to what the original text contained.

The later, editing and rewriting hand gives a more theoretical and wide-angle picture having, as McClish and Olivelle say, something of a Dharmaśāstra tendency that gives it a Brahmanical and religious aspect not found in the original. This contrasts with the more practical picture we get generally in Book 2, where we often find the content suggests the composer has spoken with practitioners to get the details, and takes the world as he finds it, without passing judgement. The most striking passage among those that have been the object of scrutiny here is undoubtedly the test of a minister, in which a forest chieftain is among the personages the passage assumes to be plausible replacements for a king said to be *adharmika*. Subsequent passages which seem to be modelled on this one continue to represent the forest chieftain as a king-like outsider who is a player in the succession to the throne of Hindu kingdoms. The place of this theme in the book history of the *Arthaśāstra* could perhaps be explained as an outcome of the effects of Hindu kingship upon the political life of forest people.

We cannot hope to find texts of ancient India in which forest people speak for themselves, and from their own point of view; our primary sources are, all of them, from an outsider's perspective, the perspective of *paṇḍita* culture, or more precisely as I am inclined to suppose, the perspective of the *laukika* brahmins and their lineages who serve the secular functions of kingship. The best we can do is to give close readings of the sources, tracking differing perspectives from which their pictures are taken. But even that little from this one text shows how very large the forest people loom in the life of Indian civilization: first, in the production of war elephants, and the royal protection of wild elephants, upon which Indian military men placed their hopes of victory, and with effects that continue into this post-war-elephant age, in the comparatively large numbers of wild elephants in India today, protected by Indian kings as a military store. Second, we see as well that military bands of forest people were a constant of Indian states both on offence and defense. We see, finally, that the forest chief was a major player in the political life of the state and the succession to kingdoms. I permit myself to think that the political development of forest people was not fixed at some past stage (that of "pre-state societies" let us say) but was greatly changed through the closeness of its interactions with Hindu kingdoms and its leaders coming to model themselves upon rajas.

The very incoherence of the multiple places that forest people occupy in the *Arthaśāstra*, post recension, is itself is a valuable datum, indicating that the places forest people occupy vis-à-vis settled agrarian kingdoms are multiple and impossible to sum into a self-consistent picture, singular and unchanging. As such it invites us to the daunting task of explaining as best we can from materials such as these what have been the roles of forest peoples in the deep history of Indian civilization. The repeated mentions of forest people in the *Arthaśāstra* assure us that their roles have been many, and critically important.[38]

---

38 This offering to Christopher Minkowski, whose wide sympathies extend from *paṇḍita* culture to elephants and my book about them, is my poor thanks for his support of the book and for imposing it upon his students.

Bibliography

Digby, Simon. *War-horse and elephant in the Dehli Sultanate: a study of military supplies.* Oxford: Orient Monographs, 1971.

Kangle, R.P., ed. *The Kauṭilīya Arthaśāstra.* Bombay: University of Bombay, 1969.

McClish, Mark. "Political Brahmanism and the state." PhD thesis, University of Texas at Austin., 2009.

Olivelle, Patrick. *King, Governance, and law in ancient India: Kauṭilya's Arthaśāstra: a new annotated translation.* Oxford: Oxford University Press, 2013. https://doi.org/10.1093/acprof:osobl/9780199891825.001.0001.

Parasher-Sen, Aloka. "Of tribes, hunters and barbarians: forest dwellers in the Mauryan period." In *India's environmental history. From ancient times to the colonial period: a reader*, edited by Mahesh Rangarajan and K. Sivaramakrishnan, 1:127–151. Ranikhet: Permanent Black, 2012.

Poncar, Jaroslav, and Thomas Maxwell. *Of gods, kings, and men: the reliefs of Angkor Wat.* Chiang Mai: Silkworm Books, 2019.

Rangarajan, Mahesh, and K. Sivaramakrishnan, eds. *India's Environmental History. From Ancient Times to the Colonial Period: A Reader.* Vol. 1. Ranikhet: Permanent Black, 2012.

Shastri, Sāhityayādarshanāchārya Tarkaratna Nyāyaratna Srī Gosvāmī Dāmodar, ed. *The Kāmasūtra by Srī Vātsyāyana Muni. Benares: Kāshi Sanskrit Series.* Kāshi Sanskrit Series 29. Jai Krishnadas-Haridas Gupta, 1929.

Singh, Upinder. *Political Violence in Ancient India.* Cambridge; London: Harvard University Press, 2017. https://doi.org/10.4159/9780674981270.

Sukthankar, V.S., ed. *Mahābhārata: critical edition.* Poona: Bhandarkar Oriental Research Institute, 1933–1971.

Thapar, Romila. "Perceiving the Forest: Early India." In *India's environmental history. From ancient times to the colonial period: a reader*, edited by Mahesh Rangarajan and K. Sivaramakrishnan, 1:105–126. Ranikhet: Permanent Black, 2012.

Thapar, Romila. "The Image of the Barbarian in Early India." *Comparative Studies in Society and History* 13, no. 4 (1971): 408–436. https://doi.org/10.1017/s0010417500006393.

Trautmann, Thomas R. "Democracy in ancient India." In *The Cambridge History of Democracy*, edited by Markku Peltonen. Cambridge: Cambridge University Press, forthcoming.

Trautmann, Thomas R. "Megasthenes on the Military Livestock of Chandragupta and the Making of the First Indian Empire." *Comparative Studies in Society and History* 63, no. 2 (2021): 339–365. https://doi.org/10.1017/s0010417521000074.

Trautmann, Thomas R. *Elephants and Kings: An Environmental History.* Chicago: University of Chicago Press, 2015. https://doi.org/10.7208/chicago/9780226264530.001.0001.

Trautmann, Thomas R. "Kauṭilya and the Arthaśāstra." PhD thesis, University of London, 1968.

Trautmann, Thomas R. *Kauṭilya and the Arthaśāstra. A Statistical Investigation of the Authorship and Evolution of the Text.* Leiden: E.J. Brill, 1971.

Zimmermann, Francis. *The Jungle and the Aroma of Meats: An Ecological Theme in Hindu Medicine.* Berkeley, Los Angeles, London: University of California Press, 1982.

PART 4

Sanskrit Intellectuals and Their Communities

CHAPTER 14

Guardians of Truth: Commentarial Interventions in Support of the Heroes of the Sanskrit Epics

Robert P. Goldman

It is both a pleasure and an honor to have been invited to offer a contribution to this Festschrift in celebration of the scholarship of my friend and colleague Professor Christopher Minkowski. I have known Chris since he was an informal student of Sanskrit here with us at Berkeley back in the early eighties, as I recall, before he returned to his alma mater for his doctoral work. Thus, on those occasions in the intervening years when we have had the good fortune to have him speak at Berkeley, we have always thought of it as a homecoming of sorts. Since those early days, we have watched his extraordinary scholarly career with pleasure and admiration as he is not only an outstanding authority on Sanskrit literature in several genres but also, as he and I are wont to joke, the veritable Dean of Nīlakaṇṭhian Studies. In fond appreciation of that connection, I would like to offer as a humble *puraskāra* to this Abhinavanīlakaṇṭha some reflections on the work of several authors of the Sanskrit commentaries on the great epic poems, the *Mahābhārata* and the *Rāmāyaṇa*, focusing most particularly on some passages from Nīlakaṇṭha Caturdhara's influential commentary on the former, the *Bhāratabhāvadīpa*.

Scholarly appreciation of the epic commentaries, their purpose, and their impact on our understanding of the epic texts themselves has been steadily increasing in the past decades. When I was a graduate student and showed some interest in looking into the commentaries on the texts we were reading in seminars, I was frequently discouraged from reading them. Instead, I was advised that it was up to us, as western philologists, to reconstruct the texts of these culturally central works and apply our own theories and methodologies alone to their interpretation and their genetic histories.

This approach seemed to me at the time to be wrongheaded in its dismissiveness of the wealth of knowledge and scholarship about the poems and their vast intertextuality with virtually the entire spectrum of vedic, literary, and śāstraic texts and also inadequate in cutting us off from the closest thing we could ever find to an intended learned audience of these great works. So, when I began reading the *Rāmāyaṇa* in earnest as I contemplated undertaking the first translation of the Baroda critical edition of the poem, I examined existing

 | DOI:10.1163/9789004536869_016

translations of both epics most of which had been obviously produced with minimal, if any, reference to their rich body of commentary.[1] This convinced me that the readers of our work should have the advantage of a dense annotation that would draw heavily on the readings, interpretations, debates, and polemics in this corpus of secondary scholarship. As a result, each volume of our translation contains notes that provide extensive excerpts from as many as a dozen commentators. As a result of this and much other material, the annotation turned out to be at least as long as the actual translation and, in some volumes, much longer. Along with the then ongoing volume-by-volume publication of the translation, I also published a piece or two on the subject.[2]

Happily, though, my collaborators and I were soon not alone in bringing to light the vital contributions of the medieval and early modern Indian scholars of the epic. There is, of course and notably, Chris's work on Nīlakaṇṭha[3] Ajay Rao's excellent studies of the Śrīvaiṣṇava commentaries on the *Rāmāyaṇa*,[4] and Vishal Sharma's work on purāṇic retellings and Mādhva commentaries on the epic and purāṇic narratives.[5] Also noteworthy in this connection is work by younger scholars, such as Sohini Pillai, on regional language versions of the *Mahābhārata*.[6] Such work has added immensely to our understanding not only of the epics themselves and their textual histories, but, importantly, of the commentators themselves as they read, interpreted, and re-interpreted the epics in keeping with the emerging schools of literary and—most critically—religious thought and practice. I myself have become increasingly interested in how the commentators, as not only scholars but practitioners of new forms of both Rāma and Kṛṣṇa *bhakti* sought to develop hermeneutic strategies to mediate between the root texts of Vālmīki and Vyāsa and the medieval and early modern sectarian versions of the epics—in both Sanskrit and regional languages—which came to replace the ancient texts in the hearts of many of their audiences.[7] With this as a background, I would like to discuss some of the strategies developed and adopted by many of the epic commentators in their efforts to read the texts in such a way as to conform to either their

1 I should note that several English translations produced in India do make occasional mentions of a commentator (rarely identified but usually Nāgeśa) or have simply silently incorporated the interpretations of a commentator into their translations.
2 Goldman 1992 and 1999.
3 Minkowski 2005, 2008, and 2010.
4 Rao 2008 and 2015.
5 Sharma 2020.
6 See, for example, Pillai 2021.
7 On this, see Goldman 2022, 883–904.

theological or social belief systems. A good example of the former type can be found throughout the Śrīvaiṣṇava commentaries on the *Vālmīki Rāmāyaṇa*.

A unique feature of this immensely influential commentarial corpus is its authors' constant concern with Vālmīki's representation of his poem's *avatāra*-hero, Rāma, as vulnerable to all manner of mental, emotional, and physical suffering and, importantly, as unaware of his divine nature. In this respect the *Rāmāyaṇa* differs sharply from most *Mahābhārata* and purāṇic characterizations of Kṛṣṇa, who is almost never shown as being discomfited, injured,[8] despondent, or ever other than fully unaware of his own divinity, which he famously reveals from time to time in the literature. More importantly, however, Vālmīki portrays Rāma as a divinity who, out of the necessity imposed by Brahmā's boon to Rāvaṇa that the latter could not be slain by any supernatural being, is compelled to not only act like a suffering mortal, but to understand himself to be one. In this, it is not only radically different from the representation of Kṛṣṇa, but—most critically—from the representation of Rāma himself at the hands of the medieval and early modern authors of more floridly devotional *Rāmāyaṇa*s whose characterizations of their hero (and heroine in some versions) have become essentially just like that of Kṛṣṇa. This is true of texts such as the *Ānanda* and *Adhyātma Rāmāyaṇa*s[9] and many regional language versions, such as Tulsīdās's important *Rāmcaritmānas*.[10]

As a result, Vālmīki's commentators found themselves in the awkward position between the Ādikavi's, "the First Poet's," representation of the hero's vulnerability and the textual milieu they themselves inhabited many centuries after the composition of the *Ādikāvya*, "First Poem." For, on the one hand, the work was invested with the reputation of its author as a divinely inspired and infallible seer, while, on the other, the later popular religious versions of the Rāma story have a radically different understanding of the nature of the Rāmāvatāra. Commentators rose to this challenge in a number of interesting ways. By exploiting *sandhi*, *vyākaraṇa*, and various *kośa*s, along with their encyclopedic knowledge of the *dharmaśāstra*s and *purāṇa*s, they either contradicted or re-interpreted the apparent meaning of the *ṛṣi*'s text in such a way as to show that the manifest meaning of passages that represent Rāma and Lakṣmaṇa as wounded, unconscious, grieving, or ignorant of their true, godly nature must be read to show the very opposite. Some of the scholiasts, not-

8 Other than, of course, at the final moment of his earthly manifestation. See *Mahābhārata* 16.5.19–25. Unless otherwise stated, all *Mahābhārata* references will be to the Pune critical edition (ed. Sukthankar et al.), hereafter abbreviated as Mbh.

9 See Goldman (forthcoming *JAOS*).

10 See Lutgendorf 1991.

ably Maheśvaratīrtha, developed a thoroughgoing system for reading the text on two distinct and mutually contradictory levels of meaning: the *spaṣṭārtha*, "the manifest level of meaning," and the *vāstavārtha*, "the real, inner meaning." In this way, for example, they re-interpret expressions of hostility towards the poem's hero or lust directed towards the poem's heroine as words of devotion. I and my collaborators have given copious examples of this sort of reading on the part of the epic commentators. I will therefore simply refer readers to the relevant passages in the annotation of our translation of Vālmīki's *Yuddhakāṇḍa*.[11] With that, let me turn from the theological imperatives of the commentators to their social and ethical concerns.

One of the social imperatives of deepest concern for the epic authors, their characters, their commentators, and, ultimately one must surmise, their audiences are those that involve social deference.[12] That is so say, deference of wives to husbands, of children to their parents and grandparents, of disciples to their gurus, of younger siblings to elder siblings, and, of course, of members of each *varṇa* to those belonging to the *varṇa*s ranked above them in the traditional *vaidika* system of *varṇadharma*. Collectively, virtually all beings, including kings and the very gods themselves, are obligated to subordinate themselves and defer to the brahman sages and seers who populate the epic world. This last point is endlessly stressed in the epics, and both texts are filled with cautionary tales of those who fail in this deference and must therefore suffer the terrible consequences. Leaving aside the ubiquitous episodes in which violators of this rule are cursed for violations of this imperative, the poems are filled with exemplary accounts of such deference, even being carried out to extremes, in familial contexts. Figures celebrated for their spectacular examples of willing subordination to their elders or husbands such as Gāndhāri, Yayāti, Bhīṣma, the four younger Pāṇḍavas, Rāma, and Lakṣmaṇa are so well-known to anyone even only generally familiar with the epics that I need not go into the matter in detail. But, aside from such paradigmatic examples, which are so commonplace as to pass largely without comment by the commentators, what I find interesting is how the scholiasts are willing to pursue this matter at length and with great ingenuity even in what might appear to be the most minor and insignificant instances in which the order of deference may appear to be violated.

---

11 Goldman and Sutherland Goldman 2009 (= Princeton Translation of the *Vālmīki Rāmāyaṇa*, hereafter abbreviated as PR). For examples in PR, see PR 6.23.1–3; 41; 6.53.2; *passim*. PR references correspond to the Baroda critical edition of the *Vālmīki Rāmāyaṇa* (eds. Bhatt and Shah), hereafter abbreviated as VR. See also Goldman, 2022, 883–914.

12 See, for example, VR 2.27.29 where Rāma tells Sītā that *dharma* itself consists of deference to one's mother and father (*eṣa dharmas tu suśroṇi pitur mātuś ca vaśyatā*).

A good example of the commentators' willingness to go to elaborate lengths to rationalize or negate the nature of a momentary interaction is to be found at 6.115.33 in the *Rāmāyaṇa*.[13] Rāma, returning in the borrowed Puṣpakavimāna to Ayodhyā at the end of his fourteen years of exile, stops at the ashram of the sage Bharadvāja and decides to wait there until Hanumān, whom he sends as a messenger, can report the news of Rāma's return to his brother Bhārata. For, in keeping with his celebrated deference to the concept of family harmony and his exemplary self-sacrifice, he wants to see how Bharata, who has been ruling as regent in his absence, will take the news of his return. If Bharata should show—by even so much as a facial gesture—any sign of displeasure at having to relinquish the rulership, then he, Rāma, would cede the throne to him in perpetuity.[14] But, in keeping with the norms of deference to one's elders, Bharata is delighted by the news and goes straightaway to the ashram with a huge retinue of wives, ministers, and soldiers to fittingly welcome his beloved elder brother home. After respectfully greeting his elder, Bharata is invited to enter the *vimāna* where he greets his own younger brother, Lakṣmaṇa, and his senior sister-in-law, Sītā. The text says simply:

> When he encountered Lakṣmaṇa and Vaidehī, Bharata, scorcher of his foes,
> delighted, reverentially saluted them and announced his name.
>
> *tato lakṣmaṇam āsādya vaidehīṃ ca parantapaḥ* |
> *abhyavādayata prīto bharato nāma cābravīt* ||[15]

Although it appears to be a simple and straightforward statement, the verse drives the commentators into virtual paroxysms of analysis and rationaliza-

13 All references to both epics are to their critical editions (VR and Mbh) unless otherwise specified. Where it is necessary for the reader to consult the original of the commentaries quoted, the first citation is to the critical edition, followed by an equal sign, which is then followed by the citation in each epic's respective Gujarati Printing Press edition, e.g., Mbh 4.3.6–11 = GPM 4.3.8–13.

14 VR 6.113.12–16 = GPR 6.127.41.

15 VR 6.115.33 =GPR 6.127.41 = Lakṣmīveṅkaṭeśvara Press edition (LPR) 6.130.42 = Kumbakonam edition (MVR) 6.130.39. Calling out one's name is one of the normative actions in a full *abhivādanam*, "respectful salutation," to one's elders, guru, etc. It is also possible that, although Bharata is their close kinsman, his brother and sister-in-law might not immediately have recognized him as the text describes him as a typical renunciant, gaunt, unkempt, filthy, and clad in bark cloth and animal skins from his fourteen years of severe asceticism. VR 6.113.26–30.

tion in an effort to find alternative ways to understand the verse. And this, of course, is because the verse seems to describe an older brother paying homage to his junior. This is a scenario that is almost as disturbing to the commentators' sense of propriety as that in which the aged king Daśaratha grovels in vain at the feet of his junior wife Kaikeyī.[16] As evidence of this, they go to great lengths in their efforts to demonstrate that this is not what is happening in the verse or, if it is, that there is some alternative way to rationalize it.

The following are some of the interpretive strategies that the commentators employ. Nāgeśa[17] and Maheśvaratīrtha[18] take the two verbal forms separately, glossing the gerund *āsādya* as "with an embrace, etc., granting esteem to Lakṣmaṇa, who was praising [him] …" (*vandamānaṃ lakṣmaṇaṃ pariṣvaṅgādinā saṃbhāvya*). In this reading, Vaidehī alone is the object of Bharata's obeisance, to which she, as the wife of his elder brother Rāma, is fully entitled. Nāgeśa and Maheśvaratīrtha justify this interpretation on the grounds of the evidence from the *Rāmāyaṇa* and the *Padmapurāṇa* indicating that Bharata is, in fact, older than Lakṣmaṇa. As an alternative, these two commentators propose that one can regard Lakṣmaṇa in some sense as "older" than Bharata and therefore deserving of his obeisance. Two arguments are put forward here. The first is:

> Others [argue] that, even though Lakṣmaṇa is junior to Bharata in chronological age, as is made clear in the section on their births, nevertheless, since he [Lakṣmaṇa] had gained greater virtue than himself [Bharata] because of [Lakṣmaṇa's] attendance on his elder [Rāma], [Bharata thinking,] "one must always practice deference to one's elders through one's own conduct," with the idea that [Lakṣmaṇa] is [Bharata's] elder/teacher (*guru*) because of [Lakṣmaṇa's] providing instruction to him, Bharata, [Lakṣmaṇa's deference to his elder Bharata] would be inappropriate.
>
> *pare tu yady api lakṣmaṇo bharatāt kaniṣṭho vayasā.* spaṣṭaṃ cedaṃ janmaprakaraṇe *tathāpi jyeṣṭhānuvartanena svāpekṣayādhikaguṇatvena svavyavahāreṇa jyeṣṭhānuvṛttir eva kartavyeti bharataṃ praty upadeśadānena ca gurutvabuddhyā tasya ca nirucitaiva.*

The second argument is that, since Lakṣmaṇa is produced from a portion of the divine *pāyasa* that had first been given to Kausalyā, he can be considered Bharata's actual elder (*kiṃca kausalyāyai prathamaṃ dattapāyasāṃśa-*

16 See VR 2.11.41.

17 Citations from Nāgeśa and Govindarāja are *ad* GPR 6.127.41.

18 Citations from Maheśvaratīrtha are *ad* LPR 6.130.42.

jatvāl lakṣmaṇasya jyeṣṭhatvam). Varadarāja Uḍāli notes simply that, although Bharata greets both Sītā and Lakṣmaṇa equally, his gesture of homage (*abhivādanam*) is reserved for Sītā alone. Otherwise, he argues, this passage would represent a contradiction to those in which it was earlier stated that Bharata is the elder of the two. (*āsādanaṃ lakṣmaṇavaidehyoḥ samānam. abhivādanaṃ tu vaidehyā eva veditavyam. anyathā pūrvāparoktaṃ bharatasya jyaiṣṭhyavacanaṃ viruddhaṃ syāt*).[19]

Govindarāja's understanding is similar to the first interpretation of Nageśa and Maheśvaratīrtha. He glosses *lakṣmaṇam āsādya* as "honoring with an embrace Lakṣmaṇa, who had made obeisance [to him]" (*lakṣmaṇam āsādya kṛtanamaskāraṃ lakṣmaṇam āliṅganena saṃbhāvya*). This, he notes, is because Lakṣmaṇa is his junior (*idaṃ ca lakṣmaṇasya kaniṣṭhatvāt*). He observes, however, that some authorities believe that, in the presence of Rāma, Bharata would have prohibited Lakṣmaṇa's expected gesture of obeisance (*lakṣmaṇena cikīrṣito namaskāro rāmasannidhau bharatena pratiṣedhita iti cāhuḥ*). Govindarāja makes an additional observation in this connection, noting that the conjunction *ca*, "and," after the name Vaidehī indicates that Bharata's homage to Rāma is included in that which he displays toward Sītā (*cakāro rāmanamaskāraṃ samuccinoti*). He then apparently feels the need to offer a strictly chronological order of the births of Bharata and Lakṣmaṇa, dismissing all discussion of the distribution of *pāyasa* to their mothers, the order of their marriages, and so on. Finally, he indicates that by the sequential nature of Bharata's greetings, first to Lakṣmaṇa and only then to Sītā (against the expected order of precedence), we are to understand that, although Sītā is in Rāma's presence, she is standing somewhat apart from him with Tārā and the other *vānara* women from Kiṣkindhā (*sītā tārādibhiḥ saha rāmasamīpa eva kiṃcid anyatra sthitety avagamyate*).

Śivasahāya offers an ingenious interpretation that is based on his reading of the verb *abhyavādayat* (v.l. for *abhyavādayata*) as both a causative and a simplex, the first taking for its object Lakṣmaṇa and the second Vaidehī.[20] The idea here is that Bharata *causes* Lakṣmaṇa to honor him, while Bharata himself offers obeisance to Sītā (*bharato lakṣmaṇam āsādyābhyavādayal lakṣmaṇena praṇāmam akārayat. hetumati ṇic. atha vaidehīm āsādyābhyavādayad avandat. atra svārthe ṇic*).

---

19 Citations from Varadarāja Uḍāli are from an undated *devanāgarī* transcription of a South Indian manuscript. The transcription, which is in the possession of the author, was graciously provided by the Rāmāyaṇa Department of the Oriental Institute, M.S. University of Baroda.

20 Citations from Śivasahāya are *ad* GPR 6.127.41.

Satyatīrtha also addresses this issue at some length.[21] Contributing a novel solution to the problem, he argues that one can take the verb *abhyavādayat* in its sense of formally declaring one's identity. Thus, according to him, Bharata simply meets Lakṣmaṇa and, upon meeting Sītā, he formally announces his name (*lakṣmaṇam āsādya prāpya. sītāṃ cāsādya nāma ca bharato 'haṃ bho abhivādayāmīti nāmābravīt*). He then enters into a discussion of the various other commentarial theories mentioned above. He more or less accepts the first argument of Nāgeśa and Maheśvaratīrtha while noting and rejecting as overly clever manipulations, those arguments that attempt to prove Lakṣmaṇa's seniority with respect to Bharata.[22]

As we can readily see from this one example out of many, the issue of social deference to elders and, more particularly, the absolute adherence to its demands on the part of the epic's exemplary characters are of extraordinary importance to the *Rāmāyaṇa* commentators. For how else could we explain the fact that all of these deeply learned scholars would expend so much intellectual energy and ingenuity on what might strike the western reader as an essentially trivial narrative verse? Put another way, what the example shows us is that when it comes to the *dramatis personae* in the epics, the commentators have a keen, almost obsessive, interest in arguing that—even if they must force the text to fit the paradigm—these figures do not in any way act counter to the social imperative.

Equal to or perhaps even more obligatory than the rigorously enjoined regimen of deference is that of truthfulness. For "truth" (*sat, satya, ṛta*, etc.) is not merely a question of a verbal act but the very force at the center of the universe. Thus, not only one's speech but, indeed, one's very life, as in the concept of the *satyakriyā*, "truth act," must, for the ideal person, conform precisely with truth.[23]

As a narrative consequence of this imperative, we see countless figures in the epics and other literature who rigorously adhere to their spoken word. Indeed, in the case of elite and highly spiritually advanced figures, one's words can bring about or alter reality itself.[24] Also, as a practical matter, one of the cul-

21 Citations from Satyatīrtha are *ad* MVR 6.130.39.

22 This example is drawn from PR 6: 1520–1522.

23 Burlingame 1917; Brown 1940, 1978, 1972.

24 Cf. *Uttararāmacarita* Act 1, vs. 10:

The speech of virtuous people in the world accords with reality.
But in the case of the primordial seers, reality itself accords with their words.
laukikānāṃ hi sādhūnām arthaṃ vāg anuvartate |
ṛṣīṇāṃ punar ādyānāṃ vācam artho 'nudhāvati ||

ture's worst transgressions is to utter words, whether in the form of a curse, a vow, or merely a casual or careless remark, and then take them back. Well-known examples from the epics are Kuntī's famous motherly advice to her sons about what they had acquired on their begging rounds, not realizing that they had brought back a bride, thus making inevitable the literature's only polyandrous marriage;[25] Rāma's thoughtless promise to put to death anyone who would interrupt his confidential conversation with Kāla, forcing him to have to "kill" his beloved brother Lakṣmaṇa through exile;[26] Arjuna's vainglorious vow to kill anyone who suggests he give his precious bow, Gāṇḍīva, to another which leads Kṛṣṇa to tell him to "kill" his eldest brother, Yudhiṣṭhira, by addressing him with the informal pronoun in violation of the imperative of deference;[27] Daśaratha's ill-advised attempt to qualify his unrestricted promise to Viśvāmitra to grant him any wish, which provokes the sage to terrifying rage;[28] and of course Arjuna's unfulfilled vow to defeat Duryodhana's army in a single day, a failure which, in the end, makes it impossible for him to enter heaven in his earthly body.[29] In short, all of these examples of the peril of retracting one's spoken word or failing to accomplish it serve to highlight Rāma's reputation as the ideal exponent and practitioner of unqualified truth.[30]

All of these episodes point out clearly the almost unparalleled valorization of truth in the construction of the ideal personality as it was conceptualized and made normative by the vedic, epic, and purāṇic texts and their commentators. But in all of this vast corpus of literature, as far as I have seen, there is only one extended passage that truly tests the linguistic, lexical, and casuistic skills of the epic commentators and confronts them with the opportunity for a bravura performance of their learning and ingenuity. The passage I am referring to is found in the opening four *adhyāyas* of the *Mahābhārata*'s fourth book, the *Virāṭaparvan*. At this point in the epic narrative, the five Pāṇḍava brothers and their wife, Draupadī, have completed their twelve years of forest exile and must now undertake the most difficult and dangerous part of their obligation. For the must now spend a thirteenth year unrecognized by anyone. Forced upon them by Yudhiṣṭhira's loss in his second gambling match with the crooked

25 Mbh 1.182.1–5.

26 VR 7.95.13–15; 96.1–3.

27 Mbh 8.49. This is an interesting example of a conflict between truthfulness and deference to one's elder; one that is resolvable only through Kṛṣṇa's suggestion that Arjuna address Yudhiṣṭhira with the informal pronoun which, to an elder, is tantamount to death.

28 VR 1.17.37–1.20.4.

29 Mbh 17.21–22.

30 See, for example, Rāma's encomium on truth at VR 2.101.11–30.

Śakuni, acting on behalf of Duryodhana, this obligation is inviolable through the strictures of truthfulness. In other words, they must live out the year in impenetrable disguises.

In these chapters, Yudhiṣṭhira asks his companions what disguise each of them will adopt in order to avoid detection and tells them of the one he will choose. Their identities thus concealed, the Pāṇḍavas and Draupadī will each take on a more or less menial position in the household of the Matsya king Virāṭa. Each of these positions corresponds roughly to some noteworthy characteristic or skill of the ones adopting them, and they announce that they will tell Virāṭa that they had served in these respective capacities in the court of Yudhiṣṭhira prior to his exile. This plan serves very well as a narrative device even as it ironically highlights the basic inability of these extraordinary figures to actually disguise themselves. Thus, in *adhyāya*s six to twelve, which follow this opening section, the brothers and Draupadī approach King Virāṭa one by one to apply for employment. The king, perceiving their ultimately unconcealable majesty, cannot really believe that they are the lowly servants they claim to be. Indeed, recognizing their innate majesty, he offers them his kingdom, etc., rather than the positions they seek. However, true to their vows, the heroes and their wife actually take up the jobs they have selected for themselves.

But here is the rub. The resumés they present to Virāṭa are, or so it would seem, evidently and, of necessity, false. And it is that which arouses the concern of the epic's commentators and provokes them to rise to the challenge the text presents. For, the tales the Paṇḍāvas and Draupadī tell the king would appear clearly to violate the categorical imperative of truthfulness. And that, in the commentators' view, is something that can never be ascribed to these godlike men and their divine wife. Therefore, they make it their mission, as it were, to demonstrate, through the deployment of their considerable, even encyclopedic, knowledge of the *kośas, vyākaraṇa*, and the broader intellectual culture of which they are defenders, that each of the Pāṇḍavas' statements about themselves and their proposed occupations are, in fact true, if phrased in such a way as to escape the understanding of all but the most learned, like them. In other words, the commentators simply cannot accept that these legendary *avatāra*-heroes, can, under any circumstances, be shown to indulge in any sort of dishonesty, prevarication, or outright falsehood.

In what follows I will take up, in the order in which they are presented in the text, each of the six character's statements about what job they will do for the king and, in Draupadī's case, his queen and closely examine the commentarial explication of the names and occupations they propose to adopt. As in the case of the *Rāmāyaṇa* example given above, some of the commentators appear to follow or conform to the interpretation of others, and in some cases they do not.[31] Each of the commentaries is interesting in its own right. However, in my *homage* to Chris, I will foreground the *Mahābhāratabhāvadīpa* of Nīlakaṇṭha, which is in any case, generally the most copious of the eight and often the most interesting. Let me now examine Nīlakaṇṭha's and the others' *explication de texte* of the Pāṇḍavas' and Draupadī's proposed job descriptions.

The passage begins in earnest at Mbh 4.1.16–18 when Arjuna questions Yudhiṣṭhira, as the eldest, as to what occupation he will adopt at Virāṭa's court in order to pass the time without being recognized.[32] Yudhiṣṭhira responds as follows:

> Now hear, you bulls among the Kurus, what work I will do when I meet that bull among kings, Virāta. I shall be a courtier of that great king. I shall be a cherished gambling master, expert with dice, a twice-born going by the name Kaṅka. I shall roll the alluring dice, made of lapis, gold, and ivory along with the sparkling nuts, the red and the black. And, should he press me, I shall tell the king that I used to be a friend, as dear as life itself, to Yudhiṣṭhira.
>
> *śṛṇudhvaṃ yat kariṣyāmi karma vai kurunandanāḥ* |
> *virāṭam anusaṃprāpya rājānaṃ puruṣarṣabham* ||
> *sabhāstāro bhaviṣyāmi tasya rājño mahātmanaḥ* |
> *kaṅko nāma dvijo bhūtvā matākṣaḥ priyadevitā* ||
> *vaiḍūryakāñcanān dantān phalair jyotīrasaiḥ saha* |
> *kṛṣṇākṣāṃl lohitākṣāṃś ca nivartsyāmi manoramān* ||
> *āsaṃ yudhiṣṭhirasyāhaṃ purā prāṇasamaḥ sakhā* |
> *iti vakṣyāmi rājānaṃ yadi māṃ anuyokṣyate* ||[33]

Passing over the definitions of the various types of dice, which the commentators spend some time elucidating, Yudhiṣṭhira's speech contains three statements that are either ambiguous or apparently false. These are, first, that he is

31 The commentaries to which I will be referring are Nīlakaṇṭha's *Bhāratabhāvadīpa*, Arjunamiśra's *Dīpikā*, the commentary of Caturbhujamiśra, the *Durghaṭārthaprakāśinī*, the *Virodhārthabhañjinī*, Sarvajñanārāyaṇa's *Bhāratārthaprakāśa*, the *Viṣamapadavivaraṇa*, and the *Lakṣābharaṇa*. All citations to these works are to the 1915 Gujarati Printing Press edition of the *Virāṭaparvan*, edited by Pt. Gangadhar Bhatta Bakre, along with eight Sanskrit commentaries (GPM).

32 Mbh 4.1.16–18 = GPM 4.1.20–22.

33 Mbh 4.1.19–22 = GPM 4.1.23–25; 27 (v.l. *priyadevanaḥ* for *priyadevitā* in 23d).

a twice-born person, second, that his name is Kaṅka, and, third, that he was a companion of Yudhiṣṭhira. The first of these, as Nīlakaṇṭha notes, is technically correct. Nonetheless, it is deliberately misleading. It is true that the term *dvija* is used generically to include individuals belonging to the three "higher" *varṇa*s of the traditional Brahmanical socio-ritual order as they alone are eligible for the *saṃskāra* of *upanayana*, the coming-of-age ritual of investiture with the *yajñopavīta*, "sacred thread." Nonetheless, when the term is used in the singular to describe a person in the epic and later literature, it refers overwhelmingly to a brahman or brahmans. Thus, Nīlakaṇṭha is willing to let it stand by virtue of its generic meaning. Arjunamiśra, however, feels the need to make the term's truthfulness explicit by noting, "[He is] a twice born by virtue of being a *kṣatriya*" (*dvijaḥ kṣatriyatvāt*). Properly parsing the alias Kaṅka, however, requires Nīlakaṇṭha to seek assistance from both the lexicons and the *veda*s. He says:

> Kaṅka by name. According to the *Viśvakośa*, "The term *kaṅka* has the following meanings: 'someone pretending to be a brahman,' 'a heron,' and 'the god of death.'" Also, since he [Yudhiṣṭhira] is the son of the god of death [Yama = Dharma] and since according to *śruti*, "You are the very self [of your father] called 'son,'" he is not different from him, and so his taking on that name does not constitute speaking falsely. That's the meaning.
>
> *kaṅko nāmataḥ. "kaṅkaś chadmadvije khyāto lohapṛṣṭhakṛtāntayoḥ" iti viśvakośāt. svasya kṛtāntasūnutvena "ātmā va putranāmāsi" iti śruteḥ tadananyatvāt tannāmno dhāraṇe mithyāvāditvaṃ na bhaviṣyatīti bhāvaḥ.*

In this way the commentator demonstrates that Yudhiṣṭhira, in obliquely acknowledging that he is a "pretend brahman" (*chadmadvija*) and the son of the god Dharma, is telling the truth for those subtle minded enough to divine it.

Regarding Yudhiṣṭhira's statement that he was a friend as dear as life itself to Yudhiṣṭhira, Nīlakaṇṭha has the following to say:

> Just as the breath of life was the animating spirit of the body that went by the name "Yudhiṣṭhira," in the same way his consciousness too, through being equal to the life breath itself, is Yudhiṣṭhira's soul, that is what is meant to be expressed.
>
> *yudhiṣṭhirākhyasya śarīrasya yathā prāṇa utthāpakas tadvac cidātmāpīti prāṇasāmyena yudhiṣṭhirātmaiva vivakṣita iti bhāvaḥ.*

On the other hand, the author of the *Virodhārthabhañjanī* remarks, "One's friend who does what's best for one becomes oneself indeed" (*svasya sakhā hitakārī svayaṃ bhavaty eva*).

Yudhiṣṭhira then turns, in due birth order, to Bhīma, asking him:

> With what work, Wolf Belly, will you amuse yourself while at Virāṭa's?
>
> *vṛkodara virāṭe tvaṃ raṃsyase kena karmaṇā* |[34]

Like Yudhiṣṭhira, Bhīma announces he will choose occupations that conform with his particular skills and predilections. In his case, these are the culinary and martial arts. He says:

> My intention is that I shall approach Virāṭa informing him that I am a master chef named Ballava. I'll make sauces for him, for I am skillful in the kitchen. I will surpass the dishes previously prepared for him by even those who are most highly trained, thus causing him delight. I will fetch even the largest stacks of firewood; and, when the king sees my prodigious work, he will be pleased. And, whether there be mighty elephants, your majesty, or immensely powerful bulls for me to tame, why, I will tame them all. Whatever wrestlers may be fighting in the arena, I'll take them all down, thereby increasing his delight. But by no means will I kill them as they fight but rather, I will throw them down in such a way that they do not die. And, if anyone should ask, I'll say I used to be Yudhiṣṭhira's cook, cattle butcher, sauce maker, and wrestler. I'll move about taking care of myself by myself, O lord of the people. I promise you that this is the way in which I will pass the time.
>
> *paurogavo bruvāṇo 'haṃ ballavo nāma bharata* |
> *upasthāsyāmi rājānaṃ virāṭam iti me matiḥ* ||
> *sūpān asya kariṣyāmi kuśalo 'smi mahānase* |
> *kṛtapūrvāṇi yair asya vyañjanāni suśikṣitaiḥ* |
> *tān apy abhibhaviṣyāmi prītiṃ saṃjanayann aham* ||
> *āhariṣyāmi dārūṇāṃ nicayān mahato 'pi ca* |
> *tat prekṣya vipulaṃ karma rājā prīto bhaviṣyati* ||
> *dvipā vā balino rājan vṛṣabhā vā mahābalāḥ* |
> *vinigrāhyā yadi mayā nigrahīṣyāmi tān api* ||

34 Mbh 4.1. 23cd = GPM 4.1.28cd. (v.l. *hetunā* for *karmaṇā*).

ye ca kecin niyotsyanti samājeṣu niyodhakāḥ |
tān apy abhibhaviṣyāmi prītiṃ tasya vivardhayan ||
na tv etān yudhyamānān vai haniṣyāmi kathaṃcana |
tathaitān pātayiṣyāmi yathā yāsyanti na kṣayam ||
ārāliko govikartā sūpakartā niyodhakaḥ |
āsaṃ yudhiṣṭhirasyāham iti vakṣyāmi pṛcchataḥ ||
ātmānam ātmanā rakṣaṃś cariṣyāmi viśāṃpate |
ity etat pratijānāmi vihariṣyāmy ahaṃ yathā ||[35]

Bhīma's multitasking skills—unique among the Pāṇḍavas—and the various positions he claims to have held under Yudhiṣṭhira provide Nīlakaṇṭha and the other commentators ample opportunity to show off their knowledge and interpretive skills. The specific professions that Bhīma claims to have mastered are *paurogava*, *ārālika*, *govikartṛ*, *sūpakartṛ*, and *niyodhaka*. The commentators take them one at a time. The first term, *paurogava*, normally "a chief cook" or "superintendent of the royal kitchens," leads Nīlakaṇṭha into an interesting and convoluted discussion of reproductive biology. He derives the term as follows:

> *Paurogava* is that among which the senses (*gāvaḥ*), that is to say, beams (*raśmayaḥ*), have precedence (*puras*), that is to say, which goes before (*purastāt*), i.e., the wind. For, when an embryo arises, the breath of life (*prāṇa*), in the form of a beam of wind, is the first thing to appear in the fluid mixture of semen and menstrual blood. After that the other senses arise, as is well known from the scientific literature. The son of that [wind/breath] is thus [called] Paurogava. But in common [usage] a *paurogava* is the superintendent of a kitchen.
>
> *paurogavaḥ puraḥ purastāt gāvo raśmayo yasya sa purogur vāyuḥ. piṇḍotpattau hi prathamaṃ rajoretasoḥ kalke vāyuraśmibhūtaḥ prāṇa āvirbhavati paścād indriyāntarāṇīti śāstraprasiddham. tasya putraḥ paurogavaḥ. loke tu paurogavo mahānasādhyakṣaḥ.*

Through this analysis Nīlakaṇṭha manages to doubly exculpate Bhīma from the offense of untruthfulness by taking the term in its normal lexical sense, as well as in its esoteric meaning of "son of the Wind God" a reference to Bhīma's self-identification as Vāyuputra. The other commentators are content to refer to the normal lexical meaning of the term.

35 Mbh 4.2.1–8 (with v.l.) = GPM 4.2.1–4ab, 6–10.

As for Bhīma's assumed name, Ballava, this also means "a cook" and so adds to the passages' various terms that have this rather overdetermined referent. Nīlakaṇṭha simply glosses "'Ballava' means 'a maker of sauces,' and so that is my name, which has the same meaning" (*ballavaḥ sūpakāraḥ etad eva mamānvarthaṃ nāma*).

The second term, *ārālika* is cited in the lexica in the senses of "a deceitful person" and "a cook." In context, the second meaning seems apposite, if a bit redundant, while the first could perhaps also be appropriate in the same way as Yudhiṣṭhira's giving himself a name that means "a pretend brahman." None of the commentators, however, interprets it in this way.

Nīlakaṇṭha first explains the term as a *taddhita* derivate of a lexeme, *arāla*, meaning "rutting elephant." He says:

> *Arālā*s are rutting elephants. One who either plays with them or masters them is an *ārālika*.
>
> *arālā mattagajās taiḥ saha krīḍati tān jayati vārālikaḥ*.

He bases this on a citation from the *Viśvakośa* that lists the following as synonyms of *arāla*: "crooked," "the resin of the Sāl tree," and "a rutting elephant" ("*arālaḥ kuṭile sarjarase vai mattadantini*" *iti viśvaḥ*). Then, interweaving his treatment of *ārālika* with his views on the term *govikartā*, discussed below, he continues, saying:

> An *ārālika* is someone who composes verses with unequal *pāda*s, someone who points out the virtues and defects in [a king's] ministers, or an elephant tamer.
>
> *viṣamaślokīkṛtas tv ārālikaḥ prakṛtīnāṃ guṇadoṣasūcako hastidamako vā*.

He also quotes Vikramāditya as saying, "*ārālika* means 'an informant,' and 'an elephant tamer'" (*tathā ca vikramādityaḥ "ārālikaḥ sūcanako hastināṃ damakas tathā"*). Nīlakaṇṭha continues, "Some say that *ārālika* can refer to a rice cook, a preparer of sauces, and a vegetable chef" (*kecit tu 'āráliko 'nnapākī syāt sūpakartā tu śākakṛt'*).

Arjunamiśra has a broader set of possible meanings. His first is consonant with the various terms for "cook" that seem to cluster around the prodigious trencherman Vṛkodara but he offers some new ones while quoting a variant of the *Viśvakośa* passage cited by Nīlakaṇṭha. Arjunamiśra says:

> An *ārālika* is a rice cook. On the other hand, [it also means] "a person engaged to proclaim the ordinances of kings" and "one who prepares drinks for children." The term is derived [from *arāla*] by substituting the *ṭhak* [*pratyāya*]. And, according to the *Viśvakośa*, "*Arāla* refers to 'an inlaid floor,' 'the resin of the Sāl tree,' and 'a rutting elephant.' It also means 'an elephant tamer.'"
>
> *ārāliko bhaktasūdaḥ. athavā rājñāṃ pramāṇaghoṣaniyuktaḥ. yad bālānāṃ pānarūpaṃ saṃskarotīti ṭhag ārālikaḥ. "arālaḥ kuṭṭime sarjarase mattahastini" iti viśvaḥ. hastināṃ damako vā.*

Arjunamiśra also offers the same quotation from Vikramāditya that Nīlakaṇṭha adduces above.

The other commentators who take up the term, generally echo all or part of Nīlakaṇṭha's and Arjunamiśra's efforts at lexicographical research or add to them additional terms referring to the culinary arts. Thus, the author of the *Viṣamapadavivaraṇa* adds to the cumulative list, glossing *ārālika* with "one who churns curds" (*dadhimanthakartā*).

It is interesting to note that most of the jobs that the commentators see as included in the broad spectrum of the term *ārālika*'s meanings, generally clustering around the work of a chef or elephant trainer, do not refer retrospectively to any particular work that Bhīma had actually performed for Yudhiṣṭhira, but prospectively to duties he will perform for Virāṭa.

It is the next term in Bhīma's "resumé," *govikartā*, that, not surprisingly, elicits the greatest attention on the part of the commentators. This, no doubt, is because the apparent meaning of the term, "cattle butcher," is so dystonic with the cultural universe of the medieval and early modern epic commentators, if not with that of the epic poets, in which cow slaughter and beef eating had become heavily tabooed.[36] The Sanskrit lexicon offers many ways of avoiding this touchy subject, as the word *gau/go* has many meanings other than "cow," such as "earth," "speech," "rays of light," and "the sense organs," to name but a few. The commentators gratefully take full advantage of the term's broad semantic range and, in the process, confer on Bhīma a number of interesting skills. These include surveying and, most notably, literary criticism.

---

36 The disparity in the reading of the term in modernity by western and Indian translators can be seen in the examples of van Buitenen (1978: 28), who renders "cow butcher," and Roy (1923: 3), who reduces Bhīma's four terms for professions to an anodyne pair, evading the issue entirely: "Formerly I was the wrestler and cook of Yudhiṣṭhira."

Nīlakaṇṭha offers a number of alternative meanings, one of which does involve cattle, albeit not their slaughter. He says:

> A *govikartā* is someone who brings about a change in cattle, even including huge oxen, changing them through training. Or, it is in keeping with his proposal in the form of his vow, "And I will tame mighty bulls."
>
> *govikartā gavāṃ mahatāṃ balīvardānām api vikartā damanena vikṛtijanakaḥ. vṛṣabhān vā mahābalān nigrahīṣyāmīty upakramāt.*

After concluding his discussion of the term *ārālika*, he continues:

> A *govikartā* is someone who embellishes *go*s, that is to say, "speech"; in other words, a user of the various forms of prose and verse .... It is also said that one who cooks fried rice (*tailānnam*) is known as a *govikartā*. Or so they say, but for that, one should consult the text.
>
> *govikartā gavāṃ vācāṃ vikartā gadyapadyādibhedānām prayokteti ... "tailānnam pacate yas tu govikartā sa ucyate" ity apy udāharanti tatra mūlaṃ draṣṭavyam.*

Arjunamiśra offers some interesting variations on the ideas Nīlakaṇṭha puts forward:

> A *govikartā* is someone who changes, that is to say, who drives back *go*s, that is to say, maddened bulls in battle. Alternatively, it refers to someone who parcels out *go*, that is, land, for villages, towns, citadels, cities, hamlets, and market towns (*kharva* v.l. for *kharvaṭa*[?]). Or the term *go* can refer to speech in the form of difficult literature, whether charming or difficult, in prose and verse in Sanskrit, Prakrit, Apabhraṃśa and Paiśācika, as well as difficult texts in logic, vedic texts, and texts in regional languages. A *kartā* of these is someone who is particularly skillful in the usage of these forms of speech, both in terms of their composition and meaning, that is to say, he is someone who is knowledgeable about the distinctions regarding the speech of others. And moreover, it is said that they say, "Bhīmasena, that subduer of his foes, is foremost among the learned." Or *go* means, "water," [so the compound means "someone who is knowledgeable about its [water's] different type"s].

govikartā gavāṃ mattavṛṣāṇāṃ yuddhe vikartā nivārakaḥ. athavā gor bhūmer grāmanagarapurapattanakheṭakharvāder vibhāgakartā. yadvā gavāṃ gadyapadyasaṃskṛtaprākṛtāpabhraṃśapaiśācikalalitaduṣkarasāhitya-duṣkaratarkavaidikadaiśikānāṃ vācāṃ viśeṇeṇārthato granthataś ca kartā prayoganipuṇaḥ paraprayogavāgviśeṣavic ca. uktaṃ ca paṇḍitānāṃ paraṃ prāhur bhīmasenam arindamam. gavāṃ jalānāṃ viśeṣajño vā.

The other commentators provide, in condensed form, much of the same analysis of the term as Nīlakaṇṭha and Arjunamiśra do.

The terms *sūpakartā* and *niyodhaka* are relatively clear in their application to Bhīma and thus do not exercise the ingenuity or the prolixity of the commentators to the degree that the previous two terms do. On the first, which normally refers to a preparer of sauces, Nīlakaṇṭha offers two, brief interpretations: "A *sūpakartā* is someone who modifies royal seals,[37] etc. Alternatively, [it has the obscure] meaning of 'someone who renders extraordinary assistance (*su+upakartā*)'" (*sūpakartā mudrādivikārakartā. pakṣe sutarām upakartā.*) Arjunamiśra, who reads the variant but synonymous *sūdakartā*, "preparer of sauces *or* broths," glosses, "*Sūdakartā*: '*sūda*' means 'sauces (*vyañjana*)'. Or the term can refer to someone who knows the art of making sauces" (*sūdakartā vyañjanaṃ sūdaḥ sūdaśāstrajño vā*). Sarvajñanārāyaṇa's interpretation is the same as the first one of Nīlakaṇṭha. Other commentators ignore the term, assuming, no doubt, that it should be taken in its apparent meaning of "a sauce cook," which fits the general context of Bhīma's association with the culinary arts, the principal area in which he will work in Virāṭa's household.

Bhīma's final term in reference to his proposed service under Virāṭa, *niyodhaka*, is the least controversial of all, referring as it does to his skills as a fighter. The only nuance in its meaning found among the commentators concerns the context and type of fighting for which the hero is celebrated. Nīlakaṇṭha, following Bhīma's statement in the text, glosses:

> The term *niyodhaka* refers to someone who fights with wrestlers but avoids killing [them]. Alternatively, [it has the obscure] meaning of someone who gives battle to a very high degree.
>
> *niyodhako māraṇavarjaṃ mallayuddhakartā. pakṣe nitarāṃ yodhakaḥ.*

37 It is not perfectly clear to which of the various meanings of *mudra/ā* Nīlakaṇṭha is referring. I have tentatively taken it in its sense of royal or administrative stamps and seals which seems to me to be most apposite in the context of courtly affairs.

Yudhiṣṭhira now turns to Arjuna to inquire about his planned disguise and occupation during their year in hiding. Interestingly, in what appears to be recognition of Arjuna's prominence in the epic, Yudhiṣṭhira departs from his usual direct question to, instead, indulge in a lengthy twenty-four-line virtual hymn of praise of the great archer, dwelling on his burning of the Khaṇḍava forest and recalling his residence in Indra's heaven. The passage also recites Arjuna's various names and epithets along with something resembling Kṛṣṇa's recital of his own *vibhūti*s in the tenth *adhyāya* of the *Bhagavadgītā*.[38] This passage, naturally, occupies the attention of the commentators until Arjuna begins to relate his plan for employment at Virāṭa's court.

He announces:

> I will swear, lord of the earth, that I am a eunuch. For these big bowstring scars of mine are hard to conceal, lord of men. I'll put fiery earrings in my ears. I'll have a braid on my head, your majesty, and I'll go by the name Bṛhannaḍā. In my feminine garb, I'll surely tell tales again and again, giving delight to the protector of the earth and the other people of the inner apartments. And, your majesty, I shall teach the women of Virāṭa's palace singing, the different kinds of dance, and the various musical instruments. I will talk a great deal about the positions of the king's subjects which they have attained through their efforts. And, son of Kuntī, I shall conceal myself by myself through this illusion. And if, scion of Bharata, I should be questioned by the king, I shall say, "I used to live in the house of Yudhiṣṭhira, as a serving maid to Draupadī." Concealed through this assumed expedient, I shall pass the time happily, lord of kings, just as did Nala.[39]
>
> *pratijñāṃ ṣaṇḍhako 'smīti kariṣyāmi mahīpate* |
> *jyāghātau hi mahāntau me saṃvartuṃ nṛpa duṣkarau* ||
> *karṇayoḥ pratimucyāhaṃ kuṇḍale jvalanopame* |[40]

38 Mbh 4.2.9–20 = GPM 4.2.11–24. Cf. Mbh 6.32.20–40 (= *Bhagavadgītā* 10.20–40).

39 In the Mbh reading, *kṛtakena yathā nalaḥ*, the reference is to the epic's well-known *Nalopākhyāna* in which the unfortunate hero, Nala, is forced to live in hiding disguised as a charioteer in the court of King Ṛtuparṇa (Mbh 3.63–64). However, GPM 4.2.32b, along with Arjunamiśra and Caturbhujamiśra, reads instead *kṛtakena yathānalaḥ*. This reading forces these commentators to understand *kṛtaka* in the sense of "ashes" giving the simile the sense of "like a fire [hidden] by ashes." Thus, Arjunamiśra glosses *kṛtakena yathānalaḥ kṛtakaṃ bhasma tenācchādito 'gnir iva*.

40 Following Mbh 22ab, a number of northern manuscripts insert a passage of one line (Mbh 4. 59* = GPM 4.2.27ab. "With conch shell bangles on my wrists, I will have assumed the third nature" (*pinaddhakambuḥ pāṇibhyāṃ tṛtīyāṃ prakṛtiṃ gataḥ*). With regard to the

veṇīkṛtaśirā rājan nāmnā caiva bṛhannaḍā ||
paṭhann ākhyāyikāṃ nāma strībhāvena punaḥ punaḥ |
ramayiṣye mahīpālam anyāṃś cāntaḥpure janān ||
gītaṃ nṛttaṃ vicitraṃ ca vāditraṃ vividhaṃ tathā |
śikṣayiṣyāmy ahaṃ rājan virāṭabhavane striyaḥ ||
prajānāṃ samudācāraṃ bahu karmakṛtaṃ vadan |
chādayiṣyāmi kaunteya māyayātmānam ātmanā ||
yudhiṣṭhirasya gehe 'smi draupadyāḥ paricārikā |
uṣitāsmīti vakṣyāmi pṛṣṭo rājñā ca bhārata ||
etena vidhinā channaḥ kṛtakena yathā nalaḥ |
vihariṣyāmi rājendra virāṭabhavane sukham ||[41]

As one might expect, Nīlakaṇṭha is unhappy with Arjuna's representation of himself as a eunuch or, as he defines the term, someone belonging to the "third nature," i.e., the neuter gender. He explains the term in a different way, playing on Arjuna's well-known friendship with Kṛṣṇa, as well as his identification with the Primal Man, Nara, and the latter's association with Viṣṇu Nārāyaṇa. Nīlakaṇṭha explains:

> "I am a *ṣaṇḍhaka*." The term *ṣaṇḍha* means "a lord of cattle," that is to say, "Śrīgopāla." His spitting image and second incarnation is [thus] a *ṣaṇḍhaka*. So, the meaning is that he is [really] saying, "I am Kṛṣṇa's friend." The alternative meaning is "a eunuch." For the *Viśvakośa* says, "*Ṣaṇḍha* can mean 'a eunuch,' 'a coward,' 'a lord of cattle,' or 'a sterile man.'"

> *ṣaṇḍhako 'smīti ṣaṇḍho gopatiḥ śrīgopalas tasya pratikṛtir avatārāntaraṃ ṣaṇḍhakaḥ kṛṣṇasakho 'smīty arthaḥ. pakṣe klībaḥ. "ṣaṇḍho varṣavare klībe gopatau vandhyapuruṣe" iti viśvaḥ.*

On this point, Arjunamiśra observes:

> "The third nature" that is to say, "embodiment as a eunuch *or* hermaphrodite." And then he was favored, through [the modification] of Urvaśī's curse [that Arjuna would become a eunuch permanently] through the boon of Dharma.

"third nature," i.e., the third or neuter gender, Nīlakaṇṭha observes, "I will have gone to, that is to say, attained, the third nature which is the state of being third with respect to the male and female [genders], in other words, a eunuch" (*tṛtīyāṃ prakṛtiṃ strīpuṃsāpekṣayā klībatvaṃ tṛtīyaṃ gataḥ praptaḥ*).

41 Mbh 4.2.21–27 = GPM 4.2.25; 26cd–32.

> *tṛtīyāṃ prakṛtiṃ napuṃsakamūrtim. urvaśyāḥ śāpam anugṛhṇāti dharmavarāc ca tadā.*

As to Arjuna's choice of a name, Bṛhannaḍā, literally (and suggestively), "Big Reed," Nīlakaṇṭha explains:

> [The term *naḍa* (v.l. *nala*)] is the same as *nara*, "man," because of the absence of any distinction between the syllables *ra* and *la*, on the one hand, and between those two and *ḍa*. Thus, [the name should be understood as] a man, and a big one at that. Also, *nara* refers to, the [eternal] companion of Nārāyaṇa, the primordial Nara. That is the meaning.
>
> *nara iti ralayor ḍalayoś cābhedān naro bṛhāṃś cāsau naraś ceti nārāyaṇasakhādyao nara ity arthaḥ.*

On the meaning of the term *ṣaṇḍhaka* and of the name Bṛhannaḍā, the other commentators, unless they are silent here, generally concur with Arjunamiśra and Nīlakaṇṭha. However, the author of the *Virodhārthabhañjanī* offers an alternative explanation of the former term. He says, "Bṛhannalā means 'like a big reed' since, as a result of his adopting the nature that is unmasculine, he lacks firmness" (*bṛhannalā bṛhannalatulyā puṃstvābhāvena sārābhāvāt*). The author of the *Viṣamapadavivaraṇa* takes Arjuna's statement that he will tell tales to the ladies of the inner apartments "again and again" as an opportunity to inject a snarky, sexist comment about Arjuna's intended audience, saying, "[His] speaking again and again is appropriate in light of the ignorance of women" (*punaḥ punar bhāṣaṇaṃ strīṇām ajñānatvād yuktam*).

The other issue of truth versus falsehood that concerns the commentators, of course, is Arjuna's statement that he was Draupadī's serving maid (*paricārikā*). This term also involves issues of gender and social hierarchy and so, again, the commentators treat it delicately.[42] Basically, *paricārikā* means "a female servant (literally, 'one who moves about')," but Nīlakaṇṭha understands that the verbal root √*car* that underlies the noun should be understood in its causative form, thus lending it the sense of someone who causes, i.e., orders, others to move about or to serve. He explains:

---

42 This term has exercised commentators in a number of contexts. See the discussion of how several of the commentators on the VR struggle with the term in their comments on the Śabarī episode of the *Araṇyakāṇḍa* in Sutherland Goldman 2018 and an analysis of Śaṅkarācārya's efforts to rationalize the term in his commentary on the *Chāndogyopaniṣad* in Robert Goldman 2019. See too, the various commentarial efforts to cope with the term in the case of Draupadī below.

> One whose habit is to instruct people to serve. Such a person is a *paricārikā*. The secret meaning is that he is Draupadī's husband.
>
> *paritaś cāriyitum śīlam yasyāḥ sā paricāriṇī saiva paricārikā, draupadyā bharteti gūḍho 'bhisandhiḥ.*

The author of the *Virodhārthabhañjanī* proposes an equally ingenious explanation of the term, saying:

> According to the *Medinī*, "The term *pari* can have the meaning of 'an embrace.'" So, a *paricārikā* is someone who roams about for the sake of embraces and thus this [statement] is not false.
>
> *"parir āliṅgane" medinī. āliṅganārthaṃ caratīti paricārikā. tena na mithyā.*

This, it would appear, is a reference to Arjuna's various wives and romances during his period of wandering while the Pāṇḍavas were living in exile in the forest. And so, we see that, as with his older brothers, nothing that Arjuna says he will be or do is in any way false, even if his words are still carefully crafted to deceive Virāṭa and his court.

Yudhisthira next turns to Nakula and Sahadeva his two junior-most brothers, the sons of Pāṇḍu's junior wife, Mādrī. Asked what role he will adopt at Virāṭa's court, Nakula says:

> I shall be King Virāṭa's horse wrangler for I like that kind of work, and I shall go by the name Granthika. I am skilled in training horses as well as in equine medical treatment. I have always loved horses, king of the Kurus, just as you have. And to any of the people of Virāṭa's city who may call me out on this, I shall tell them, "This is how I pass the time."[43]

43 Following Mbh 4.3.4, a number of *devanāgarī* manuscripts insert a passage of one line (Mbh 4. 76 * = GPM 4.3.6ab): "Formerly, dear sir, the Pāṇḍava placed me in charge of his horses" (*pāṇḍavena purā tāta aśveṣv adhikṛtaḥ purā*). Nīlakaṇṭha proposes a particularly ingenious way of validating this statement. He says, "The word *aśveṣu* [should be read as follows]. Those for whom there is no tomorrow (*śvaḥ*) are heroes. For, in battle they value their lives no more than blades of grass. Among them I was placed at the head, that is to say, I was the foremost. That is the secret meaning. The meaning of the alternative reading is clear." (*aśveṣu nāsti śvo yeśāṃ te śūrās samare tṛṇīkṛtaprāṇās teṣv adhikṛto mukhya iti gūḍhāśayaḥ. pakṣe spaṣṭo 'rthah*).

aśvabandho bhaviṣyāmi virāṭanṛpater aham |
granthiko nāma nāmnāhaṃ karmaitat supriyaṃ mama ||
kuśalo 'smy aśvaśikṣāyāṃ tathaivāśvacikitsite |
priyāś ca satataṃ me 'śvāḥ kururāja yathā tava ||
ye māṃ āmantrayiṣyanti virāṭanagare janāḥ |
tebhya evaṃ pravakṣyāmi vihariṣyāmy ahaṃ yathā ||[44]

The commentators are particularly interested in parsing Nakula's alias, Granthika. The term, derived from the root √*grath*, "to tie up *or* string something together," would seem to fit the occupation of a horse handler. But, of course, the term *grantha* is widely used to refer to texts which are "strung together *or* composed." Thus Nīlakaṇṭha, playing on this meaning and on the fact that the Aśvins are Nakula's and Sahadeva's divine fathers, says:

> The term 'texts' refers to the fact that he knows the *Āyurveda* and the *Ādhvaryava* (*Yajurveda*). He is one who understands [these texts] by virtue of his being a son of the Aśvins. For, as the *śruti*, says, 'Verily the Aśvins are the physicians and the *adhvaryus* of the gods.'
>
> *granthān āyurvedaṃ ādhvaryavaṃ ca vettīti granthiko 'śvinoḥ sutatvāt. "aśvinau vai devānāṃ bhiṣajāv aśvināv adhvaryū" iti śruteḥ.*

Sarvjñanārāyaṇa suggests that the name is appropriate because, "One who knows the texts on evaluating horses is called a *granthika*" (*aśvaparīkṣāgranthān vettīti granthikaḥ*). The author of the *Virodhārthabhañjanī* proposes the puzzling explanation that someone who plays with a text is a *granthika*, because of knowing the technical texts on horses (*granthena dīvyati granthikaḥ. aśvaśāstrajñānāt*). The author of the *Viṣamapadavivaraṇa* presents a number of alternative interpretations of the name. He says:

> 'Going by the name, namely (*nāma nāmnā*) of Granthika, I ....' A *granthika* is someone who adopts a false identity in order to defeat his enemy. Crookedness, deception, and dishonesty [are synonyms]. Or, if we accept the variant reading, 'Going by the name of Granthika, I, with a rope,' then it means 'Granthika by name, a fastener of ropes,' that is to say, 'I know how to tie the ropes of horses.'

44 Mbh 4.3.2–4 = GPM 4.3.3–5.

> *granthiko nāma nāmnāhaṃ iti grathnāti śatruvijayāya kuṭilabhāvaṃ karotīti granthikaḥ. grathi vaki kauṭilye. yadvā granthiko nāma dāmnāham iti pāṭhe dāmnā saha granthiko nāma dāmagranthika iti. aśvarajjūṃ grathituṃ jāne.*

Yudhiṣṭhira now turns to Nakula's twin brother, Sahadeva, who, in response to the same question as was asked of the others, replies:

> I shall be the lord of the earth Virāṭa's enumerator of cattle, their milker. And the one who keeps them corralled. I am skillful in enumerating cattle. You should know that I will be known by the name Tantipāla. I will conduct myself skillfully; so give up your anxiety. Earlier, you always placed me in charge of the cattle, and it was in that post that I learned how to do the job skillfully, lord of the people. Their characteristics, behavior and auspicious markings, all of that I know very well and more besides, lord of the earth. I know about bulls, your majesty, the ones with highly valued characteristics by merely sniffing the urine of which even barren cows can give birth. So that is how I will occupy myself, and I will find constant delight. Moreover no one other [than us] will recognize me. May this please you, lord of the earth.

> *gosaṃkhyātā bhaviṣyāmi virāṭasya mahīpateḥ |*
> *pratiṣeddhā ca dogdhā ca saṃkhyāne kuśalo gavām ||*
> *tantipāla iti khyāto nāmnā viditam astu te |*
> *nipuṇaṃ ca cariṣyāmi vyetu te mānaso jvaraḥ ||*
> *ahaṃ hi bhavatā goṣu satataṃ prakṛtaḥ purā |*
> *tatra me kauśalaṃ karma avabuddhaṃ viśāṃpate ||*
> *lakṣaṇaṃ caritaṃ cāpi gavāṃ yac cāpi maṅgalam |*
> *tat sarvaṃ me suviditam anyac cāpi mahīpate ||*
> *vṛṣabhān api jānāmi rājan pūjitalakṣaṇān |*
> *yeṣāṃ mūtram upāghrāya api vandhyā prasūyate ||*
> *so 'ham evaṃ cariṣyāmi prītir atra hi me sadā |*
> *na ca māṃ vetsyati paras tat te rocatu pārthiva* ||[45]

As usual, the commentators are particularly interested in parsing the brothers' various aliases in an effort to demonstrate that they have a hidden meaning that truthfully describes each of them. Nīlakaṇṭha begins with an analysis of

45 Mbh 4.3.6–11 = GPM 4.3.8–13 (with v.l.).

the name Tantipāla, which evidently refers to the practices of the traders at a cattle market. He says:

> *Tanti* means a long and thick rope to which the traders attach a smaller rope to which oxen are tied, which is useful for their purpose. But according to its secret meaning, *tanti* means "speech" and according to the *śruti*, "Its cords have the inner [secret] name of 'speech.'" Thus, he is a guardian of speech [and is saying,] 'I will carry out the import of your words.' That is the meaning.
>
> *tantipālaḥ balīvardā yasyāṃ dīrghasthūlarajjvāṃ vaṇigbhir alpair dāmabhir badhyante sā prakṛtopayoginī, gūḍhābhisaṃdhau tu tantir vāk "tasya vāk tantir nāmāni dāmāni" iti śruter vākpālaḥ. yuṣmadvacanārthānuṣṭhātety arthaḥ.*

Sarvajñanārāyaṇa gives the same interpretation as Nīlakaṇṭha as far as the discussion of the cattle ropes goes.

The author of the *Virodhārthabhañjanī* takes a different tack. He argues that "*Tantī* means that as he is possessed of the knowledge of the highest teachings of the technical treatises on cattle, he guards them and thus is a *tantipāla*" (*tantī gośāstrasiddhāntajñānī san pālayati tantipālaḥ*). But the author of the *Viṣamapadavivaraṇa* offers other, different interpretations. First, he comments, "He stretches out the expanse of time. That is, he adheres to it, so he is a *tantipāla*" (*tanoti samayavistāraṃ pālayatīti tantipālaḥ*). Alternatively, he proposes that we read the variant *tantrapāla*, understanding the term *tantra* to refer to a traditional kingdom with its seven elements. He protects it because of his wide-ranging knowledge (*yadvā tantrapala iti pāṭhe tantraṃ saptāṅgarājyaṃ pālayatīti tantrapālaḥ sarvaveditvāt*). Finally, he notes that the conventional referent of the term is someone who corrals cattle (*saṅketaśabdena gavāṃ nivārakaḥ*).

Finally, Yudhiṣṭhira, filled with anxiety about how Draupadī will manage in Virāṭa's court as a solitary woman without a protector, turns to her and asks the familiar question he had posed to his brothers, but phrasing it in the third person. She replies as follows:

> In the world, there are independent (literally, unprotected) serving women known as *sairandhrī*s, scion of the Bharatas. No other women can move freely about as they do, for such is the fixed opinion among the people. I will say that I am a *sairandhrī* who is skilled in dressing hair. And since you are asking me, I would say that I move about under my own

protection. I shall serve the king's wife, the illustrious Sudeṣṇā. Once I am with her, she will protect me. So, you should not have any such anxiety.

sairandhryo 'rakṣitā loke bhujiṣyāḥ santi bhārata |
naivam anyāḥ striyo yānti iti lokasya niścayaḥ ||
sāhaṃ bruvāṇā sairandhrī kuśalā keśakarmaṇi |
ātmaguptā cariṣyāmi yan māṃ tvam anupṛcchasi ||
sudeṣṇāṃ pratyupasthāsye rājabhāryāṃ yaśasvinīm |
sā rakṣiṣyati māṃ prāptāṃ mā te bhūd duḥkham īdṛśam ||[46]

As usual, the commentators are at pains to explain the character's names and job descriptions as meaning virtually anything other than the standard lexical ones. In the case of Draupadī, as with Arjuna, gender is added to the problem of self-identification and proposed job category. The commentators are keen, first off, to deal with the term *sairandhrī*, which, it appears, Draupadī will use as a job category, as a caste name, or, perhaps, as a personal name as well. In the lexica, the term is said to mean "a female servant in the *antaḥpura*," "a skilled female servant who performs a specific duty," or "the name of a mixed caste or tribe said to be born of the union of a *dasyu* male and an *ayogava* woman (a woman born of the *pratiloma* union of a *śūdra* male and a *vaiśya* woman)."[47] Any of these meanings could apply in the context of Draupadī's faux self-identification and the terms are, of course, not mutually exclusive in a social environment in which class or caste identity is frequently associated with occupation. Nonetheless, a number of the commentators add several other definitions or variations on these three. They appear to struggle with some of the terms in their effort to show that Draupadī is cunningly telling the truth about herself.

Nīlakaṇṭha offers as his first alternative, a definition that seems unique to him:

Sairandhrī: *sīra* means "a curtain or screen," that is to say, "an interior curtain." A woman by or for whom it is held up is called a *sīrandhrā*. She is herself a *sairandhrī*. The term is a secondary nominal derivative with the

46 Mbh 4.3.16–18 = GPM 4.3.18; 20–21. Following Mbh 4.3.17ab northern manuscripts insert a passage of two lines Mbh 4.96* = GPM 4.3.19:

And, scion of Bharata, if I am questioned by the king, I shall say, "I used to live in Yudhiṣṭhira's household as Draupadī's servant."
yudhiṣṭhirasya gehe 'smi draupadyāḥ paricārikā |
uṣitāsmīti vakṣyāmi pṛṣṭā rājñā ca bhārata ||

47 On the caste names terms *ayogava* and *sairandhra/sairindhra*, see Apte 2003, ad loc and *Manusmṛti* 10.32.

same meaning as its nominal base. So, the term can refer to either a female servant of the inner apartments or the wife of a king. The *Viśvakośa* says, "*Sairandhrī* is a feminine noun for a craftswoman who lives in the house of another. She is the product of a union between a woman born as a result of the mingling of the social classes and a Mahillaka." *Medinī*, however, has the reading *-matallikayoḥ*, "an excellent (?) man" [for *mahillakayoḥ*]. In the first reading, *mahillakā* means "a king's wife." In the second, *matallikā* means "a praiseworthy wife." In reference to a servant, we should break up the words so as to read "unprotected," that is to say, "independent."

sairandhrīti. sīraṃ pratisīrāntaḥpaṭaḥ dhriyate 'nayāsyai veti sīrandhrā saiva sairandhrī svārthe taddhitaḥ. antaḥpuraparicārikā rājabhāryā vā. "sairandhrī paraveśmasthā śilpakṛt svavaśā striyām. varṇasaṃkarasaṃbhūtastrīmahillakayor api" iti viśvaḥ. "matallikayoḥ" iti medinīpāṭhaḥ. ādye mahillikā rājadāraḥ. dvitīye matallikāḥ praśastastriya iti. dāsīpakṣe 'rakṣitā iti cchedaḥ svavaśā ity ukteḥ.

With the exception of the author of the *Viṣamapadavivaraṇa*, none of the other commentators quoted in the GPM add anything substantial to the meanings given the term by Nīlakaṇṭha. Instead, they generally mention one or two of the options he mentions. The *Viṣamapadavivaraṇa* however, does add something different. It says:

*Sairandhṛī*s in the world are protected. [The word] *svataḥ*, "by themselves," should be added. These *sairandhrī*s are thus those who live as they please (*svairam*). Here, through the elision of the letter *v*, they are called *sairandhrī*s ... The meaning is that other women, that is to say, ordinary women, cannot enter the king's palace as they please in the way these *sairandhrī*s do. They are not servants of anyone and are independent.

sairandhryo rakṣitā loke svata iti śeṣaḥ. svairam ātmānaṃ dharanti tāḥ svairandhryaḥ. atra vakāralopāt sairandhryaḥ ... yathaitāḥ sairandhryo rājabhavanaṃ yadṛcchayā yānti tathānyāḥ prākṛtastriyo nety arthaḥ. tā adāsyaḥ kasyacin na svādhīnāḥ.

The highly charged and related questions of Draupadī's sexual vulnerability and how she will be protected—issues that, as seen above, deeply concern Yudhiṣṭhira—are of great interest to the commentators as can be seen in their treatment of two of the terms she uses in her first statement: *arakṣitā*, "unprotected," and *bhujiṣyā*, "female servant." Interestingly the terms are situated in phonological environments that permit them to be read as either positive or negative terms. In other words, they can also be read as *rakṣitā*, "protected," and *abhujiṣyā*, "not a servant." The commentators, in grappling with the terms, make full use of their ambiguity.

As noted above, Nīlakaṇṭha uses the ambiguity presented by the *sandhi* to read either *rakṣitā*, "protected" or *arakṣitā* "unprotected" applying either to elite women or their servants, respectively. Thus, again, he observes:

> If we take the term to refer to such servant women, we should break up the words to read the term as "unprotected" as it says that they are independent. If we take it to refer otherwise [to noblewomen] we should read, "protected."
>
> *dāsīpakṣe 'rakṣiteti cchedaḥ svavaśety ukteḥ. pakṣe rakṣitety eva.*

As for the second term, he says similarly:

> "A *bhujiṣyā* is either a woman who is to be protected or a protectress. For we can take the verbal root √*bhuj* in the senses of protecting or bringing food, such is its nature."
>
> *bhujiṣyā pālanīyā pālayitrī ca. bhujapālanābhyavahārayor ity asya rūpam.*

After defining the term *sairandhrī*, Arjunamiśra notes:

> These women ... are "unprotected," that is, by another person. If one reads the text as "protected women," then they are protected by another person. *Bhujiṣyā*s are servant women who arrange their protection by themselves and are devoted to a single husband. That is the way they are. These women do not leave their man for another, even if he has the appearance of a superior man, as other people do because of their base nature. For such is the fixed opinion among the people. Therefore, [she says,] "I shall move about protected by myself." Those [women] whose vow is in the form, "[I have but] one husband," are like that. The hidden meaning is that "The princes, starting with the horse driver and the chief cook, as well as Indra, etc., will protect me."
>
> *sairandhryah ... arakṣitānyeneti. rakṣitastrīkā iti pāṭhe pararakṣitāḥ santi. bhujiṣyāḥ paricārikā etā ātmanaiva kṛtarakṣā ekapatnvratāś ca. tad evaṃprakārā bhavanti. tatra striyo manuṣyād uttamapuruṣānurūpam anyaṃ tā*

na yānti yathetare janāḥ kṣudratvād eveti loke niścayaḥ. tad evātmaguptā cariṣyāmīti. ekapatir eva vrataṃ yāsāṃ tās tathā. sārathisūdādhyakṣādayaḥ kumārā indrādayaś ca māṃ rakṣiṣyantīti gūḍharthaḥ.

Caturbhujamiśra proposes an alternative reading of the phrase *naivam anyāḥ striyaḥ*, "no other women ... as they do" in the second line of Draupadī's speech. He breaks up the sequence as *naiva mānyāḥ striyah* to read "no respectable women." He says:

> Even though they are servants, *bhujiṣyās* are respectable women. They are attractive but they do not go out, that is to say, they do not go forth seeking men. Therefore, they are quite protected. That is the syntax. "They go about protected" means "being protected, they pass the time." Thus, *bhujiṣyās* are not [really] servants.
>
> *bhujiṣyāḥ dāsyo'pi satyo mānyāḥ abhigamyā nābhiyānti nābhigacchantīti pumāṃso 'rthyante atas tā ve(ra?)kṣitā evety anvayaḥ. rakṣitā yānti rakṣyamāṇāḥ kālam ativāhayanti. bhujiṣyā dāsyo na tathā.*

Sarvajñanārāyaṇa, however, is more definite in his judgement. He says, tersely, "A *bhujiṣyā* is a female servant" (*bhujiṣyā dāsī*).

In the following *adhyāyas*,[48] the Pāṇḍavas and Draupadī enter the Matsya kingdom one by one and approach King Virāṭa requesting positions at court consistent with their earlier statements of intention. Once again, the commentators exert themselves in an effort to rationalize their claims regarding their skills and prior service at Yudhiṣṭhira's court in Indraprastha. But much of this repeats their comments on the earlier chapters where they display their greatest grammatical knowledge and lexical ingenuity to show, with almost fierce determination, that these noble figures, so central to the culture's image of elite manhood and womanhood, cannot possibly be guilty of violating the strict code of truthfulness.

On the basis of the texts treated above, I believe that my argument here has been amply confirmed. Although the medieval and early modern commentators on the *Rāmāyaṇa* and the *Mahābhārata* are well known for their theological analyses, debates, and rationalizations concerning the divinity of the epics' respective *avatāra*-heroes, Rāma and Kṛṣṇa—no doubt under the influence of purāṇic and regional-language devotional versions of the epic

48 Mbh 4.6–12=GPM 4.7–12.

stories—they are also deeply concerned with showing that, despite the apparent surface meaning of their texts, its heroes and heroines do not deviate from two central social imperatives, deference to their elders and unwavering adherence to the truth. In this task they deploy to the fullest their lexical knowledge and grammatical skills, displays of learning and ingenuity.

Abbreviations

GPM Gujarati Printing Press edition of the *Mahābhārata*
GPR Gujarati Printing Press edition of the *Rāmāyaṇa*
LPR Lakṣmīveṅkaṭeśvara Press edition of the *Rāmāyaṇa*
Mbh Poona Critical Edition of the *Mahābhārata*
MVR Kumbakonam Edition of the *Rāmāyaṇa*
PR Princeton Translation of the *Rāmāyaṇa*
VR Baroda Critical Edition of the *Rāmāyaṇa*

Bibliography

Apte, Vaman Shivaram. *The Practical Sanskrit-English Dictionary*. New Delhi: Motilal Banarsidass Publishers, 2003. Reprint, revised and enlarged of the 1890 original: *The Practical Sanskrit-English Dictionary*.

Bākre, Mahādeva Gaṅgādhara Bhaṭṭa, ed. *Śrīman Mahābhāratam: Virāṭa Parva* IV. Bombay: Gujarātī Printing Press, 1915.

Bhatt, G.H., and U.P. Shah, eds. *The Vālmīki Rāmāyaṇa: Critical Edition*. 7 vols. Baroda: Oriental Institute, 1960–1975.

Brown, W. Norman. "Duty as Truth in Ancient India." *Journal of the American Philosophical Society* 116, no. 3 (1972): 252–268. https://www.jstor.org/stable/986119.

Brown, W. Norman. "The basis for the Hindu Act of Truth." *Review of Religions* 5 116, no. 3 (1940): 36–45.

Brown, W. Norman. "The Metaphysics of the Truth Act (*Satyakriyā*)." In *Mélanges d'Indianisme à la mémoire de Louis Renou. 40e anniversaire de la fondalion de l'Institut de Civilisation Indienne de l'Université de paris 1967*, 171–177. L'Institut de Civilisation Indienne. Série in 8°, Fasc. 28. Paris: Éditions E. de Boccard, 1968.

Buitenen, J.A.B. van, ed. *The Mahābhārata. Vol. 3, The Book of Virāṭa and The Book of Effort*. University of Chicago Press, 1978.

Burlingame, Eugene Watson. "The Act of Truth (*Saccakiriyā*): A Hindu Spell and its Employment as a Psychic Motif in Hindu Fiction." *Journal of the Royal Asiatic Society*, 1917, 429–467.

Goldman, Robert P. "Acquainted with Grief: Rāma, Kṛṣṇa, and Medieval Debates on the Suffering of God." *Journal of the American Oriental Society*, 2022.

Goldman, Robert P. "Creating Context: Ādiśaṅkarācārya's Reading of Three Ākhyāyikās of the Chāndogya Upaniṣad." In *Self, Sacrifice and Cosmos: Late Vedic Thought, Ritual and Philosophy*, edited by Lauren Bausch, 161–181. New Delhi: Primus, 2019.

Goldman, Robert P. "Gods in Hiding: The Mahābhārata's Virāṭa Parvan and the Divinity of the Indian Epic Hero." In *Modern Evaluation of the Mahābhārata: Professor R.K. Sharma Felicitation Volume*, edited by Satya Pal Narang, 73–100. New Delhi: Nag Publishers, 1995. Reprinted in *Purāṇa* 41 (1999): 95–131.

Goldman, Robert P. "Translating Texts Translating Texts: Issues in the Translation of Popular Literary Texts with Multiple Commentaries." In *Literary Studies East and West: Cross-cultural Perspectives; Selected Conference Papers*, edited by Cornelia N. Moore and Lucy Lower, 93–106. Honolulu: University of Hawaii; the East-West Center, 1992.

Goldman, Robert P., and Sally J. Sutherland Goldman, eds. *The Rāmāyaṇa of Vālmīki: An Epic of Ancient India. Vol. 6, Yuddhakāṇḍa.* Princeton, NJ: Princeton University Press, 2009. With introduction, annotations, and translation by Robert P. Goldman, Sally J. Sutherland Goldman, and Barend van Nooten.

Goldman, Sally J. Sutherland. "Women at the Margins: Gender and Religious Anxieties in Vālmīki's Rāmāyaṇa." *Journal of the American Oriental Society* 138, no. 1 (2018): 45–72. https://doi.org/10.7817/jameroriesoci.138.1.0045.

Kane, P.V., ed. *Uttararāmacarita of Bhavabhūti.* Translated by C.N. Joshi. Delhi: Motilal Banarsidass, 1971.

Krishnacharya, T.R., and T.R. Vyasacharya, eds. *Srimad Vālmīki Rāmāyaṇa: A Critical Edition with the commentary of Govindarāja and Extracts from Many Other Commentaries and Readings*. Bombay: Nirṇaya Sāgar Press, 1911–1913.

Lutgendorf, Philip. *The Life of a Text: Performing the Rāmcaritmānas of Tulsidas*. Berkeley: University of California Press, 1991.

"*Mahābhārata* of Vyāsa (2nd Century Bce–1st Centruy Ce)," n.d.

"*Manusmṛti* of Manu (2nd Century Bce—2nd Century Ce)," n.d.

Minkowski, Christopher. "Nīlakaṇṭha and His Historical Context." *Vostok* (*Oriens—Moscow Academy of Sciences*) 4 (2008): 37–49.

Minkowski, Christopher. "Nīlakaṇṭha's Mahābhārata." *Seminar* 608 (2010): 32–38.

Minkowski, Christopher. "What Makes a Work 'Traditional'? On the Success of Nīlakaṇṭha's Mahābhārata Commentary." In *Boundaries, Dynamics and Construction of Traditions in South Asia*, edited by Federico Squarcini, 225–252. Florence: Firenze University Press, 2005.

Mudholkar, Shastri Shrinivas Katti, ed. *Rāmāyan of Vālmīki. With three commentaries called Tilaka, Shiromani, and Bhooshana.* 7 vols. Bombay: Gujarātī Printing Press, 1914–1920.

Nārāyaṇ, Rām Āchārya, ed. *The Manusmṛti: With the Commentary of Kullūka*. Tenth. Bombay: Nirṇaya Sāgar Press, 1946.

Pillai, Sohini Sarah. "Blessed Beginnings: Invoking Viṣṇu, Kṛṣṇa, and Rāma in Two Regional Mahābhāratas." In *Many Mahābhāratas*, edited by Nell Shapiro Hawley and Sohini Sarah Pillai, 257–278. Albany, NY: SUNY Press, 2021.

"*Rāmāyaṇa* of Vālmīki (4th Century Bce–1st Centruy Ce)," n.d.

Rao, Ajay. *Re-figuring the Rāmāyaṇa as Theology: A History of Reception in Premodern India*. Abingdon, UK: Routledge, 2015.

Rao, Ajay. "Theologising the Inaugural Verse: Śleṣa Reading in Rāmāyaṇa Commentary." *Journal of Hindu Studies* 1, nos. 1–2 (2008): 77–92.

Roy, Pratap Chandra, trans. *The Mahabharata of Krishna-Dwaipayana-Vyasa: Translated into English Prose from the Original Sanskrit Text. Virata Parva*. Calcutta: Datta Bose & Co, 1923.

Sharma, Vishal. "An Argument with God: Bhagavannindā in the Mahābhārata Tradition." PhD thesis, University of Oxford, 2021.

Śrīkṛṣṇadāsa, Gaṅgāviṣṇu, ed. *Śrīmadvālmīkirāmāyaṇam. With the commentaries of Govindarāja, Rāmānuja, and Maheśvaratīrtha and the commentary known as Taniślokī*. 3 vols. Bombay: Lakṣmīveṅkaṭeśvara Mudraṇālaya (Veṅkaṭeśvara Steam Press), 1935.

Sukthankar, Vishnu Sitaram, ed. *The Mahābhārata: For the First Time Critically Edited*. 24 vols. Poona: Bhandarkar Oriental Research Institute, 1933–1970. With the cooperation of Shrimant Balasaheb Pant Pratinidhi and Others.

"*Uttararāmacarita* of Bhavabhūti (8th Century Ce)," n.d.

Vira, Raghu, ed. *The Virāṭaparvan: Being the Fourth Book of the Mahābhārata: The Great Epic of India*. Poona: Bhandarkar Oriental Research Institute, 1936.

Vivekatilaka of Varadarāja Uḍāli (Uḍāri). Baroda: Oriental Institute, 1964. Undated Devanāgarī transcription of manuscript of Rāmāyaṇa commentary, transcript no. 25664, copied April 6, 1964.

CHAPTER 15

Disruptive Readings: Sanskrit Literary *Commentary* as *Translation*

Deven M. Patel

Students of Sanskrit regularly discover that learning to translate and interpret a text are indistinguishable acts. To translate and to interpret means to solve problems, to make decisions, to locate context for words and sentences, and to read between the lines to make sense of what is metaphorical and unexpressed. While grammar is famously indispensable to make sense of words (*na ca antareṇa vyākaraṇam laghunā upāyena śabdāḥ śakyāḥ jñātum*),[1] most readers of Sanskrit literature (who are not solely using a dictionary) have come to not only rely on the ubiquitous *ṭīkā* but also to warmly embrace its existence. In fact, the *ṭīkā*—often technically understood as limited to an interlinear "gloss" to clear up semantic difficulties but can be more broadly understood in Sanskrit tradition as "commentary" itself—is arguably *the* signature of learned and literate culture in India for the past one thousand years, much as *glossa* was in the European Middle Ages.[2] Prior to its recognizably settling into a familiar textual form, the *ṭīkā* presumably "moves" (*ṭīk*) or disturbs the source in some way, or rather it moves through the source, a "running" commentary, as it were. Perhaps one may think of it as moving the reader closer to the text or, conversely, moving the text closer to the reader. As an *effect*, the *ṭīkā* undoes some knotty word or passage or shakes loose an embedded sense. Before exploring in this essay a provisional logic that prioritizes framing the *ṭīkā* as "translation" (or pre-translation) rather than "commentary," it is important to mention that inspiration for this line of thinking comes directly from a set of groundbreaking essays on the *ṭīkā* by scholars of the Sanskrit Epic, foremost among them being Christopher Minkowski, Robert P. Goldman, and Sally Goldman, all of whom took the *ṭīkā* form on *itihāsa* and *kāvya* seriously before it became an

1 *Vyākaraṇamahābhāṣya* (*Paspaśāhnika* 3).

2 See Copeland for the parallel case with Latin *glossa*. See Ganeri for a detailed typology of the commentary genres in Sanskrit philosophical cultures. On the *ṭīkā* specifically, he writes: "A commentary whose function is only to elucidate obscure or otherwise tricky words in the text is styled a *ṭīkā* … The term *ṭīkā*, again like gloss, is also used in a more general sense, as a synonym then of [other forms of commentary such as] *vṛtti* or *vivaraṇa*." (Ganeri 2008: 109).

 | DOI:10.1163/9789004536869_017

integral part of contemporary studies on both genres. Collectively, these essays not only helped to historicize textual practices, like commentary-writing and translation, but also to think in new ways about *what* we are reading and, more importantly, *how* and *how we could* read the ubiquitous literary networks of source-and-commentary to form our own (mental) translations of complex works.

In addition to being the standard designation for "commentary" on a source text, the word *ṭīkā* (and here the term is used to stand in for any of the multiple terms and types of commentary-writing in Sanskrit) in some modern Indian languages takes the pejorative sense of (unsolicited) criticism, fault-finding, or nitpicking. Complaints about the *ṭīkā* are ancient too, of course, as well as a staple of modern Indological criticism of "native" habits of reading. Much of the anxiety about the *ṭīkā* emanates from the fact that it consistently seems to underperform the expectations that are put onto it. Usually, when a *ṭīkā* behaves as something other than a faceless, interchangeable resource to help ease semantic difficulty, or a fully coherent and satisfying interpretation of the source text, it incurs the familiar charge of being either useless or unproductive. Bhoja's famous grousing (in his *Yogasūtra* commentary [on 1.6]) about the *ṭīkā*'s "usefulness" is an oft-cited example: "Whatever is rather difficult to understand, that they avoid by saying 'It is clear'. With respect to the clear meanings, they over-elaborate with useless analysis of compounds, etc. The *ṭīkā* seems to confuse issues for readers by aimlessly prattling on unhelpfully at the wrong places."[3] In other places, if a *ṭīkā* takes up an interpretive function, and again fails to meet expectations, it is often labeled as outlandish, far-fetched, or "traditional." To give an example, Christopher Minkowski outlines, in his early essays on Nīlakaṇṭha Caturdhara's virtuoso *ṭīkā* of the *Mahābhārata*, that the helpfulness of the *ṭīkā* was appreciated by Indologists of an earlier era until it transgressed being "more than a minimal gloss or explanation":

> When Nīlakaṇṭha really asserts himself as a commentator, offering more than a minimal gloss or explanation, that is, when readers come up against what Nīlakaṇṭha really wants to do with his text, his views are often discounted, and not gently. Even the most grudging of his readers

3 *durbodhaṃ yad atīva tadvijahati spaṣṭārtham ity uktibhiḥ spaṣṭārtheṣv ativistṛtiṃ vidadhati vyarthaiḥ samāsādikaiḥ / asthāne 'nupayogibhiś ca bahubhir jalpair bhramaṃ tanvate śrotṝṇām iti vastu-viplava-kṛtaḥ prāyeṇa ṭīkākṛtaḥ*. Among the early Indologists, Bopp voices a similar complaint (see fn. 23 in Minkowski 2011). See also Karen Preisendanz' review of perspectives about "unproductive" philosophical commentaries as being either "incidental … collections of glosses or annotations" or "continuous expository commentaries that combine various types of annotation into one text." (Preisendanz 2008: 609, 611).

> acknowledge that he is occasionally useful in specific passages, but when it comes to his intellectual project, he is not only set aside; he is even deplored.
>
> MINKOWSKI 2011: 230

Minkowski's cataloguing of Indological complaints takes in not only the *ṭīkākāra*'s exegetical deficiencies but also his editorial decisions in literally producing an influential edition of the *Mahābhārata*. In the case of Nīlakaṇṭha's "intellectual project," Minkowski argues that the mistaken expectation of modern readers was that the *Bhāratabhāvadīpā* was the "traditional" understanding of the great epic. He ameliorates the discrepancy between "native" and "Indological" approaches in terms of misguided expectations and a "misdiagnosis" of Nīlakaṇṭha's "traditionalism" among a range of modern readers of the Epic:

> Modern complaints about Nīlakaṇṭha's commentary found their source here: expectations arose that were ill-adapted to what Nīlakaṇṭha's work was for. Those expectations were inevitably disappointed. How is it that Indologists took Nīlakaṇṭha's work to be 'traditional'? There was, first of all, the murky conception among Indologists of what being 'traditional' amounted to. Moreover, they would have encountered Nīlakaṇṭha's text and commentary in widespread use among 'native' readers. This fact, coupled with the predisposition to see Sanskritic literature as part of a timeless and changeless universe of discourse, would have given rise to a judgment of Nīlakaṇṭha's work as typical of the 'traditional' understanding of the text.
>
> MINKOWSKI 2011: 233

Momentarily setting aside the question of what Nīlakaṇṭha's *ṭīkā* was for, which is examined below, and the tension that binds conceptions of "native" with "non-native" reading horizons—which are certainly significant for unmasking the politics of philological practice—this passage significantly signals that the very same markers of distinction for variable commentarial and editorial acts hold for forms of *translation* as well: choices are made on the basis of the reader's intellectual project (historicist, pragmatic, semiotic, etc.), their attitudes toward the text (fixed and semantically invariant, somewhat unstable and polysemous, totally open and reader-dependent), and the hermeneutic strategies found to be acceptable and useful to enact a specific interpretive vision.

While seeing the *ṭīkā* as a special kind of translational act would certainly not dissolve complaints about its minimalist approach and deficient results, it may attenuate charges of unproductiveness and inadequacy. In fact, exploring the ways in which the *ṭīkā*—as incomplete commentary and inadequate translation—may offer some clarifying context for historicizing ancient and medieval reading practices as well as mentoring another way to read today. The explanatory inadequacy, semantic incoherence, and fragmentary form of the *ṭīkā*, widely viewed in terms of complaint, could be seen as a purposefully sufficient and pedagogically enlightened approach to reading a source, whose potential for significance is eminently spacious. The *ṭīkā*, in this light, proves to be not a "commentary" nor a "translation" but rather an in-between genre, marked by very specific lexical and methodological strategies to think about commentary, translation and interpretation as ultimately providing, at turns, the framework to disrupt, to harmonize, or occasionally to resist the heterogeneous possibilities of the source text.

1 Incomplete Commentaries and Inadequate Translations

In a pioneering essay on Epic commentaries, Robert Goldman tantalizingly suggests that we take the several dozen Epic commentaries on the *Rāmāyaṇa* to "collectively constitute ... a massive translation project" but immediately interrogates the suitability of calling these works "translations":

> These texts are not, of course, translations in the commonly accepted sense of the term. For one thing they are in the same language—more or less—as their sources. Then, too, they are not coherent narratives but rather complex series of glosses, re-phrasings, grammatical exegeses, rhetorical analyses, and hermeneutics of the poems on a verse-by-verse basis which treat passages as their authors deem it appropriate to do so. The complexity of these documents is considerably enhanced by the fact that 1.) they are in many cases powerfully and explicitly intertextual among themselves, and 2.) their authors not only cite the opinions of other commentators with which they may agree or disagree, but very frequently provide as many as five or six alternative explications of their own.
>
> GOLDMAN 1992: 93; 96

This passage summarizes the key features of most Sanskrit *ṭīkā*: intralingual (Sanskrit-to-Sanskrit); narratively incoherent; organized as a complex series of philological acts (gloss, rephrasing, grammatical exegeses, rhetorical analysis, and an author and reader-sensitive "hermeneutics"); and finally, an explicit exploration of polysemy. One might here amplify the point about "incoherence" to say that most *ṭīkā*—especially those on *kāvya*—consistently strike the reader as incomplete and unfinished, especially when it comes to expectations about interpretation, or meaning-making, that burden the genre among contemporary readers. One surmises that the hesitation in calling these works *actual* translations stems not only from the obvious fact that they do not meet the equivalency expectations of what is commonly understood to be translation *proper* (the interlingual translation)[4] but also that they are visibly fragmentary and, like notes or passing glosses, only guide or inform or suggest a translation and not actually perform one.

While little to no emic theorization on the literary and epic *ṭīkā* exists[5] (nor, for that matter, for translation), ancient authors of philosophical *ṭīkā* have more explicitly theorized their reading practices and, more recently, modern scholars have attempted to insightfully reconstruct emic views on translation practices from the fifteenth century until the present moment.[6] In his survey of words for "translation" in Indian languages, for example, Brian Hatcher rightfully dismisses the neologisms attributed to the practice (*anuvāda*, *bhāṣāntara*, and so on) as generally imprecise. None of these terms accompanies a tradition that theoretically reflects on transferring meaning within or between languages, or from genre to genre.[7] Regarding the relationship of comment-

4 Roman Jakobsen formulated a now-canonical tripartite division of translation as follows: intralingual translation or *rewording* is an interpretation of verbal signs by means of other signs of the same language; interlingual translation or *translation proper* is an interpretation of verbal signs by means of some other language; intersemiotic translation or *transmutation* is an interpretation of verbal signs by means of signs of nonverbal sign systems. (Jakobson 1959/2000: 114).

5 Sheldon Pollock's statement concisely frames the "unthematized" nature of the literary commentary, despite scholarship on the "divergent functions" and "various discursive formats": "[We] know next to nothing about how their three most important tasks of philological commentary were executed: text constitution (commentators were also editors); emendation (they were also text critics); and analysis (they were also interpreters). These matters have been unstudied because they are unthematized in the tradition itself; the data to understand them were never assembled in any work of systematic thought but only exist dispersed in the commentaries themselves .... [W]e must infer their precepts from their practices." (Pollock 2015: 118).

6 On thirteenth, fourteenth, and fifteenth century literary translations (intra- and interlingual) of Sanskrit literary classics, see Patel 2011; on translation practices among seventeenth-century Jain communities, see Cort 2015; for post-nineteenth century studies of translation (and translation theory relating to Indian literature), see especially Hatcher 2017, Merrill 2009, Devy 1999, Chandran 2014, Gopinathan 2019, Choudhuri 2010, and Mukherjee 1994.

7 Cort summarizes some valiant attempts at naming the practice: "As alternatives, one might apply a widespread tripartite distinction used by modern authors in all the North Indian

ary and translation, scholars usually focus on the ways in which medieval and modern translations from a source to target language utilize, reflect, or mimic the commentary, or how the translation can be seen as a "commentarial reading." Hatcher unpacks this point in discussing the translation methodology of the nineteenth century writer, scholar, and translator Vidyāsāgara:

> At the same time, [Vidyāsāgara] refers directly to the challenges attendant upon achieving equivalence (*aikya*) between the original text (*mūlagrantha*) and the translation. In this connection, he says that in many cases he found it necessary to follow the lead of earlier Sanskrit commentators in order to make sense of the meaning (*artha*) and the purport (*tātparya*) of certain passages. Thus, having introduced the ideal of the accurate translation (*avikala anuvāda*), he seems to fall back on another understanding of *anuvāda* as commentarial reading.
>
> HATCHER 116

From an earlier era, several centuries before Vidyāsāgara, authors were already capturing lexical and syntactic elements from Sanskrit commentaries on canonical works to populate their own retellings of the Sanskrit classics, which amounted to something akin to a translation but not quite in the conventional way that translation is understood today or in other premodern contexts.[8] The example of the seventeenth-century poet Banarsidas' Hindi *Nāmamāla*, nominally based on a Sanskrit work of the same name, is illustrative. While Banarsidas apparently studied the work with his Sanskrit teacher, Cort writes that in whatever manner the Hindi *Nāmamāla* is conceived as a translation, "to characterize Banarsidas's *Nāmamāla* as a translation clearly stretches any definition of this genre near to breaking point" (Cort 2015: 81), underscoring the point that the Sanskrit commentaries and the Sanskrit pandits who either taught through them or embodied their process in oral pedagogy continued to inform the vernacular work even though no "direct reference to a Sanskrit original" was made.

The oral pedagogy that constitutes a tacit communique that binds generations of teachers and students in shared literary culture seems to manifest as

---

languages, among translations as *shabdarth* (literal, word-for-word translation), *bhavarth* (translating the underlying deeper message and intention), and *chhaya* (free adaptation or "transcreation") … Another set of distinctions was employed by Saroj Agraval in her book on the *Prabodhacandrodaya* tradition in Hindi, in which she distinguished among *anuvad* (translation), *rupantar* (adaptation), *svatantra* (texts influenced by or in the style of the original), and *anshatah* (texts partially influenced by the original)." (Cort 93, fn. 103).

8 See Patel (2011) and Cort (2015).

a formal dialogue of sorts within the *ṭīkā* format. Goldman aptly suggests that the *ṭīkā* on the *Rāmāyaṇa* seem to stage a dialogue within its borders and, in turn, invite dialogue from the outside:

> [T]aking the view that the commentators share with us a fundamental desire to "translate" this ancient poem in all its richness and ambiguity so that it can be more fully understood by a contemporary audience—whether in medieval India or twentieth century America—we have attempted to enter into a series of "dialogic" exchanges with the commentators and with other modern translators in every case in which the meaning of a passage in the text is either significantly contested or subject to a multiplicity of readings.
>
> GOLDMAN 1992:96

Outside of a centrally shared "struggle to disambiguate the meaning(s) of a passage and also seek to creatively reambiguate the text in pursuance of a set of larger agendas" (Goldman 1992:97), the differences *in function* between the commentary and translation, conventionally understood, invite reflection. To begin with, being in the same language as the source, the commentary comes across as a process, as that which precedes and informs an effort to translate. Outside of the occasional *anvaya* (syntactical reordering), the *ṭīkā* leaves behind reordered lexical and semantic puzzles for the reader to mentally piece together for a finished glimpse. It occasionally articulates ambiguity, juxtaposing options, implicitly (usually) weighing them with a series of "or else's" (*yadvā*) or some other such formula. The final product here—if this form of translation can be viewed as a product—is necessarily unfinalized in the commentary. Translation *proper*, on the other hand, *only* emerges as a final product, having shed all of the hermeneutic scaffolding, as it were, upon which it was built.

And yet, for all that distinguishes the two activities—if, in fact, we insist on seeing them as separate activities (distinct from being separate *products*)—much is shared. First, the standard specter of fidelity haunts them both. Second is the suspicion of competency, both textual and contextual.[9] Here there is an important difference, however. Whereas rarely will the traditional Sanskrit exegete derive a starkly incorrect surface reading/meaning, grumbles about interpretation/propriety or "usefulness" may surface from time to time; for the

9 Goldman (1992: 95) cites and translates a traditional Sanskrit maxim on the importance of context when "translating" the source for oneself: "when in doubt (about the meaning of the word), derive the correct sense from the context" (*sandigdhe vākyaśeṣān nirṇayaḥ*).

translator, anxieties about faithfulness—to word, meaning, and significance—and "adequacy" seemingly hang around the neck through every step of the process. While wary of removing the scare quotes and implying that *ṭīkā* are functionally *translations*, it may not be too disagreeable to see that both activities—the literary commentary and translation—are essentially hermeneutic, especially for genres like *kāvya* (and Veda), where linguistic-literary effect underwrites not only its value for the author and audience that respectively produce and consume it but also gives it distinction as a discrete cultural form.

In resisting a hasty conflation of *ṭīkā* with translation and bracketing the convention to assume it to be an unproductive commentary or merely a helpful gloss, a provisional position might understand the *ṭīkā* as delivering a close-to-the-text reading that makes more elaborate readings possible. As such, on the one hand, it is a pre-commentary, or a pre-translation, a service to those who would draw from this preliminary practice to interpret or retell (if not create an equivalency with) a source. On the other hand, for those uninterested in formally reconstituting the semiotic possibilities of the source into a true expository commentary (*bhāṣya*) or interpretive essay (*nibandha*), or a translation *proper*, the *ṭīkā* still functions as a standalone reading for informal understanding and pedagogical utility. In this regard, the *ṭīkā* is a work complete in its incompleteness, mediating a familial relationship between the commentarial and translational act. While it does not, in itself, directly transfer meaning between two languages or two cultures—the necessary condition most would require to call it a "translation"—it does constitute the very same process that precedes all translation, in its marshalling of the lexical, intertextual, and hermeneutic substrata to complete the act; and furthermore, it factors in the very same scope and purpose a translation would: the capacity of the reader's comprehension, the temporal distance that has elapsed between source and reader, references and allusions in need of being supplied, and the level of elaboration necessary to meet the specific needs of the audience.

That the majority of Sanskrit texts are commentarial, the majority of which are *ṭīkā* with its characteristically limited elaboration, their sufficiency and usefulness for the Sanskrit literary tradition may lie beyond the value they provide as information bearers. They carry forward, rather, a hermeneutic, pedagogical, and material logic that validates their numerical magnitude. The significance of writing a *ṭīkā* on literary texts for pedagogical reasons and for justifying credentials in institutional contexts has been briefly treated elsewhere;[10] however,

10 See Patel (2014: 59–65) for a brief discussion about the relationship of commentary writing with institutions of scholastic culture.

the potency of the *ṭīkā* form in literary contexts that is consciously understood as a prerequisite for creative and interpretive work has been relatively under-explored. While the fact that emic theorization of the literary *ṭīkā* is virtually non-existent,[11] and sustained deliberation into Indian practices of translation missing until relatively recent times, the closest place to locate a potential hermeneutic logic for the literary *ṭīkā* is in the so-called "scriptural" commentary performed, and occasionally theorized, in religious contexts.

2 Religious Reading Paradigms

Modern scholarship on the philosophical and broadly religious or "scriptural" commentary in Sanskrit has bloomed in the past several decades. Sustained philological labor has made available to non-specialists the kind of exegetical practice—and nascent theoretical meta-reflection—that Jain, Buddhist, and Mīmāṃsaka authors conducted for centuries. As prelude to explaining the stunningly thoughtful practices of *anuyoga* and *nikṣepa* in Jain hermeneutics, for instance, scholars of Jain commentary have drawn attention to that tradition's need for a "special hermeneutics for accessing a level of meaning that is beyond the [lexical, syntactical, and semantic] level of language" (Jyväsjärvi 2010:134) since the ancient Jain sūtra "without some sort of accompanying commentarial explication is equivalent to somebody who is asleep" (Dundas 1996: 74).[12] The use of the phrase "beyond" language here locates significance outside of the text, whereas the expression of significance lying "beneath" language (asleep, perhaps) suggests a different metaphor of access and transference. Nevertheless, the same holds for both the Jain *sūtra*, the Vedic *mantra*, and *buddhavacana*: that the access point is still the various linguistic levels and that the attitude taken to the source is still either one of a hidden box in need of skillful opening or a sleeping person to be adeptly awakened.

More interesting than the urge to extract a hidden, or "sleeping," meaning—which is common to many charismatic *ṭīkā* authors across genres—is the particular process Jain exegetes map for themselves. Two particular practices stand out in Śvetāmbara hermeneutics for what Jyväsjärvi (2010: 135) describes as

11 See Pollock (2015: 124–127) for the interesting example of the fourteenth-century commentator Aruṇagirinātha, who articulates the types of readers for whom his *ṭīkā* on the *Kumārasambhava* is composed.

12 Dundas explains that the eleventh century commentary Abhayadeva Sūri "claims that 'sūtra' can also be derived from *supta*, "asleep" on the grounds that scripture is effectively unawakened when without a commentary." (Dundas 1996: 84).

their emphasis on "commentary-writing as a form of thinking, articulating, and debating ideas." One of them, called *anuyoga*, "conjoins" scriptural text "with its broadest connotative context [to bring it] into full association with the complexity of reality" (Dundas 1996:76). Like the word *ṭīkā*, or *vyākhyā*, or indeed "commentary" and "translation," *anuyoga* seems to be a philological cluster concept that groups together a range of hermeneutical procedures that sequentially climbs from a pre-discursive level of words and referents to access points that take in wider questions of meaning and doctrine.[13] In conjunction with the more general *anuyoga* method is the interpretive technique called *nikṣepa*, where "each of the significant words of the root text are subjected to a systematic analysis from a codified set of angles—four, six, or more ... examined from the points of view of name or designation (*nāma*), representation or effigy (*sthāpana*), the concrete, material aspect or substance (*dravya*), and the inner reality or moral, spiritual, essential aspect (*bhāva*)" (Jyväsjärvi 2010:149). Most significant about this technique is that it implies that the sūtra—the source—is intrinsically open-ended, partial, and unfinished:

> [Nikṣepa allows] the commentator to explore all the facets and connotations of a word before settling on one, which he then determines to be the topic (*pagati, ahigāra*) ... [W]ords can make only a partial meaning explicit, while indicating more through synecdoche. In order to be able to interpret the rules correctly, one has to become a connoisseur of words. Synonymy serves a similar function, as it explores the connections between signifier and signified from the opposite perspective: different names for a single phenomenon, or words belonging to the same semantic field.
>
> JYVÄSJÄRVI 2010: 151

Whereas *anuyoga* emphasizes a more universalizing gesture of mechanically reproducing a philological practice common across genres and texts, a *nikṣepa* approach to the sūtra seems to actively embody, and thus endorse, a willfully provisional practice without which the source does not optimally function

13 Jyväsjärvi identifies a late fifth-century text that unpacks the range of methods: "The principles of *anuyoga* were laid down in the *Anuyogadvārasūtra*, a late (fourth-fifth century CE) canonical text, which lists four *anuyogadvāra*, 'doors' or 'points of access' of disquisition: *upakrama* or introduction, *nikṣepa*, *anugama*, or further explanation, and the seven *nayas* or standpoints ... Thus, exactly what is included under the term anuyoga turns out to be somewhat ambiguous [Saṅghadāsa: combined presentation (*niyoga*), exposition (*bhāṣā*), detailed exposition (*vibhāṣā*), and exhaustive exposition (*vārttika*)." (Jyväsjärvi 2010:147).

(even though it may *exist*) for the individual reader, the intimate community of readers, and the cultural and pedagogical institutions that sustain and proliferate it. Special attention here might be given to the consciously unfinished nature of the interpretive act according to *anuyoga* and *nikṣepa* procedures, resistant to the temptation to make static or fix an invariant meaning. It should be noted, however, that both of these practices operate on a "microlevel of interpretation" that appear to be "skirting the topic," something Jain commentators like the sixth-century Saṅghadāsa are well aware of, even if they do not care to expatiate on what that awareness entails.[14] Saṅghadāsa's silence may imply that effective pedagogy often stops short of providing a translation and certainly stops short of providing an interpretation. However, something *akin* to a translation is still necessary and the *ṭīkā*'s microlevel gloss and allusion maintains the integrity of the original while simultaneously laying the groundwork for interpretation. The author and the reader/translator are co-creators of the meaning in this scenario, especially the underlying meaning.

Furthermore, the resultant incompleteness of the *ṭīkā*—disturbing the source out of its slumber and yet itself remaining sketchy and elusive—need not indicate its inadequacy nor necessarily even a quaint plurivocal humility that takes in all perspectives; rather, the in-process semblance of a translation might intimate a creative potency that engenders an escape from the tempting leash of fixed reading agendas. Speaking now about the overwhelming presence of *ṭīkā* literature in Sanskrit, one surmises that the inability to easily *use* documents such as these for historicist, culturalist, or even literary-critical purpose—as they rest comfortably in their unfinishedness—may account for much of the annoyance or, more commonly, indifference that this massive archive provokes. Yet, they centrally exemplify "traditional" Sanskrit philological practice: rooted in oral pedagogy and tending toward granting capacious interpretive latitude to the student or reader who recognizes the teacher/commentator's back-and-forth between the source and himself as a reflexive method that effectively disrupts direct communication with the source while also transforming that communication and, with it, the source itself. The reading then becomes as much creative as it is analytical.

As with the Jain religious readers, so too seems to be the attitude toward source and exegesis among some Buddhist and Mīmāṃsaka commentators.[15] Both of these traditions view their omniscient source, *buddhavacana* and Veda

---

14 In Saṅghadāsa's words, "the masters of *anuyoga* preach while leaving out the [apparent] topic." [*mottūṇam ahigāraṃ aṇuyogadharā pabhāsanti* (*Bṛhatkalpabhāṣya* 3025)] (Jyväsjärvi 2010:150).

15 See Eltschinger (2018) and Gornall (2020).

respectively, as infinitely generative and draw on methods of reading them by staying as close to the language of that source. Pāli and Sanskrit scholars in the middle of the first millennium CE explored ways in which the omniscient teaching of the Buddha could be mediated through exegetical procedure to not only understand but also to interpret, imitate, and ultimately pedagogically perform *buddhavacana* itself though commentaries. These interpretive practices first entailed an intensely focused and immediate encounter with the words of the sources themselves. Maria Heim, for example, explains the ways in which Buddhaghosa distinguishes the multiple approaches one takes with the source text—restating it concisely or with elaboration, working out the semantics, phrasing, conventions, ambiguities, contexts, and so—ultimately seeing these practices as "descriptions of the Buddha's words that then provide interpretative cues for the commentator" (Heim 2021: 60). The elusive and quiet provocation of *buddhavacana*—culminating in "interpretative cues for the commentators"—shapes not only the commentary's content but also informs, Heim suggests, the commentarial *practice* itself which, following the Buddha's example, guides the devoted student to discover rather than receive insight.[16] As Heim succinctly states the process: "[G]rasping *how* the Buddha taught was prerequisite to grasping *what* he taught." (Heim 2021: 6). Most literary *ṭīkā*—and the oral pedagogy it represents—seem to relate to their source in a similar fashion, inviting the reader to engage in a direct relationship with the source text so as to enter its semantic and affective world with as little mediation as possible.

Similar to Buddhaghosa's example is the approach of Vasubandhu, whose *Vyākhyāyukti* gives examples for how philosopher-commentators should proceed in encountering *buddhavacana* (which Vasubandhu apparently sees as

16 Heim thus reads the Buddha's "omniscience" to give "a name to features of the textual corpus that he found immeasurably and infinitely generative. This is to say, perhaps the Buddha's omniscience was something that commentarial work with the texts *discovers*, rather than (only) presumes. For example, the modular nature of texts, passages, and categories that Buddhaghosa encounters and enacts can be seen as one way in which texts generate new meaning and continued to speak to multiple (or perhaps infinite) contexts. What we might call rampant intertextuality was for him the unlimited application of the teachings of an omniscient being." (Heim 2021: 15). While arguing for a similar view of the source-commentary relationship as essentially unitary, Eliot Deutch explains the process as more a "recovery" rather than a "discovery": "In making this argument I am not claiming that we have some completed original text as such on which diverse commentaries were written, but rather that what constitutes the text in Indian thought is precisely the *sūtra* (or *karika*) and/or other authoritative sources, together with the ongoing exegetical work. In Indian philosophy we have as the basic unit what we might call the "tradition text": the philosophical content of a 'school', in the best sense of the word." (Deutsch 1988: 169).

partly commentarial in its own right) rather than systematically following, as Richard Nance explains, "general hermeneutic principles"; consequently, the "art of commentary," for Vasubandhu, "is best learned by imitation" (Nance 2015: 228).[17] The conclusion reached by Nance correlates well with pedagogical impulses suggested elsewhere in the Sanskrit exegetical tradition, especially the Vedic exegetical tradition and the other modes of commentary that stem from it.[18] Seeing the imitative pedagogical processes of commentary-writing and the underlying desire to explain the source as belonging to a shared praxis analysis marks a return to an earlier point made that interpreting and translating go hand-in-hand, as for the Buddhas (or any source), teaching and commenting proceed without distinction. Thus, the pedagogical function is constitutive in the commentarial function and, as such, is hyperconscious of the range and multiplicity of content and the manner of delivering that content to a student or uninitiated reader. To an extent, therefore, the source is looked upon as both a complete repository of meaning and yet a perpetually unfinished product, in a manner similar to the way in which Deutsch describes a "tradition text" in Indian philosophy as "a kind of authorless truth … that becomes something new with each vital engagement with it" (Deutsch 1988: 170). Usually, after the initial disturbance that a *ṭīkā* initiates on the received source, restating and augmenting its language, any creative

17 On the commentarial nature of *buddhavacana* itself, Nance writes: "Vasubandhu is thus concerned not only to provide commentary on the fragment, but also to read the fragment as itself, in part, a commentary—a move suggesting that he understands the speech of the Buddhas to be partly commentarial … When speaking, Buddhas do not typically signal a shift from teaching to explication; the distinction between teaching and explication is left unmarked in the scriptural passage itself" (Nance 2015: 230). See also Gornall (2020: 88–118) for a comprehensive study of Pāli commentaries on Abhidhamma handbooks by scholar-monks in twelfth-century Sri Lanka that "acknowledged a separate reality of for scripture beyond its exact wording," adopting exegesis-related methodology of works like the *Vyākyāyukti*.

18 Kiyotaka Yoshimizu describes the importance of the communique shared between teacher and student in influencing the method and practice of Vedic exegesis: "The expounders of the Veda, however, do not spontaneously decide on which word the structure of a Vedic text is focused. A student has to learn how to interpret a Vedic text from his teacher, who also previously come up as a student, learned the interpretation from the teacher's own teacher. Thus, every expounder has a preceding expounder from whom he inherits the method of how to interpret the text. Therefore, it is possible to say that the intention that a person signifies in expounding a Vedic text is, contrary to the intention of the speaker in everyday discourse, not the cause that establishes the meaning of the text, but the effect evoked by the self-determining text of the Veda. Expounders may make an earnest effort to inform their students of the sentence meaning, which was, however, already established within the text in advance." (Yoshimizu 2008: 57).

engagement that might follow stems from this initial engagement that leaves something to the imagination, a sort of pre-commentary or pre-translation that is both emptied of the burden of fixed significance but also capable of manufacturing it. Writers of scriptural commentary generally display both of these processes in their work—restating the source and reconfiguring it for wider significance—while many *ṭīkā* works on literature remain content to provide the pre-translational support without an accompanying exposition of significance.

The approaches taken by Buddhaghosa, Vasubandhu, Saṅghadāsa, among others, as well as most commentary writers in all genres, surely resonate most deeply with the longstanding Mīmāmsaka approach to the Veda, the omniscient source *par excellence* in need of explication and interpretation. The hermeneutic procedures by which individual words, sentences, and coherent text units of the Veda are theorized to be meaningful and, especially in ritual contexts, unified and purposeful have been widely influential for Sanskrit textual practices in all subsequent genres.[19] Unlike the Veda, however, or *buddhavacana* and the canonical Jain *sūtra*, applications of these methods for *kāvya* are not governed by the ultimate requirement for the source to ultimately signify an authoritative truth or even foundational logic for moral action. As such, unless a choice is made to reconfigure the linguistic units to *explicitly* arrive at a set of meanings that can then be directed to a singular interpretation, the commentary on *kāvya* can satisfy pedagogical and readerly expectations by simply adopting a position of neutrality toward the text, akin to what the fifteenth-century commentator Mallinātha claimed to be doing when he wrote in each of his commentaries on Sanskrit *mahākāvya* that he was not presenting anything that was not in the source (*nāmūlaṃ likhyate kiṃcit*). This emphasis on the linguistic elements does not preclude higher-level interpretations nor arguments

19 McCrea (2000: 449) offers a concise statement of the Mīmāṃsaka's approach to the Vedic text: "The job of the Mīmaṃsaka is, by a careful analysis of the functional connections between such a sentence and other surrounding ones, to settle upon one of these many possible meanings as contextually appropriate; one must determine which elements of the sentence are predicated (*vidhi*) and which form the subject of predication (*anuvāda*), which are primary and which are subordinate, which are qualified and which are qualifiers, which are 'intended to be expressed' (*vivakṣita*) and which, if any, are 'not intended to be expressed' (*avivakṣita*) .... [T]he process by which sentence-meanings mutually specify one another to produce a specific textual meaning is analogous to that by which individual word meanings within a sentence mutually specify one another so as to refer to particular entities." See also Tubb and Boose (2007) and Patel (2014) for examples of the way in which Mīmaṃsā-inflected approaches to textual meaning-making influence *kāvya* commentary writing.

over the correctness of one interpretation over the other. Without insinuating that *only* the source is needed, the deftly governed mediation the *ṭīkā* provides reorients the readers' gaze toward the source without forcibly tethering them to it. The very format of the interlinear *ṭīkā* allows for an in-betweenness that, by staying close to the text's words, resists being immediately assimilated into another discourse.

Among some of the scriptural commentators discussed above, there seems to be a shifting, sometimes oscillating, emphasis from the doctrinaire aspects of a commentary to its method for delivering doctrine. For scriptural sources, the twin aspects of commentary (form and doctrinal truth) ultimately seem to coalesce. For the literary source, however, with its intrinsic significance not tied to any truth-affirming tradition, the method—the commentary itself on the level of verse and complete work—can perhaps remain singularly divorced from any *siddhānta* or even an ambition to reach such a culminating interpretation. The *ṭīkā* nevertheless provides the raw material to construct and reconstruct readings that either uncover a text's meaning or partially transcend the work to connect up with other semiotic systems. This process—in all of its stages—is well-represented, for example, in Nīlakaṇṭha's oeuvre.

3 Nīlakaṇṭha's Disruptive Reading

When the inclination to domesticate the source toward some familiar theological or literary-critical position arises for the philosopher, essayist, or commentator (as is the case with Nīlakaṇṭha), the process invariably commences with the bare-bones *ṭīkā*. Veda as the ultimate source that harbors all commentary seems to be at the heart of Nīlakaṇṭha's readings of the *Mahābhārata*. Christopher Minkowski's painstaking reconstruction of Nīlakaṇṭha's diverse portfolio of work—commentary and independent essays—demonstrates a method that is somewhat consonant with that of the religious readers described above and firmly in a tradition that the seventh-century Mīmāmsaka philosopher-commentator Kumārila describes in his *Tantravārttika*, where *itihāsa-purāṇa* and Veda are inextricably tied together as delivering consistent and authoritative knowledge.[20] The method rests on an interpretive structuring initiated

20 Parimal Patil, drawing from Kumārila's *Arthavādādhikaraṇa* of the *Tantravārrtika* 1.2.7, explains: "According to Kumārila, the Sanskrit epics—the *Mahābhārata* and the *Rāmāyaṇa*—were composed by Vyāsa and Vālmīki in strict accordance with Veda. They are, therefore, completely consistent with it. Each contains injunctions and an incredible

by standard *ṭīkā* practices that sequentially surge toward a greater complexity and ingenuity that ultimately culminate in a so-called *vedicizing* and *vedānticizing* of the *Mahābhārata*, which aims to draw the reader to see the interrelated patterns of shared significance written into both Veda and *Mahābhārata*. Although sometimes understood as an idiosyncratic mode of commenting, which it certainly is in Nīlakaṇṭha's case, drawing Veda and *Mahābhārata* together in a shared interpretive space has an established pedigree.[21]

Minkowski traces the arc of Nīlakaṇṭha's commenting process across the various textures of his *ṭīkā*—from restated glosses of earlier commentators like Devabodha to a unique brand of analytical ones—to his occasionally remarkable readings on certain passages (especially the *Harivaṃśa*) that flow into his more second-level allegorical, intertextual, and intersemiotic injections of his *ṭīkā* into Vedic and Purāṇic passages. Most of these works, Minkowski explains, take shape in a genre that brings out the secrets of Vedic mantras (*mantrarahasyaprakāśa*), which Nīlakaṇṭha produced "after the commentary on the *Mahābhārata* was largely complete":

> In these highly idiosyncratic works, Nīlakaṇṭha assembled verses taken from the *Ṛgveda*, and constructed an interpretation of the verses in such a way as to elicit from them the whole narrative of the *Rāmāyaṇa*, or the Kṛṣṇa story from the *Bhāgavatapurāṇa*, or other works. It is this mode of assembling and interpreting Vedic verses that I am calling *vedicizing*. It is characterized by a running, word-by-word commentary on the Vedic verse, using a great deal of interpretive latitude, in order to elicit from

---

number of purely narrative episodes. Both epics are also said by Kumārila to be universally studied and accepted by the learned, and to be relevant for nearly everyone. Kumārila explains that many of the injunctions found in the epics are directly based on those found in the Veda and so are absolutely authoritative with respect to dharma. The epics are so consistent with Veda that Kumārila says that we can even reason that the epic injunctions that are not found in Veda itself are still authoritative, since they must be based on Vedic statements that were known to the learned authors of the epics but are no longer extent and so no longer known to us." (Patil 2006: 50).

21 Arguing against views that reading *Mahābhārata* and Veda together constitutes an "innovation," Adluri and Bagchee write: "The Indologists' insistence on an "epic" tradition in contradistinction to the Vedic is especially puzzling when one recalls that the Mahābhārata's self- proclaimed title, *itihāsa purāṇa*, is a Vedic genre, occurring for the first time in Atharvaveda 15.6.11. The Brāhmaṇas provide for the narration of *itihāsa* and *purāṇa* narratives between certain rites (see, for example, Śatapatha Brāhmaṇa 11.5.6– 7 and 14.5.4 and 10). Chāndogya Upaniṣad 7.1.2, 7.1.4, 7.2.1 and 7.7.1 and Bṛhadāraṇyaka Upaniṣad 2.4.10 also refer to *itihāsapurāṇa*. Chāndogya Upaniṣad 7.1.2 specifically adds that the *itihāsapurāṇa* is *pañcama veda*, the fifth Veda." (Adluri and Bagchee 2018: 160).

> the Vedic words the narration of an epic or Purāṇic episode … In the *Mantrarahasyaprakāśa* works, such episodes are then strung together to create continuous narratives.
>
> MINKOWSKI 2005: 416

As a source in need of a "special hermeneutics" to uncover hidden meaning, the *Mahābhārata* is generally not put on the same level as the omniscient Jain *sūtra*, *buddhavacana*, or Veda. *Kāvya* even less so. Critical conversations of "non-scriptural" Sanskrit works such as these probably circulated, as they seem to today, in informal contexts that nevertheless utilize *ṭīkā* within direct instructional or communal reading contexts. As Minkowski noticed about Nīlakaṇṭha's work, however, in contrast to many Indologists before him, the *Bhāratabhāvadīpa* represents a handful of works that emerge after the sixteenth century that extend the traditional practices of *ṭīkā* writing to occasionally enact what Goldman calls "bravura readings" in the *ṭīkā* space itself.

Most well-known to students of the *Mahābhārata* are Nīlakaṇṭha's stunning commentarial moves in the *Dīpā* on *Mahābhārata* 3.187 (in Nīlakaṇṭha's 'Bombay' edition; 3.185 in the Critical Edition) on the flood narrative[22] and the exhilarating readings of *Virāṭaparvan* passages, where, as Goldman explains, Nīlakaṇṭha "exploits not just the peculiarities of Sanskrit, its multiple homonymity, and the flexibility made possible by differing ways of analyzing compounds, but the larger 'lexicon' of cultural clichés to bring out—or does he create?—the speaker's 'hidden intention'" (Goldman 1992: 103) to either locate unexpected meanings in the source (note the genuine ambivalence about whether Nīlakaṇṭha "bring[s] out" or "creates") or to suppress an unwanted meaning, as in the example of Nīlakaṇṭha's (and Arjunamiśra's) exegetical gymnastics to make Bhīma's occupation of *gokartā* to mean anything but a "cow butcher" (Goldman 1992:103). Again, as neither a translator nor an independent essayist, Nīlakaṇṭha's ability or, as both Minkowski and Goldman describe, his "ingenuity" to work within the linguistic and semiotic systems of Sanskrit to either intentionally widen the text's scope (toward Vedāntic allegory or "to exculpate the Pāṇḍavas of the sin of mendacity"), or to willfully read against the grain, circumvents altogether the problem of fidelity to "translating" (and editing) the text in a certain prescribed way since the *Mahābhārata* and *kāvya*, for that matter, are not held to the standard of Veda in terms of stability and sacrality.

22 See Minkowski 2011:226–227.

Furthermore, as Goldman concludes: "There is a humbling indeterminacy about texts such as these and we must be very cautious about rejecting an interpretation, even a seemingly forced one, without trying to understand exactly what these kinds of interpretations mean and what their relationship to the contextualization of the epics really is" (Goldman 1992:104–105). If the Sanskrit commentaries are to be taken as preliminary forms of a perpetually incomplete intralingual Sanskrit translation—one that expresses the source "in other words" without formally organizing them into a fixed form—we may begin to consider the value such an exercise confers to the reader and interlingual translator. The first advantage, of course, is that by merely presenting one or more options, it attenuates the challenge of synonymy for the intralingual translator and the challenge of equivalence for the interlingual translator. Simultaneously, however, it may trigger the independent essayist to find pregnant semantic and semiotic potencies in the source language itself, both expanding the scope of interpretation and also suggesting a narrower entry point. Nīlakaṇṭha, as Minkowski demonstrates in several of his essays (especially in his 2002 essay on the *Mantrakāśīkhaṇḍa*), extends glossing practices to discover *mantrarahasya* readings in the source.[23] The case of the *Dīpā* on the *Harivaṃśa* is especially instructive about the diachronic development of more intricate readings from the *ṭīkā* springboard. The *Dīpā* on the main *Mahābhārata* books seems to have encouraged Nīlakaṇṭha to compose four independent *mantrarahasyaprakāśa* works which, in turn, nourished a mature approach to the *Dīpā* on *Harivaṃśa*. Minkowski recaps the ways in which this section of the *Dīpā* is "more ornate than the commentary on the *Mahābhārata* proper":

> Each chapter's comment begins with a versified synopsis of the contents of the chapter; there are numerous Pāṇinian analyses of forms, and copious citations of Vedic and Smṛti sources. These latter two feature are much less in evidence in the epic commentary proper, and the versified synopses are missing altogether. Thus, the text of the *Dīpa* on the *Harivaṃśa* incorporates the development of Nīlakaṇṭha's thoughts and interests over a long part of his literary career, from those of a rather doctrinaire nondualist to those of an author who is extending himself to epic and mythological subjects in unusual ways, and who is developing a distinctive, even idiosyncratic style of commentary.
>
> MINKOWSKI 2005: 417

23 For a detailed discussion of Nīlakaṇṭha's Vedānta and Vaidika readings of the *Kāśikhaṇḍa* of the *Skandhapurāṇa*, see Minkowski 2002: 329–344.

In describing the expansion of Nīlakaṇṭha's use of the *ṭīkā*, to restate the original source in creative ways, Minkowski notes that to "unlock the verse's concealed narrative," Nīlakaṇṭha homes in on "salient words" that both the Epic and Vedic text share, even when there is no transparently shared meaning between them" (Minkowski 2005: 422). The anachronistic reading (*vedicizing*), however "improbable as a literary approach, and peripheral to the understanding of the ancient epic text" (Minkowski 2005: 431), nevertheless manifests Nīlakaṇṭha's ability to maximally extract the semantic possibilities of the Epic's lexicon to extrapolate a dynamic hermeneutic reading from a gloss that never threatens the original word's semantic range. Perhaps Nīlakaṇṭha's readings seem idiosyncratic because he himself completes one of the many generative interpretive routes that a *ṭīkā* on a source text can inspire, while most authors of *ṭīkā* rarely exceed their pedagogical goals to merely make available those readings.

To take the example of Nīlakaṇṭha's *Bhāratabhāvadīpā* as an archetypal *ṭīkā* is, of course, a stressed proposition. He energetically demonstrates an ability to synthesize familiar *ṭīkā* practices—which take pains to provide access points for understanding without violating the source's expressive incommensurability—with a broader agenda to reach agreement between seemingly distinctive discourses of Veda and *itihāsapurāṇa*, and ultimately to produce new knowledge about both works. He does this by expanding his own synonymic glosses and sundry marginal notes to draw the connections between them without unmooring his project from the specific language of the sources. To call his project an interpretation, a translation, or a commentary does not quite seem to capture what he actually creates. For a historicist interpretation to understand why he might have undertaken the project to read the *Mahābhārata* in the transformative way he did, Chris Minkowski's sustained and ongoing attempts to historicize Nīlakaṇṭha's oeuvre on multiple fronts offer compelling arguments.

4 Inconclusion

Multiple perspectives, including a distance-reading of vast corpora of *ṭīkā* production through multiple paradigms, especially historicist ones, would be necessary to really muster the assurance to deliver a general theory of the practice. A provisional approach might be to take the *ṭīkā* form as a complete genre whose internal logic is to remain preliminary and perform a consciously unfinished act of translation—which is simultaneously also an act of interpretation—in a continuum that may, further on, lead to a reconfigured act of intralingual translation (which may be called adaptation, transcreation, etc.) or an interlingual translation, where the provision of synonymy and syntactic structures can be directly implanted into the target language along with the discovered sense, or it can be part of a specialized creative or essay project (like Nīlakaṇṭha's), or it can stand alone as a pedagogical event that delivers understanding without any further reworking.

Ultimately, its intention as a pedagogical strategy seems to be to empower the reader. While retaining ambiguities and ambivalences within the source and also subtly suggesting a strategy to read them, the *ṭīkā* strengthens an individual's understanding by instigating further investigation. In withholding statements of interpretation and providing glosses, it furthers pedagogical objectives to hone skills and not merely confer instruction. While many may find it unfounded, and thus justifiably read against it, the initial faith one puts in the commentator's implicit translation of the source structures the reader's capacity to receive and assess the reading it prompts. The success of the *ṭīkā*, then, may not lie in the coherence or completeness it delivers, nor even whether it satisfies the reader's initial expectations, but rather in its ability to lead the reader to reconfigure the source's language in a manner that may even exceed the (presumed) interests of the commentator and *auctor*. [In Nīlakaṇṭha's case, one may posit that, as his intellectual project unfolded over many works, he even exceeded his own *ṭīkā*.]

Without overstating a theoretical relationship that all *ṭīkā* might uniformly share with its sources (especially relatively open literary sources), the point of drawing commentary and translation together can be conceived as two-fold. A first instinct draws attention to the shared surface form engaged in locating or creating equivalencies to transfer a fixed meaning. The second marks the shared hermeneutic function between the two philological genres to suggest that while scholars notice commentarial elements in translation,[24] relatively less attention is paid to the intrinsic structural and performative logic of the commentary being to ultimately foster and facilitate future translation and not

24 Medieval vernacular translations of Sanskrit classics, for example, certainly relied on the language and logic of Sanskrit commentaries they knew intimately and seemingly admired (see Patel 2011 for examples), but even translators who had reservations about the conclusions of Sanskrit commentaries nevertheless incorporated many of their readings into their translations. Minkowski (2011: 230, fn. 26), for example, reiterates Holtzmann's observation that Ganguli, the English translator of the "Bombay" (or Vulgate) *Mahābhārata*, used many of Nīlakaṇṭha's readings in his translation even though he criticized the author of the Bhāvārthadīpa for his "inveterate predilection for a needless subordination of the expressions of the poet to purely brahmanical ideas."

solely to ameliorate difficult words and passages. If a *ṭīkā* is taken to be unfinished, reliant on an attitude that likewise views the source to be capable of yielding multiple ports of significance, then any other form of textual activity that relies on these nodes in the network must also be understood as ultimately incomplete, including the translation (intra- or interlingual), the essay, and any other such critical or interpretive work.

It is perhaps no coincidence that a philosophical text which has received enormous commentarial attention in the *ṭīkā* form, such as the *Bhagavadgītā*, has also received countless translations and essays. It is unlikely that the recent proliferation of dozens of partial or complete *Gītā* translations into English stems from any new insights about rendering the Sanskrit's *surface* lexicon and syntax. It suggests, rather, that the process of renewing the *Gītā* through commentarial engagement and translation is as much, if not more, conjoined with a process of intralingual reworking from *English* to English, or any other such concatenation of non-Sanskrit source-and-target languages, with their attendant (and hidden) hermeneutic and aesthetic inflections to create a dynamic equivalence for a changing readership. This modern process is not unlike what Cort describes for early-modern vernacular-language translations of the *Bhāgavata-purāṇas* by translators who did not know Sanskrit but "were either simply vernacular retellings or else 'translations' from the same language, much in the spirit of Robert Bly's popular 'translations' of Kabir and Mirabai, in which he worked from existing English translations" (Cort 2015: 91).

For many readers of Sanskrit literature and the Sanskrit epics setting out to translate (or even reading earlier translations alongside the original source) would be a joyless chore—and sometimes filled with the dread of getting it totally wrong—were it not for guidance gained from the experience of reading the multiple *ṭīkā* works available for the Sanskrit classics, along with the numerous works (broadly, translations) they have inspired in multiple languages and media over the centuries. Often it is easier to read the Sanskrit source through a Sanskrit translation, just as it is easier to teach the source by translating into other Sanskrit words. Even when an ancient or modern *ṭīkā* is unavailable for a particular work, recollection of the pedagogical and interpretive process they embody, in a constant goalless motion, is often sufficient to empower the intrepid reader to confidently encounter and disrupt the alien work from its apparent untranslatability.

Acknowledgements

I thank member-colleagues from the Philadelphia Area Buddhist Studies Group (PABST) for valuable discussion about the provisional arguments put forth in this essay. Many of their excellent suggestions will be taken up as these ideas evolve with more evidence.

Bibliography

Adluri, Vishwa, and Joydeep Bagchee. *Philology and Criticism: A Guide to Mahabhārata Text Criticism*. London: Anthem Press, 2018.

Cardona, George, ed. *Vyakaranamahabhasya (Mahabhasya) of Patañjali: Adhyāya 1*, n.d. http://gretil.sub.uni-goettingen.de/gretil/1_sanskr/6_sastra/1_gram/pmbhassu.htm. Published online. Formatted by Masato Kobayashi. Based on the edition by Franz Kielhorn (Bombay 1880–1885), revised by K.V. Abhyankar (Poona 1972–1996). With additional references of the edition Gurukuljhajjar, Rohatak (Rohtak): Hariyana sahitya samsthan, 1961–1963, 5 vols.

Choudhuri, Indra Nath. "Towards and Indian Theory of Translation." *Indian Literature* 54, no. 5 (2010): 113–123.

Copeland, Rita. "Gloss and Commentary." In *The Oxford Handbook of Medieval Latin Literature*, edited by Ralph Hexter and David Townsend, 1–23. Oxford: Oxford University Press, 2012.

Cort, John E. "Making it Vernacular in Agra: The Practice of Translation by Seventeenth-Century Jains." In *Tellings and Texts: Music, Literature and Performance in North India*, edited by Francesca Orsini and Katherine Butler Schofield, 62–105. Cambridge: Open Book Publishers, 2015. https://www.openbookpublishers.com/books/10.11647/obp.0062.

Deutsch, Eliot. "Knowledge and the Tradition Text in Indian Philosophy." In *Interpreting Across Boundaries*, edited by Gerald James Larson and Eliot Deutsch, 165–173. Princeton: Princeton University Press, 1988.

Devy, Ganesh. "Translation and literary history—an Indian view." In *Postcolonial Translation: Theory and Practice*, edited by Susan Bassnett aand Harish Trivedi, 182–188. New York: Psychology Press & Routledge Classic Editions, 1999.

Dundas, Paul. "Somnolent Sūtras: Scriptural Commentary in Śvetāmbara Jainism." *Journal of Indian Philosophy* 24, no. 1 (1996): 73–101.

Eltschinger, Vincent. "From Commentary to Philosophy, or Lectio and Disputatio in Indian Buddhist Commentarial Literature." In *L'espace du sens. Approches de la philology indienne/The Space of Meaning. Approaches to Indian Philology*, edited by Silvia d'Intino and Sheldon Pollock, 271–303. Paris: Collège de France (Publications de l'Institut de Civilisation Indienne), 2018.

Ganeri, Jonardon. "Sanskrit Philosophical Commentary." *Journal of the Indian Council of Philosophical Research* 25 (2008): 107–127.

Goldman, Robert P. "Translating Texts Translating Texts: Issues in the Translation of Popular Literary Texts with Multiple Commentaries." In *Translation East and West: A Cross-Cultural Apporach*, edited by Cornelia Niekus Moore and Lucy Lower, 93–105. Honolulu: University of Hawai'i, 1992.

Gopinathan, G. "Translation, Transcreation and Culture: The Evolving Theories of Translation in Hindi and other Modern Indian Languages." London: SOAS, 2016. http://www.soas.ac.uk/literatures/satranslations/Gopin.pdf Published online.

Gornall, Alastair. *Rewriting Buddhism: Pali Literature and Monastic Reform in Sri Lanka, 1157–1270*. London: University College London, 2020. https://www.jstor.org/stable/j.ctv13xprn9.

Hatcher, Brian A. "Translation in the Zone of the Dubash: Colonial Mediations of "Anuvāda"." *The Journal of Asian Studies* 76, no. 1 (2017): 107–134.

Heim, Maria. *Voice of the Buddha: Buddhaghosa on the Immeasurable Words*. Oxford: Oxford University Press, 2018. https://doi.org/10.1093/oso/9780190906658.001.0001.

Jacobson, Roman. "On linguistic aspects of translation." In *The Translation Studies Reader*, edited by Lawrence Venuti, 113–118. London: Routledge, 1959–2000.

Jyväsjärvi, Mari. "Retrieving the Hidden Meaning: Jain Commentarial Techniques and the Art of Memory." *Journal of Indian Philosophy* 38, no. 2 (2010): 133–162.

McCrea, Lawrence. "The Hierarchical Organization of Language in Mīmāṃsā Interpretive Theory." *Journal of Indian Philosophy* 28, no. 5/6 (2000): 429–459.

Merrill, Kristi. *Riddles of Belonging: India in Translation and other Tales of Possession*. New York: Fordham University Press, 2009.

Minkowski, Christopher. "Nīlakaṇṭha Caturdhara's Mantrakāśīkhaṇḍa." *Journal of the American Oriental Society* 122 (2002): 329–344.

Minkowski, Christopher. "Nīlakaṇṭha's Vedic Readings in the Harivaṃśa Commentary." In *Epics, Khilas, and Purāṇas: Continuities and Ruptures*, edited by Petteri Koskikallio, 411–433. Zagreb: Croatian Academy of Sciences and Arts, 2005.

Minkowski, Christopher. "What Makes a Work 'Traditional'? On the Success of Nīlakaṇṭha's Mahābhārata Commentary." In *Boundaries, Dynamics and Construction of Traditions in South Asia*, edited by Federico Squarcini, 226–252. London: Anthem Press, 2011.

Mukherjee, Sujit. *Translation as Discovery*. Hyderabad: Orient Longman, 1994.

Nance, Richard F. "Mindsets and Commentarial Conventions among Indian Buddhists." *Journal of the American Academy of Religion* 83, no. 1 (2015): 210–235.

Patil, Parimal. "Gloss and Commentary." In *Conflict between Tradition and Creativity in Indian Philosophy: Text and Context* [*Proceedings of the Seventh International Conference Studies for the Integrated Text Science*], edited by Toshihiro Wada, 47–60. Nagoya: Nagoya University, 2006.

Pollock, Sheldon. "What Was Philology in Sanskrit?" In *World Philology*, edited by Sheldon Pollock, Benjamin A. Elman, and Ku-ming Kevin Chang, 114–136. Cambridge, MA: Harvard University Press, 2015.

Preisendanz, Karin. "Text, Commentary, Annotation: Some Reflections on the Philosophical Genre." *Journal of Indian Philosophy* [*Special Issue: Theory and Method in Indian Intellectual History*] 36, no. 5/6 (2008): 559–618.

Tubb, Gary, and Emery Boose. *Scholastic Sanskrit: A Manual for Students*. New York: American Institute of Buddhist Studies, 2007.

Yoshimizu, Kiyotaka. "The Intention of Expression (*vivakṣā*), the Expounding (*vyākhyā*) of a Text, and the Authorlessness of the Veda." *Zeitschrift Der Deutschen Morgenlandischen Gesellschaft* 158, no. 1 (2008): 51–71.

CHAPTER 16

Debating God in the Delta: *Trimūrti*, Transcendence and Hierarchy in Late Advaita Vedānta

Jonathan Duquette

1 Introduction

Genealogies of modern Hinduism have been deeply concerned with problematizing the notion of an essential unity underlying the diversity of traditions known under the umbrella term of 'Hinduism'. In his compelling book *Unifying Hinduism*, Nicholson discards the notion that Hinduism, as a unified set of religious beliefs and practices, is an invention of nineteenth-century Orientalists and English-educated Indians. Instead, he argues that the idea of 'Hindu unity' emerged earlier in the writings of late medieval thinkers, such as Vijñānabhikṣu (sixteenth century?), who 'began to treat as a single whole the diverse philosophical teachings of the Upaniṣads, epics, Purāṇas, and the schools known retrospectively as the 'six systems' (*ṣaḍdarśana*) of mainstream Hindu philosophy' (Nicholson 2010: 3). The unification of the so-called 'orthodox' (*āstika*) philosophical schools—Sāṃkhya, Yoga, Nyāya, Vaiśeṣika, Vedānta and Mīmāṃsā[1]—promulgated by Vijñānabhikṣu and his contemporaries, together with the composition of compendia and doxographies portraying these schools as partial approximations of the truth taught in Vedānta, would have laid the basis for the formation of modern Hindu thought. Nicholson holds that it is based on such hierarchical and inclusivistic tendencies that 'neo-Advaita' thinkers such as Sarvepalli Radhakrishnan came to regard Advaita Vedānta as the culminating point of Hinduism, and indeed 'religion itself in its most universal and deepest significance' (Radhakrishnan 1954: 23).

1 The classification of the *ṣaḍdarśana* into six 'orthodox' (*āstika*) schools probably postdates the twelfth century. Not all doxographies classified schools in this manner, and some of the early Jaina doxographies included seven schools, not six; see Gerschheimer 2007: 239–240 on this point. To give an example from the period with which we are concerned in this article: in his *Ṣaḍdarśanīsiddhāntasaṃgraha*, Rāmabhadra Dīkṣita (middle of seventeenth century), an influential scholar whom I shall refer to later on, combines Nyāya and Vaiśeṣika into one school and adds grammar to his list of six schools.

 | DOI:10.1163/9789004536869_018

However, we know that the situation was more complex on the ground. As Nicholson himself points out, Vijñānabhikṣu—in his view, the representative par excellence of this new inclusivistic attitude—not only rejected 'heterodox' (*nāstika*) schools of Buddhism and Jainism as his predecessors had done, but also regarded Advaita Vedānta itself as a *nāstika* school because of this tradition's defence of the illusory nature of the world. Vijñānabhikṣu, indeed, belonged to a different branch of Vedānta (a variant of realist *bhedābheda*), which did not accept the implications of Advaita's pure non-dualist stance. In fact, debate and rivalry, even across *āstika* schools, were just as prominent during and after the period in which Vijñānabhikṣu flourished. Proponents of Nyāya, Vedānta, and Mīmāṃsā continued to actively debate with one another on matters of doctrine and religious practices, while also turning a critical gaze to the canonical figures of their own traditions. Schools of thought diversified internally while new schools emerged that attempted to dislodge the older ones. Refutations and polemical tracts were written and circulated across wide networks of household and renunciate brahmins who were active in *maṭhas*, *agrahāras*, and temples; the number of polemical works actually seems to have increased during the early modern period.[2] This process did not slow down with the onset of the modern period. Despite an increasing colonial presence in India during the eighteenth and nineteenth centuries, and the resulting loss of numerous traditional sources of scholarly patronage, Sanskritic scholarship continued to flourish even in the nineteenth century. Scholars carried on debating, producing, and exchanging sophisticated works in Sanskrit, prominently in the field of Vedānta.

While there is certainly something to be said about philosophical attempts to unify disparate theological commitments on the eve of the modern era, to conceptualise Hinduism principally in terms of 'philosophical schools' and their unity can be misleading.[3] Scholars writing in Sanskrit were not only proficient across several *śāstras*, but also belonged to devotional traditions that deeply informed their philosophical stances. From the medieval period onwards, schools of Vedānta with distinctively Śaiva and Vaiṣṇava leanings developed, whose proponents challenged on a sectarian basis the presumed

---

2 A quick survey of Potter's *Encyclopedia of Indian Philosophies Bibliography* reveals that the number of works with titles ending in *-bādhaka, -bhaṅga/bhaṅgī, -bhañjana/bhañjanī, -dhvaṃsa/dhvaṃsana, -dhikkāra, -dūṣaṇa/dūṣaṇī, -gadā, -khaṇḍana, -nirāsa, -mardana*, etc., all of which convey the polemical nature of the work, dramatically increases from the seventeenth century onwards.

3 McCrea 2008 rightly problematises using 'schools' or 'systems' as basic units of analysis in the study of Indian philosophy and religion.

homogeneity of the Purāṇas, epics and Upaniṣads. If, for instance, Śrīvaiṣṇava theologians of Vedānta rejected the validity of Śaiva-leaning Purāṇas and Upaniṣads, their Śaiva co-religionists argued that the epics were Śiva-centred.

In order to contextually understand influential, neo-Advaita accounts of a unified Hinduism qua world religion, what we need, first and foremost, is an intellectual and social history that addresses the rich array of sectarian debates and innovative theological thinking that developed across networks of scholars of Vedānta writing in Sanskrit in the immediately preceeding and simultaneous period. Uncovering this 'alternative history' of Vedānta, in which pandits—as I shall also refer to them in this article—play a more prominent role, will enable us to complicate and better situate the popular, unified undertandings of Hinduism that were set forth outside traditional networks during the colonial period. While the assumptions, methods and contexts of the scholars of Advaita Vedānta operating during and after the seventeenth century certainly differed from those of neo-Advaitins, it is, in fact, in the works of these understudied scholars—not in those of Vijñānabhikṣu, Madhusūdana Sarasvatī or other late medieval scholars—that we find the first genuine attempts to elaborate an 'all-inclusivism' akin to that of neo-Advaita, particularly in relation to the notion that there is a supreme reality underlying gods and goddesses.[4]

There is perhaps no more widespread notion in modern Hindu thought than the idea that Hindu deities—Brahmā, Viṣṇu, Śiva, Śakti and others—embody different aspects of the same absolute godhead. This notion is prominently represented in the figure of the *trimūrti*: the 'three forms' consisting of Brahmā, Viṣṇu and Śiva, which respectively account for the cosmic functions of creation, preservation and destruction. Writing about the *trimūrti*, Sarvepalli Radhakrishnan held that the three deities 'are not to be conceived as independent persons [for] they are the threefold manifestations of the one Supreme' (Radhakrishnan 1968 [1953]: 815). To be sure, Radhakrishnan's comment has long antecedents in the history of India. To take a single example from Kālidāsa's *Kumārasaṃbhava*, verse 7.44:

> *ekaiva mūrtir bibhide tridhā sā*
> *sāmānyam eṣāṃ prathamāvaratvam* |
> *viṣṇor haras tasya hariḥ kadācid*
> *vedhās tayos tāv api dhātur ādyau* ||

4 While I lay the basis for this argument in the present article, I intend to explore in greater details its ramifications in a forthcoming publication.

> There is just one form that divided in three ways.
> Each of them might appear first or last,
> sometimes Śiva before Viṣṇu,
> sometimes Viṣṇu before Him,
> sometimes Brahmā before both,
> and sometimes both before Brahmā.

Reading the works of Radhakrishnan and other neo-Advaita thinkers, however, we may perhaps be left with the impression that this conception of God is inherent to Vedānta and even to 'Hindu thought'. After all, if Advaita Vedānta—the paradigmatic representative of Hindu religious thinking in their view—regards Brahman as the non-dual (*advaita*) reality underlying everything, it is natural to suppose that all deities are equal expressions of this Brahman. Once again, the reality on the ground is more complex, for behind this seemingly simple idea lies a long history of sectarian debates involving Vedānta scholars from diverse backgrounds, affiliations, and worldviews. Are deities truly different manifestations of the same single encompassing reality? Or are deities simply equal to one another in terms of their qualities and powers? Does Brahman transcend the *trimūrti*, and how do deities relate to Brahman in this respect? We will see that such questions continued to preoccupy pandits of Advaita Vedānta writing on the eve of the modern era.

This article explores some of the ideas defended by four scholars of Advaita Vedānta active betwen the seventeenth and nineteenth centuries, who belonged to a dynamic intellectual network centred around the Kāverī delta in South India. Their scholarly contributions, mostly polemical in nature, reveal that the question of God was far from a settled matter. While indications of polemical debates on the nature of deities and their relative superiority are already found in the writings of early Vaiṣṇava theologians of Vedānta, it is during the early modern period that this question started to be debated with greater interest. A key transformative moment occurred in the sixteenth century when the prominent scholar Appaya Dīkṣita (1520–1593) authored numerous Śaiva works challenging the claim that Viṣṇu is the absolute Brahman, arguing instead that Śiva is supreme insofar as He transcends the *trimūrti*. Appaya's Śaiva oeuvre brought about a kind of revolution in the realm of vedāntic thinking about God. It left a clear mark in the works of the prominent pandits who form the scope of this study and many others.

2 A Turning Point: The Śaiva Oeuvre of Appaya Dīkṣita

The sixteenth century constitutes a turning point in the history of Vedānta. Until then, vedāntic theology had been mostly dominated by Vaiṣṇava scholars. In the aftermath of Śaṅkara's *Brahmasūtrabhāṣya*, four major Vaiṣṇava-leaning commentaries on the *Brahmasūtra*s were written, namely by Rāmānuja, Madhva, Nimbārka and Vallabhācārya, the first two of which led to the formation of systematic schools of thought—Viśiṣṭādvaita Vedānta and Dvaita Vedānta—and generated an important amount of commentarial literature. At the time Appaya was writing, the Viśiṣṭādvaita Vedānta tradition had already secured a prominent position. Scholars of Vedānta still read and commented on the works of Rāmānuja (ca. 1077–1157; see Carman 1981: 27), Vātsyavarada (late twelfth / early thirteenth century), Sudarśanasūri (late thirteenth century) and Veṅkaṭanātha (trad. dates 1268–1369), and also wrote new works engaging with specific doctrinal aspects of the system. These scholars of Viśiṣṭādvaita Vedānta all shared the view that the supreme deity referred to in the Upaniṣads and in the *Brahmasūtra*s as Brahman is identical to Viṣṇu.

It is in this context that Appaya Dīkṣita emerged as a leading defender of Śaivism, a religious tradition centred on the god Śiva to which he belonged by birth. A scholar of polymathic erudition, Appaya wrote profusely in a range of Sanskrit disciplines prominent in his day—especially poetic theory (*alaṃkāraśāstra*), scriptural hermeneutics (*mīmāṃsā*) and theology (*vedānta*)—and spent most of his prolific career writing about Śaivism. His Śaiva oeuvre spans nearly three decades, and includes polemical works arguing for Śiva's supremacy over Viṣṇu as well as Advaita-leaning works that led to the foundation of the first Śaiva school of Vedānta: Śivādvaita Vedānta.[5] In his magnum opus of Śivādvaita Vedānta, the *Śivārkamaṇidīpikā*, Appaya articulates the view that the *Brahmasūtra*s understand Śiva as the conceptual and semantic equivalent of Brahman, the absolute reality eulogized in the Upaniṣads. In his Śaiva works, Appaya responds on various grounds to arguments put forth by Rāmānuja and his successors.

If he argues against a Śaiva-leaning interpretation of Upaniṣads in his *Vedārthasaṃgraha* and elsewhere, Rāmānuja himself does not engage with the claim that Śiva is superior to Viṣṇu (or other deities, for that matter). However, we know that there were ongoing debates on this question around his time since Vātsyavarada, in his *Tattvanirṇaya*, refutes the *pūrvapakṣa* that Rudra (a form

5 For a detailed study of Appaya's Śivādvaita Vedānta and of his broader Śaiva oeuvre, see Duquette 2021.

of Śiva) is the supreme reality referred to in scriptures. Not long after, in his monumental subcommentary on the *Śrībhāṣya*, the *Śrutaprakāśikā*, Sudarśanasūri refutes a Śaiva position that appears to have been prominent in his time, namely that Śiva is identical to Brahman and as such transcends the *trimūrti*:

> Because [the word *brahman*] can technically mean [Viṣṇu], the fact that [the three deities of the] *trimūrti* are equal, that [the three deities of the] *trimūrti* are one and the same, and that Brahman is superior to [the three deities of the *trimūrti*] is refuted. The following position is also rejected: 'Īśvara alone is Brahman; this [Brahman] is Sadāśiva [i.e., Śiva] [and] transcends the *trimūrti*; the three [deities of the *trimūrti*] beginning with Brahmā are not the Lord'.[6]

Sudarśanasūri rejects the claim that Śiva is identical to Brahman and that He transcends the *trimūrti*. It is not clear whether this position was held by a specific Śaiva author or shared by a greater number of Śaivas. Significantly, it is precisely the same view that Appaya will defend in his Śaiva works a few centuries later, thus generating a significant critical response from Vaiṣṇava scholars of Vedānta. As we shall see, pandits of Vedānta continued debating this very same question as we enter the modern period.

In the passage above, Sudarśanasūri refutes three distinctive views concerning the *trimūrti*:

- the three deities of the *trimūrti* are equal (→ *sāmya*)
- the three deities of the *trimūrti* are one and the same (→ *aikya*)
- there is a supreme godhead, Brahman, that transcends (→ *uttīrṇa*) these three deities.

Though Sudarśanasūri does not elaborate upon these three views in his subcommentary, we can get more clarity on this point from the *Rahasyatrayasāra* authored by the great Śrīvaiṣṇava scholar Veṅkaṭanātha. In the introductory verse of the sixth chapter on the nature of the supreme deity (*paradevatādhikaraṇa*), Veṅkaṭanātha says:

> There is a single self; all deities are one [and the same]; the three [deities of the *trimūrti*] are, when properly understood, equal [in terms of their knowledge, powers, etc.]; the three [deities of the *trimūrti*] are one [and the same]; something else [than these three deities] is the Lord: good

6 *rūḍhiśaktyā trimūrtisāmyaṃ tadaikyaṃ brahmaṇas taduttīrṇatvaṃ ca niṣiddham, 'īśvara eva brahma, sa trimūrtyatiriktaḥ sadāśivaḥ, brahmādayas trayo 'py anīśvarāḥ' iti pakṣaś ca vyudastaḥ* (*Śrutiprakāśikā* 2006: 6).

> people disregard such statements made by naive people. Based on Upaniṣads, in agreement with each other [on this point], and on statements by Manu, Vyāsa and others that agree [with these Upaniṣads], it is our view that the revered Nārāyaṇa [i.e., Viṣṇu] [alone] is the Lord who embodies everything, [that He alone] bestows liberation and [that He alone] is enjoyed by those who have achieved liberation.[7]

Veṅkaṭanātha rejects the view—likely to have been defended by proponents of Advaita Vedānta—that there is no difference whatsoever between Brahman, deities and individual selves (*jīva*). Although the same supreme deity 'rules within' (*antaryāmin*) every entity, whether it is human or divine, there are essential differences between these entities: deities differ in nature and functions, the individual self is not the same as Brahman, and selves differ from each other. In particular, the three deities of the *trimūrti* are not equal (*tulya*, *sama*) because their knowledge, powers, qualities, etc. vary; nor are they one and the same (*eka*) because they stand in relationship to each other. It is equally not possible to hold that there exists some other godhead superior to them. Just like Sudarśanasūri, Veṅkaṭanātha rejects that there is a divine state transcending the *trimūrti*—a 'fourth state' (*turīya*) as Appaya and other authors will later refer to it. In his view, Viṣṇu, the member of the *trimūrti*, *is* Brahman, although, unlike Brahmā and Śiva, He is not subject to the laws of *karma* and endowed with qualities (*guṇa*) that normally constitute worldly entities: He is the eternal and omniscient Lord who incarnated as Viṣṇu out of His own will. In Viśiṣṭādvaita terminology, Brahmā and Śiva are 'modes' (*prakāra*) of Viṣṇu, or His 'surbordinates' (*śeṣa*), insofar as they cannot exist without Him.

In the introductory verse above, Veṅkaṭanātha refutes two other views in addition to the three already mentioned by Sudarśanasūri, namely that there exists a single self (*ātmaikya*) and that all deities are one and the same (*devataikya*). Together, these five views form the object of the *Pañcamatabhañjana* ('A Refutation of the Five Views'), a Viśiṣṭādvaita Vedānta treatise written by Raṅgarāmānuja in the late seventeenth century. Significantly, the last view

7 *ātmaikyaṃ devataikyaṃ trikasamadhigatā tulyataikyaṃ trayāṇām anyatraiśvaryam ityādyanipuṇaphaṇitīr ādriyante na santaḥ. trayyantair ekakaṇṭhais tadanuguṇamanuvyāsamukhyoktibhiś ca śrīmān nārāyaṇo naḥ patir akhilatanur muktido muktabhogyaḥ* (*Rahasyatrayasāra*: 130). In the *Nyāyasiddhāñjana*, Veṅkaṭanātha likewise claims that the following four positions must be rejected: 1) the view that the three deities of the *trimūrti* are equal; 2) the view that they are one and the same; 3) the view that there exists something beyond them; and 4) the view that another entity is supreme: *trimūrtisāmyaikyottīrṇavyaktyantaraparatvapakṣāś catvāro 'pi nirastāḥ veditavyāḥ* (*Nyāyasiddhāñjana*: 284). Note that the wording here is close to that found in Sudarśanasūri's passage.

stated in Veṅkaṭanātha's verse—namely, that there exists some other deity beyond the *trimūrti* and that this deity is the Lord (*īśvara*)—is understood to refer to Śiva in the *Pañcamatabhañjana*. Indeed, the entire work is directed against Appaya Dīkṣita and his view that there is a 'fourth' reality transcending the *trimūrti*, and that this reality is Śiva, not Viṣṇu. Raṅgarāmānuja felt compelled to address Appaya's views, even a century after they were ushered. He was one among many in his time, and not the last.

2.1 *Śiva Transcends the* trimūrti

Appaya expounds his position on the *trimūrti* in the *Śivatattvaviveka* ('Inquiry into the Śiva-principle'), his auto-commentary on the *Śikhariṇīmālā* ('Garland of [Verses in the] *Śikhariṇī* [Metre]'), a hymn of praise to Śiva comprised of sixty verses in the *śikhariṇī* metre. Appaya explains in the beginning that he wrote this work to put an end to the blabbering (*pralāpa*) of evil-minded people (*kumatikula*) and 'heretics' (*avaidika*) who have no respect for Śiva; it becomes clear as we read on that Appaya targets Śrīvaiṣṇava theologians of Vedānta. It is in his commentary on verses 12 and 13 that Appaya puts forth his position on the *trimūrti*. In verse 12, he states that Śiva, although endowed with limitations (*upādhi*) and qualities (*guṇa*) on a phenomenal level, is devoid of limitations (*upādhivinirmukta*) in reality; He is in truth the non-dual Brahman consisting of existence, consciousness and bliss (*saccidānanda*), the inner self that is revealed once all dualities have vanished.[8] Appaya pursues the same thought in verse 13, highlighting here that Śiva is the controller (*niyantṛ*) of all beings, *including the deities of the trimūrti*—Brahmā, Hari (= Viṣṇu) and Hara (= Śiva). In his elaborate commentary on this important verse, Appaya elaborates on Śiva's transcendence and His relationship to the *trimūrti*. The verse reads as follows:

---

8 *anantaṃ yat prāhur niratiśayasaccitsukhavapuḥ parābhūtadvaitaprakaraviśadaṃ pratyaguditam | akhaṇḍārthair vedyaṃ vacananikarair aupaniṣadair mahādeva jyotis tad asi paramātmeti viditam* || 12 ||—'That which [the sages] have declared to be infinite, to consist in unsurpassed existence, consciousness and bliss, to be pure once all dualities have vanished, to manifest inwardly, to be the [reality] cognized through the several upaniṣadic statements that convey an unbroken meaning—That You are, O Mahādeva, the Light known as the supreme self' (*Śikhariṇīmālā*: 10). When employing the term *akhaṇḍārtha* to describe the type of upaniṣadic statements through which the non-dual reality of Śiva is revealed, Appaya uses a typical Advaitic terminology. The doctrine of *akhaṇḍārtha* is indeed of vital importance to Advaita Vedānta, as it explains how certain linguistic utterances can reveal an impartite object (Brahman). Unlike ordinary sentences construed in terms of a logical subject-predicate relation, these sentences of 'unbroken meaning' express the simple identity of a thing with itself. The most famous example of such statements are the *mahāvākyas* of the Upaniṣads such as *tat tvam asi* ('You are That'), which have the linguistic power to convey the relation of identity between the individual self and Brahman.

> Though You transcend qualities [of the phenomenal world], You are mixed with *māyā* and appear to be endowed with qualities. Adorned with Ambā, a body, a dark throat and three eyes, You are widely known by the names of Śiva, Bhava and others. You are victorious, the controller of [all] people as well as Hari, Hara and Brahmā.[9]

Although Śiva transcends all phenomenal qualities (*guṇātīta*), He manifests as if He were endowed with qualities (*saguṇa iva*). As Appaya explains in his commentary, Śiva's attributes—His dark throat, His three eyes, etc.—are not ultimately real but only imagined (*kalpita*) through the mirific power of *māyā*, just as a rope is falsely imagined to be a snake due to limiting factors (*upādhi*) (e.g., a fearful experience in the past, lack of light, etc.) which distort perception. In contrast, the three deities of the *trimūrti*—including Śiva the Destroyer (Hara), which Appaya regards as an emanation (*vibhūti*) of the supreme Śiva—are *really* endowed with qualities and thus belong to the phenomenal world. Hence Śiva is fundamentally different from the *trimūrti*. Since He is beyond the interplay of *guṇa*s, He also undoubdtedly transcends the *trimūrti*.[10]

Appaya and Veṅkaṭanātha approach the problem of the *trimūrti* in very different ways. Unlike Veṅkaṭanātha, who argues that the personified Viṣṇu in the *trimūrti* is the supreme deity itself, Appaya does not hold that Hara, the embodied form of Śiva who functions as the Destroyer in the *trimūrti*, is the same as the supreme Śiva. This divergence is rooted in key metaphysical differences between Viśiṣṭādvaita Vedānta and the Śaiva Vedānta upheld by Appaya. Appaya's views align closely with the non-dualist position of Advaita Vedānta and its doctrine of levels of reality: while the supreme Śiva belongs to the ultimate level of reality (*pāramārthika*), Hara belongs to a lower level of reality (*vyāvahārika*) that vanishes upon the realisation of His non-dual nature (just like the false perception of the snake disappears upon the realisation that it is a rope). For tenants of Viśiṣṭādvaita Vedānta, however, Viṣṇu, though also non-dual in essence, is 'qualified' (*viśiṣṭa*) by the phenomenal world and its entities; that is, it is in a real relation with these entities. While form (*mūrti*, *rūpa*) is real for them, it is not for Appaya who holds difference (*bheda*) to be a condi-

9 *guṇātīto 'pi tvaṃ saguṇa iva māyāśabalitaḥ pariṣkurvan sāmbaṃ vapurasitakaṇṭhaṃ trinayanam | abhikhyābhiḥ khyātaḥ śivabhavamukhābhir vijayase niyantā lokānāṃ sahariharapaṅkeruhabhuvām* || 13 || (*Śikhariṇīmālā*: 11).

10 Appaya says earlier in the *Śivatattvaviveka*: *guṇamūrtibhyaḥ paraṃ māyopādhikaṃ jagatkāraṇaṃ brahmeti vajralepāyate*—'[The view that] Brahman [i.e., Śiva], the cause of the world, which has *māyā* as its limiting factor, is beyond forms endowed with qualities [i.e., beyond the *trimūrti*], is written in stone' (*Śikhariṇīmālā*: 13).

tional product of *māyā*. Accordingly, Appaya denies that Viṣṇu is the supreme principle and that He occupies a special place in the *trimūrti*. He concludes his commentary on verse 13 of the *Śikhariṇīmālā* as follows:

> In ascertaining the [nature of the] Viṣṇu-principle, two views have been put forth. The best scholars should consider and accept whatever is correct on this subject. In both views, it was established that Brahmā, Viṣṇu [and] Hara are different parts of the blessed Śiva—Ambikā's lover, the first among all manifested forms of Brahman [and] the Lord of the world—just like Bhairava and other [deities are different parts] of Hara, and just like Matsya and other [deities are different parts] of the Lord of Śrī [i.e., Viṣṇu]. This being the case, it is proven that the Lord whose crest-jewel is the moon crescent [i.e., Śiva] is superior to all deities.[11]

According to Appaya, only two views can be defended as far as Viṣṇu's nature as a supreme principle is concerned: 1) Viṣṇu is the 'fourth state' (*turīya*) insofar as He transcends His own emanations (*vyūha*)—the divine beings Saṃkarṣaṇa, Pradyumna and Aniruddha who are endowed with various qualities such as knowledge, power, etc., and who are active on a phenomenal level; or 2) Viṣṇu belongs to the 'fourth domain' (*turīyakoṭi*) on par with Śiva. However, none of these two views, argues Appaya, entails the superiority of Viṣṇu over Śiva: in both, Śiva retains His superior position. Just as Matsya and other incarnations (*avatāra*) of Viṣṇu are parts (*kalā*) of Viṣṇu, and just as Rudra is a part of Hara (Śiva the Destroyer), so are the three deities of the *trimūrti*, including Viṣṇu, parts of the supreme Śiva. Even if Viṣṇu were to transcend the *trimūrti*, He would remain in this position subordinate (*upasarjana*) to Śiva in the way Śiva's consort is.[12]

---

11 *matadvayam upanyastaṃ viṣṇutattvavinirṇaye* |
yad atra yuktaṃ tad grāhyaṃ vimṛśya vibudhottamaiḥ ||
matadvaye 'pi bhagavān ambikāramaṇaḥ śivaḥ |
aśeṣabrahmamūrtīnām ādibhūto jagatpatiḥ ||
bhairavādyā harasyeva matsyādyāḥ śrīpater iva |
brahmaviṣṇuharās tasya kalābhedā iti sthitam || 13 ||
evaṃ ca yady apīśasya candrakhaṇḍaśikhāmaṇer utkarṣaḥ sarvadevebhyas tata eva prasidhyati (*Śikhariṇīmālā*: 21).

12 In his subcommentary on the *Śivatattvaviveka*, Tyāgarāja Śāstrī (one of the pandits studied in this article) glosses *upasarjana* with *guṇabhūta*. It might be useful to think here of the subordinate relationship of Viṣṇu to Śiva in Mīmāṃsā terms. In the injunction 'One who desires sovereignty should sacrifice', the sacrificial act is subordinate (*guṇabhūta*) to the result, sovereignty, which is the predominant (*pradhānabhūta*) aim. Without the sacrificial act, the aim of sovereignty cannot be achieved. What is subordinate is therefore

At least six rebuttals to the *Śivatattvaviveka* were written before the middle of the eighteenth century. The famous Mādhva scholar of Dvaita Vedānta, Vijayīndra (ca. 1514–1595), wrote no less than two treatises dealing with this work: the *Paratattvaprakāśikā* and the *Turīyaśivakhaṇḍana*. As its title suggests, the second work deals exclusively with Appaya's thesis that there is a 'fourth Śiva' (*turīyaśiva*) transcending the *trimūrti*. Vijayīndra does not argue, like Śrīvaiṣṇavas do, that there is no such thing as a 'fourth entity' beyond the *trimūrti*, but simply that Śiva is not that entity: Viṣṇu is. In his wake, Nārāyaṇācārya (ca. 1600–1660) wrote the *Viṣṇutattvaviveka*, also a rebuttal to Appaya's *Śivatattvaviveka*. Not long after, a similar response came from another Vaiṣṇava school of Vedānta, Śuddhādvaita Vedānta. A key commentator of this school, Puruṣottama (ca. 1657/1668–1725), attacks Appaya's thesis in his *Prahastavāda*. Like Vijayīndra, Puruṣottama denies that the 'fourth entity' in question is Śiva, arguing instead that the upaniṣadic and purāṇic passages cited by Appaya in support of his position really highlight a *śeṣa–śeṣin* relationship: only Viṣṇu, not Śiva, stands in such a relationship with Brahman. It is around the same period that the Śrīvaiṣṇava Raṅgarāmānuja wrote the aforesaid *Pañcamatabhañjana*, most of which is directed against Appaya's 'fifth view' that Śiva transcends the *trimūrti*. A century or so after Raṅgarāmānuja, another Śrīvaiṣṇava scholar, Paravastu Vedāntācārya (eighteenth century?), wrote an influential work, the *Vedāntakaustubha*, in which he refutes theses from the *Śivatattvaviveka* and other Śaiva works. It is in direct response to this work that Tyāgarāja Śāstrī, one of the pandits discussed in this article and an illustrious descendant of Appaya, wrote a subcommentary on the *Śivatattvaviveka*, the *Śivatattvavivekadīpikā*, thus reviving the debate in the modern era.

2.2 *Viṣṇu Is Divine, Just like Śiva*

The composition of the *Śivatattvaviveka* marked a new beginning in the history of brahminical debates on God. The long-standing view that Viṣṇu is

---

a necessary means to an end. Likewise, Viṣṇu, although inferior to Śiva, has an important and, in effect, necessary role to play in the functioning of the world. In his Śivādvaita Vedānta work, notably in the *Ratnatrayaparīkṣā* (see below), Appaya repeatedly describes Viṣṇu as a part (*bhāga, avayava*) or aspect (*aṃśa*) of Śakti, Śiva's consort, insofar as He is the material cause (*upādānakāraṇa*) of the world. In this scenario, both Śakti and Viṣṇu are subordinate to Śiva, and yet without them Śiva could not manifest the phenomenal world. Hence Viṣṇu's subordination to Śiva does not merely convey His inferiority to Śiva, but also the fact that Viṣṇu accomplishes a task of key importance for Śiva. This contrasts with the role ascribed to Śiva in Viśiṣṭādvaita Vedānta, which is essentially one of service to Viṣṇu: Śiva is an 'accessory' (*śeṣa*) to Viṣṇu (the 'principal' or *śeṣin*), which means that He is totally subservient to Viṣṇu.

supreme was now being challenged and tentatively replaced with another view—unprecedented in the realm of vedāntic theology both in terms of its scope and influence—namely that Śiva is supreme insofar as He transcends the *trimūrti*. The *Śivatattvaviveka* is not the only polemical work of Appaya arguing along those lines; other works of his such as the *Śivakarṇāmṛta*, the *Brahmatarkastava* and the *Bhāratasārasaṃgrahastotra* also seek to uphold Śiva's supremacy based on a creative exegesis of passages from Purāṇas, epics and Upaniṣads. In his later Śivādvaita Vedānta works, however, Appaya begins to display a more tolerant attitude towards Viṣṇu and His worship. More heavily aligned with Advaita Vedānta, the Vedānta tradition which Appaya cherishes above all others,[13] these works insist on Viṣṇu's divinity in accordance with the belief that all deities are ultimately manifestations of the same non-dual Brahman. This position is key, for instance, to Appaya's *Ratnatrayaparīkṣā* ('An Examination of the Three Jewels'), a short devotional hymn (*stotra*) comprised of eight verses with a self-authored commentary in which he argues that Śiva, Śakti and Viṣṇu are 'three jewels' (*ratnatraya*) who share the same divine nature of Brahman. In verses 3 and 4, Appaya says:

> The *dharmin* is the agent of the five [activities] beginning with the creation of the world. The *dharma* assumes, in the form of a Man, the state of the material cause of the entire world. In the form of a Woman, the divine queen belongs to the primordial creator in whom She resides. According to those who know scriptures, the [two] divisions of *dharma* just mentioned [i.e., the Man and the Woman] are in the domain of Brahman, just like the *dharmin*.
>
> That *dharmin* is Śambhu, and is well known as Śiva, the supreme Brahman. She is known as Ambikā or Umā, and said to be the undivided queen of this *dharmin*. As for the highest Man, the material cause of this world, He is well known under the name 'Nārāyaṇa'. [All this] is understood from the words of the blessed Kūrma and Nārada as well.[14]

13 Appaya made significant contributions to Advaita Vedānta. His *Parimala* on Amalānanda's *Kalpataru* (a commentary in the Bhāmatī tradition), his *Siddhāntaleśasaṃgraha* (an analytical survey of various Advaita views on different topics) and his *Nyāyarakṣāmaṇi* (a study of the first *adhyāya* of the *Brahmasūtras*) are remarkable works of Advaita Vedānta, still studied today. Appaya remained a great admirer of Śaṅkara and of his *Brahmasūtrabhāṣya* throughout his entire career. While we do perceive a leaning towards Advaita in his early Śaiva works, it is in his mature works of Śivādvaita Vedānta that his reliance on the doctrines and hermeneutical methods of this school is most prominent.

14 *kartṛtvaṃ tasya dharmī kalayati jagatāṃ pañcake sṛṣṭipūrve dharmaḥ puṃrūpam āptvā sakalajagadupādānabhāvaṃ bibharti | strīrūpaṃ prāpya divyā bhavati ca mahiṣī svāśra-*

Appaya had stated in the opening verse that the non-dual consciousness of Brahman assumed two forms in manifesting the world: the property (*dharma*) and the property-possessor (*dharmin*). In the verses above, he explains what these terms refer to. The *dharmin* is Śiva and it is identical to the supreme Brahman in essence as well as in function in terms of His primary role in creation. The *dharma*, on the other hand, has two aspects: a masculine aspect called Nārāyaṇa, which functions as the material cause of the world, and a feminine aspect called Ambikā or Umā. While this scheme still emphasises Śiva's supremacy, it also affirms that both Viṣṇu and Śakti belong, *along with Śiva*, to the 'domain of Brahman' (*brahmakoṭi*). However, the identity (*abheda*) between the 'three jewels' does not entail that all deities are 'equal' or 'identical' to one another: each 'jewel' has a specific, irreplaceable and hierarchical function in the overall workings of the world. Though Śakti and Viṣṇu are one (*eka*) with Śiva from an absolute standpoint, they do play different roles from a phenomenal standpoint and thus remain subordinate to Śiva.

In another Śivādvaita work, the *Ānandalaharī* ('Waves of Bliss'), Appaya explores in greater depth the question of whether Viṣṇu is an individual self (*jīva*) or a deity sharing Brahman's divine nature. In his commentary on verse 23, he introduces an opponent who argues that Viṣṇu is an individual self. Though some passages in the *Bhagavadgītā* and the *Viṣṇupurāṇa* unequivocally describe Viṣṇu as a supreme being, other passages are found in Purāṇas that affirm the supremacy of Brahmā and Rudra. Just as the claim that Brahmā and Rudra are supreme should be understood in view of their respective roles in the creation and destruction of the world, so should the claim of Viṣṇu's supremacy be understood in view of His role in the preservation of the world. In other words, concludes the opponent, such claims do not entail that Viṣṇu is supreme but only that He plays an important role in the functioning of the world. Appaya significantly rejects the opponent's position, holding that Viṣṇu shares the nature of Brahman and is not merely an exalted individual self. Later, he begins his commentary on verse 35 with a very personal defence of Viṣṇu's divinity:

> My tongue could not move to assert, on the basis of a few *mantras*, explanatory passages in the Vedas and passages from Purāṇas found somewhere in a corner, that the revered Nārāyaṇa—who is unanimously worshipped

---

yasyādikartuḥ proktau dharmaprabhedāv api nigamavidāṃ dharmivad brahmakoṭī || 3 || *yo 'sau dharmī sa śaṃbhuḥ sa śiva iti paraṃ brahma ceti prasiddho yā coktā dharmiṇo 'syāvighaṭitamahiṣī sāmbikometi vittā* | *yaś copādānabhūtaḥ para iha puruṣaḥ saiṣa nārāyaṇoktyā khyātaḥ śrīkūrmavākyair adhigamitam idaṃ nāradasyāpi vākyaiḥ* || 4 || (*Ratnatrayaparīkṣā*: 6, 8).

as the supreme Brahman by all virtuous people [and] whose nature as the supreme Brahman has been ascertained in the blessed *Mahābhārata*, the *Viṣṇupurāṇa*, and other works that have been composed, for the sake of supporting the Vedas, by the blessed Vedavyāsa, an expert in the interpretation of all Vedas [and] the crest-jewel among all-knowing [scholars], who descended [to earth] precisely in order to classify the Vedas; by the blessed Parāśara, whose [own] knowledge of the highest truth about deities was granted by Pulastya's boon; [and] by [other] great sages—is an individual self (*jīva*). If I did so, my head would burst into a hundred pieces and I would be [guilty] of treachery towards the Vedas, the rishis and the deities. Therefore, my view is that Nārāyaṇa is in the domain of the supreme Brahman.[15]

To call Viṣṇu a *jīva* based on a few scriptural passages would not only be unfair, Appaya holds, but a treachery (*droha*) against the Vedas, the rishis and the deities, a treachery for which he would not forgive himself. Unlike Vaiṣṇavas before him, who had unanimously claimed that Śiva is an individual self (*jīva*) devoid of any divine status, Appaya holds in his mature works that Viṣṇu is not a *jīva* but a deity as 'divine' as Śiva. How should we then understand the repeated claim in the *Śivatattvaviveka* that Viṣṇu is an inferior deity? I have argued elsewhere that Appaya held this position in his early polemical works to *defend* the worship of Śiva against claims of Śiva's inferiority put forth by Śrīvaiṣṇava theologians of Vedānta. This is suggested by his systematic engagement with Śrīvaiṣṇava positions held before him, and by the fact that he explicitly demands at times that his Vaiṣṇava opponent acknowledges the divinity of Śiva just as he acknowledges Viṣṇu's.[16] As an Advaitin, Appaya, in the end, is uninterested in whether one prefers to worship Śiva or Viṣṇu, or whether Śiva or Viṣṇu is the central object of the *Brahmasūtra*s and the Upaniṣads. Since the supreme reality, according to him, consists of Brahman without attributes (*nirviśeṣa*, *nirguṇa*), both theistic stances must be respected as legitimate ways of gradually approaching the truth of pure non-duality.

15 *vedavibhāgārtham evāvatīrṇena sakalavedatātparyābhijñena sarvajñaśikhāmaṇinā bhagavatā vedavyāsena pulastyavaradānalabdhadevatāpāramārthyavedanena śrīparāśareṇānyaiś ca maharṣibhir vedopabṛṃhaṇārthaṃ praṇīteṣu śrīmahābhārataviṣṇupurāṇādiṣu niścitaparabrahmabhāvasya sadbhiḥ sarvair apy avibhāgena parabrahmety eva pūjitasya śrīnārāyaṇasya kvacit koṇe niviṣṭamantrārthavādapurāṇavacanādileśam avalambya jīvabhāvaṃ vaktuṃ nāsmajjihvā pravartate. tathā cen mūrdhā ca śatadhā bhavati vedavaidikadroho devatādrohaś ca jāyate. ato nārāyaṇaḥ parabrahmakoṭir ity evāsmākaṃ siddhāntaḥ* (*Ānandalaharī*: 66).

16 See Duquette 2021, Chapter 4, for my detailed argument on this point.

Despite Appaya's unprecedented proclamation of Viṣṇu's divinity, several Vaiṣṇava scholars of Vedānta after him argued that Appaya's sole motivation in his polemical works was to initiate a dispute between Śaivas and Vaiṣṇavas. Some Śaiva scholars responded vehemently to such accusations by writing polemical tracts in further support of Appaya's early polemical positions, thus generating further acrimonious debates with Vaiṣṇava scholars. A priori, it might be supposed that pandits writing in the tradition of Advaita Vedānta should not be concerned with such polemics. Whether one worships Viṣṇu or Śiva, as Appaya himself claimed, should be of no importance given their core belief that Brahman is the non-dual reality of everything. But were they really indifferent? If these pandits considered Appaya an authority on Advaita, how did they engage with his Śaiva work, and more generally with the claim of Śiva's supremacy? How did Appaya's work impact on their conception of God? It is striking that Appaya retains a central importance for Advaitins writing from the seventeenth century onward. Whether they sided with him or not, they all felt compelled to address his views.

3 Hari, Hara, Harihara: Bodhendra Sarasvatī (ca. 1610–1692)

Our story begins a few decades after Appaya's demise with a scholar named Bodhendra Sarasvatī. Bodhendra was the 59th pontiff of the Śaṅkarite *maṭha* in Kāñcīpuram, and the disciple of Viśvādhika Sarasvatī, to whom he pays homage in all of his extant works.[17] A devotee of Rāma, Bodhendra is known for works in the *nāmasiddhānta* tradition, such as the *Nāmāmṛtarasāyaṇa* in which he defends the view that the singing of God's name is soteriologically efficacious. He is also known for a number of Advaita Vedānta works. Among the latter, the *Hariharādvaitabhūṣaṇa* and the *Hariharabhedadhikkāra* are two extensive works in which Bodhendra defends the view that there is essentially no difference between Viṣṇu and Śiva and that both deities are aspects of the same reality—a reality he refers to as Harihara.[18] He begins the *Hariharādvai-*

17 Viśvādhika Sarasvatī was the 58th pontiff of the same *maṭha*; see *Hariharādvaitabhūṣaṇa*: iv–v for more details on his identity.

18 Claims of identity between Viṣṇu and Śiva are fairly common in epic literature. In the *Harivaṃśa*, for instance, the sage Mārkaṇḍeya claims that he sees no difference whatsoever between Viṣṇu and Śiva, and then sings the praises of Harihara (see 1.37.40 ff.). The identification between the non-dual Brahman, on the one hand, and Śiva and Viṣṇu, on the other hand, is also found in works of Advaita Vedānta, notably in the *Vivekacūḍāmaṇi* traditionally ascribed to Śaṅkara: *nārāyaṇo 'haṃ narakāntako 'haṃ purāntako 'haṃ puruṣo 'ham īśaḥ | akhaṇḍabodho 'ham aśeṣasākṣī nirīśvaro 'haṃ nirahaṃ ca nirmamaḥ*

tabhūṣaṇa with a sweeping statement about their underlying unity. Not only are both deities equal (→ *sāmya*) insofar as they are both endowed with all possible auspicious qualities (*kalyāṇaguṇa*); they are also one and the same (→ *aikya*) since they both govern the activities of world creation, maintenance and destruction. Anyone desirous to achieve the highest good needs to acknowledge their identity, and Bodhendra sets as his task to demonstrate that the intent (*tātparya*) of the Upaniṣads is precisely to communicate that both Śiva and Viṣṇu have sovereignty over everything (*sarveśvaratva*).[19] To do this, he proceeds in three steps. In the first section of the work, he establishes the supremacy of Viṣṇu (*haripāramya*) and refutes all claims to the effect that Śiva is supreme; here Bodhendra engages substantially with Appaya's views. In the second section, he establishes the opposite conclusion, namely the supremacy of Śiva (*harapāramya*). In the third and concluding section, he establishes the identity of both deities (*hariharādvaita*) on this basis.

Bodhendra begins the first section by examining various arguments advanced against Viṣṇu's supremacy on the basis of upaniṣadic passages, including the position that Viṣṇu is an individual self (*jīva*) and the concomitant view that His expressed sovereignty (*īśvaratva*) is only figurative (*lākṣaṇika*). One other such argument concerns the controversial interpretation of the *mantra* '*nārāyaṇaparaṃ brahma*' in the *Mahānārāyaṇa Upaniṣad* (or *Mahopaniṣad*, as Bodhendra calls it). The interpretation of this *mantra* had engaged the attention of Śaiva and Vaiṣṇava theologians from the early medieval period onwards: Haradatta Śivācārya (twelfth / thirteenth century), Vātsyavarada and Sudarśan-

||—'I am Nārāyaṇa, the slayer of Naraka. I am the slayer of [Tri]pura, I am the Puruṣa, I am the Lord [Śiva]. I have unbroken consciousness, I am the witness of everything, I have no ruler, and I am devoid of the ideas of "I" and "mine"' (v. 494 in *Vivekacūḍāmaṇi*: 214). The *Nāmāmṛtarasāyaṇa*, *Hariharabhedadhikkāra* and *Hariharādvaitabhūṣaṇa* all begin with a praise of Rāma (or Rāma's name). Though he begins the first work with such praise, Bodhendra insists there on the efficacy of using both the names 'Śiva' and 'Viṣṇu' to reach the aims of human existence (*puruṣārtha*) (Venkatkrishnan 2015: 121). In his view, Rāma is an embodiment of Śiva and Viṣṇu (see, e.g., v. 2 of the *Hariharādvaitabhūṣaṇa*: *śrīśagaurīśvarābhinnarūpaṃ rāmam aham bhaje* |). Bodhendra wrote the *Hariharabhedadhikkāra* after the *Hariharādvaitabhūṣaṇa*, for the latter mentions the former (see, e.g., *Hariharādvaitabhūṣaṇa*: 107). In the opening statement of the *Hariharabhedadhikkāra*, which is still unpublished as far as I know, Bodhendra declares that it is necessary for anyone desirous to achieve any aim of existence to discard any thought of difference between Viṣṇu and Śiva: *iha khalu dharmārthakāmamokṣākhyacaturvidhapuruṣārtheṣu yatkiṃcitpuruṣārthārthināṃ sarveṣāṃ viṣṇuśivayor bhedabuddhirāhityam āvaśyakam evety āśayena* ... (Adyar 1952: 479).

19 *tasmān viṣṇau śive ca tatsāmyaikyayoḥ siddhisādhanarūpasya tayoḥ sarveśvaratvasyopaniṣattātparyaviṣayatvaṃ krameṇa darśayiṣyate* (*Hariharādvaitabhūṣaṇa*: 107).

asūri discussed it early on, and so did Śrīkaṇṭha, Appaya, Vijayīndra and Puruṣottama (among others) later on.[20] The point of controversy lies in the fact that the compound *nārāyaṇaparam* can be analysed in two ways, leading to two opposite interpretations: a) by taking *nārāyaṇa* as a separate word with an elision of the nominative case ending, thus expressing the fact that Nārāyaṇa (i.e., Viṣṇu) *is* the supreme Brahman (i.e., *nārāyaṇa paraṃ brahma*); or b) by taking *nārāyaṇa* in composition with *param*, and reading the compound as an ablative *tatpuruṣa* expressing the fact that Nārāyaṇa is 'other' than Brahman (i.e., *nārāyaṇāt paraṃ brahma*). Bodhendra refutes the second interpretation. In his view, this interpretation is not compatible with the various 'contextual identifiers' (*tātparyaliṅga*[21]) in this Upaniṣad indicating that Viṣṇu rules over everything and is the cause of the world. One of these *tātparyaliṅga*s states that there should be concordance between the initial and final theses of a text (*upakramopasaṃhāraikya*), and that in case of disagreement, the thesis stated in the introduction (*upakrama*) should prevail.[22] Now, if we accept the aforesaid interpretation of the *mantra* and claim that Viṣṇu is not the supreme Brahman, the initial thesis of the *Mahānārāyaṇa Upaniṣad*—namely that Viṣṇu is supreme because He created the world—is contradicted. To make sure that the principle of concordance is maintained, we need to interpret the *mantra* as conveying Viṣṇu's supremacy.[23]

---

20 I discuss debates around the interpretation of this *mantra* in Duquette 2021, Chapter 2.

21 The six 'contextual identifiers' (*tātparyaliṅga*) allow one to ascertain the underlying intention (*tātparya*) of a text or textual passage. They are: 1) the concordance between the initial and final theses (*upakramopasaṃhāraikya*); 2) the repetition or recurrence (*abhyāsa*) of the subject under discussion; 3) the novelty (*apūrvatā*) of the subject (i.e., typically the subject should not be knowable through other *pramāṇa*s than *śruti*); 4) the clear mention of an independent result (*phala*) associated with the subject; 5) the mention of subordinate statements of praise (*arthavāda*) related to the subject; and 6. the logical justification (*upapatti*) of the subject.

22 The claim that the initial thesis should always prevail over the final thesis was challenged prominently by theologians of Dvaita Vedānta. As an Advaitin, Bodhendra believed in the primacy of the initial thesis. For more details on this debate, see Duquette 2016, and more recently, Bronner & McCrea 2021.

23 *upakrama eva nārāyaṇasya jagatkāraṇatvoktyā sarveśvaratvam uktam eveti tadvirodhe upakramādhikaraṇavirodhaś ca. evaṃ cāsyā upaniṣado nārāyaṇe sarveśvaratāparatvaṃ ṣaḍvidhais tātparyagrāhakaliṅgair avagamyate*—'The fact that Nārāyaṇa rules over everything is indeed taught [in the *Mahānārāyaṇa Upaniṣad*] because it is mentioned right at the beginning [of this Upaniṣad] that Nārāyaṇa is the cause of the world. If this is contradicted [i.e., if we interpret the *mantra* as expressing that Nārāyaṇa is not Brahman], [the principle stated in the] *upakramādhikaraṇa* [i.e., *Mīmāṃsāsūtra* 3.3.1–8] is contradicted. And thus, the fact that this Upaniṣad is intent on teaching Nārāyaṇa's sovereignty over everything is understood by means of the six contextual identifiers' (*Hariharādvai-*

In order to ascertain Viṣṇu's supremacy, Bodhendra must reject arguments advanced in support of Śiva's supremacy, for if we accept Śiva's sovereignty, it naturally follows that Viṣṇu cannot rule over everything. One of these arguments, defended by Appaya and others, is that Śiva is the 'fourth' (*turīya*) entity transcending the *trimūrti* as well as the single creator of the world (*jagatkāraṇa*). Bodhendra holds that if some purāṇic and upaniṣadic passages do seem to support this interpretation, other passages suggest that Viṣṇu is the 'fourth' and that He created the world. How are we to resolve this? According to Bodhendra, the debate about which deity has primacy over the other is not relevant because both positions are found expressed across the entire scriptural corpus. Instead, we should accept that both deities designate the *same* supreme Lord (*parameśvara*), that is, the 'reflected consciousness' (*bimbacaitanya*) of Brahman that governs the creation, preservation and destruction of the world. If both deities differ in terms of their attributes on the phenomenal level of the *trimūrti*, they are one and the same on the level of the Lord.[24] Bodhendra repeatedly stresses this point in his work.

According to Bodhendra, previous debates on the nature of God were grounded merely in an attempt to 'strengthen the sovereignty' (*sarveśvaratvadṛḍhīkaraṇa*) of one's own preferred deity (*iṣṭadevatā*), not in an attempt to understand the nature of deities based on sound hermeneutical principles. That there were indeed partisan debates in Bodhendra's time—and that Appaya's

---

tabhūṣaṇa: 4). The principle that the initial thesis must prevail over the thesis stated in the conclusion of a text or textual passage is discussed by Śabara in his *bhāṣya* on *Mīmāṃsāsūtra* 3.3.1–8, a section sometimes referred to as the *upakramādhikaraṇa*. For more details, see Duquette 2016: 5–7.

24 *tathā cobhayos tatsarveśvaratvanirvāhāya bimbacaitanyātmakasarveśvararūpeṇa tayor aikyaṃ sarvaiḥ śreyo 'rthibhir avaśyam abhyupagantavyam*—'And likewise, all those who wish to achieve the highest good should accept, in order to explain that both [deities are described in scriptures as having] sovereignty over everything, that they are both one and the same in the form of the Lord of everything, the reflected consciousness [of Brahman]' (*Hariharādvaitabhūṣaṇa*: 12). Bodhendra will say later on: *evaṃ ca śivanārāyaṇayor ubhayor api bimbacaitanyarūpeṇaikyaṃ śrīmūrtirūpeṇa bhedaś ca siddhyatīti tayos tāratamyagandhasyāpi nāvakāśaḥ*—'And thus it is established that both Śiva and Nārāyaṇa are one and the same in the form of the reflected consciousness [of Brahman] and different in their embodied form [in the *trimūrti*]. There is no place for even a trace of gradation between them' (*Hariharādvaitabhūṣaṇa*: 19). In the Vivaraṇa tradition of Advaita Vedānta, *bimbacaitanya* precisely defines the Lord (*īśvara*): the Lord is the consciousness of Brahman conditioned (*upahita*) by objective ignorance (*ajñāna, avidyā*). See Madhusūdana Sarasvatī's concise definition in the *Siddhāntabindu*: *ajñānopahitaṃ bimbacaitanyam īśvaraḥ* (*Siddhāntabindu*: 29). Incidentally, this suggests that Bodhendra's training in Advaita might have been in the Vivaraṇa tradition.

Śaiva theses were central to them—is clear from what he says later in the first section of the *Hariharādvaitabhūṣaṇa*:

> It is entirely fair that [Appaya] Dīkṣita, who is regarded as a foremost devotee of Śiva, said—[though] he knew perfectly well that it is inconsistent—that bad things should be said to those who blame Śiva, whether these things are consistent or not. Nevertheless, some very unfortunate people tormented by fate, although knowledgeable in all *śāstras*, nowadays hold that Dīkṣita said that—although it is entirely inconsistent—based on his own investigation of the meaning of the Vedas, [and] that this was his final position. [Although they] resort to Śiva, [these people] hate Viṣṇu and [will therefore] go to hell without even obtaining the fruit of worshipping Śiva. This is really astonishing to me.[25]

Arguably, Bodhendra is addressing here Śaiva scholars in his own time and place who held that Appaya believed in Viṣṇu's inferiority and plainly encouraged Śaivas to attack Vaiṣṇavas. Bodhendra wonders how these 'unfortunate people tormented by fate', though good scholars in their own right, have misunderstood Appaya's position (*siddhānta*) to that extent. In his view, Appaya knew very well that denying the supremacy of Viṣṇu is not consistent with scriptures; we have seen that this is exactly what Appaya says in the *Ānandalaharī*. Although he rejects some of the claims made by Appaya in the *Śivatattvaviveka*,[26] Bodhendra believes that he holds the same essential position as Appaya, namely that Śiva and Viṣṇu are one and the same and that both deities share the same fundamental nature as the reflected consciousness (*bimbacaitanya*) of Brahman.

In support of this, Bodhendra notably mentions a passage from the *Siddhāntaleśasaṃgraha*, Appaya's early compendium of philosophical positions in Advaita Vedānta, where the nature of the liberated state is discussed. Appaya

25 *yady api saṃgatam asaṅgatam vā śivanindakānām aniṣṭaṃ vaktavyam iti śivabhaktāgragaṇyaiḥ dīkṣitair idam asaṅgatam iti jñātvaivoktaṃ tad ucitam eva, tathāpīdānīṃtanāḥ sudurbhagā daivena nirdagdhāḥ sakalaśāstrapāṭhayitāro 'pi kecanātyantāsaṃgatam api tam arthaṃ dīkṣitā vedārthaṃ nirṇīyoktavantas teṣāṃ siddhānta evāyam artha iti gṛhṇantaḥ śivam āśritya viṣṇuṃ nindantaḥ śivabhaktiphalam apy aprāpya narakaṃ gacchantītīdam atyadbhutaṃ bhāti.* (*Hariharādvaitabhūṣaṇa*: 16).

26 For instance, Bodhendra refutes the argument in the *Śivatattvaviveka* that passages from the *Matsyapurāṇa* describe Śiva as manifesting in the form of Nārāyaṇa, the deity of the *trimūrti*, thus indicating the former's superiority: *etena ca śivatattvaviveke etanmahopaniṣadupabṛṃhaṇatvenodāhṛtamatsyapurāṇavacanānāṃ tātparyābhidhānaprastāve 'bhihitam* [...] *ityādikaṃ nirastam* (*Hariharādvaitabhūṣaṇa*: 38).

argues here that those who experience the nature of the Lord (*īśvarabhāva*) in liberation do not experience suffering. The question arises from the fact that even an incarnation (*avatāra*) of the Lord such as Rāma is known to have experienced suffering, such as when he was cursed by the sage Bhṛgu. Appaya's response is that Rāma's suffering was not real, but a mere 'show' put up by the Lord. Since the Lord, being eternally liberated, cannot really experience suffering, it must be the case that He does so through His incarnations 'like an actor', merely to preserve the rule (*maryādā*) that the curse of sages should be useful and relevant.[27] Bodhendra's point here is that Appaya's reference to the 'incarnations of the Lord' such as Rāma in this passage from the *Siddhāntaleśasaṃgraha* implies that he did not identify the 'Lord' solely with Śiva, but also with Viṣṇu.[28] Thus Bodhendra is able to hold that Appaya truly believed that Śiva and Viṣṇu are both supreme and he thereby condemns Śaivas who believe that Appaya emphatically denied Viṣṇu's supremacy.

In the second section of the *Hariharādvaitabhūṣaṇa*, Bodhendra argues along similar lines that Śiva rules over everything and that He is the sole cause of the world. He rejects counter-arguments—quite possibly advanced by contemporary scholars—to the effect that Śiva is a *jīva*, etc. Like in the first section, he argues that the arguments put forth in support of the supremacy of one

27 The relevant passage of the *Siddhāntáleśasaṃgraha* is the following: *na ca parameśvarasya raghunāthādyavatāre tamasvitvaduḥkhasaṃsargādiśravaṇān muktānām īśvarabhāve punarbandhāpattiḥ, tasya vipraśāpāmoghatvādisvakṛtamaryādāparipālanāya kathaṃcid bhṛguśāpādisatyatvaṃ pratyāyayituṃ naṭavad īśvarasya tadabhinayamātraparatvāt*—'And it does not follow from the fact that scriptures say that incarnations of the supreme Lord like Rāma and others are associated with ignorance, suffering, etc., that liberated people remain bound [to *saṃsāra*, and thereby suffer] once they [have attained the liberated] state of the Lord. The reason is [that scriptures] only intend [to express] that the Lord, just like an actor, pretends [to be associated, through His incarnations, with ignorance, suffering, etc.] in order to preserve the rule, established by Himself, that the curse of sages [like Bhṛgu] should not be in vain, etc., [and] to make [people] understand in some way the truth behind the curse of Bhṛgu, etc'. (*Siddhāntaleśasaṃgraha*: 157).

28 Bodhendra concludes: *evaṃ śivasyāpi bimbacaitanyarūpatvaṃ teṣāṃ vivakṣitam eveti bimbacaitanyarūpeṇa viṣṇuśivayor abhedaḥ śrīmūrtirūpeṇa bhedaś ca dīkṣitānām abhimata ity etat tatrāpi siddham eva*—'And the following has been clearly established with respect to that point too: since he [i.e., Appaya] really intends to express the fact that Śiva too is the reflected consciousness [of Brahman] in this way, it is therefore Dīkṣita's intention that Viṣṇu and Śiva [be regarded as] non-different in the form of the reflected consciousness [of Brahman] and that they are different [solely] in terms of their embodied form [in the *trimūrti*]' (*Hariharādvaitabhūṣaṇa*: 20–21). As we have seen, this is exactly Bodhendra's own view. Note, however, that Appaya narrates the same episode in his polemical *Rāmāyaṇatātparyasārasaṃgrahastotra* to establish Viṣṇu's inferiority to Śiva. Was Bodhendra familiar with this work?

deity can also apply to the other deity, for scriptures contain plenty of statements in support of the supremacy of both. This leads Bodhendra to the third section, where he makes the point again and again that both deities are united in the reflected consciousness of Brahman. If we do not accept that both Viṣṇu and Śiva rule over everything, there is no way to reconcile scriptural texts that have conflicting 'marks' (*liṅga*) concerning the nature of the supreme self.[29] Bodhendra pays attention to upaniṣadic passages where the supreme Lord is referred to with various names: Viṣṇu, Śiva, Āditya, etc. All these names must refer to the same entity because it is impossible that several deities impel all beings: only a single entity can play this role. Bodhendra also argues in this context that Śaṅkara defended the identity between Śiva and Viṣṇu in some of his commentaries on Upaniṣads. Although not all of his arguments are necessarily convincing,[30] it is obviously important for Bodhendra to prove that this major figure of the Advaita Vedānta tradition also upheld the unity of both deities.

It is equally important for Bodhendra to show that Appaya also shared this view. Towards the end of the *Hariharādvaitabhūṣaṇa*, Bodhendra reiterates the point that Appaya's views about Śiva and Viṣṇu have been deeply misunderstood. Not only has he proclaimed in various ways their unity in the *Śivatattvaviveka*, *Siddhāntaleśasaṃgraha* and elsewhere, but he also made his position very clear in this respect in the *Ratnatrayaparīkṣā* (see above, § 2.2). As we recall, Appaya argues in this work that the 'three jewels' (*ratnatraya*) of Viṣṇu, Śiva and Śakti all share the divine nature of Brahman, and yet play different roles on a phenomenal level. Bodhendra introduces an opponent who highlights the gradation between deities that is implied on this level. Interestingly, though it is very clear in the *Ratnatrayaparīkṣā* that Śiva is the property-possessor (*dharmin*) and that Viṣṇu is only Śiva's *dharma*, the opponent argues that Appaya actually held the opposite view, namely that Viṣṇu is the *dharmin* and Śiva the *dharma*. It is not clear how the opponent reaches this conclusion, but the reply of Bodhendra is revealing:

---

29 *evaṃ cātra viṣṇuśivayor bimbacaitanyātmakasarveśvararūpeṇābhedo 'vaśyam abhyupagantavyo. anyathā paramātmani viṣṇutvabodhakabahuśrutiliṅgānāṃ vā śivabodhakabahuśrutiliṅgānāṃ voparodhāpatteḥ*—'And it must be accepted that in this [Upaniṣad] both Viṣṇu and Śiva are non-different insofar as they rule over everything as [Brahman's] reflected consciousness. Otherwise, it would undesirably follow that the marks [found in] numerous scriptural passages that indicate Viṣṇu to be the supreme self interfere with the marks [found in] numerous scriptural passages that indicate Śiva to be [that same] supreme self' (*Hariharādvaitabhūṣaṇa*: 108).

30 Some of the passages he mentions to prove his point are taken from minor Upaniṣads on which it is doubtful that Śaṅkara wrote a commentary.

> [You should not say this] because if it were the case, it would undesirably follow that the description given of Viṣṇu by this same Dīkṣita in the *Siddhāntaleśasaṃgraha*—namely, [his description of Viṣṇu as an entity] reached by those who desire liberation, having the nature of the reflected consciousness [of Brahman], unrestrained, independent, unsurpassed, etc.—is incongruous. For this reason, it is established that [Dīkṣita] considered that the aforesaid non-difference between both [Viṣṇu and Śiva] is intended in the Vedas. As explained earlier, his [occasional] statements to the effect that Viṣṇu is inferior [to Śiva] were made with the intention that something bad should be said in a way or another to those who hate Śiva.[31]

It is striking that Bodhendra reiterates his defence of Appaya in his concluding statement. If Appaya made dubious statements about Viṣṇu, it is simply because he needed to respond to those who denigrated Śiva. Whether we agree or not with Bodhendra's reading of Appaya's polemical views, it is clear that Bodhendra wished to show that Appaya shared his own liberal views about God. If Bodhendra relies at times on the authority of other Advaitins such as Śaṅkara, Sureśvara, Nṛsiṃhāśrama and Madhusūdana Sarasvatī to support his claims,[32] it is really Appaya's opinion that matters to him in the end. The *Hariharādvaitabhūṣaṇa* not only confirms that polemical debates about the nature of the godhead were actively taking place a few decades after Appaya's demise, but also that Appaya's Śaiva works continued to occupy a significant place in them.

31 *tathātve siddhāntaleśasaṃgrahe dīkṣitair eva viṣṇor mumukṣuprāpyabimbacaitanyarūpatvaniravagrahasvātantryasamābhyadhikarāhityādyabhidhānasyāsaṃgatyāpatteḥ. atas tayor uktābheda eva vaidikatvena teṣāṃ tātparyaviṣaya iti siddham. tair viṣṇunikarṣābhidhānaṃ tu śivanindakānāṃ yathā kathaṃcid aniṣṭaṃ vaktavyam ity āśayeneti prāg evoktam* (*Hariharādvaitabhūṣaṇa*: 126–127).

32 *nṛsiṃhāśramacaraṇair advaitadīpikāyām ayam apy abhedaḥ kaṇṭharaveṇoktaḥ. anyeṣāṃ apy advaitināṃ mahatām ayam abhedo 'bhimata evety etat tatkṛtagrantheṣu vyaktam eva*—'This non-difference [between Śiva and Viṣṇu] was also explicitly taught by the venerable Nṛsiṃhāśrama in the *Advaitadīpikā*. Other great Advaitins too intended [to teach] this non-difference; this is [made] entirely clear in the works they composed' (*Hariharādvaitabhūṣaṇa*: 127).

4 All-Inclusivism: Ayyaṇṇa Dīkṣita (ca. 1750)

Ayyaṇṇa Dīkṣita is a scholar who was active in Tiruviśainallūr around the middle of the eighteenth century.[33] He is fondly remembered for his playing of the *vīṇā* and for an original treatise of Advaita Vedānta, the *Vyāsatātparya-nirṇaya*. Not unlike Bodhendra before him, in the *Vyāsatātparyanirṇaya* Ayyaṇṇa argues for the identity of Śiva and Viṣṇu, but he does so primarily with a focus on the *Brahmasūtras*. As an Advaitin, Ayyaṇṇa holds that the true intention of Vyāsa (*vyāsatātparya*) when writing the *Brahmasūtras* was to teach pure non-dualism (*kevalādvaita*). However, we will see that Ayyaṇṇa has his own vision of what 'pure non-dualism' stands for. Not only will he argue that *all* deities are equal manifestations of the non-dual Brahman, but also that *all* philosophical schools implicitly teach pure non-dualism. Hence Ayyaṇṇa dedicates his work to anyone—whether trained in the Vedic tradition (*vaidika*) or not (*laukika*)—who is 'free from partiality' (*pakṣapātavinirmukta*), i.e., anyone who is free from attachment to a specific religious or philosophical position. While Bodhendra begins the *Hariharādvaitabhūṣaṇa* by celebrating Rāma as an embodiment of Śiva and Viṣṇu, Ayyaṇṇa opens the *Vyāsatātparyanirṇaya* with the statement that the non-dual Brahman has the form of both deities:

> He ought to be served as the Lord in the form of the Lord of Śrī [i.e., Viṣṇu] and in the form of the moon-crowned one [i.e., Śiva]; [He ought to be served] as the person who serves [this Lord], and [also ought to be served] as the entire world, which is the place for [this person's] experience. I pay homage to Him, Brahman, whose innate nature is existence, bliss and consciousness.[34]

The verse conveys that Brahman is non-dual because He pervades equally the Lord (*īśa*), the individual self or person (*jīva*) who worships the Lord and the world (*viśva*) where this experience takes place. Ayyaṇṇa describes the Lord

---

33 Ayyaṇṇa Dīkṣita (also known as Ayyaṇṇa Kavi or Ayyaṇṇa Sūri) was the son of Veṅkateśa Kavi and the pupil of Śrīdhara Veṅkateśa, a well-known religious figure to whom Ayyaṇṇa pays homage at the beginning of his *Vyāsatātparyanirṇaya* (… *tasmai śrīdhara-veṅkaṭeśagurave kurve praṇāmāyutam*). Śrīdhara Veṅkateśa, also known as Ayyavāḷ, was patronised by the Marāṭha king Śāhajī of Thanjavur, who reigned between 1684 and 1711. Hence Ayyaṇṇa probably flourished around the middle of the eighteenth century. See the entry on Ayyaṇṇa in the *New Catalogus Catalogorum* (vol. 1, p. 366).

34 *śrīśākṛtīndumakuṭākṛti copasevyam īśātmakaṃ tadupasevitṛjīvarūpam | viśvātmakaṃ tad-upabhogapadaṃ yad īḍe tad brahma satyasukhabodhanijasvarūpam* || (*Vyāsatātparya-nirṇaya*: 1).

as having the form of both Śiva (*indumakuṭa*) and Viṣṇu (*śrīśa*). We will see that for Ayyaṇṇa the non-dual reality of Brahman extends beyond this dual form to encompass Brahmā, Śakti and the various 'emanations' (*vibhūti*) and 'incarnations' (*avatāra*) of Śiva and Viṣṇu, respectively. Significantly, he comes to this conclusion by means of defending Appaya's stance on deities against opponents who claim that Appaya was a staunch Śaiva merely interested in polemics.

The *Vyāsatātparyanirṇaya* is divided into two sections (*pariccheda*): the first deals with the true import (*tātparya*) of the *Brahmasūtras* and what this implies with regard to other philosophical schools; the second is a detailed examination of the thesis, a priori associated with Appaya, that Śiva is the 'fourth' entity transcending the *trimūrti*. Ayyaṇṇa begins by noticing that the intention of the *Brahmasūtras* is 'not entirely clear' (*anatisphuṭāśāya*), as a result of which various commentators (*bhāṣyakāra*)—Śaṅkara, Rāmānuja, Śrīkaṇṭha, Madhva, Vallabha and others—have come up with very different interpretations of the *Brahmasūtras*.[35] In his view, all these commentaries have some validity (*prāmāṇya*) as far as determining the true import of the *Brahmasūtras* is concerned, for all are rooted in the thinking of an authoritative figure and make use of valid means of knowledge (*pramāṇa*). However, if it is true that no commentary can be entirely discarded for these reasons, it is not possible to affirm, on the other hand, that only one of these commentaries teaches exactly what Vyāsa intended. The truth is that every commentator tends to align his interpretation along the tenets of the Vedānta school to which he belongs. None of the logical and exegetical arguments raised by exponents of these schools or those of interpreters who do not belong to a specific school of Vedānta, are entirely neutral.[36] How, then, are we to ascertain Vyāsa's true teaching in the *Brahmasūtras*?

35 Ayyaṇṇa's claim that the *Brahmasūtras*' intention is not entirely clear contrasts with the position Appaya takes in the *Śivārkamaṇidīpikā*. Appaya holds there that the *Brahmasūtras* have a clear (*prasanna*) meaning and an innate clarity (*svābhāvikaprasāda*) that can be revealed through a good commentary, notably Śrīkaṇṭha's *Brahmamīmāṃsābhāṣya*.

36 *tasmād yuktimātrabalena vā śrutimātrabalena vā yuktisahakṛtaśrutibalena vā madhyasthavidvatsaṃnidhānabalena vā tattadvādimātrasaṃpratipannakūṭasākṣivākyasadṛśasūtasaṃhitābodhāyanavṛttyādyanugṛhītabhagavatpādarāmānujādibhāṣyagranthaparyālocanayā vā ayam evārtho vyāsatātparyaviṣaya iti kenāpy avadhārayituṃ na śakyate*—'For this reason, it is not possible for anyone [who belongs to a Vedānta school] to ascertain—whether through reasoning alone, or through [relying on] scriptures alone, or through [relying on] scriptures with the help of reasoning, or through being in contact with an impartial scholar, or through investigating [by oneself] the commentaries of Śaṅkara, Rāmānuja, etc., [commentaries] that are respectively backed by the *Sūtasaṃhitā*, the

Ayyaṇṇa's way of circumventing this difficulty is clever. He appeals to the only truly neutral (*madhyastha*) party, namely the exponents of traditions—Ayyaṇṇa refers to them generically as *tantras*—who *oppose* (*prativādin*) Vedānta, namely Nyāya, Vaiśeṣika, Sāṃkhya, Yoga, Pūrvamīmāṃsā, Pāśupata (in which he includes Śaiva Siddhānta) and Pāñcarātra. Ayyaṇṇa argues that when exponents of these traditions engage with Vedānta on the whole, they focus *only* on the pure non-dualist position held by Advaita Vedāntins. In his view, the fact that these traditions target the position of Advaita Vedānta in their refutations proves that *only* Śaṅkara's commentary—the only commentary on the *Brahmasūtras* that defends *kevalādvaita*—is authoritative on the question of determining the import of the *Brahmasūtras*. Accordingly, since *kevalādvaita* is the true teaching of the *Brahmasūtras*, it must be the case that all the Upaniṣads—the meaning of which the *Brahmasūtras* intend to clarify—expound this same non-dual position. On this account, the *Bhagavadgītā*, Purāṇas and all schools that rely, explicitly or not, on the authority of these foundational texts also, ultimately, expound non-dualism. Ayyaṇṇa's grand thesis echoes Appaya's claim in the *Śivārkamaṇidīpikā* that *all* religious texts ultimately teach non-duality:

> Even though the Upaniṣads and [other] religious scriptures, as well as all [other] compositions, including the several Purāṇas and *smṛti* texts such as the *Mahābhārata* and so forth, culminate in non-duality alone; [even though] the *Brahmasūtras* too shine forth, for those who have discernment, as having their ultimate rest [in that same non-duality]; [and even though] nothing but that [non-duality] was accepted by the best of ancient teachers, such as Śaṅkara and others—nonetheless, it is only through the grace of the one who has the young moon as His crest-jewel [i.e., Śiva] that the inclination towards non-duality appears in people, not otherwise.[37]

However, Ayyaṇṇa's inclusivism differs from Appaya's in two significant ways. First, Ayyaṇṇa does not argue that one ought to worship Śiva in order to develop

---

Bodhāyanavṛtti, etc., and that are, as it were, false testimonies insofar as they are accepted only by the exponents of their respective [schools]—that a given teaching [and not any other] was intended by Vyāsa' (*Vyāsatātparyanirṇaya*: 8).

37 *yady apy advaita eva śrutiśikharagirām āgamānāṃ ca niṣṭhā sākaṃ sarvaiḥ purāṇasmṛtinikaramahābhāratādiprabandhaiḥ | tatraiva brahmasūtrāṇy api ca vimṛśatāṃ bhānti viśrāntimanti pratnair ācāryaratnair api parijagṛhe śaṅkarādyais tad eva* || 6 || *tathāpy anugrahād eva taruṇenduśikhāmaṇeḥ | advaitavāsanā puṃsām āvirbhavati nānyathā* || 7 || (*Śivārkamaṇidīpikā*: 2).

a natural inclination towards the gnostic realisation of non-duality. Though Ayyaṇṇa acknowledges the importance of worship (see below), he does not give any precedence to Śiva since all deities, on his view, are expressions of the same non-dual Brahman. Secondly, and more radically, Ayyaṇṇa holds that *all schools*, even the so-called 'heterodox' (*nāstika*) traditions of Buddhism, Jainism and materialism, have pure non-dualism as their 'ultimate teaching' (*paramasiddhānta*). Although Ayyaṇṇa does acknowledge that the validity of such traditions is denied, at times, in scriptures, he holds that these systems are indirectly (*paramparayā*) helpful (*upayogin*) to the study of Vedānta. Appaya, on the other hand, not unlike Advaitins before him, rejects the teachings of these traditions on the basis that they deny the authority of the Vedas; so did Vijñānabhikṣu, Madhusūdana Sarasvatī and others after him. Ayyaṇṇa's all-inclusivist thesis is to my knowledge unprecedented. It differs from the quasi-inclusivist thesis propounded by Vijñānabhikṣu and his contemporaries, and in a way prefigures the universalistic trends of later neo-Advaita thinkers who hold that all views culminate in pure non-duality.

The second section of the *Vyāsatātparyanirṇaya* takes a stand against those who have recourse to the *Śivatattvaviveka* and other polemical works by Appaya to promote the idea of 'Śiva the fourth' (*turīyaśiva*)—that is, Śiva as an entity distinct from and transcending the *trimūrti*. If he holds the view that the true nature of Brahman is unqualified (*nirviśeṣa*), Ayyaṇṇa maintains that it is only through the grace (*prasādād eva*) of the supreme Lord (*parameśvara*), or the embodied Brahman (*saguṇabrahman*), that we can achieve the understanding of His non-dual nature. The worship of the supreme Lord purifies the mind from its dualities and thus enables the worshipper to receive the Lord's grace, which alone can lead him to the realisation of the attributeless Brahman. The question is then: what is the nature of that supreme Lord? As we recall, theologians of Vedānta had defended diametrically opposed views on this question. While Vaiṣṇava theologians of Vedānta had argued that Viṣṇu is supreme, Appaya and other Śaiva scholars had denied this, arguing instead for Śiva's supremacy. Ayyaṇṇa rejects both positions and distinctively argues that not only Śiva and Viṣṇu, but *all deities* are the supreme Lord. It is worth noting that Ayyaṇṇa's key argument is a direct outcome of the conclusion he reached in the first section of the *Vyāsatātparyanirṇaya*: if we accept the view of non-difference (*abheda*) taught in pure non-dualism—arguably the underlying teaching of the *Brahmasūtras*—the very deliberation (*vicāra*) as to whether Śiva *or* Viṣṇu is Brahman is simply not possible, for both deities must be ontologically non-different from each other.[38] It might be argued against

38 *abhedapakṣe śivasya nārāyaṇasya vā jagatkāraṇatvam iti vicāra eva na saṃbhavati,*

this that though these deities do not differ from each other in terms of their essential nature (*svarūpa*), they nevertheless possess different features (physical attributes, powers, etc.) due to the operation of limiting factors.[39] Ayyaṇṇa's response is that these mutual differences do not ultimately matter in the context of worship (*upāsanā*) because the worshipper's aim is to fully identify with his preferred deity. In this state of union, differences cease to exist. Some may also argue, on the basis of various scriptural passages, that either Śiva or Viṣṇu possesses features of an individual self (*jīva*), and that only the other deity is supreme. Like Appaya, Ayyaṇṇa denies that any deity can be a *jīva*, and holds that every deity of the *trimūrti* has sovereignty (*īśvarabhāva*). Just like Bodhendra had said, the arguments used to defend the supremacy of a deity can also be used to support the other deity's supremacy. Since we cannot really decide between the two alternatives, it must be the case that the Lord is an embodied entity of our choosing, non-different from Śiva, Viṣṇu and other deities.[40]

Ayyaṇṇa then examines in detail the claim that 'Śiva the fourth' (*turīyaśiva*) is supreme, an argument advanced by those who seek to reinvigorate Appaya's polemical views. If the *turīyaśiva* is truly the cause of the world, He must be endowed with a 'body' (*vigraha*) for this purpose, given that the creative act requires a connection to *māyā* and its epiphenomena. Ayyaṇṇa then asks: what is the relation between the *turīyaśiva* and this *vigraha*? Does the *turīyaśiva* create the world while being 'conditioned by a body' (*vigrahopahita*) or while being 'accidentally designated by a body' (*vigrahopalakṣita*)? In the first case, the body functions as a limiting factor (*upādhi*) for the *turīyaśiva*; it is produced by *māyā* and persists as long as the *turīyaśiva* exists, just as a red object functions as a limiting factor for a piece of glass while it is placed in front of

---

ubhayor bhedābhāvāt—'If we hold the view of non-difference, the very deliberation as to whether Śiva or Nārāyaṇa is the cause of the world is impossible, for there is no difference between both [deities]' (*Vyāsatātparyanirṇaya*: 53).

39 A standard example for the operation of a limiting factor (*upādhi*) is the space enclosed in a pot: the pot is an *upādhi* of space because it appears to 'limit' space in its enclosure. This limitation, however, does not affect space's true nature (*svarūpa*), namely that it is omnipresent and homogeneous. In the present case, the limiting factor is *māyā*: although deities are ultimately non-different from each other, they are different on a phenomenal level because they are endowed with different features conditioned by *māyā* (*māyopahita*).

40 *tasmād vinigamanāvirahena śivaviṣṇvādyabhinno 'smadabhimatamāyopahitamūrtaparameśvara eva sarvakāranam ity abhyupetam*—'Therefore, since there is a lack of determining argument [in support of a deity's supremacy], we must admit that the cause of everything is the [form of the] embodied supreme Lord that we prefer, which is [in essence] non-different from Śiva, Viṣṇu and other [deities], and which has *māyā* as its limiting factor' (*Vyāsatātparyanirṇaya*: 57).

it. In the second case, the body is merely an accidental feature (*upalakṣaṇa*) of the *turīyaśiva* insofar as it 'designates' it accidentally, just as crows temporarily perched on a house 'accidentally designate' that house. Ayyaṇṇa rejects the first option on two grounds. First, the body in question, even if it belongs to the 'fourth' state transcending the *trimūrti*, must have parts (*sāvayava*) and must therefore be an effect (*kārya*) of something else (the underlying inference being that everything 'produced' is made of parts). But this entails the fault of infinite regress (*anavasthā*) since *another cause* having a body as its limiting factor would need to be invoked to explain the creation of the *turīyaśiva*'s body. Secondly, since in this scenario the body functions as a limiting factor that is *produced* by *māyā*, the body cannot be beginningless (*anādi*), thus implying the same fault of infinite regress. Ayyaṇṇa leans toward the second option: only a form that is 'accidentally designated' by a body can be the cause of the world, for only such a form can be the product of a consciousness conditioned by *māyā* (*māyopahitacaitanya*). According to Ayyaṇṇa, if the creator of the world is embodied, it takes the form that *we* choose (*asmadabhimata*)—whether Śiva, Viṣṇu or another form. This is possible only if consciousness itself participates in this experience by having *māyā* as its limiting factor, not if the body is itself a limiting factor. In this scenario, the body merely 'designates' the form of the deity that appears in our consciousness, and this deity can therefore be *any* deity.

The direct implication here is that no deity can be superior to others. Ayyaṇṇa spends much of the second section refuting the view that Śiva is superior to Viṣṇu or to any other deity for that matter. Just like Bodhendra, Ayyaṇṇa holds that if we claim that Śiva is superior to Viṣṇu based on some scriptural passages, we must reconcile those passages with other passages that affirm Viṣṇu's superiority. The only way out is to accept, in line with the doctrine of *kevalādvaita*, that there is no hierarchy or gradation (*taratamabhāva*) whatsoever between the various deities worshipped—that is, no 'fourth form' (*turīyamūrti*) transcending other forms—but only a single supreme Lord 'woven together' (*anusyūta*) by our consciousness in the form of any given embodied deity (*guṇamūrti*). Significantly, Ayyaṇṇa supports his position with the help of Appaya's authority. To an opponent claiming that the superiority of Śiva to Viṣṇu is clearly defended by Appaya in the *Śivatattvaviveka* and other works, Ayyaṇṇa replies:

> [It is not the case that Śiva is superior to Viṣṇu] because, even if the author of these works, the venerable Appaya Dīkṣita, taught that Viṣṇu is inferior to Śiva in order to oppose his [Vaiṣṇava] opponents, in reality it was not his intention to do so: he said this, making use of the *na hi nindā* prin-

> ciple,[41] with the intention that the worship of Śiva should [also] be performed [just like that of Viṣṇu]. [...] Therefore, those who desire to be free should worship only the undivided and unqualified supreme Brahman—[whose nature is] the existence, consciousness and bliss, the pure non-duality intended by Vyāsa [in the *Brahmasūtras*]—in His embodied form conditioned by *māyā*, as the cause of the creation, etc. of the world, and in the form of Brahmā, Viṣṇu, Rudra, Rāma, Kṛṣṇa and [other deities and their various manifestations]. It is only through His grace that one can acquire the knowledge of Brahman. All of this is agreeable.[42]

Interspersed in this passage are three quotations taken from Appaya's Śivādvaita and Advaita works in which his liberal views about Viṣṇu are made explicit. In Ayyaṇṇa's view, Appaya's defence of the *turīyaśiva* exemplifies the *na hi nindā* principle of interpretation, according to which blaming (*nindā*) a given deity is only done for the sake of praising another deity. In other words, Appaya's claims about Viṣṇu's inferiority are only meant to praise Śiva. Backed by Appaya's authority, Ayyaṇṇa concludes not only that all schools implicitly teach the pure non-dualism intended by Vyāsa in the *Brahmasūtras*, but also that *all* deities—the *trimūrti* and their various avatars—can be worshipped in order to realise the non-dual Brahman.

5 Śiva Alone Is Supreme: Tyāgarāja Śāstrī (1815–1903)

It should be clear that the widespread idea that all Hindu deities are equal manifestations of the non-dual Brahman is not inherent to Advaita Vedānta. Up until the modern period, this notion did not meet the approval of all Advaita Vedāntins, as evident from the polemical debates fueled by Appaya's Śaiva works. We have seen that Bodhendra and Ayyaṇṇa develop their inclusivistic approaches by drawing on Appaya's liberal views, while other scholars emphas-

41 This principle of Mīmāṃsā hermeneutics is posited in Śabara's commentary on *Mīmāṃsāsūtra* 2.4.20: *na hi nindā nindyaṃ ninditum prayujyate, kim tarhi ninditād itarat praśaṃsitum*—'Blame is not employed in order to blame something that must be blamed, but to praise something that is different from what is blamed'.

42 *tadgranthakartṛbhir appayyadīkṣitacaraṇaiḥ prativādinirākaraṇāya viṣṇoḥ śivāpekṣayā nikarṣapratipādane 'pi vastutas tatra tātparyābhāvena tatpratipādanasya na hi nindā nyāyena śivopāsanāvaśyakartavyatāparatvāt.* [...] *tasmād vyāsābhimatakevalādvaitarūpasaccidānandākhaṇḍanirviśeṣaparabrahmaṇa eva māyopahitāmūrtarūpeṇa jagajjanmādikāraṇatvarūpeṇa brahmaviṣṇurudrarāmakṛṣṇādirūpeṇa ca mumukṣūpāsyatvaṃ tatprasādād eva brahmajñānaprāptiś ceti sarvaṃ ramaṇīyam* (*Vyāsatātparyanirṇaya*: 79).

ise Appaya's pro-Śaiva positions in order to promote their own sectarian position. What retains the attention of Bodhendra and Ayyaṇṇa is not Appaya's arguments about Viṣṇu's inferiority and Śiva's supremacy, but his understanding of the shared divinity of all deities. Notwithstanding Bodhendra's and Ayyaṇṇa's remarkably inclusivistic stances, the question of God's nature, particularly in relation to the supremacy of either Viṣṇu or Śiva and their respective status in the *trimūrti*, continued to be debated in the following decades. In the late eighteenth century, Tryambaka Śāstrī (ca. 1770?—1835), an Advaita Vedāntin of impressive erudition hailing from Kalyāṇadurga (Karnataka), wrote several polemical works and defended his positions in numerous debates. One such debate took place in the court of Kṛṣṇarāja Voḍeyar III together with Rāmaśeṣa Śāstrī, then the leading scholar of the court; the debate, which 'raged for three days', pertained to whether the word *nārāyaṇa* means Viṣṇu or Śiva (Ravishankar 2018: 176, 180). A similar debate took place in Kāñcīpuram not long after, involving one of Tryambaka's own students, Kuṇigala Rāmaśāstrī (1807–1860). In line with Appaya Dīkṣita, who had allegedly first made this argument, Rāmaśāstrī argued that the word *nārāyaṇa* means Śiva. Hundreds of Vaiṣṇava scholars met to confront him on this question, as a result of which Rāmaśāstrī wrote the *Navakoṭi*, a work 'advocating Advaita and Śaivism and criticising Vaiṣṇavism' (ibid.: 254). This work attracted at least two rejoinders: the *Daśakoṭi* by Aṇṇaṅgarācārya, a scholar of Viśiṣṭādvaita Vedānta belonging to the well-known Śeṣa family (Raghavan 1957: 185), and the *Navakoṭibhañjanavāda* by another prominent Viśiṣṭādvaita Vedānta scholar, Kṛṣṇatātācārya.

Tyāgarāja Śāstrī (1815–1903), also known as Rāju Śāstrī, was active around the same period. Hailing from Mannārguḍi in the Tamil country, Tyāgarāja was a prolific scholar of Advaita Vedānta and a direct descendant of Appaya. He authored more than thirty works on Advaita Vedānta and also a number of treatises criticising Viśiṣṭādvaita Vedānta theses. Foremost among these works is the *Nyāyenduśekhara*, which is a rejoinder to the *Nyāyabhāskara* by the Śrīvaiṣṇava scholar Anantācārya (alias Anantālvār, 1822–1862). The *Nyāyenduśekhara* is a response to the *Advaitasiddhicandrikā*, which is a commentary by Gauḍa Brahmānanda Sarasvatī (beginning of eighteenth century) on Madhusūdana Sarasvatī's *Advaitasiddhi*. Following in the footsteps of his prestigious ancestor, Tyāgarāja was a staunch defender of Śaivism. He wrote the only extant commentary on Appaya's *Śivatattvaviveka*, the *Śivatattvavivekadīpikā*, in order to 'increase the devotion [of his audience] to the lotus-feet of Śiva'.[43] Just like the *Nyāyenduśekhara*, this work too responds to Śrīvaiṣṇava

43 The first half of the sixth introductory verse of the *dīpikā* reads: *śivatattvavivekadīpikaiṣā śivapādāmbujabhaktivṛddhihetuḥ* (*Śivatattvavivekadīpikā*: 1).

criticisms: it is primarily a rejoinder to the *Vedāntakaustubha* authored by the eighteenth-century Śrīvaiṣṇava scholar Paravastu Vedāntācārya. Paravastu was a prominent critic of Appaya. He wrote the *Mahābhāratatātparyarakṣā* to refute Appaya's position, developed in his *Bhāratasārasaṃgrahastotra*, that the *Mahābhārata* highlights Śiva's supremacy. The *Vedāntakaustubha* is principally directed against the *Śivatattvaviveka*.[44]

In his *Śivatattvavivekadīpikā*, Tyāgarāja follows Appaya's text closely, glossing nearly every word and providing additional explanations when necessary. Given this approach, the *dīpikā* was perhaps intended as a starting-point for those eager to familiarise themselves with Appaya's masterpiece of Śaiva polemics. Tyāgarāja, however, also takes time to address complex counter-arguments, typically at the end of his commentary on the verses. Sometimes he responds directly to arguments found in the *Vedāntakaustubha*, and, at other times, to arguments from scholars presumably contemporaneous with him. An example of the latter is found at the end of his commentary on verse 13 where, as we have seen earlier, Appaya argues that Śiva transcends the *trimūrti*. At the end of his auto-commentary on this verse, Appaya remarks that despite the clarity of his arguments, some people remain confused. Tyāgarāja explains that such people misunderstand Appaya not only because of their foolishness but due to their pride in having studied texts belonging to a non-Vedic tradition: the Pāñcarātra tantric tradition dear to Śrīvaiṣṇavas. At this point, Tyāgarāja introduces a first key objection. The objector is evidently a Śrīvaiṣṇava scholar who has read the *Vedāntakaustubha* and relies on it to make his point:

> The author [of the *Śivatattvaviveka*] set out [to explain] all this [namely, that Śiva transcends other deities, etc.] as if it was already established that there exists a fourth state [above the *trimūrti*] (*turīyamūrti*), [but] this remains unproven because [this thesis] was seized by the fault-finding arguments presented in the *Kaustubha*. After quoting the following passage from the *Śivatattvaviveka*:
>
> > In [the following passage from the] *Brahmopaniṣad*, the supreme reality consisting of all deities, designated by the word *akṣara*, is under-

44 The editor of the *dīpikā* edition I have at hand confirms this: *śivatattvavivekagranthasya khaṇḍanātmako 'yaṃ viśiṣṭādvaitamatānuyāyinā kenacit praṇīto vedāntakaustubhākhyo grantha iti tannirāsārtham evaitadvyākhyārambha iti*—'The work titled *Vedāntakaustubha*, written by some follower of the Viśiṣṭādvaita school, is a rejoinder to the *Śivatattvaviveka*. It is precisely in order to refute it that [Tyāgarāja] undertook [the composition] of this commentary [on the *Śivatattvaviveka*]' (*Śivatattvavivekadīpikā*: 1).

> stood to be a different entity than Brahmā and other [deities]: 'Now, this Puruṣa has four seats: the navel, the heart, the throat and the head. In these shines forth Brahman with four parts, namely Brahmā in the waking state, Viṣṇu in the dream state, Rudra in the dreamless state and the Imperishable (*akṣara*), which is the fourth state [transcending the three other states]. He is also Āditya, Viṣṇu and Īśvara',
>
> [the author of the *Kaustubha*] raised the following counter-argument: since [Appaya] conveys [in this passage] that the one Brahman has four parts (*amśa*) beginning with Brahmā, etc., he understands that [these four parts], referred to by the names of Brahmā, Viṣṇu, Rudra and the Imperishable, constitute a single consciousness. [For this reason, Appaya could not reasonably hold that there is a fourth state beyond the *trimūrti*] because parts (*aṃśa*) and whole (*aṃśin*) naturally designate something unitary.[45]

In verse 13, Appaya had argued that Śiva is identical to Brahman as the 'fourth state' transcending the *trimūrti*, whereas Rudra, an emanation of Śiva, belongs to the *trimūrti*. According to the opponent here, Paravastu convincingly showed in his *Vedāntakaustubha*, in line with the general Śrīvaiṣṇava position on the matter, that the very idea of 'another entity' (*tattvāntara*) transcending the *trimūrti* is problematic. The quoted passage from the *Brahmopaniṣad* suggests that Appaya actually understands the *trimūrti* along with the Imperishable (*akṣara*) as four parts (*aṃśa*) of a single whole (*aṃśin*), which is the single consciousness of Brahman—not as a hierarchy comprising the *trimūrti* at the bottom and the Imperishable (identified with Brahman) above it. In other words, according to this passage, there is no 'fourth state' transcending Brahmā, Viṣṇu and Rudra. It is therefore not possible to argue, like Appaya does, that Śiva is that 'fourth state'. At first sight, Paravastu's interpretation of Appaya's thesis comes close to how Bodhendra and Ayyaṇṇa both understand Appaya's true intention (*tātparya*) in their respective works: Appaya did not believe that there is a deity superior to others because for him what really exists is only

---

45 *nanu sarvam idaṃ turīyamūrtivyavasthāpanasiddhavat kāreṇa pravṛttam. tadvyavasthāpanam eva kaustubhoktadūṣaṇagrastatvād asiddham. tathā hi 'brahmopaniṣadi athāsya puruṣasya catvāri sthānāni bhavanti nābhir hṛdayaṃ kaṇṭho mūrdhā ca, tatra catuṣpādaṃ brahma vibhāti jāgarite brahmā svapne viṣṇuḥ suṣupte rudras turīyam akṣaram ādityo viṣṇuś ceśvaraś ca iti brahmādibhyaḥ param akṣaraśabditaṃ sarvadevātmakaṃ tattvāntaraṃ darśitam' ityādiśivatttvavivekagrantham anūdya tad dūṣaṇam uktam—ekasyaiva brahmaṇaḥ caturmukhādyaṃśacatuṣṭayavattāṃ pratipādayatā 'nena brahmaviṣṇurudrākṣaraśabdoktānām ekacetanasyaivāvagamāt aṃśāṃśivyapadeśasya aikya eva svarasatvād iti* (*Śivatattvavivekadīpikā*: 105–106).

the single consciousness of Brahman. However, there is a key difference in that Paravastu uses the terms *aṃśa* and *aṃśin* to describe the relationship between Brahman and the deities: these terms are typical of the 'partitive theology' of Viśiṣṭādvaita Vedānta, not of the pure non-dualism of Advaita Vedānta defended by Bodhendra and Ayyaṇṇa.[46]

In his response, Tyāgarāja highlights the different ways in which Viśiṣṭādvaitins and Advaitins understand the concept of difference. He does not think that the objector's argument hits the mark. While it is true that Appaya believes in the fundamental unity of Brahman's consciousness—as an Advaitin, he upholds the unity of the self (*ātmaikya*)—he does not hold that there is no difference whatsoever between the four deities in question. Though they emphasise non-difference (*abheda*) as the ultimate feature of reality, Advaitins do acknowledge the existence of conditional divisions (*aupādhikabheda*) superimposed on the unity of Brahman through the mirific power of *māyā*. Hence Brahmā, Viṣṇu, Rudra and the Imperishable are not merely 'parts' (*aṃśa*) of Brahman: they are conditioned—that is, ultimately unreal—divisions of Brahman that differ from each other in accordance with their respective qualities or attributes (*guṇa*). This is the reason why it is possible for Advaitins like Appaya to defend unity while upholding that there is a 'fourth state' beyond the *trimūrti* subject to the interplay of *guṇas*. For a Viśiṣṭādvaitin like Paravastu, however, the sense given to unity (*aikya*)—that of an organic whole comprising a 'part-possessor' (*aṃśin*) and its qualifying parts (*aṃśa*)—is entirely different, for it is based on the presupposition that the parts in question are real (*vāstava*), not accidental (*aupādhika*). In fact, argues Tyāgarāja, Paravastu himself is not in a position to claim that all deities are fundamentally one, for the 'parts' constituted by the deities of the *trimūrti* are *really* different from their 'part-possessor', Brahman. In fact, to say that the relationship between part and part-possessor

---

46 I borrow the expression 'partitive theology' from Lipner (1986: 85). According to Rāmānuja, Brahman is 'partite' insofar as it consists of two 'parts' (*aṃśa*)—conscious beings (*cidaṃśa*) and non-conscious beings (*acidaṃśa*)—of which He is the part-possessor (*aṃśin*). The relation that stands between *aṃśa* and *aṃśin* is the same as that between a qualifier (*viśeṣaṇa*) and the thing it qualifies (*viśeṣya, viśiṣṭa*). The qualifier (*viśeṣaṇa*) is not hierarchically inferior to what it qualifies but inseparably connected (*apṛthaksiddha*) to it, and thus on the same ontological level. Insofar as *aṃśa* qualifies the *aṃśin*, it does not exist independently of the latter although it has its own distinctive nature. Thus the relation between *aṃśa* and *aṃśin* highlights the 'difference-in-identity' relation that is central to Viśiṣṭādvaita Vedānta theology. Scholars of Viśiṣṭādvaita Vedānta use these terms (and others) to describe the relationship between the body and the self (*ātman*) as well as between the individual self (*jīva*) and Brahman. See *Śrībhāṣya* ad *Brahmasūtra* 2.3.45 for Rāmānuja's definition of *aṃśa* as qualifier: *viśiṣṭasyaikasya vastuno viśeṣaṇam aṃśa eva.*

naturally entails unity (*aikya*) only displays how ignorant Paravastu is of his own tradition (*svamatājñānavijṛmbhita*). As a matter of fact, Viśiṣṭādvaitins do not even accept the unity of the individual self (*jīva*) and the Lord (*īśvara*). For them, the self always retains its individuality, even in the liberated state. Furthermore, even if we concede that the *trimūrti* along with the Imperishable are really four parts (*aṃśa*) of the single consciousness of Brahman, we must accept that Brahmā and Rudra too are divine, just like Viṣṇu; this is a consequence which the author of the *Kaustubha* would certainly reject.[47]

The objector also attempts to downplay the existence of a 'fourth state' transcending the *trimūrti* by drawing a parallel with Viṣṇu's incarnations (*avatāra*), such as Rāma and Kṛṣṇa. Just as we designate these incarnations as 'parts' (*aṃśa*) of Viṣṇu, so should we regard the three deities of the *trimūrti* as 'parts' of Brahman. The idea is that Viṣṇu does not transcend His incarnations but forms a unity, as it were, with them: Rāma, as Viṣṇu's *aṃśa*, inheres in Viṣṇu and is not 'inferior' to Him in any way. The relationship between *aṃśa* and *aṃśin* is not hierarchical but one of expression or manifestation: just like the rays of the sun manifest the sun's nature, so do the incarnations of Viṣṇu manifest Viṣṇu's divine nature. Likewise, there is no need to invoke the existence of an entity transcending the *trimūrti* if we realise that the deities of the *trimūrti* are not inferior to Brahman, but rather inhere in Brahman as His parts, thus expressing His nature as pure consciousness. Why then is there the need, on proving the existence of a 'fourth state' of consciousness (*turīyacetana*)? For Tyāgarāja, the comparison with Viṣṇu and His incarnations is not adequate. The existence of a 'fourth state' is in fact compatible with the scheme of parts and part-possessor if we understand their relationship as resulting from limiting factors. Once again, the problem comes down to how Advaitins and Viśiṣṭādvaitins approach differently the concept of metaphysical difference (*bheda*).

Another problem for Tyāgarāja is the claim made by Viśiṣṭādvaitins to the effect that Brahmā and Rudra are *different* from Brahman. How can we reconcile this claim with their thesis that all deities partake of the single consciousness of Brahman? This is not a problem for Advaitins who understand differences as conditioned by *māyā* (*māyopahita*), and therefore as ultimately unreal. The situation is different for Viśiṣṭādvaitins, for they do not accept that

47 *aṃśacatuṣṭayavato brahmaṇo vāstavaikatvopagame ca brahmarudrayor api viṣṇor iva brahmakoṭipraveśāpatteḥ*—'Because if you accept that Brahman is endowed with four parts [and is nevertheless] really unitary, it would undesirably follow [for you] that Brahmā and Rudra too, like Viṣṇu, fall into the domain of Brahman' (*Śivatattvavivekadīpikā*: 106).

differences are merely accidental. The objector featured in the *dīpikā* expects such an objection and responds to it by highlighting the fact that Brahmā and Rudra are endowed with different defining attributes (*dharma*) than Brahman. This is fully attested in scriptures which declare that Brahmā and Rudra are created, have the nature of individual selves, are subject to *karma*, etc. Brahman, on the other hand, has opposite attributes: eternality, not being subject to *karma*, etc. However, while such attribute-based differences (*bheda*) are real—which allows the claim that Brahman is different from Brahmā and Rudra—they do not preclude identity at a fundamental level. This makes sense within the metaphysical framework of difference-in-identity accepted in Viśiṣṭādvaita Vedānta. Tyāgarāja does not reject these premises. The problem for him lies in the fact that the objector only ascribes such attributes to Brahmā and Rudra, and not to Viṣṇu, though scriptures also claim Viṣṇu to be created, etc. This is a case of 'inconclusiveness' (*anaikāntikatā*), a logical fallacy involving the reason (*hetu*) in an inference. A reason is inconclusive if it does not co-exist in a unique way with what the inference sets out to establish (*sādhya*). In the present case, the inference seeks to establish that Brahmā and Rudra are different from Brahman on the basis that they have attributes such as being created, etc. Since the reason here—i.e., possessing attributes, such as being created, etc.—also applies to Viṣṇu, it is inconclusive. To support his claim, Tyāgarāja cites passages from scriptures that assert that Viṣṇu is created by Prajāpati, that the three deities of the *trimūrti* are individual selves, that Viṣṇu's incarnations are subject to the laws of *karma*, etc. After responding in detail to a few more objections, Tyāgarāja concludes his commentary on verse 13 by reasserting what Appaya had declared two centuries before: Śiva alone is the 'fourth entity' transcending the *trimūrti*.

6 In Defence of Viṣṇu: Rāmasubrahmaṇya Śāstrī (1840–1922)

Tyāgarāja's defence of Appaya's Śaiva theses quickly garnered attention. What had started as an argument with a Viśiṣṭādvaitin surprisingly turned into a dispute with another prolific Advaitin: Rāmasubrahmaṇya Śāstrī from Tiruviśainallūr. Like Tyāgarāja, Rāmasubrahmaṇya wrote several independent works of Advaita Vedānta as well as polemical tracts attacking other schools of Vedānta. Prominent polemical pieces include the *Nyāyabhāskarakhaṇḍana*, a refutation of the aforesaid *Nyāyabhāskara*, and the *Candrikākhaṇḍana*, in which Rāmasubrahmaṇya defends Śaṅkara's commentary against criticisms advanced by the Mādhva scholar Vyāsatīrtha in his *Tātparyacandrikā*. Rāmasubrahmaṇya was a devoted Vaiṣṇava, and his Vaiṣṇava affiliation closely informed his approach of Advaita. While his aim in the *Bhāṣyatātparyasaṃgraha*, for instance, is to reassert the intention behind Śaṅkara's commentary in order 'to put an end to the delusion of Śaivas',[48] his goal in the *Nyāyenduśekharadoṣaghātana* is to attack the Śaiva theses propounded in the abovementioned *Nyāyenduśekhara* by Tyāgarāja Śāstrī. Rāmasubrahmaṇya may well have written other polemical works directly targeting Tyāgarāja's theses, and he wrote at least one treatise critically engaging his illustrious ancestor, Appaya Dīkṣita: the *Viṣṇudveṣakara-mahāśaivamatamardana* ('Crushing the Position [held by the] Great Śaiva who Hates Viṣṇu').[49]

In the *Viṣṇudveṣakaramahāśaivamatamardana*, Rāmasubrahmaṇya makes his intention behind the work very clear: to defend Viṣṇu against Śaivas who claim that Viṣṇu is not supreme. He traces the source of the problem to Southerners (*drāviḍa*) such as Haradatta Śivācārya and others—among whom he certainly includes Appaya too—whom he deems responsible for having first distorted the meaning of *śāstra*s. Rāmasubrahmaṇya, indeed, composes his *Mardana* in order to redress this situation.[50] He complains that although the

48 This work is still unpublished; see entry 1546 in Hultzsch (1896: 149): *śaṅkarāryakṛte bhāṣye yat tātparyam avasthitaṃ | tac cāhaṃ saṃgrahīṣyāmi śaivadurmohaśāntaye* ||—'I shall sum up the intention of the commentary written by Śaṅkara in order to put an end to the delusion of Śaivas'.

49 In 2005, Dominik Wujastyk met with descendants of Rāmasubrahmaṇya and was told the following by a family member: 'She recounted how Rāmasubrahmaṇya participated in an intellectual conflict with a scholar who lived in Maṉṉārguḍi. Although Rāmasubrahmaṇya was a devoted Vaiṣṇava, he had no serious argument with Śaivas, and had even been into a Śaiva temple on some occasions, though he prayed to Rāma. When the scholar Mahāmahopādhyāya Rāju Śāstrī, *alias* Tyāgarājamakhin, of Maṉṉārguḍi, a descendant of Appayadīkṣita, wrote a pro-Śaiva tract, Rāmasubrahmaṇya wrote a counter-blast. A counter-counter-blast came from Maṉṉārguḍi, and so forth for about four exchanges, of which the *Viṣṇudveṣakara-mahāśaivamatamardanam* is one' (Wujastyk 2012: 244). I surmise that the *mahāśaiva* referred to in the title of this work is Appaya, not Tyāgarāja, for two reasons: 1) most of the arguments dealt with in this work can be traced to Appaya's Śaiva polemical works; 2) Rāmasubrahmaṇya refers to Appaya as a *mahāśaiva* once in the text (*mahāśaivāppayyadīkṣitair bhāratatātparyasaṃgrahe* ...; *Mahāśaivamatamardana*: 21). Nonetheless, since Tyāgarāja is a descendant of Appaya who defended the latter's views, this work may also be considered a rejoinder to Tyāgarāja's Śaiva work.

50 *drāviḍaharadattādyaiḥ śaivaiḥ śāstrārthaviplavo janitaḥ | tam ahaṃ dūrīkartuṃ karomi rāmasya pādayor namanam* || 1 ||—'The distortion of the meaning of the *śāstra* originated with Śaivas from the South such as Haradatta and others. In order to get rid of this confusion, I pay homage to Rāma's feet' (*Mahāśaivamatamardana*: 1). Rāmasubrahmaṇya hailed from Tiruviśainallūr in the Tamil country, but considered himself a descendant of so-called 'northern' (*audīcya*) brahmins; see the first verse in the last portion of the *Mahāśaivamatamardana* where he describes his lineage: *tadgrāmasthā viprā vatsās tv audīcyākhyās teṣāṃ madhye* ... (*Mahāśaivamatamardana*: 68).

scriptures all concur in saying that Brahman is non-dual, yet people 'devoted to bad doctrines' (*durvādaniṣṭha*) maintain that Viṣṇu's supremacy has no place in the *śāstra*s:

> Although it is possible even for someone like me to say that there is not any harm in [holding] that Brahman, insofar as He is spoken of in hundreds of scriptures as being clearly non-dual, is Śiva or Viṣṇu, [nevertheless] some people devoted to bad doctrines adopted a wrong interpretation when writing in their works that there is no place in *śāstra*s for the supremacy of Viṣṇu. I refute this view here [in this work].[51]

This assertion strongly echoes Appaya's own attitude towards Vaiṣṇavas. Like Appaya, Rāmasubrahmaṇya does not wish to engage in acrimonious debates on the nature of God—the fact that God is the non-dual Brahman is overtly clear enough—but he feels compelled to do so in order to address and effectively reject stances denying Viṣṇu's supremacy. But if Appaya sought to defend Śiva against claims made by tenants of a different branch of Vedānta, namely Viśiṣṭādvaita Vedānta, Rāmasubrahmaṇya responds here to Appaya and indirectly to his descendant Tyāgarāja Śāstrī, an Advaitin who lived not far from Tiruviśainallūr and around the same time as Rāmasubrahmaṇya.

Several of the arguments invoked in support of the supremacy of a deity rely on Purāṇas that have a devotional leaning towards that deity. Hence Rāmasubrahmaṇya begins his treatise, quite remarkably, by rejecting *en bloc* the legitimacy of Purāṇas. Purāṇas are neither scriptures (*śruti*) nor texts that support the meaning of these scriptures (*upabṛṃhaṇa*). Thus they are of no use in ascertaining the meaning of the two main sections of the Vedic corpus—namely, the *jñānakāṇḍa* represented by the Upaniṣads, which contain salvific knowledge about Brahman, and the *karmakāṇḍa* represented in Vedic *mantra*s and *brāhmaṇa*s, which pertain to ritual action. Since neither Jaimini nor Bādarāyaṇa—the authors of the foundational *sūtra*s on these two *kāṇḍa*s, namely the *Mīmāṃsāsūtra*s and the *Brahmasūtra*s, respectively—considered Purāṇas in their respective investigations (*mīmāṃsā*), these texts should not be accepted by learned people (*śiṣṭa*). Rāmasubrahmaṇya further argues that though it is correct to assume that all Upaniṣads are 'Vedic' and as such authoritative, not all are helpful towards achieving the knowledge of Brahman. In his view, only

51 *yady apy advaitam eva śrutiśatagaditaṃ brahma tat syāc chivo vā viṣṇur vā nātra kācit kṣatir api gadituṃ śakyate mādṛśaiś ca | kecid durvādaniṣṭhā hariparabhavanaṃ śāstrasaṃghe na cāstīty evaṃ granthaṃ likhitvā kugatim upagatās tanmataṃ khaṇḍyate 'tra* || 3 || (*Mahāśaivamatamardana*: 1).

the ten principal Upaniṣads (*daśopaniṣad*)—namely, *Īśā*, *Kena*, *Kaṭha*, *Praśna*, *Muṇḍaka*, *Māṇḍukya*, *Taittirīya*, *Aitareya*, *Chāndogya*, *Bṛhadāraṇyaka*—should be included in the *jñānakāṇḍa*. Sectarian Upaniṣads such as the *Śvetāśvatara*, *Kaivalya*, *Atharvaśikhā*, *Atharvaśiras* and others should not; nor do these Upaniṣads have any relevance as far as the performance of ritual actions is concerned. In other words, these Upaniṣads have no authority whatsoever. While he acknowledges the authority of such Upaniṣads for people who are not engaged in one of the four 'stages of life' (*āśramabāhya*), he denies it for all the others—effectively, then, all brahmins, whether they are *brahmacārin*s, *gṛhastha*s, *vānaprastha*s or *saṃnyāsin*s. Significantly, these 'sectarian' Upaniṣads are precisely those which are amenable to a Śaiva interpretation, and which form the scope of Appaya's exegesis in the *Śivatattvaviveka*.[52]

Following this line of argumentation, in the *Viṣṇudveṣakaramahāśaivamatamardana*, Rāmasubrahmaṇya refutes the view that Śiva is the central object of the *Mahābhārata*. This bold thesis was first propounded by Appaya in his *Bhāratasārasaṃgrahastotra*, and Rāmasubrahmaṇya directly challenges claims advanced in that work. One of the arguments he addresses is found in Appaya's self-authored commentary on verse 8. Here Appaya argues that passages in the *Mahābhārata* that describe Śiva as being inferior to Brahmā and Viṣṇu do not actually refer to Śiva, but to His emanations (*vibhūti*)—i.e., the various forms of Rudra the Destroyer, one of the three deities of the *trimūrti*.[53] Names such as Viśālākṣa and Sthāṇu, or expressions such as the 'one who is born from Brahmā'

52 Not all Advaitins would agree with Rāmasubrahmaṇya's claim that only the ten principal Upaniṣads have relevance for gaining the knowledge of Brahman. A few decades before Rāmasubrahmaṇya, another South Indian Advaitin, Upaniṣadbrahmayogin, famously wrote commentaries on 108 Upaniṣads. Likewise, not all theologians of Vedānta with a Vaiṣṇava affiliation would agree with Rāmasubrahmaṇya's quick dismissal of the *Śvetāśvatara Upaniṣad* and other similar 'sectarian' Upaniṣads. Prominent Viśiṣṭādvaita Vedāntins, including Yāmuna, Rāmānuja, Vātsyavarada and Veṅkaṭanātha, acknowledged their authority and tried to demonstrate that they proclaim Viṣṇu's supremacy (Stark 1992).

53 The verse in question reads: *viśālākṣaḥ sthāṇur druhiṇatanayaḥ saṃharaṇakṛt sarojākṣakrodhaprabhava iti vaiśeṣikapadam | vibhūtitvajñaptyai kim api nihitaṃ nātha niyataṃ nikarṣoktis tābhyāṃ bhavata iti buddhiṃ nirasitum* ||—'The particular names [and expressions] Viśālākṣa, Sthāṇu, son of Brahmā, Destroyer, the one who was born out of the anger of the lotus-eyed one [i.e., Viṣṇu], are given to [Śiva] in order to convey in some way that [Śiva] has emanations [namely, Rudras], [and] to eliminate the notion that there are sayings that convey that You, O Lord [Śiva], are inferior to them [i.e., Brahmā and Viṣṇu]' (verse 8, *Bhāratatātparyasaṃgraha*: 65). Appaya opens his commentary on this verse by claiming that there are several Rudras that function as particular emanations (*vibhūtiviśeṣa*) of Śiva—he refers to Śiva here as Sāmbaśiva—who is the cause of the *trimūrti* (and so transcends it), the controller of *māyā* and the embodied Brahman

or the 'one born out of Viṣṇu's anger', all convey some sort of defect or inferiority on the part of Śiva. On Appaya's account, these names and expressions refer to forms of Rudra (*vibhūtirudra*), not to the transcendent Śiva. In truth, the central deity worshipped in the *Mahābhārata* is not any form of Rudra but the transcendant Śiva (*turīyaśiva*). Rāmasubrahmaṇya examines in detail multiple passages from the *Mahābhārata* cited by Appaya in support of this view, and concludes, like other detractors of Appaya had done before him, that the *Mahābhārata* contains no suggestions whatsoever that there exists 'another' Śiva transcending the *trimūrti* who is devoid of the attributes (*dharma*) conveyed in the aforesaid names and expressions.

Rāmasubrahmaṇya makes an interesting comment about Appaya's own textual sources. He draws attention to two passages from the Śāntiparvan that alledgedly demonstrate, according to Appaya, that a 'fourth Śiva' transcends the *trimūrti*. According to Rāmasubrahmaṇya, these passages cannot be accepted as evidence in support of Appaya's thesis, for two reasons. Firstly, these passages are not found in the 'Northern' edition of the *Mahābhārata* commented upon by Nīlakaṇṭha Caturdhara; this alone is sufficient ground for these passages to be discarded (*prakṣipta*). Secondly, since these same passages are found in a 'Southern' edition but not in the 'Northern' edition as well, it is legitimate to doubt whether or not these passages belong at all to 'the *Mahābhārata*' (Rāmasubrahmaṇya presumably has in mind an Ur-*Mahābhārata*). This doubt makes it impossible to take these passages as solid evidence (*niścitapramāṇa*) in support of Appaya's thesis.[54] Incidentally, this argument suggests that Rāmasu-

(*saguṇabrahman*): *santi trimūrtikakāraṇasya māyādhiṣṭhātṛsaguṇabrahmarūpasya śrīsāmbaśivasya vibhūtiviśeṣarūpā rudrāḥ* (ibid.).

54 *na ca* [...] *iti vacanaṃ ca trimūrtyuttīrṇaturīyaśive pramāṇam iti vācyam, nīlakaṇṭhamiśravyākhyāte uttaradeśamahābhāratakośe tvadudāhṛtavacanādarśanena tadvacanānāṃ prakṣiptatvāt dākṣiṇātyakośasattvottaradeśakośāsattvābhyāṃ tadvacanānām astitvanāstitvaśaṅkākalaṅkitatvena niścitapramāṇatvābhāvāc ca* (*Mahāśaivamatamardana*: 25). The quoted passages (*vacana*) in question are found in Appaya's auto-commentary on verse 7 of the *Bhāratasārasaṃgrahastotra*. When making this argument, Rāmasubrahmaṇya addresses the opponent directly (***tvadudāhṛtavacanādarśanena***). At first sight, this might suggest that he is addressing a contemporary opponent. However, the fact that the argument is found in the *Bhāratasārasaṃgrahastotra*, suggests that he is addressing Appaya here in the second person; alternatively, he might be indirectly referring to Tyāgarāja insofar as he belongs to Appaya's lineage. 'Nīlakaṇṭha Miśra' refers here to Nīlakaṇṭha Caturdhara, the famous commentator on the *Mahābhārata* who lived in Banaras in the second half of the 17th century. The *uttaradeśamahābhāratakośa* probably refers to the Bombay lithograph edition (Gaṇapatakṛṣṇājī Press) of the *Mahābhārata*, which was published, together with Nīlakaṇṭha's commentary, in 1863. The *dākṣiṇātyakośa* could be inter-

brahmaṇya was one of the very first Indian scholars to acknowledge and make use of the fact that there are differences between 'Southern' and 'Northern' textual versions of the *Mahābhārata*, roughly at the same time Moriz Winternitz (1863–1937) and Heinrich Lüders (1869–1943) were proposing to produce a critical edition of the *Mahābhārata* based on different recensions.[55]

After arguing, against Appaya, that the *Mahābhārata* in fact extols the supremacy of Viṣṇu,[56] Rāmasubrahmaṇya attempts to undermine Appaya's skills in Mīmāṃsā. He pays attention to Appaya's interpretation of a scriptural statement from the *Aitareya Brāḥmaṇa* in which Viṣṇu is praised as the highest of the gods.[57] In the *Śivatattvaviveka*, Appaya argues that the inferiority of Agni stated in this passage is only vis-à-vis a particular deity (*yatkiṃciddevāpekṣayā*), not vis-à-vis all deities, as the genitive plural *devānām* first seems to indicate. Otherwise, it would undesirably follow that the other scriptural passages claiming that Agni is supreme *along with* another deity (e.g., Soma) would be contradicted. If this is the case, then the same must apply to Viṣṇu's superiority since the word 'Viṣṇu' is mentioned together (→ *samabhivyāhāra*) with the word 'Agni' in the same sentence. Appaya's point is that Viṣṇu's superiority is only vis-à-vis a single deity, and that this deity cannot be Śiva. Rāmasubrahmaṇya rejects this interpretation in toto because its underlying argument is

---

preted as a reference to the Kumbhaghoṇam edition completed in 1910 by Krishnacharya, twelve years before Rāmasubrahmaṇya passed away; the passages in question are found in this edition. It is quite possible that Rāmasubrahmaṇya, who lived less than ten kilometers away from Kumbhaghoṇam, would have had access to this edition. Rāmasubrahmaṇya's use of the word *kośa* suggests that these 'editions' were either mass-produced manuscripts (like the lithograph Bombay edition) or that he simply used the word to refer to what we now call a 'printed edition'. The fact that the passages in question are *prakṣipta* on the mere basis that they are not found in the 'northern' edition interestingly suggests that Rāmasubrahmaṇya accepted this edition to be more authoritative than the 'southern' edition. I wish to thank Christopher Minkowski and Vishal Sharma for drawing my attention to these two editions of the *Mahābhārata* and for productive exchanges on this passage. I am also grateful to Camillo Formigatti for shedding additional light on the meaning of *kośa* in this context.

55 Winternitz drew attention to the need for a critical edition of the *Mahābhārata* back in 1897 and published a first fascicle introducing formally the idea in 1908 together with Lüders. The collating of manuscripts continued but came to a halt with the First World War. In 1918, the project of a critical edition was resumed by scholars from the Bhandarkar Oriental Research Institute in Poona. I am grateful to Christopher Minkowski for pointing this information out to me.

56 *mahābhāratasya viṣṇusarvotkarṣa eva paramatātparyam* (*Mahāśaivamatamardana*: 30).

57 The passage is: *agnir vai devānām avamo viṣṇuḥ paramas tadantarā vai sarvā devatāḥ*—'Agni is in truth the lowest of the gods, Visnu the highest. All other gods are in truth in between' (*Aitareya Brāhmaṇa* 1.1.1).

contradicted by a stronger evidence (*prabalapramāṇa*), namely the fact that the Aitareya passage claims *explicitly* that Viṣṇu is supreme (*viṣṇuḥ paramaḥ*) and that *all* deities (*sarvā devatāḥ*) stand between Viṣṇu and Agni. Rāmasubrahmaṇya invokes here the Mīmāṃsā hermeneutical principle formulated in the well-known *śrutiliṅgasūtra*.[58] This *sūtra* states, in order of importance, the six ways of determining the meaning of a text or textual passage. The first item in the list is *śruti*, that is, what is explicitly said in a given scriptural passage. When it comes to interpreting a scriptural sentence, *śruti* dominates over the proximity (*sthāna*) of given elements in that sentence. In the present case, this means that the proximity of the word 'Viṣṇu' to the word 'Agni' constitutes weaker evidence (*durbalapramāṇa*) than what the Aitareya explicitly says. In Rāmasubrahmaṇya's view, such an argument could only have been put forward by someone who is unfamiliar with the *śrutiliṅgasūtra* and therefore unskilled in Mīmāṃsā. Hence he exclaims: 'Please, [Appaya,] look at Jaimini's *śrutiliṅgasūtra*!'[59]

Overall, Rāmasubrahmaṇya takes a very different approach to Appaya's Śaiva works than Bodhendra and Ayyaṇṇa. While these two scholars labourously demonstrate that Appaya is not concerned with sectarian disputes about the nature of God, Rāmasubrahmaṇya holds Appaya to be a 'Great Śaiva' who hates Viṣṇu. Thus he refutes at length several of his key Śaiva arguments while also attacking the legitimacy of the scriptures he relies on in his Śaiva exegesis. In the *Mardana*, Rāmasubrahmaṇya is not interested in trying to promote the supremacy of Śiva on par with Viṣṇu. Although he is a proponent of Advaita Vedānta, he does not concede, in this work, that both Śiva and Viṣṇu are manifestations of the same non-dual Brahman. Nonetheless, Rāmasubrahmaṇya displays a more liberal approach in his *Anubhāṣyagāmbhīryagrantha*, wherein he defends Śaṅkara's *Brahmasūtrabhāṣya*. While he upholds a Vaiṣṇava approach to Śaṅkara's commentary,[60] he also claims that the three principal schools of Vedānta—Dvaita, Viśiṣṭādvaita and Advaita—agree on basic tenets, such as the nature of the embodied Brahman, the individual self, and the nature

58 *Mīmāṃsāsūtra* 3.3.14: *srutilingavākyaprakaraṇasthānasamākhyānāṃ samavāye pāradaurbalyam arthaviprakarṣāt*.

59 *śrutiliṅgasūtraṃ jaiminīyaṃ paśyatu bhavān* (*Mahāśaivamatamardana*: 40–41).

60 For instance, Rāmasubrahmaṇya argues in this work that Śaṅkara believed in the orthodoxy of the Pāñcarātra teachings and the superiority of Nārāyaṇa over other deities: *pāñcarātrasya prāyeṇa vedamūlakatvasya taduktanārāyaṇotkarṣatadārādhanāder bhagavatpādābhimatatvam ityetat svīkāryaṃ*—'It should be accepted that the Bhagavatpāda [i.e., Śaṅkara] upheld that Pāñcarātra [teachings] are for the most part grounded in the Vedas [and] that he accepted the superiority of Nārāyaṇa, His worship, etc. taught in [Pāñcarātra]' (*Anubhāṣyagāṃbhīryagrantha*: 21).

of liberation as bliss; they only differ on details. On this account, despite their doctrinal differences, all Vedānta schools are, as a matter of fact, authoritative.[61] In taking this broad view of Vedānta, Rāmasubrahmaṇya comes very close to Appaya. Indeed, he seeks support for his own views in Appaya's doxography of Vedānta schools, the *Caturmatasārasaṃgraha*, a work in which Appaya likewise asserts his acceptance of all Vedānta schools. This liberal take on Vedānta suggests that Rāmasubrahmaṇya's position in the *Mardana* is not his ultimate view—as he makes clear in the beginning, Brahman is non-dual and it does not matter whether we call it Śiva or Viṣṇu—and that his aim is merely to 'crush' those who hate Viṣṇu.

7 Conclusion

Advaita Vedāntins were not unanimous on the question of God as we entered the modern period. Although they all belonged to the same 'school', and thus shared the same basic metaphysical premises about the non-dual nature of the world, self, and Brahman, the four pandits examined in this study did not agree on the ultimate nature of the godhead nor on which deity, if applicable, ought to be considered supreme. Bodhendra and Ayyaṇṇa defend an inclusivist position in which both Śiva and Viṣṇu—and, specifically in the case of Ayyaṇṇa, all Hindu deities and their incarnations—represent different aspects of the same non-dual consciousness of Brahman. Tyāgarāja and Rāmasubrahmaṇya, however, adopt a more sectarian approach. Tyāgarāja, on the one hand, argues that only Śiva is the supreme Brahman while Rāmasubrahmaṇya, on the other hand, holds that Viṣṇu is the supreme Brahman. While these four pandits represent only a small sample of the scholars of Advaita Vedānta who were active in this period, their varied writings reveal that the widespread idea that all Hindu deities are 'manifestations of the one Supreme' is not inherent to Advaita Vedānta. It is an idea that came about through a gradual process of individual and collective thinking that took place both within and without the Advaita tradition, and against the backdrop of a long history of sectarian debates going back to the beginning of the second millenium. These debates, often polemical in nature, not only revolved around the nature of God, but also entailed

61 See *Anubhāṣyagāṃbhīryagrantha*: 30 ff. Rāmasubrahmaṇya concludes: *tasmāt pradhānāṃśe matatrayasyāpi sāmyād upasarjanāṃśeṣu vailakṣaṇye 'pi matatrayaṃ nirmatsarāṇām upādeyam eva*—'Therefore, since the three [Vedānta] schools agree on essential matters and differ only on details, [scholars] free from envy should entirely accept these three [schools]' (*Anubhāṣyagāṃbhīryagrantha*: 31).

wider considerations about consciousness, transcendence, sovereignty, identity, as well as the authority of scriptures, epics and Purāṇas.

A paradigm-shifting moment occured in the sixteenth century when Appaya Dīkṣita challenged the dominant theological model of the period, Viśiṣṭādvaita Vedānta. For the first time, the long-standing view that Viṣṇu is the supreme Brahman was tentatively replaced with another view—unprecedented in the realm of vedāntic theology—namely, that Śiva, not Viṣṇu, is supreme insofar as He transcends the *trimūrti*. As his career progressed, however, Appaya shifted his focus away from sectarian polemics to embrace a more comprehensive view of God. In his later works, Appaya displays a remarkably tolerant attitude towards Viṣṇu and His worship, even insisting on His divinity on par with Śiva. However, the ambivalence implicit in Appaya's Śaiva writings fueled various polemical debates in the centuries that followed. It is, indeed, significant that the four Advaita Vedāntins discussed here all feel compelled to engage with Appaya's writings. Both Bodhendra and Ayyaṇṇa formulate their inclusivist view of God in response to opponents who have arguably misunderstood Appaya's claims concerning Viṣṇu's inferiority. A descendant of Appaya himself, Tyāgarāja revives his illustrious ancestor's pro-Śaiva views to counteract some of his contemporary Viśiṣṭādvaita opponents. Desirous to settle the matter with Tyāgarāja, Rāmasubrahmaṇya criticises Appaya's work while also adopting a parallel approach when defending the supremacy of Viṣṇu against claims of Śiva's superiority. It is perhaps fair to say that without Appaya and the debates his Śaiva writings generated, the Advaitin position on God would not have been drawn out with the same clarity and vividness. We could go even further and argue, as I shall do in greater detail in a future publication, that Appaya's work radically transformed Advaita Vedānta, ushering in a kind of 'post-sectarian' movement. The seventeenth century onward indeed bears witness to an increasing number of attempts to move beyond sectarian disputes and to formulate a more unified vision of Vedānta.

This study also serves as a reminder that socio-religious concerns can play a key role even in what appears to be purely philosophical matters. Despite the fact that Advaita Vedānta is primarily concerned with the attributeless (*nirguṇa*) nature of Brahman, modern pandits of Advaita Vedānta debated vehemently, like their predecessors had done, on whether Śiva or Viṣṇu ought to be identified with Brahman as the 'fourth' reality transcending the *trimūrti*. At odds with Radhakrishnan's understanding of Advaita Vedānta as 'transcending all religious approaches', the writings of these pandits foreground Advaita as a religious philosophy embedded in a diversified intellectual, social and religious environment. The scholars discussed here belonged to a network of scholars centred around the Kāverī delta in South India, not far from the important cultural and religious centre of Thanjavur. Bodhendra was based in Kāñcīpuram, and was an older contemporary of Śrīdhara Veṅkateśa, Ayyaṇṇa's own teacher. Śrīdhara, whose well-known religious tolerance[62] might well have inspired Ayyaṇṇa's own inclusivistic tendencies, was one of the first members of a remarkable academic community founded in Tiruviśainallūr around 1693 by the Marāṭha king Śāhajī of Thanjavur (regn. 1684–1711; see Wujastyk 2012: 236).[63] Ayyaṇṇa lived in this community along with his teacher, and so did Rāmasubrahmaṇya a century later. As for Tyāgarāja, he lived in Mannārguḍi, only fifty kilometres away from Tiruviśainallūr. Right from the beginning, this community was meant to attract scholars from a variety of intellectual but also religious backgrounds. While Śrīdhara was himself a Śaiva, another important founding member, the celebrated scholar Rāmabhadra Dīkṣita (fl. 1693), was a devotee of Rāma.[64] Likewise, although they were all proponents of Advaita Vedānta, the four pandits discussed in this study shared distinct religious affiliations and lineages that shaped, in different ways and degrees, their philosophical stance as Advaitins, and thereby the sort of debates that were taking place. The fact that they all belonged to the same intellectual network further facilitated these exchanges. To a significant extent, the history of Advaita Vedānta in the modern era is the history of scholars—most of whom remain either unknown or unstudied—who worked in similar eclectic environments. To better understand how and why Advaita Vedānta evolved as it did in this period requires that we pay attention not only to textual sources reaching back to Appaya Dīkṣita and even earlier, but also to the local networks in which pandits lived, wrote, and debated.

---

62 Śrīdhara is still remembered today in South India for his religious tolerance and generosity, and also for writing devotional songs (*bhajans*) of enduring appeal (see Wujastyk 2012: 239–240).

63 The foundation of this academic community involved the donation of land, tax-free and in perpetuity, to a group of forty-six scholars and their descendants. The scholars who settled in this village and its surroundings were experts in various areas of *śāstras* ranging from grammar and Vedic hermeneutics to theology, medicine, and law. For more details, see Sastri 1904.

64 Śrīdhara wrote Śaiva devotional works such as the *Śivabhaktikalpalatikā* and the *Śivabhaktalakṣaṇa* (Sastri 1904: 187). Rāmabhadra Dīkṣita was a devotee of Rāma as shows the fact that he begins his works with an invocation to Rāma (ibid.: 137).

Acknowledgements

I wish to thank the editorial committee for their feedback on this article, and Dominik Wujastyk for sharing materials on Rāmasubrahmaṇya Śāstrī. I also wish to thank James Madaio for fruitful exchanges on late Advaita, and for his detailed feedback on the article. This study was made possible thanks to the support of a research grant awarded by the John Templeton Foundation (*Global Philosophy of Religion Project*, ID # 61613, University of Birmingham).

Bibliography

Primary Sources

Ānandalaharī of Appaya Dīkṣita. *Anandalahari, Sivamahimakalika Stuti, Sivakarnamritam, Sri Ramayana Tatparyasarasamgraha Stotra, Bharatasarasangraha Stotra of Srimad Appayya Deekshitendra*, ed. Pandit Sri K. Ramachandra Sarma. Hyderabad: Srimad Appayya Deekshitendra Granthavali Prakasana Samiti, 1989.

Anubhāṣyagāṃbhīryagrantha of Rāmasubrahmaṇya Śāstrī. *Anubhāṣyagāṃbhīryagrantha*. Cennapuri: Veṅkaṭeśvarasabhādhyakṣa, 1912.

Bhāratatātparyasamgraha of Appaya Dīkṣita. *Bhāratatātparyasaṃgrahaḥ*. Śrīraṅgam: Vani Vilas Press, 1929.

Hariharādvaitabhūṣaṇa of Bodhendra Sarasvatī. *Hariharādvaita Bhūsanam (with Kārikā) by Bodhendrasarasvatī*, ed. T. Chandrasekharan. Madras: Superintendent Government Press, 1954.

Mahāśaivamatamardana of Rāmasubrahmaṇya Śāstrī. *Viṣṇudveṣakaramahāśaivamatamardanam*, ed. Piccu-Ayyar-Ākhya-prabhu. Tañjore: Śrīkṛṣṇavilāsamudrālaya, 1912.

Nyāyasiddhāñjana of Veṅkaṭanātha. *Nyaya Siddanjana by Srimad Vedanta Desika with two old commentaries*, ed. Śrīnivāsatātācārya. Madras: Ubhayavedāntagranthamālā, 1976.

Rahasyatrayasāra of Veṅkaṭanātha. *Śrīmadrahasyatrayasāraḥ*. Srirangam: Sri Poundarikapuram Swami Asramam, 2000.

Ratnatrayaparīkṣā of Appaya Dīkṣita. *The Ratnatrayapariksha by Sri Appayya Dikshita with his commentary*. Mysore: Seshadri Press, 1952.

Siddhāntabindu of Madhusūdana Sarasvatī. *Siddhāntabindu of Madhusūdana with the Commentary of Purushottama*, ed. and transl. Prahlad Chandrashekhar Divanji. Gaekwad's Oriental Series LXIV. Baroda: Oriental Institute, 1933.

Siddhāntaleśasaṃgraha of Appaya Dīkṣita. *The Siddhāntaleśasaṃgraha of Appayya Dīkṣita*, ed. S.S. Suryanarayana Sastri. Vol. 2: Roman and Sanskrit texts. Madras: University of Madras, 1937.

Śikhariṇīmālā of Appaya Dīkṣita. *Śikhariṇīmālā tatkṛtaśivatattvavivekākhyavyākhyāsahitā*, ed. Harihara Śāstrīn. Kumbhaghoṇa: Śrīvidyāmudrākṣaraśālā, 1895.

Śivatattvavivekadīpikā of Tyāgarāja Śāstrī. *Sikharini mala (1 to 14 slokas) (with commentary Siva Tatva Vivekam) of Srimad Appayya Dikshitar. Commentary in Sanskrit Siva Tatva Viveka Dipika by Mahamahopadhyaya Sri Tyagarajamakhivara*, ed. Mani Dravid Sastry. Chennai: Mannargudi Periava Seva Samithi, 2014.

Śivārkamaṇidīpikā of Appaya Dīkṣita. *The Brahma Sūtra Bhāṣya of Śrīkaṇṭhācārya with the Commentary Śivārkamaṇi Dīpikā by the Famous Appaya Dīkṣita*, ed. R. Halasyanatha Sastri, Vol. 1. New Delhi: Nag Publishers, 1986 (reprint of 1908).

Śrutaprakāśikā of Sudarśanasūri. *Brahmasutra Sribhashyam of Sri Ramanuja (with Srutaprakasika commentary of Srisudarsanasuri and Bhavaprakasika of Srirangaramanujamuni)*. Vol. 1. ed. K.E. Devanathacharya. Chennai: Sri Nrisimhapriya Trust, 2006.

Vivekacūḍāmaṇi ascribed to Śaṅkara. *Vivekachudamani of Sri Sankaracharya*, ed. and transl. Swami Madhavananda. Almora: The Advaita Ashrama, 1921.

Vyāsatātparyanirṇaya of Ayyaṇṇa Dīkṣita. *Vyāsatātparyanirṇayaḥ*, ed. Vidwan N. Ranganatha Sharma. Bangalore: Udaya Prakashana, 1995.

Secondary Literature

Adyar, *Descriptive Catalogue of Sanskrit Manuscripts in the Adyar Library*. Vol. 9: Mīmāṃsā and Advaita Vedānta. Adyar: The Adyar Library and Research Centre, 1952.

Bronner, Yigal and Lawrence McCrea, *First Words, Last Words: New Theories for Reading Old Texts in Sixteenth-Century India*. Oxford: Oxford University Press, 2021.

Carman, John B., *The Theology of Ramanuja: An Essay in Interreligious Understanding*. Bombay: Ananthacharya Indological Research Institute (Indian reprint), 1981.

Duquette, Jonathan, 'On Appaya Dīkṣita's Engagement with Vyāsatīrtha's *Tarkatāṇḍava*', *Journal of Indological Studies* 28: 1–23, 2016.

Duquette, Jonathan, *Defending God in Sixteenth-Century India: The Śaiva Oeuvre of Appaya Dīkṣita*. Oxford University Press, Oxford Oriental Monographs Series, Oxford, 2021.

Gerschheimer, Gerdi, 'Les 'six doctrines de spéculation' (*ṣaṭtarkī*). Sur la catégorisation variable des systèmes philosophiques dans l'Inde classique', in K. Preisendanz (ed.), *Expanding and Merging Horizons: Contributions to South Asian and Cross-Cultural Studies in Commemoration of Wilhelm Halbfass*, Vienna: Österreichische Akademie der Wissenschaften (Philosophisch-Historische Klasse, Denkschriften 351; Beiträge zur Kultur- und Geistesgeschichte Asiens 53), 2007, pp. 239–258.

Hultzsch, E., *Reports on Sanskrit Manuscripts in Southern India*. Madras: The Superintendent, Government Press, 1896.

Lipner, Julius J., *The Face of Truth: A Study of Meaning and Metaphysics in the Vedāntic Theology of Rāmānuja*. London: The Macmillan Press Ltd., 1986.

McCrea, Lawrence, 'Playing with the System: Fragmentation and Individualization in Late Pre-colonial Mīmāṃsā', *Journal of Indian Philosophy* 36: 575–585, 2008.

New Catalogus Catalogorum, Vol. 1, ed. V. Raghavan., Madras: University of Madras, 1968.

Nicholson, Andrew, *Unifying Hinduism: Philosophy and Identity in Indian Intellectual History*. Columbia University Press, South Asia Across the Disciplines Series, New York, 2010.

Radhakrishnan, Sarvepalli, *The Hindu View of Life*. Allen and Unwin, London, 1954.

Radhakrishnan, Sarvepalli, *The Principal Upaniṣads*. Allen and Unwin, London, 1968 (second impression). First impression in 1953.

Raghavan, V., 'Sanskrit Literature c. 1700–1900', *Journal of the Madras University* (Section A. Humanities). Madras: University of Madras. Vol. XXVIII, No. 2, January 1957, pp. 175–204.

Ravishankar, Chinya V., *Sons of Sarasvatī: Late Exemplars of the Indian Intellectual Tradition*. State University of New York Press, Albany 2018.

Sastri, T.S. Kuppuswami, 'Ramabhadra-Dikshita and the Southern Poets of His Time', *The Indian Antiquary*, Vol. XXXIII, 1904. Bombay: Bombay Education Society's Press, Byculla., pp. 126–142 and pp. 176–196.

Stark, Sylvia, '*Śvetāśvataropaniṣad* 3.8–3.11 in the Light of Viśiṣṭādvaita (with special reference to the interpretation of Vātsya Varadaguru)', *Wiener Zeitschrift für die Kunde Südasiens* 36, 1992, pp. 187–194.

Venkatkrishnan, Anand, *Mīmāṃsā, Vedānta, and the Bhakti Movement*. Ph.D. Dissertation, Columbia University, 2015.

Wujastyk, Dominik, 'Rāmasubrahmaṇya's Manuscripts: Intellectual Networks in the Kaveri Delta, 1693–1922', in *Aspects of Manuscript Culture in South India*, ed. Saraju Rath. Leiden-Boston: Brill, 2012, pp. 235–252.

CHAPTER 17

Gotmāī's Suit: A Brahman Woman of Property in Seventeenth Century Western India

Rosalind O'Hanlon

Viewed from the perspective of a social historian rather than a fellow Sanskritist, Chris Minkowski's wonderful scholarship has always struck me as having something of Dr Who's Tardis about it. When I first met Chris in Oxford in 2007, two years after he had himself moved there from Cornell, my impression was of intellectual interests that were sharply focussed and compact in scope. Sharply focussed was certainly right. But on closer acquaintance, I quickly learned that his work in the field closest to my own—the working lives of small but strategically placed communities of Maratha Brahman pandits in Mughal India—actually opened for us a strikingly broad and complex new field for intellectual discovery.

Chris's interest in these communities has been just one of many fields of Sanskrit scholarship to which he has contributed with such distinction. During my first year as his colleague in Oxford I discovered, as so many others had done before me, his extraordinary intellectual range, running from the most ancient of Vedic studies to the scientific revolution of the seventeenth century as viewed from the perspective of Sanskrit scholars. With my own early modern interests, Chris's studies of seventeenth century pandit lineages were particularly compelling and revelatory. He explored their constitution out of entwined forms of intellectual and family filiation, and his approach to their wider social networks brought to light a delicate interplay between social, religious and intellectual history approaches.

His work on the Banaras pandit and polymath Nīlakaṇṭha Caturdhara in particular—as scholar of astrology, astronomy, the magic of numbers, philology, geography and the arts of war, commentator on the Mahābhārata, student of Vedānta—tell us us much about Chris's own wide-ranging intellectual interests.[1] 'Histories of the book' approaches formed the framework with which

1 Christopher Minkowski, 'Nīlakaṇṭha's Teachers and Gurus, Part 1: Lakṣmaṇa Paṇḍita' *Wiener Zeitschrift für die Kunde Südasiens*, 55, 2013–2014, 33–76; 'Nīlakaṇṭha's Mahābhārata,' *Seminar* 608, 2010, 32–38; 'What makes a work "traditional"? On the Success of Nīlakaṇṭha's Mahābhārata Commentary,' in Fr. Squarcini, ed., *Boundaries, Dynamics and Construction*

 | DOI:10.1163/9789004536869_019

he explored the lives of pandit astrologers, and the manuscript libraries that were their essential their working tools.[2] Studies of these and other Sanskrit intellectuals enjoying the patronage of royal courts in Mughal India, alongside their European counterparts, enabled him to illustrate in remarkably close detail the interaction between the Mughal court and the world of Sanskrit learning.[3] All of these studies made him a valued associate of Sheldon Pollock's 'Sanskrit Knowledge Systems' project. Like so many others, I have hugely enjoyed and benefitted from Chris's range of skills and insight, and look forward to much more of it in the years to come.

In other ways for me too, Chris has been a very generous colleague. It was very much the early opportunity that Chris kindly gave me to collaborate together on a project on seventeenth century pandit communities in their migration between Maharashtra and Banaras, that helped to set my own research directions in the years that followed. With remarkable patience, he tolerated my clumsy attempts to work out the connections between their lineages and the wider social histories of caste, gender and household with which I was more familiar. He even forgave an early, and foolhardy suggestion on my part that we might dispense with diacritical marks. Without his experience and inspiration, my own years in Oxford would have been much the poorer.

I hope therefore that Chris might enjoy this small essay offered in gratitude for his kindness and intellectual generosity. The essay deals with themes that echo, albeit in very different social-historical settings, those in Chris's work. A Brahman household and lineage is at the centre of the essay, although a lineage not of intellectuals, but of local revenue officials based in Maharashtra's Konkan littoral. Nonetheless, their stores of paper records also mattered greatly to them, both as the working tools of their trade, and as evidence of their rights to their offices. Complex and sometimes bitter relationships within the lineage also feature here, and the importance of sons to a family's survival. Here also we see the political connections that could link the most local of Konkani family conflicts to relationships at the heart of Aurangzeb's Mughal empire. Like

of Traditions in South Asia. Florence: Firenze University Press, 2005, 225–252. 'Nīlakaṇṭha's Instruments of War: Modern, Vernacular, Barbarous', *Indian Economic and social History Review* 41, 3, 2004, 365–385.

2 Christopher Minkowski, 'Sanskrit Scientific Libraries and their Uses: Examples and Problems of the Early Modern Period' in F. Bretelle and C. Proust, eds., *Looking at it from Asia: The Processes that shaped the sources of the History of Science*. Boston Studies in Philosophy of Science 206, 2010, 81–114.

3 Christopher Minkowski, 'Learned Brahmins and the Mughal Court: The Jyotiṣas', in Vasudha Dalmia and Munis D. Faruqi, *Religious Interactions in Mughal India*. Delhi: Oxford University Press, 2014, 102–134.

Nīlakaṇṭha, a single individual of the later seventeenth century is the focus of the essay, and one remarkable in her determination and skill. As an unlettered Brahman widow, however, her field of action did not lie in literary exegesis or learned debate, but rather in navigating the complex interface between the household and the public spheres of temple and provincial governor's office.

Gotmāi Kāle was a widow of a Brahman family based in the *taraf* or district subdivision of Chiplun town in the subāh or province of Dabhol in Maharashtra's Konkan littoral. She lived from the middle decades of the seventeenth century through into the early decades of the eighteenth. This was a prosperous region, with trade from the Persian Gulf and Indian ocean landing at the ports of Dabhol and Cheul, before moving up through the passes of the western ghats and into the Deccan. As the last surviving direct descendant of the family, Gotmāi found herself drawn into a prolonged struggle to retain control of the family's prestigious hereditary office as district accountants, against distant relatives and outsiders bent on usurping the office and the valuable rights attached to it.

An important question for any scholar interested, as Chris has been, in the nature of family and lineage in early modern India, concerns their nature as complex corporate entities, blending scribal skills, social and marital interdependencies and material resources, and dependent for their stability above all on the cohesion of their members. For the seventeenth and eighteenth centuries, indeed, it is particularly difficult to gain insights into the mundane and local contexts in which women outside the world of courtly and state elites sought to control their family rights to property and office. Nor do we have very much understanding of women's ability to negotiate the internal family relationships upon which such control often depended in the early modern centuries.

From a range of different intellectual perspectives, this is an important omission. As I have argued elsewhere, Brahman scholarly labour depended overwhelmingly on the resources and cohesion of the Brahman household, with Brahman wives and mothers at its centre. Intellectual prestige was intimately connected with the reputation of the family, both as a provider of teaching for new generations of scholars, and as a vital guarantor of the purity of biological kinship. The largest seventeenth century gathering of named Brahman scholars that we know of assembled together not in intellectual debate but to assure themselves of the purity of their marriage relationships.[4] In this setting, where

---

4 Rosalind O'Hanlon, 'Brahman Lineages Beyond the Mughal Court', in Richard Eaton and Ramya Sreenivasan, eds., *The Oxford Handbook of the Mughal World*. Oxford University Press, forthcoming, and O'Hanlon, 'Letters Home: Banaras pandits and the Maratha Regions in early modern India', *Modern Asian Studies*, 44, 2, 2010, 201–240.

intellectual and biological filiation ran through the same skeins of family connection, what forms of agency were open to women?

Women's agency in other kinds of early modern household matters too. As Samira Sheikh, Indrani Chatterjee and others have recently observed, our understanding of women's ability to control capital and their access to the political networks on which such control often depended, is a field still very much under development.[5] How did women of small landholder families, or local office-holders with ancestral lands attached, regard the rights that their families had accumulated? How did they seek to protect them from predatory neighbours or rivals, negotiate with local judicial institutions, handle the documentary culture that accompanied property rights in many parts of India during this period, invoke the protection of religious authorities or agencies of the state? How far did the family itself, so often the key arena within which women's life chances were determined, also shape their ability to control capital and resources?

Local and apparently obscure as it was, Gotmāi's long struggle offers us some sharp insights into many of these questions. She waged her protracted campaign in the intimate surroundings of her home, amid fraught family relationships, in the public space of the local temple, in the neighbourhoods of Chiplun town where she sought refuge in the homes of relatives, and in the office of the provincial governor. Some understanding of the way in which women experienced life at this prosaic interface between family, community institutions and agencies of the state might have much to tell us about women's scope for independent action in early modern India, their local political competencies, their attitudes to authority and their ability to protect such access to property rights as they had. It might also illuminate wider social attitudes in this milieu to women themselves, when they sought to become actors beyond the bounds of the domestic sphere. Developing our appreciation of these forms of agency also contributes, of course, to our understanding of their changing history with the coming of colonialism and its new regimes of property right and family law.

5 Samira Sheikh, 'Jibhabhu's Rights to Ghee: Land Control and Vernacular Capitalism in Gujarat, c. 1803–1810', *Modern Asian Studies* 51, 2, 2017, 350–374; Indrani Chatterjee, *Forgotten Friends: Monks, Marriages, Memories of Northeast India*. Delhi: Oxford University Press, 2013; Farhat Hasan, 'Property and Social Relations in Mughal India: Litigations and disputes at the Qazi's Court in Urban Localities', *Journal of the Economic and Social History of the Orient* 61, 5–6, 2018, 851–877; Ramya Sreenivasan, 'Drudges, Dancing Girls, Concubines: Female Slaves in Rajput Polity, 1500–1850', in Indrani Chatterjee and Richard M. Eaton., eds, *Slavery and South Asian History*. Bloomington and Indianapolis: Indiana University Press, 2006, 137–161; Shadab Bano, 'Women and Property in Mughal India', *Proceedings of the Indian History Congress*, vol. 68, 2007, 406–415.

Let us look first, then, at this local milieu itself, with its judicial institutions, agencies of the state, bureaucratic culture and long history of recorded property rights. Unlike Bengal, for example, where most written records of landed rights dated back to the Mughal period at the earliest, the Marathi-speaking regions had a much earlier and continuous recorded history of rural settlement, cultivation and lands gifted by local states in return for service of different kinds. As elsewhere in peninsular India, the village headman, *pātīl* or *mukaddam* took a leading role in opening up land for cultivation, and the hereditary cultivator's right of *mirās* derived from family participation in such early pioneering. *Pātīls* were responsible for maintaining cultivation in the village and for overseeing collection of revenue and its regular payment to the state. Rights to offices, such as that of *kulkarṇi* or village accountant, with *inām* lands attached, could likewise constitute a village *mirās*. At the supra-village level of the *parganā*, a geographical unit of anything between 20 and 100 villages, the offices of *deśamukh* and *deśakulkarṇi* carried responsibility for maintaining cultivation and the flow of revenues, and for local justice and security. The *deśakulkarṇi*, the office of district-level accountant under contention in Gotmāi's family, oversaw revenue collection and the framing of accounts for the *parganā*. Expected to work alongside the *deśamukh*, the *deśakulkarṇi* was a state appointment, remunerated from village revenues and often tax free land given in *inām*. Literacy was essential to both of their offices, and as senior state officials, both were expected to attend the periodic assemblies of local officials and holders of hereditary landed rights which were convened whenever these rights were in dispute. The *havāldār* was again a state appointment, carrying a wide remit of responsibility for security, military affairs, justice and revenue collection over several *parganās*. The administrative division of the *subhā* or province represented a higher level still, and a *subhedār* carried the same broad remit as a *havāldār*, but over an entire *subhā*, although the size and boundaries of a *subhā* could vary significantly with changes of regime and nomenclature.[6]

This was a setting in which the hereditary property rights of such families were extremely important and tenaciously maintained. Local memory, as well as written records in stone, copper plate and increasingly from C14th, on paper, secured these rights and provided evidence that they had been legitimately conferred, either by other *mirāsidārs* acting collectively, or by the state in return for agreed services. Local judicial assemblies of the caste or of holders of *mirās* rights in the locality, with officers of the state present in important cases, could

6 For these contexts, see Hiroshi Fukazawa, *The Medieval Deccan: Peasants, Social Systems and States*. Delhi: Oxford University Press 1991, 1–48.

be convened to hear disputes, examine evidence and witnesses, and determine outcomes.[7]

Our source for Gotmāi's life and struggle lies in a *karīnā*, the kind of lengthy factual statement usually made to a *majālis* or judicial assembly meeting to hear a local dispute. We have access only to an early twentieth century printed copy of the original Marathi *karīnā*, collected in the course of the nationalist historian V.K. Rajwade's tours around western India in search of family papers that otherwise might not survive the colonial state's neglect of private Marathi collections.[8] The form of the document is worth noting. Its opening lines tell us that it was written in the year 1116 of the south Indian regnal year of *Śuhū r*, and the *Manmath* or 29th year of the Jovian *saṃvatsara* calendar, which brings us to the CE year of 1715.[9] The long history that it describes stretches back into the early seventeenth century. It consists of sections of narrative, interspersed with the testimony of the principal disputants in the case, and lists of names of all those who attended the judicial assembly convened to hear the case. Gotmāi's active part in the history took place well before 1715. Only two dates emerge in the *karīnā* with any certainty. One is the date of its compilation in 1715. It is not quite clear what occasioned its production in that year since, as we shall see, the history that it traced concluded with the period of the Mughal invasions of the Maratha country in the 1690s. It is possible that the Kāle family took steps to produce the *karīnā* with the return of stability to the region in the second decade of the eighteenth century, as Shivaji's grandson Shahu consolidated his authority in Satara, and appointed his first peshwa Bālājī Viśvanāth, in 1713. The document recorded Gotmāi as still living at this time, but now an old woman recounting events and a contest that had taken place many decades earlier. The other date we know for certain is 1659. We are able to determine this date because, as we shall see, Gotmāi's principal antagonist took the opportunity of the Bijapur general Afzal Khān's entry into the Maratha country to obtain a crucial piece of documentary evidence about the Kāle family rights. We know that Afzal Khān set out from Bijapur to take on Shivaji's insurgent Maratha forces in May 1659.[10]

---

7 For this institutional milieu see in particular Sumit Guha, *History and Collective Memory in South Asia, 1200–2000*. Seattle: University of Washington Press, 2019, 88–109.

8 The *karīnā* is published in *Marāṭhyāncya Itihāsācī Sādhane, Dvitīya Sammelan Vṛtta*, 1914, 52–67. Hereafter MIS *karīnā*. For the historian Rajwade, see Prachi Deshpande, *Creative Pasts: Historical Memory and Identity in Western India, 1700–1960*. New York: Columbia University Press, 2007, 130–135.

9 I am most grateful to Shailendra Bhandare for his help in working out this date.

10 Gajanan Bhaskar Mehendele, *Shivaji: His Life and Times*. Parama Mitra Publications, Thane, 2011, 203.

The *karīnā* is not in her words, but in those of the scribe responsible for taking down the evidence and collating it into a narrative. However, Gotmāi's voice comes through the narrative very clearly. Her account formed one of its major sources, and it contains immediate and striking details of her character, her strategies, her local reputation, and her ultimate success in protecting her family rights. The concluding paragraph of the document, moreover, tells us that 'Gotmāi related the whole of the present story out of her own mouth'.[11]

Before moving on to the narrative, it is worth reflecting on the assembly that the *karīnā* recorded as having met earlier to hear the case. Described towards the end of the document, this assembly was very large, reflecting the seriousness with which many local leaders and state officials evidently regarded the dispute.[12] Some 138 individuals were named as witnesses, grouped according to their status, place of origin and profession. There were *subhā* level officials, the *subhedār* himself, the *havāldār*, then a range of *subhā* level record-keepers: two *daftardārs*, a *majūmdār*, a *chiṭnavīs* and a *potnīs*. Their equivalents from the local administration of Chiplun pargaṇā were also present, some further six officials. There were in addition various groups of Brahmans: astrologers, priests, *vaidya* medical practitioners, and learned scholar Brahmans from elsewhere resident in Chiplun, amounting to about 28 people. Some twenty-four *śarāfs* and *seṭias* represented the interests of local money-changers and bankers. Then there were 13 principal local office-holders of Chiplun itself, the *pātīl*, the *kulkarṇī*, the *chaudharī*, the *potdār* and so on. Representatives from key communities were there: goldsmiths, oil-makers and weavers. Some 4 *nāyakvāḍis*, chiefs amongst the local Dalit Mahār and Berāḍ communities were present, responsible for village watch and ward. There was a large further group of some 50 named individuals, whose titles indicate office or landed rights. Representatives of families who were party to the dispute, some 14 individuals, also attended. This, then, was a huge gathering of local and regional administrative and landed elites, reflecting the importance of the office at stake, and the family issues surrounding it.

Let us turn, then, to the narrative. Its opening is a little difficult to follow because some phrases are missing from first lines of the printed text. Its opening lines tell us that the office of *deśakulkarṇi* of Chiplun in the subāh of Dabhol had been bestowed on one Harī Krishna. From that time to the date of the *karīnā*'s writing in 1715, he and his descendants had continuously enjoyed the office. After a short break in the text, we learn that Kālopant, the husband of

11 MIS *karīnā*, 66.

12 MIS *karīnā*, 57–60.

Gotmāi, had been one of these descendants, until he fell victim to a violent murder. Kālo and his wife Gotmāi had a son, Antājī Kālo, a boy of great acumen, who left Chiplun to work at the court of Bijapur. This had long been, of course, a well-trodden career path amongst scribal elites in the western Deccan.[13] After his father's murder, the talented son Antājī returned from to Chiplun with his family, presumably to take over the *deśakulkarṇī* office. But then he himself died, leaving Gotmāi alone with her daughter-in-law, now also herself a widow.

At this point, Gotmāi found herself having to look for a *mutālīk*, someone to manage the *deśakulkarṇī* office on her behalf. She set about this, the narrative suggests, 'citānarūpa', which means according to her own intelligence and liking—with the implication, perhaps, that she had her own ideas as to the kind of manager she was looking for.[14] There was a male relative who could have done the job, her husband's cousin Bālo Vināyak, but he, the *karīnā* tells us, 'had no education', *vidyā nāhī*. So it was settled instead by a local tribunal that Harbājī Dātīr, the father of Gotmāi's now widowed daughter in law, should be appointed as *mutālīk*. Dātīr took on the role of manager, and held it for several years.

But then Gotmāi and Dātīr fell out.

> In order to make a quarrel with him, she began to say that owing to some defect of his daughter, her own son had died, and so she began to slander him and to look for a different *mutālīk* more to her liking, but without success. Their quarrel became notorious, and people predicted disaster unless it could be calmed.[15]

At this point two new characters came on the scene, Timājī Bhāskar and his brother Gangājī. They had been working as *kulkarṇīs* or simple village accountants in the nearby town of Helvāk, but then lost their position, and came to Chiplun with their families looking for opportunities. Finding a place to stay proved difficult, so they took up residence at the house of Gotmāi's paternal uncle, Timājī Nāik Deuskar. Once ensconced, and enjoying the good opinion of the local community, they suggested to Gotmāi that Timājī should take on the role of manager. After some days, Gangājī went back to Helvāk to carry on the work of the *kulkarṇī*. Timājī had no son to help him, so Gangājī gave him his own son, Harī Trimal in adoption. Timājī and Harī Trimal secured the agree-

13 Richard M. Eaton, *A Social History of the Deccan: Eight Indian Lives*. Cambridge: Cambridge University Press, 2005, 144–145.

14 MIS *karīnā*, 52.

15 MIS *karīnā*, 52–53.

ment of the local panchāyat or judicial assembly that they should do the work. At first they treated Gotmāi well, and enjoyed the good opinion of everyone in the district.

Matters then took a darker turn, as Timājī and his adopted nephew evidently decided to exploit their position. 'They poisoned their own good fortune. They began to neglect Gotmāi, keeping her short of food and clothing, and treating her with contempt. They set out to make themselves masters and proprietors'. Wider political circumstances at this time, which this point in the narrative enables us to date to the late 1650s, were in their favour. Assailed by Mughal and Maratha forces alike, the Sultanate state of Bijapur was in a state of near political collapse, its dowager queen Badi Sāhiba effectively exercising power on behalf of her young son Ali Adil Shah II (r. 1656–1672) and many of its Irani nobility looking to their own futures. In September 1659, frustrated by the Maratha warrior leader Shivaji's raids into Bijapur's territories, the Bijapur court despatched its military commander Afzal Khān to engage Shivaji's forces.

The narrative of the *karīnā* referred to this despatch of Afzal Khān, and described the opportunity that his arrival opened up for Timājī. Timājī learned that Afzal Khān's powerful political adviser had brought along with him a supply of the Bijapur Sultan's personal *ḍaula*. These were the pre-prepared blank paper orders, often complete with seals and signatures, that could be filled out with the names of those to be granted offices or estates.[16] Such orders would have been a valuable asset in Afzal Khān's coming campaign. As soon as he heard that Afzal Khān's retinue had a supply of such orders, the *karīnā* continued, Timājī made contact with one Kāśī Diyānatrāo. We know from other sources that Kāśī Diyānatrāo was another Maratha Brahman in the service of the Bijapur court. He hailed originally from a family of revenue officials near Pune and had survived many contests for revenue rights similar to Timājī's own. His star had risen quickly in Bijapur. He became a member of the royal council of advisers to the sultans Muhammad and Ali Adil Shah II, and confidential adviser to Khāwas Khān, one of the leaders of the Dākhaṇi party at the Bijapur court. His services brought him the title of 'Diyānatrāo', evidently in reference to his piety, and he appears in contemporary records as Kāśī Diyānatrāo, or simply Diyānatrāo.[17]

---

16 For usages of this kind of document, see Shireen Moosvi, 'Reforming Revenue Administration II: Aurangzeb's Farman to Rasikdas, 1665', in Shireen Moosvi, *People, Taxation and Trade in Mughal India*. Delhi: Oxford University Press, 2008. I am very grateful to Nandini Chatterjee for this reference.

17 G.H. Khare, 'Some records from a Deshpande family of Wai', *Proceedings of the Indian Historical Records Commission*, vol. XLVI, Aurangabad, 1979, 16–19.

Timājī may well have known him, or known of him, through the networks that connected Brahman families of revenue collectors in the western Deccan. The *karīnā* described Timājī's exploitation of the connection without equivocation.

> Timājī got his name inserted onto one of these documents by ingratiating himself with Kāśī Diyānatrāo, and also by promising to supply 300 boats to Bijapur. By this means, he got Gotmāi's name thrown off the document, and without saying anything, returned to Chiplun.[18]

It is likely that Diyānatrāo's interest in boats arose in the context of Shivaji's own early experiments in naval warfare at the end of the 1650s. These early experiments were reported in a letter from the Portuguese governor of Goa in August 1659, as part of the Maratha leader's drive to control Bijapur's territories on the Konkan coast, as well as to strengthen his position against the English and Portuguese factories there.[19]

Evidently, however, Timājī realised that without a properly attested *mahzar*, the document of right usually confirmed by a local *majālis* assembly, his own order of appointment meant little, and that any such assembly would have to be packed with influential and supportive witnesses.[20] The *karīnā* identified two such witnesses. 'It would be important at the very least to obtain the favourable testimony of the respectable merchants Keśāvśet and Gāgśet, and when Timājī got a chance he went to ask them'.[21] But the two merchants were having none of it. Instead, they went behind Timājī's back, to warn Gotmāi.

> They told her that Timājī had got his name inserted into one of the Padishah's orders, and now was looking for witnesses who would help him obtain a confirmatory mahzar issued with his name on it, and that Timājī had visited them to ask them to bear witness in his favour. They asked her, did you have any idea this was going on?[22]

---

18 MIS *karīnā*, 53.

19 P.R. Kakodkar, ed and tr., *The Portuguese and the Marathas: Translation of articles of the late Dr. Pandurang S. Pisurlenkar*. Bombay: State Board for Literature and Culture, 1975, 3–4.

20 For these judicial forms and their associated documents, see Rosalind O'Hanlon, 'In the Presence of Witnesses: Petitioning and judicial 'publics' in western India, c. 1600–1820', *Modern Asian Studies* 53, 2, 2019, pp. 52–88; Nandini Chatterjee, 'Mahzar-namas in the Mughal and British Empires: the uses of an Indo-Islamic Legal form', *Comparative Studies in Society and History* 58, 2, 2016, 379–306.

21 MIS *karīnā*, 53.

22 MIS *karīnā*, 53–54.

Gotmāi was deeply distressed at this news, and confessed that she had known nothing about it. The merchants asked her whether she still had the family's documents of property right in her possession, and she assured them that she had. They urged her to guard the the documents carefully, and not to let anyone else see them. Gotmāi took them at their word, but her precautions, the narrative, suggested, were not very sophisticated.

> Being a woman, she did not arrange to keep the documents in another place. Instead, she bundled them up and hid them in a pillow that she kept nearby her, so that every day when she rose from her bed, she could press down on the pillow, and check that the documents were still there.[23]

Meanwhile, the would-be usurper Timājīpant decided that if his quest was to succeed, he would have to search Gotmāi's house and get hold of the documents that she possessed confirming the Kāle family's rights.

> So he went to see her and appealed to her heart by talking to her of his joys and sorrows. As they talked, he noticed that her hand kept going to the pillow, and he guessed that the documents were hidden within it. Eventually, he saw his opportunity. He made up a bundle of documents with his own scrawls on them, thrust them into the pillow, and pulling out her documents, made off with them. In the days that followed, Gotmāi felt the pillow to confirm that her documents were still there, and believed that they were safe.[24]

After some days, the *karīnā* went on, she began to have her doubts, and so she opened the pillows to look at the documents. But she could not read for herself, and no-one else in the household could read the documents either. So she took them to her father, Krishnājī Deuskar, who examined them. 'He became very agitated, and said, they are nothing but scraps of scribble.'[25] He stormed off to Timājī's house and in a great rage accused him of taking his daughter's family name out of the document. Timājī was at the time still living in Krishnājī's brother Timājī Deuskar's house. On learning of the episode, Timājī Deuskar threw the miscreant Timājī Bhaskar out.

Thwarted thus in his attempt to appropriate the family office, Timājī's relations with Gotmāi went from bad to worse. Gotmāi strove mightily to have

23 MIS *karīnā*, 54.
24 MIS *karīnā*, 54.
25 MIS *karīnā*, 54.

Timājī removed as manager and to have someone else appointed, fearing that her battle might very well end in her losing her life. But without anyone nearby to take her side, she turned to her maternal nephew, one Krishnājī Somnāth, who was then working as an official for the Subhedār of Dabhol, and who used to visit her to ask after her from time to time. 'She said to him, you are my son, but misfortune has befallen me, and my trusted *mutālīk* has turned against me, wiping out my name and pushing his own in instead. But her nephew took little heed of her entreaties'.[26] At this point, Gotmāi turned to seek refuge in the home of one relative after another, her insistence on the rightness of her cause and her refusal to brook any contradiction leading to frequent quarrels. Rumours and finger-pointing began to follow her, and hostility from every relative in whose home she sought refuge.

> They said, she carries a bag of gold around with her, and all she cares about is to protect it. So saying, each made difficulties for her, and threw her out of their homes. Her fate pursued her, and she found no-where to live.[27]

Eventually, Gotmāi reached a dramatic decision. 'I must either flee the country, or drink poison to end my life'.[28]

With a fine sense of theatre, however, Gotmāi did neither. She retired to the prominent temple of Paraśurām in Chiplun, took up a seat in front of the god, and embarked on a public fast. She impressed the temple priest, Nārambhaṭ, who asked her why she had come to the temple. 'She said to him, "I have no place to go, wherever I go those devils in the guise of Brahmans follow me. So I have come here to die near the god"'.[29] The priest's heart melted, the *karīnā* went on, and he invited Gotmāi to come and stay at his home.

Her public protest caused a wider stir. It brought another local official, Antājī Bāvājī, the *mazūmdār* or keeper of revenue accounts at Chiplun, to enquire into her situation. After hearing her story, he issued a stern rebuke to the would-be usurper Timājī Bhāskar, ordering him to build a separate house for Gotmāi, and to keep her well supplied with food and clothing. But the struggle and bitter exchanges between the two continued, even worse than before. After some further time, the same Antājī Bāvājī, now promoted to be *subhedār* or governor of the nearby district of Rajapur, visited Chiplun. He remembered Gotmāi's

26 MIS *karīnā*, 54.

27 MIS *karīnā*, 55.

28 MIS *karīnā*, 55.

29 MIS *karīnā*, 55.

history, and enquired after her. The *karīnā* described her response in touching detail. We can perhaps again hear Gotmāi's own voice as she compares herself to Draupadī, heroine of the epic Mahābhārata, rescued by the god Krishna himself from the malign Kaurava prince Dushashana as he attempted to disrobe her in public.

> She was as happy as if literally her very family god had taken human form and come to meet her. She laid her account before him. He said "If this goes on, you will be killed, and the office will fall into his hands. So, if you have any relative, bring them, and ask them to take up the office, do not delay". She said, "I am lying helpless at the Swami's feet. Apart from the Swami I have no-one. You yourself are my father and mother, brothers, all my family and relations. I am completely helpless. You came to me just like Sri Krishna came in answer to Draupadī's loud wailing. I have nothing left, please, I beg you to come to my aid".[30]

In the face of this compelling personal drama that Gotmāi was able—perhaps very deliberately—to act out, the *subhedār* of Rajapur wrote for her a personal letter to Mahādjī, the *subedhār* of Dabhol, asking him to help her. So she took the letter to Dabhol, where the *subhedār* also asked her if she did not have any relatives. She said that indeed she did—her maternal nephew Krishnājī Somnāth, who was then employed in Dabhol, and to whom, as we saw, she had earlier appealed for help. So the *subhedār* summoned Krishnājī and tried to persuade him to take on the family office. Braving the *subhedār's* wrath, Krishnājī refused again, and left the place. Then, by a stroke of fate, Vināji Balāl, son of Gotmāi's unlettered kinsman Bālo Vināyak, returned to Chiplun after a long absence. Gotmāi was delighted, and took him to meet Mahādjī, who was relieved at this discovery of a relative. But because Vināji was unlettered, he called Krishnājī Somnāth back, introduced him to Vināji, and the two recognised each other. Despite the mediation of Gotmāi's uncle Timāji Deuskar, who came from Chiplun, Krishnājī still declined to take on the position. He suggested instead that his son, Harīpant, should take on the office, working alongside Gotmāi's unlettered kinsman, Vināji. Documents were prepared to confirm Vināji in the office, and thus Gotmāi gave Vināji into Krishnājī's care.

But there remained the question of the revenues collected while Timājī and his adopted nephew had control of the *deśakulkarṇī* office, and of the records maintained during this period. Here again, we are reminded of the practical

30 MIS *karīnā*, 55.

realities of the Brahman scribal household, its deep involvement with local finances, and the key role of family relationships in sustaining these arrangements. By the time the arrangement with Krishnājī and Vinājī had been made, the *karīnā* continued, Timājī himself had passed away, and Gangājī excluded from the business and sent away to Chiplun. Wishing to check on the *deśakulkarṇī's raṇābandh*, the undertaking given to make payments to the state revenues following its own local collections, Krishnājīpant went into the *deśakulkarṇī's* office, met the clerks and other staff, and set about the work. He summoned Gangājīpant and demanded that he produce the old records and the lists of collections from the province. But Gangājī protested his inability to produce any records, and instead recited his own long tale of woe, which was evidently received with some scepticism.

> He began to weep, and cried, "I used to have these papers, but I kept them in the cowshed. Then one day the cowshed burned down, and they were all destroyed. No a scrap of them remains". So he started playacting in this way. "My brother Timājī died. My son Harī Trimal has also died. I will soon be dead myself". With this piteous complaining, he fell at their feet and pleaded piteously to be given an honourable discharge from the job.[31]

The implication was that no proper accounts had been rendered for the *pargaṇā*, which might well explain why Krishnājī Somnāth had been so reluctant to take on the job in the first place. When news of this performance reached the *śeṭias*, who would likely have had some considerable stake in the correct payment of revenues, they declared that he should not be let off lightly in this way. So Gangājī was taken to the *subhedār* Mahādjī in Dabhol, who also pressed him for the accounts, but he repeated the same story. He followed it up with a sworn written statement, that all the papers he was aware of were destroyed in a fire before his very eyes, and in any case Krishnājī himself had no right to press him for information about the profits and losses connected with the office. Evidently unimpressed, Mahādjī the *subhedār* demanded a payment of the fine thousand lāris said to be outstanding. Various respectable people tried to intercede for Gangājī, but the *subhedār* pressed his demand regardless. At this point, Krishnājīpant seems to have decided that perhaps Gangājī deserved a second chance in the family, and might be allowed to take on the role of manager of the *deśakulkarṇī's* office, so long as he treated Gotmāi well. The narrator described his evident contrition. 'He began to treat Gotmāi not just

31 MIS *karīnā*, 56.

like a mother, but like a goddess; the cruelties of the past were forgotten, and they lived in amity'.[32]

Still Gotmāi was not quite content. Perhaps she felt that only a son of her own could offer security, both for herself, and for the family's control of the *deśakulkarṇī* office. 'So she went to her nephew Krishnājī, and said, "You have three sons. May they live as long as the sage Markāndeya". She entreated him earnestly. "Please, allow me to adopt one of them"'.[33] This posed a puzzle for the local pandits, who looked into the matter, and found that as a woman, she was not able herself to adopt a son; only a man could adopt in this way. So it was agreed that her unlettered kinsman Vināji Balāl, the cousin of her dead husband, should adopt one of Krishnājī's sons, Janārdan. We now approach the end of Gotmāi's role in this story. Though Vinājī had a son of his own, he wrote a letter before witnesses that he would treat them both as his own family, and they would all live together in amity, and do the work of the *deśakulkarṇī*' office, and enjoy its perquisites, from generation to generation. The two sons were put to study so that they could carry out their duties efficiently, and an oath was taken on holy water. Thereafter the work of the office, and relations within the family, were conducted in a very satisfactory way.[34]

This part of the narrative then listed all of the important people, described above, who attended the assembly to hear the case. From the account of the *karīnā*, we can deduce that this assembly probably took place in the late 1680s, before the decade of the Mughal invasions in the 1690s. Before this era of warfare, a further challenge to the Kāle family's office developed, which Gotmāi's maternal nephew Krishnājī Somnāth had been forced to contest. The narrative of the *karīnā* continued: 'After Krishnājī's account was decided to be true, the lawless plundering of the Mughlai occurred, and at that time these two conflicting accounts emerged. Gotmāi related the whole of the present story out of her own mouth'.[35] Gotmāi seems to have taken no part in this later phase, this time her nephew Krishnājī taking the lead in successfully defending the family's interests. The outcome of this second stage of the judicial narrative drew to a close. Gotmāi was still living, and in amity with one of Krishnājī's sons, Harīpant. The narrative then concluded with some reflections on the importance of giving true testimony in any judicial case involving family rights.

Here, then, I think, there is much to interest us in the details of Gotmāi's suit, conducted in these mundane but often fiercely competitive lower reaches

32 MIS *karīnā*, 57.
33 MIS *karīnā*, 57.
34 MIS *karīnā*, 57.
35 MIS *karīnā*, 66.

of state administration. What we see here—the attempt by a manager to appropriate a significant office as his own property, aided by forgery and the manipulation of local justice—was actually quite a familiar situation in these settings. What is remarkable, though, is the degree of traction that the unlettered, but determined Gotmāi was able to gain in the face of such a challenge. The conflict seems to have been resolved without recourse to formal judicial procedures until the challenge which occasioned the great assembly of the late 1680s, when Krishnājī Somnāth defended the family's office. Rather, Gotmāi conducted her earlier struggle to protect the the family's rights through informal social means. She staged a major drama in the local temple, which brought the *subhedār* to her aid, and then a further public drama in his office, replete with classical references, as she endeavoured to persuade him to help her by dispensing with the services of her nephew.

She was not herself literate, and with its stock of male members so depleted, the Kāle family had not diversified into scholarly alongside administrative and revenue management roles. Nonetheless, it was Gotmāi who kept custody of the family's documents of property right, and she knew what to do when she had been the victim of deception. Her lack of literate skills made it impossible for her to exercise the bureaucratic power of the *deśakulkarṇī* in her own right, and in this she was unlike widows who wielded political authority or mercantile influence on the basis of their connections and experience. Yet she was clearly determined to find a manager that she liked, and willing to quarrel with those she saw as unsuitable or disrespectful to her, dragging in the death of her son to strengthen her hand. Ultimately what she wanted, of course, was a male relative to take on the office so that it could be preserved within her family, and she was assertive enough to suggest that she should herself be allowed to adopt one. Her goal was preservation of her family's rights and its name, but to do this she was willing to flout convention, to make herself unpopular, to test the limits of local community tolerance, to take over a temple for the staging of her personal drama, and to put great pressure on state officials.

Her suit also gives us some idea of contemporary attitudes to women and widows such as herself. The scribe's narrative suggests that many people regarded her as a 'difficult' woman, quarrelsome by nature and determined to use her deficient woman's judgement in the management of her affairs. She was clearly seen as a soft target by predatory local Brahmans eager to appropriate the family's office. Her nephew was clearly very reluctant to come and work with her, and possibly not just because he may have suspected that the *parganā* accounts were in disorder, with substantial funds missing. Yet her testimony, told now in what must have been her old age, was clearly regarded as crucial to a correct understanding of the history of the quarrel, and it was presented to an extraordinarily large and distinguished assembly of state officials, local notables, community heads, merchants and family members.

Although the Kāle family itself shared much in common with many Maratha Brahman families of the period, what we have in this *karīnā* is the fragment of an archive, rather than anything like a larger individual corpus or family history. It is difficult for us to tell how far other women in Gotmāi's position might have sought to protect family property and to shape family roles and relations so tenaciously. Such histories may be useful, however, not only to reconstruct something of the household relationships of Maratha Brahman families more widely in this period, but to add to our understanding of the kind of agency that women of such families were able to wield as they negotiated the difficult terrain between household, extended family, local community and agencies of the state.

Acknowledgements

I am extremely grateful to participants at the various seminars where versions of this essay were presented: the Oxford South Asian History seminar, the 'Unstable Archives' webinar series organised by Megan Robb and Sneha Krishnan in September 2020, and the 'Women of Capital' panel held at the Madison South Asia conference in 2015. As ever, I am most indebted to Shailendra Bhandare, who has been kind enough in the preparation of this essay once more to share his deep expertise in Marathi language and his breadth of knowledge about Maharashtra in the early modern era.

Bibliography

Bano, Shadab. "Women and Property in Mughal India." *Proceedings of the Indian History Congress* 65 (2007): 406–415.

Chatterjee, Indrani. *Forgotten Friends: Monks, Marriages, Memories of Northeast India*. New Delhi: Oxford University Press, 2013.

Chatterjee, Nandini. "Mahzar-namas in the Mughal and British Empires: the uses of an Indo-Islamic Legal form." *Comparative Studies in Society and History* 58, no. 2 (2016): 397–406.

Deshpande, Prachi. *Creative Pasts: Historical Memory and Identity in Western India, 1700–1960*. New York: Columbia University Press, 2007.

Eaton, Richard M. *A Social History of the Deccan. Eight Indian Lives*. Vol. 1.8. The New Cambridge History of India. Cambridge: Cambridge University Press, 2005.

Fukazawa, Hiroshi. *The Medieval Deccan. Peasants, Social Systems and States*. New Delhi: Oxford University Press, 1991.

Guha, Sumit. *History and Collective Memory in South Asia, 1200–2000*. Seattle: University of Washington Press, 2019.

Hasan, Farhat. "Property and Social Relations in Mughal India: Litigations and disputes at the Qazi's Court in Urban Localities." *Journal of the Economic and Social History of the Orient* 61, no. 5/6 (2018): 851–877.

Kakodkar, P.R. *The Portuguese and the Marathas: Translation of Articles of the late Dr. Pandurang S. Pisurlenkar*. Bombay: State Board for Literature and Culture, 1975.

Khare, G.H. "Some Records of a Deshpande Family of Wai." *Proceedings of the Indian Historical Records Commission* 46 (1979): 16–19.

Mehendale, Gajanan Bhaskar. *Shivaji: His Life and Times*. Thane: Parama Mitra Publications, 2011.

Minkowski, Christopher. "Learned Brahmins and the Mughal Court: The Jyotiṣas." In *Religious Interactions in Mughal India*, 102–134. New Delhi: xford University Press, 2014.

Minkowski, Christopher. "Nīlakaṇṭha's Instruments of War: Modern, Vernacular, Barbarous." *Indian Economic and Social History Review* 41, no. 3 (2004): 365–385.

Minkowski, Christopher. "Nīlakaṇṭha's *Mahābhārata*." *Seminar* 608 (2010): 32–38.

Minkowski, Christopher. "Nīlakaṇṭha's Teachers and Gurus, Part 1: Lakṣmaṇa Paṇḍita." *Wiener Zeitschrift Für Die Kunde Südasiens* 55 (2013): 33–76.

Minkowski, Christopher. "Sanskrit Scientific Libraries and their Uses: Examples and Problems of the Early Modern Period." In *Looking at it from Asia: The processes that shaped the sources of the History of Science*, edited by Florence Bretelle-Establet, 81–114. Boston Studies in the Philosophy of Science No. 265. Dordrecht, Heidelberg, London, New York: Springer Science & Business Media, 2010.

Minkowski, Christopher. "What makes a work 'traditional'? On the Success of Nīlakaṇṭha's Mahābhārata Commentary." In *Boundaries, Dynamics and Construction of Traditions in South Asia*, edited by Federico Squarcini, 225–252. Florence: Firenze University Press, 2005.

Moosvi, Shireen. *People, Taxation and Trade in Mughal India*. New Delhi: Oxford University Press, 2008.

O'Hanlon, Rosalind. "In the Presence of Witnesses: Petitioning and judicial 'publics' in western India, c. 1600–1820." *Modern Asian Studies* 53, no. 2 (2019): 52–88.

O'Hanlon, Rosalind. "Letters Home: Banaras pandits and the Maratha Regions in early modern India." *Modern Asian Studies* 44, no. 2 (2010): 201–240.

O'Hanlon, Rosalind. In *Brahman Lineages Beyond the Mughal Court*, edited by Richard M. Eaton and Ramya Sreenivasan. New York: Oxford University Press, forthcoming.

Rajwade, V.K. "The Origins of the Chitpavans," Marāṭhyāncya Itihāsācī Sādhane, Dvitīya Sammelan Vṛtta (1914): 52–66.

Sheikh, Samira. "Jibhabhu's Rights to Ghee: Land Control and Vernacular Capitalism in Gujarat, c. 1803–1810." *Modern Asian Studies* 51, no. 2 (2017): 350–374.

Sreenivasan, Ramya. "Drudges, Dancing Girls, Concubines: Female Slaves in Rajput Polity, 1500–1850." In *Slavery and South Asian History*, edited by Indrani Chatterjee and Richard M. Eaton, 137–161. Bloomington; Indianapolis: Indiana University Press, n.d.

Index

Note: Entries for texts are, generally, listed under their respective authors. Individual Parvans and Kāṇḍas are listed under the Mahābhārata and the Rāmāyaṇa, respectively. Individual deities are listed under 'Deities.'

www.ingramcontent.com/pod-product-compliance
Lightning Source LLC
LaVergne TN
LVHW012052011026
849162LV00018B/190